ELECTRICIAN'S MATE
3 & 2

Prepared by

BUREAU OF NAVAL PERSONNEL

NAVY TRAINING COURSES

NAVPERS 10546

UNITED STATES
GOVERNMENT PRINTING OFFICE
WASHINGTON : 1959

For sale by the Superintendent of Document)!, V. S. Government Print inn Office Washington 25. D. C. - Price $2 75

THE UNITED STATES NAVY

GUARDIAN OF OUR COUNTRY

The United States Navy is responsible for maintaining control of the sea and is a ready force on watch at home and overseas, capable of strong action to preserve the peace or of instant offensive action to win in war

It is upon the maintenance of this control that our country s glorious future depends, the United States Navy exists to make it so

WE SERVE WITH HONOR

Tradition, valor, and victory are the Navy's heritage from the past To these may be added dedication, discipline, and vigilance as the watchwords of the present and the future

At home or on distant stations we serve with pride, confident in the respect of our country, our shipmates, and our families

Our responsibilities sober us. our adversities strengthen us

Service to God and Country is our special privilege We serve with honor

THE FUTURE OF THE NAVY

The Navy will always employ new weapons, new techniques, and greater power to protect and defend the United States on the sea. under the sea. and in the air

Now and in the future, control of the sea gives the United States her greatest advantage for the maintenance of peace and for victory in war

Mobility, surprise, dispersal, and offensive power are the keynotes of the new Navy The roots of the Navy lie in a strong belief in the future, in continued dedication to our tasks, and in reflection on our heritage from the past

Never have our opportunities and our responsibilities been greater

ACTIVE DUTY ADVANCEMENT REQUIREMENTS

' Recommendation of potty afHcort, officer* and approval by commanding officer roquirod for all advancemontt.

INACTIVE DUTY ADVANCEMENT REQUIREMENTS

'Recommendation of patty officert, officerj and approval by commanding officer required for all advancements. #Active duty periods may be substituted for drills and training duty.

READING LIST

NAVY TRAINING COURSES

Basic Electricity , NavPers 10086

Basic Hand Tool Skills (metal working skills only),

NavPers 10085 Blueprint Reading and Sketching (chapters 1-9), Nav-

Pers 10077-A

OTHER PUBLICATIONS

Bureau of Ships Manual , chapters 4; 6; 22; 31; 34; 45, section 1; 60; 61, section I; 62; 63; 64; 66; 81, section III; 82; 83; 88, section III, part 2.

U.S. Navy Safety Precautions (chapter 18), OpNav 34P1

USAFI TEXTS

United States Armed Forces Institute (USAFI) courses for additional reading and study are available through your Information and Education Officer.* The following is a partial list of those courses applicable to your rate:

SELF TEACHING

Number Title
MA 784 Electric Wiring
MB 290 Physics I (Mechanics)
MB 785 Electrical Measuring Instruments
MB 858 The Slide Rule
CORRESPONDENCE
CB 290 Physics I (Mechanics)
CB 785 Electrical Measuring Instruments
CB 858 The Slide Rule

♦"Members of the United States Armed Forces Reserve components, when on active duty, are eligible to enroll for USAFI courses, services, and materials, if the orders calling them to active duty specify a period of 120 days or more, or if they have been on active duty for a period of 120 days or more, regardless of the time specified on the active duty orders."

Vl

PREFACE

This training course is written to aid enlisted men in the U. S. Navy and Naval Reserve in preparing for advancement to the rate of Electrician's Mate 3 or 2.

The subjects with which the striker or EM3 must be familiar before he can qualify for advancement to Electrician's Mate 3 or 2 are outlined in the Manual of Qualifications for Advancement in Rating, listed in appendix n of this book. This training course contains information on each examination, and, insofar as practicable, information on each practical factor. Because examinations for advancement in rating are based on these qualifications, interested personnel should refer to them for guidance (the LATEST qualifications should always be consulted).

The Electrician's Mate 3 & 2 training course was prepared by the U.S. Navy Training Publications Center, which is a field activity of the Bureau of Naval Personnel. Technical assistance was provided by the Bureau of Naval Personnel, the Bureau of Ships, and the Electrician's Mates Schools (Class A and Class B), U. S. Naval Training Center, Great Lakes, Illinois.

CONTENTS

Chapter Page
1. ORGANIZATION 1
Electrician's Mates 1
Assignment of personnel 9
Engineering department 11
Records 21
Publications 28
2. ELECTRICAL CABLE AND
ACCESSORIES 35
Cable designations 35
Installation 41
Cable maintenance 99
3. ELECTRIC LIGHTING 117
Principles of light 117
Light sources 125
Lighting fixtures 139

Navigational lights 147
Lighting equipment ; 157
Lighting distribution systems 166
Maintenance 168
Safety precautions 173
4. SEARCHLIGHTS 179
24-inch carbon-arc searchlight 179
12-inch incandescent searchlight 212
8-inch, 60-cycle sealed-beam searchlight . 222
5. MAINTENANCE OF MOTORS AND
GENERATORS 232
Introduction 232
Cleaning motors and generators 233
General maintenance 234
Sleeve bearings 236
Chapter Page
Ball bearings 241
Brushes 249
Commutators and collector rings 255
Armatures 261
A-c rotors 266
Field coils 268
A-c stator coils 271
Motor and generator air coolers 272
Periodic tests and inspections 273
6. MAINTENANCE AND REPAIR OF
BATTERIES 280
Dry batteries 280
Storage batteries 285
Maintenance 296
Troubles 309
Repairs 315
Survey 323
Safety precautions 325
7. PROTECTIVE DEVICES 331
Fuses 332
Troubleshooting electrical circuits 336
Control and protective accessories 348
Circuit breakers 388
8. CONTROL DEVICES 395
Introduction 395
D-c contactors 403
A-c contactors 425
Maintenance 438
Periodic tests and inspections 439
Safety precautions 440

9. ELECTRICAL SYSTEMS IN SMALL CRAFT 447
Engine starting system 447
Lighting system 500
Chapter Page
10. ELECTRIC POWER DISTRIBUTION 507
A-c electric power systems 508
D-c electric power systems 519
Switchboards 520
Preparing generators for operation 546
Operation of generators 547
Operation of electric plants 557
Operating procedures 562
Maintenance 566
11. ELECTRIC AUXILIARIES 579
Degaussing installations 579
Anchor windlass 615
Electric elevators 621
Electrohydraulic steering gear 626
Electric galley equipment 634
12. SATURABLE REACTORS 645
Introduction 645
Basic half-wave circuit 650
Basic full-wave circuit 658
Transfer characteristics 667
Full-wave circuit without control circuit
rectifiers and without control
circuit e^. 669
Conclusion. . 676
Appendix
I. Answers to quizzes 687
n. Qualifications for advancement in rating . . 704 Index

ELECTRICIAN'S MATE 3 & 2

CHAPTER

ORGANIZATION

ELECTRICIAN'S MATES

The United States naval, military, and civilian organizations, including ships, aircraft, submarines, bases, yards, factories, and supply lines all together add up to the world's greatest seapower. Under emergency conditions thousands of men are processed into this astonishingly large organization within months and weeks. The methods and facilities used to convert civilians into Navy men and Navy men into competent seamen and technicians are as varied as the number of jobs to be accomplished. Training facilities include self-study courses, textbooks, training aids, and practical training in operating and maintaining the actual equipment in service schools as well as aboard ship. The average enlisted man of today is a young man with a few years of service. No matter how complex the organization and ships become, he is still the most important factor in the effective operation and maintenance of the equipments and the ships of the Navy. This training course is designed to meet the needs of the Navy electrician.

Electrician's Mates perform both military and professional duties. The military duties are the same as those of other petty officers irrespective of the professional or specialty Ratings. The professional duties include a variety of tasks that require many specialized skills and \'7btechniques. In order to accomplish these specialty duties, Electrician's Mates must have a good

working knowledge of the basic principles of electricity and electronics, as well as a working knowledge of practical mathematics.

Requirements for Advancement

The military requirements and the professional qualifications for all ratings are listed in the Manual of Qualifications for Advancement in Rating (Revised), Nav-Pers 18068, commonly known as the Quals Manual. The Quals Manual is periodically revised to reflect organizational and procedural changes in the Navy that affect the rating structure, and to incorporate additional skills and techniques required by the development and installation of new equipment.

PROFESSIONAL QUALIFICATIONS.-A reprint of the professional qualifications for advancement for the Electrician's Mates rating is presented in appendix n of this training course. These qualifications are current through change 1 dated 1 June 1953. Personnel preparing for any examination subsequent to the date of this change should refer to the latest revision of the Quals Manual.

The rating of the Electrician's Mate consists of one general service rating and two emergency service ratings. General Service Electrician's Mates (EM) stand watch on ship's service generator and distribution switchboards and other electrical equipment; operate and maintain electrical auxiliaries; maintain and repair electric power and lighting circuits, fixtures, and equipment; and repair and rebuild electrical equipment in an electrical shop.

The two emergency service ratings consist of Power and Lighting Electrician's Mates (EMP), and Shop Electrician's Mates (EMS). Electrician's Mates in the emergency service ratings are not responsible for all of the qualifications in the general service ratings. Emergency Service Power and Lighting Electrician's Mates (EMP) operate and maintain power and lighting circuits and equipment aboard ship, and stand watch on generator and distribution switchboards, electrical propulsion motors, and control equipment. Emergency Service Shop Electrician's Mates (EMS) repair and rebuild electrical equipment in an electrical shop.

The qualifications that apply to the particular rates and grades are indicated in the applicable rates column of the excerpted qualifications in appendix II. This column indicates the lowest rate for which a particular qualification applies. The qualifications for advancement to the higher rates also include those of the lower rates.

The qualifications in appendix II are separated into two primary divisions consisting of PRACTICAL FACTORS and EXAMINATION SUBJECTS. The individual qualification items under these primary divisions are further divided into appropriate subject matter areas that indicate the required skills and knowledges.

The PRACTICAL FACTORS are qualifications that include particular tasks performed on the job and tested by practical demonstrations with materials, tools, and equipment. The EXAMINATION SUBJECTS are qualifi-f cations that include the minimum knowledges required for

the work performance and tested by a written examination.

A man qualifying for the next higher rate must complete all of the military and professional practical factors I before he can be recommended to take the advancement

examination. As the candidate shows proficiency in each practical factor a record of its completion is entered on the RECORD OF PRACTICAL FACTORS, NavPers 760 (EM), which is a standard form used for this purpose.

The Record of Practical Factors provides a standard checkoff list of both military and professional practical factors required to be demonstrated in each rate as a prerequisite for advancement. The supervising officer initials and enters the date of completion of each practical factor in the appropriate column provided on the form. Each division maintains a Record of Practical Factors for each enlisted man in pay grades E-2 through E-6 (apprentice through first class). When an enlisted man \ is transferred, the signed copy of the form is inserted in

the correspondence side of the Enlisted Service Record, which is forwarded to the man's new duty station. In this way, the man's record is kept up to date and used on a ' continuing basis as he progresses in his rating.

MILITARY REQUIREMENTS.-The Electrician's Mate, in addition to his technical duties performs military i duties. Underway he must man general quarters stations

which may include (1) being a member of a repair party or

(2) standing watch on a switchboard in the steering engine room or in the elevator control room. In port he may be assigned such military duties as shore patrol, security watches, or taking charge of a draft of men and being responsible for their safe transportation and delivery.

As previously mentioned, the military requirements, as well as the professional qualifications are listed in the Manual of Qualifications for Advancement in Rating (Revised), NavPers 18068. The military duties are not included in this training course. They are discussed in Basic Military Requirements. NavPers 10054; in Military Requirements for Petty Officers 3 and 2, NavPers 10056; and in Military Requirements for Petty Officers 1 and Chief NavPers 10557.

REFERENCE MATERIAL.—The EM striker or the EM 3 in preparing for advancement in rating must study certain publications in addition to this training course. The Reading List in the front of this book is especially useful as supplementary study material. The references listed under the headings, Navy Training Courses and Other Publications, are of particular importance because questions on the examination for advancement may be based on material contained in these courses and publications as well as on material in this training course.

These references are taken from Training Publications for Advancement in Rating, NavPers 10052-F, which is an annual bibliography published by the Bureau of Naval Personnel. This bibliography lists the current Navy training courses and other publications that have been prepared for the use of all enlisted personnel concerned with advancement in rating examinations. This bibliography is used by examining authorities in preparing military and professional examinations for advancement in rating and also by personnel preparing to take these examinations.

In addition to the basic Navy training courses contained in the Reading List, Mathematics, Vol. 1, NavPers 10069-B and Basic Electronics (portions of chapters 1, 2, and 3), NavPers 10087 should be included as supplementary study material. Mathematics, Vol. 1 will help you to acquire the necessary knowledge of shop mathematics.

Navy training courses can be obtained by application to your Information and Education Officer. He is also in

a position to help you acquire other publications that may not be readily available. If you need any study material in preparing for your advancement, do not hesitate to inform your I and E officer.

LEADERSHIP.—In satisfying the military requirements for advancement in rating you are required to study Military Requirements for Petty Officers 3 and 2, Nav-Pers 10056, which contains a chapter entitled, Military Command and Leadership. After reading that chapter you should begin to think of ways to apply the information to your duties as an EM 3 or EM 2.

When you become a petty officer you become a link in the chain of command between the officers of your division and your men. Your responsibilities are more than just giving orders and seeing that work is accomplished. You also have a responsibility for sharing your knowledge with others. When the Navy promotes you it expects you to bear some of the burden of training others.

Many books have been written on the subject of leadership, and many traits have been listed as a necessary part of the makeup of a leader. Whether you are a successful leader or not will be decided, not by compiled lists of desirable traits, but for the most part by the success with which you stimulate others to learn and to perform.

Self-confidence is one of the keys to leadership, but it must be supported by enthusiasm and especially by knowledge. For example, if you are supervising a group in performing preventive maintenance on a piece of electrical equipment you should not only know the necessary procedures thoroughly, but also be ready to pitch in and help do the job yourself if necessary. Your men will respect you as a man who has demonstrated his know-how and skill in his profession.

A cooperative attitude is a requirement of leadership. Do not let knowledge of your job techniques make you unreasonable and overbearing with lower rated men whom you may have to instruct. Your attitudes will have a definite effect on the attitudes and the actions of these men.

Be competent in your instruction of others; the opportunity to acquire knowledge and to master new skills was not given to you solely for your own benefit, but also for the benefit of the Navy as a whole. As new types of tools and equipment for use in your rating are made available, you should be the first to learn about their operation and maintenance.

WARRANT AND LIMITED-DUTY OFFICER.-The paths of advancement from enlisted ratings to warrant officer categories and/or limited-duty officer (LDO) classifications are difficult to set forth because they are so varied. In the warrant category, there are four grades ranging from warrant officer (WO-1) through warrant officer (WO-4). In the LDO classification, the commissioned ranks follow from ensign through commander.

How to Study

The general methods of study are the same for everyone, but the real art entails discovery of the methods that are most advantageous for the individual. It is always best to study about a particular equipment while working on it. With a piece of equipment available, the student should study the technical manual and relate the physical location and size of the component. On the job, he should learn by doing.

PLAN OF STUDY.—When studying theory or operational material, it is very important to set up some plan of study. Study must be made a habit. It must be done under conditions and surroundings that will not distract the student. It is important that learning be done in an orderly fashion so that the acquired bits of knowledge will serve as stepping stones in the process of learning. The material at hand should be read and studied with as much concentration as possible.

RULES OF STUDY.-Some basic rules for studying are:

1. A comfortable, quiet, and well-lighted location should always be used if possible. With pencil and paper handy for recording notes, the student should start to read.

2. A portion of a chapter and the number of pages to be studied should be decided upon,

depending upon the subject.

3. The material should be read quickly in order to get the main point of the subject.

4. Then the material should be reread carefully.

5. When the material has been reread, the book should be put aside.

6. The main points should be listed.

7. With the book open, the student should check the main points.

8. The material should then be reread more slowly. This time the student should try to remember the details and connection of each part.

9. When he has finished reading, the student should write a detailed summary of what he has learned, using the book only if necessary.

10. When the details of the material have been thor-▶ oughly digested, the student should turn to the end of the

chapter and answer as many questions as possible without referring to the text.

11. The answers should be checked and corrections should be made.

This general method should be of great benefit to those who find it difficult to learn and retain what they have read. It should be remembered that electricity cannot be learned in a hurry. However, a consistent application of effort over a period of time will bring a man to his goal sooner than he thinks.

READING WITH UNDERSTANDING.-Technical matter should not be read with the idea of covering a specific number of pages or chapters without regard for the complexity of the subject matter. It is better to read a small amount of material and digest it thoroughly than to cover a large number of pages and have only a rough idea of what is going on. Basic material should be read in order to get a thorough background before proceeding to more difficult material. It is easier to grasp new knowledge with a good background of fundamentals. In order to work out problems that are not fully clarified, the student should always have pencil and pad handy while studying.

Quizzes should be used to give an indication of the amount of information retained. Oftentimes textbooks have questions and problems at the end of each chapter. Answering these questions is a good way to review the chapter. Another suggested way is to read the chapter or section, close the book, and then try to summarize it.

The student should keep a notebook on all the publications that deal with the Electrician's Mate and the petty

n

officer In general. He should list the title of the publication, its short title, the bureau or civilian agency responsible for publishing it, its location aboard ship, and a summary of its contents. This record will be invaluable in assisting personnel to readily find information on a given subject.

Scope off EM 3 and 2 Training Course

The EM 3 and 2 training course is designed to aid the EM striker in preparing for advancement to EM 3 and likewise to aid the EM 3 in preparing for advancement to EM 2. The text is written to cover the examination subjects and, where practicable, the practical factors in the Electrician's Mates Rating (group VII) of the Manual of Qualifications for Advancement in Rating (Revised), Nav-Pers 18068.

The first chapter of this training course explains the qualifications and requirements for advancement and lists useful reference material. It also includes the organization of the electrical division In a large ship and a resume of the duties and responsibilities of Electrician's Mates.

The second chapter describes the types, uses, and installation of shipboard electric cable and accessories. It includes the various tests and inspections necessary for the proper maintenance and repair of electric cables.

Chapter 3 explains the principles of light to familiarize the EM striker or EM 3 with the fundamentals and terminology of electric lighting. It describes the types of lamps and lighting fixtures used on naval vessels, and the practical methods of calculating illumination requirements. Lighting distribution systems are explained, including the maintenance of lighting installations and the safety precautions to be observed when working with this equipment.

The fourth chapter covers the types of Navy searchlights. It describes the construction, operation, and maintenance of these searchlights including the safety precautions involved.

Chapter 5 deals with the maintenance of motors and generators. It acquaints the Electrician's Mate with the preventive and corrective maintenance procedures used by the Navy to obtain optimum performance of these machines.

Chapter 6 presents the maintenance and repair of batteries. It describes the types, characteristics, and installation of dry batteries and storage batteries used in the Navy. The maintenance, troubles, and repair of storage batteries are presented in detail, Including the safety precautions to be observed when working with batteries.

The seventh chapter covers the protective devices used in electrical circuits and equipment. It includes a description of the construction and function of the various types of fuses, switches, relays and circuit breakers. A description is also given of troubleshooting single-phase and three-phrase circuits.

Chapter 8 presents control devices, which include rheostats, electric brakes, and contactors. It includes a description of the construction, operation, and maintenance of these devices. Also included is an analysis of circuit diagrams of d-c and a-c motor starters.

Electrical systems in small craft are discussed in chapter 9. This discussion includes the engine starting systems installed in gasoline-engine and diesel-engine powered boats and a description of the lighting systems.

Chapter 10 covers the electric power distribution systems installed in naval vessels including the methods of interconnecting the various switchboards and the operation of automatic bus transfer equipment. It describes cross-plant (parallel) and split-plant operation and operation under casualty conditions.

Chapter 11 describes electric auxiliaries, which includes degaussing installations, anchor windlass, electric elevators, electrohydraulic steering gear, and electric galley equipment.

The principles of saturable reactors (magnetic amplifiers) are covered in chapter 12. The discussion includes the components that comprise a magnetic amplifier and an analysis of the basic half-wave and full-wave circuits with various values of control voltage.

ASSIGNMENT OF PERSONNEL

The complement and allowance of personnel of Navy ships are established by both the Chief of Naval Operations and the Chief of Naval Personnel. The Chief of Naval Operations determines the number of men required for a

particular ship for certain jobs, and the Chief of Naval Personnel determines the ranks and ratings of these officers and men.

Complement

The complement of a combat ship is based on the number of personnel required to (1) man the stations for battle,

(2) perform the basic administrative requirements, and

(3) maintain the continuous watches required under wartime conditions of readiness. The complement is, therefore, a fixed number based on the mission of the ship and its installed equipment.

Allowance

The allowance of a combat ship is based on a percentage of the complement necessary to maintain and operate the ship under peacetime conditions. The allowance is a flexible component in personnel administration based on the national policy and budgeting limitations. It should be understood that a ship with peacetime allowance is still a very effective fighting unit. However, in the event of an emergency the prevailing allowances of ships are expanded to wartime complements as quickly as possible.

To Sections

For wartime organization, each division is divided into three approximately equal sections, each being adequate to maneuver and fight the ship under emergency conditions. The section is the primary organization unit of the ship for administration of condition watches, watch standing, liberty, and messing and berthing. Each section should include an adequate number of qualified rated and nonrated personnel to man all required stations in emergencies, including those stations for getting underway and proceeding to sea for limited operations as may be required by the weather, surprise hostile activity, or other emergency situations.

Two primary considerations in the assignment of personnel in the organization structure are the number and qualifications of the available personnel that must be

employed in the various battle, watch, and administrative billets to effectively fulfill the mission of the ship, department, or division.

Enlisted Assignments

The assignment of enlisted personnel is accomplished through the use of divisional, sectional, watch, and billet number assignments. Assignments to the various billets prescribed by the Ship's Battle Bill, Ship's Watch Organization, Ship's Organization Bills, and departmental and divisional administrative and watch organizations are published in the division Watch, Quarter, and Station Bill (fig. 1-1) and supplementing watch and duty lists.

Ships seldom experience the ideal conditions presented by the Personnel Allocation List because of unavoidable fluctuations in the ranks and ratings onboard and because of differences in (1) capabilities of individuals, (2) material resulting from improvements and alterations, and (3) operating conditions. Hence, division officers will be required to modify assignments of personnel to stations and duties in places where these inconsistencies occur. Necessary appraisals and revisions must be made continuously of the various assignments to achieve maximum operational efficiency and optimum utilization of personnel.

The division officer's notebook and individual watch, quarter, and station cards are used to advantage for controlling, recording, and disseminating information on such assignments. The procedures for the use of these techniques are published in Shipboard Procedures, NWP50 (Naval Warfare Publication) and the division officer's guide.

ENGINEERING DEPARTMENT

Navy ships are operated under standard administrative and battle organizations to facilitate quick expansion (without major change) from peacetime to wartime status. This organization divides the ship's personnel into the (1) operations, (2) navigation, (3) gunnery, (4) engineering, (5) supply, (6) medical, and (7) dental departments. Aircraft carriers and seaplane tenders have, in addition, an air department, and repair ships have a repair department.

The scope of this training course does not permit a description of the entire ship's organization, which is published in United States Navy Regulations and more specifically in the Ship's Organization and Regulations Manual. However, a brief description of the engineering department is included with particular emphasis on the electrical division and the duties and responsibilities of Electrician's Mates.

The engineering department is under the direct supervision of the engineer officer. It normally consists of five divisions, as illustrated by the organizational chart in figure 1-2. The machinery and boiler divisions are under the supervision of the main propulsion assistant; the electrical division is under the electrical officer; and the repair and auxiliary divisions are under the damage control assistant. These officers are charged primarily with the operation, maintenance, and repair of the machinery and equipment allotted to their divisions, and they also act as assistants to the engineer officer in the performance of his duties.

The maintenance and repair of all electronic equipment, except as assigned to another department, is the responsibility of the operations officer.

Administrative Organization

The electrical division is under the supervision of the electrical officer. In large ships the electrical division consists of several groups (fig. 1-2), each of which is under the direct supervision of an EM 1 or EMC. The Electrician's Mates assigned to these groups are responsible for the operation, maintenance, and repair of the specific electrical equipment and circuits included in the respective groups.

ELECTRICAL OFFICER.-The electrical officer is responsible, under the engineer officer, for the organization, administration, and operation of the electrical division and its assigned personnel and material in support of the over-all mission of the ship. He is responsible for the operation, maintenance, and repair of the electrical machinery and systems throughout the ship, except those assigned to another department.

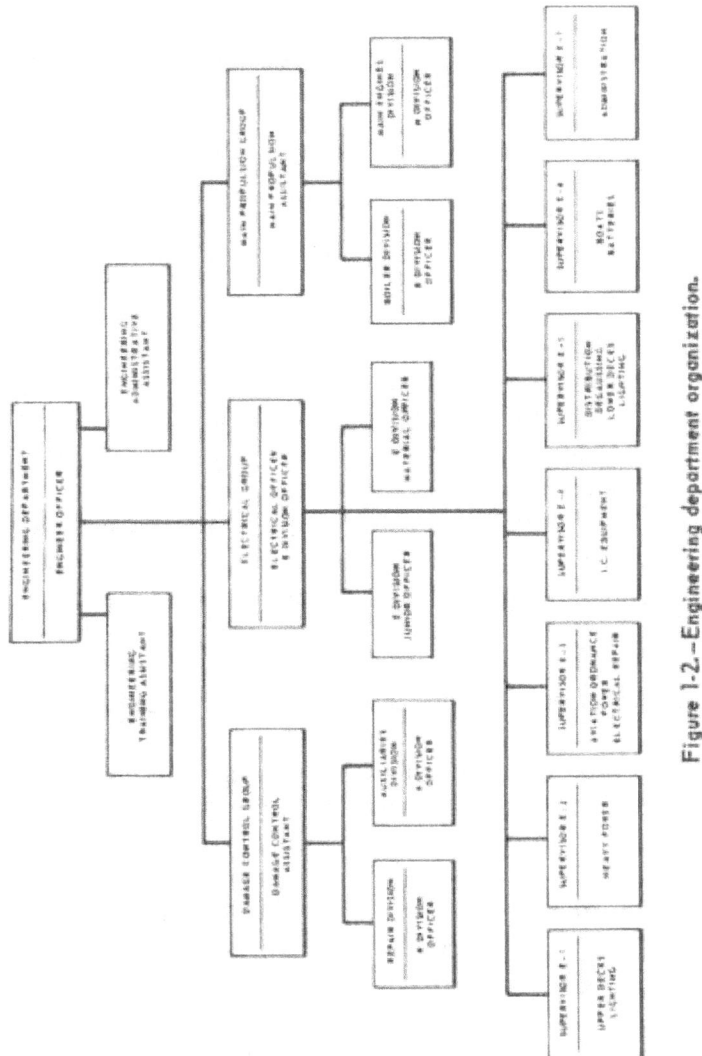

Figure 1-2.—Engineering department organization.

JUNIOR DIVISION OFFICER.—The junior division officer assists the division officer in coordinating and administering the functions of the division. He must develop a thorough understanding of the functions, operation, organization, and equipment of the division so that he can assume the duties of the division officer.

MATERIAL OFFICER.—The material officer is responsible, under the electrical officer, for the readiness of all assigned electrical equipment and the administration of the electrical material maintenance program.

Watch Organization

When the watch organization is established, extreme care is used to ensure that all personnel thoroughly understand their duties, responsibilities, authority, and organizational relationships. Personnel assigned to watchstanding duties are entrusted with the safety of the ship, machinery, and personnel. Confusion or conflict among watch personnel concerning the responsibilities or authority could result in a disaster, such as collision, grounding, or even loss of the ship. In some instances of naval disasters, certain watch personnel were held accountable for failing to take proper action because they were not aware of their duties and responsibilities or because they did not think they had the authority to act. Conversely, in many instances, serious damage and loss of life have been averted by the timely action of watch-standers working together as a coordinated and integrated team.

An effective watch organization is based on sound organizational principles applied to tactical and operational requirements and published in the form of operational charts for the guidance and control of the personnel who will stand the watches. Because of the major differences in the functions and relationships of watch officers and other watchstanders when underway and in port, the watch organization is considered separately for the two situations.

Watches in port or at sea are normally of four hours duration. The day's duty in port or at sea extends from 0800 to 0800 the following day, unless otherwise directed by the engineer officer. The change from IN-PORT to

UNDERWAY conditions occurs at the time the steaming watch is set. The change from UNDERWAY to IN-PORT conditions occurs at the time the auxiliary watch is set.

UNDERWAY.—The main propulsion assistant prepares the underway watch list (fig. 1-3) and submits it to the engineer officer for approval.

The engineering officer of the watch (EOOW) is in charge of the underway engineering watch organization and is primarily responsible for the operation of the main propulsion plant and auxiliaries during his watch. He is responsible for the operation of all other engineering department machinery and equipment in general and for the progress of the engineering department routine.

The engineering junior officer of the watch (EJOW) when assigned, assists the engineering officer of the watch in the performance of his duties and must be prepared to assume the duties of the engineering officer of the watch.

The underway key watchstanders consist of the (1) engineroom supervisors, (2) boilerman of the watch, (3) electrical petty officer of the watch, (4) interior communications petty officer of the watch, (5) damage control petty officer of the watch, and (6) petty officers in charge (fig. 1-3).

The electrical petty officer of the watch is stationed at the control distribution switchboard. He exercises control of the electrical distribution system and all distribution boards and operating generators through the petty officers in charge in each such space.

The interior communications petty officer of the watch is stationed in the designated interior communications room. He exercises control of the operating interior communications systems and equipment through the petty officer in charge in each such space.

Watch personnel in performing their duties shall:

1. Promptly obey all orders issued to them by the engineering officer of the watch or other competent authority.

2. Carry out applicable provisions of the Bureau of Ships Manual, Ship's Organization and Regulations Manual, Ship's Instructions and Notices, Engineering Organization Manual, Engineering Instructions and Notices, Machinery Operating Instructions and Safety Precautions, and all other directives issued by competent authority.

3. Not leave their posts without being properly relieved.

s!z _ " n *

4.IL

..o

*5

u »

3 5

is

4. Ensure that gages and meters are read correctly; that no loose rags, tools, or other material are adrift that might fall into machinery, cause an accident, or create a fire hazard; that proper settings of valves, switches, and safety devices are maintained; and that oil flow through bearings is uninterrupted.

5. Be able to detect any unusual sound or vibration of machinery, and investigate any abnormalities immediately.

6. Immediately investigate smells of hot oil, burning insulation, and smoke.

7. Visualize casualties and emergencies, which might occur and visualize the steps that should be taken to remedy them.

8. Know all safety precautions, operating instructions, and casualty control procedures of their assigned station.

IN PORT.—The engineer officer prepares the duty list for in-port watches (fig. 1-4). The in-port duty list must cover the routine administration of main propulsion, damage control, and electrical matters as well as emergency situations, such as emergency getting underway, fire, and rescue and assistance.

The engineering department duty officer is designated by the engineer officer to stand a day's duty and supervise the routine of the engineering department during a particular day in port. While on duty and during the absence of the engineer officer, he shall act on behalf of the engineer officer in all routine departmental matters and is responsible for the security and proper functioning of the department.

The engineering department duty petty officer is responsible to the engineering department duty officer for the security of departmental spaces and the proper functioning of engineering department personnel during a particular day in port. This watch shall be stood by a senior petty officer in each duty section (when his section has the duty) as a day's duty.

The engineering supervisory watch is responsible to the engineering department duty officer for the operation of departmental machinery and equipment, and the proper functioning of assigned personnel. This watch shall be stood by qualified petty officers, as designated by the

5 5t5 >- r ». *;

• - * ■ N M O •* *

-o ■

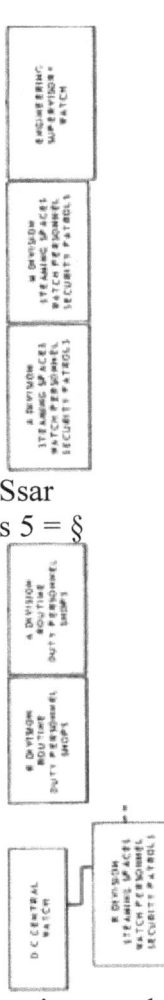

Ssar

s 5 = §

main propulsion assistant and scheduled by the engineering department administrative assistant.

RECORDS

Naval vessels are required to maintain certain records, which are an important part of your job as an Electrician's Mate. They provide an effective means of keeping the engineer officer posted on the status of the material in all parts of the plant and on the performance of all divisions of the department.

Revised Individual Allowance List

The revised individual allowance list (RIAL), which supersedes the machinery index, is a listing of all machinery and equipment except electronic equipment installed aboard a naval vessel. The RIAL for each item of equipment includes the (1) material group number, (2) complete nameplate data, (3) manufacturers'instruction book number, and (4) location in the ship. This data is required by the Bureau of Ships to provide adequate repair parts, battle damage components, and replacement equipment for the forces afloat.

The electronic installation record serves the same purpose for electronic equipment as the RIAL serves for all other units.

Material History

The engineer officer is responsible for maintaining the ship's material history. The material history, which supersedes the machinery history and hull repair record books, consists of cards filed in loose-leaf binders. The following four types of cards form the basis of the ship's material history:

1. Machinery History Card (NavShips 527).
2. Material History Card—Electrical (NavShips 527A).
3. Electronic Equipment History Card (NavShips 536).
4. Hull History Card (NavShips 539).

The purpose of these cards, when properly used, is to provide a comprehensive record of the items concerned. They must be kept up to date and available for inspection

at all times and are Integrated Into preventive maintenance programs, such as the Current Ship's Maintenance Project (CSMP).

MATERIAL HISTORY CARD-ELECTRICAL (Nav-Ships 527A).—The electrical officer Is responsible for maintaining the Material History Card—Electrical, which is the basic maintenance history card for electrical equipment (fig. 1-5). It provides for recording failures and other information pertaining to electrical equipment. An appropriate card is filled In for each item in the RIAL.

... , »r. • . . ,> jfW Ch

if Hp tt imtn »vw tyrt-M^ tw"«i

■it w off

fkLrt. K<»lt»»< mictr. »ju> tprt njU, HjHi.MA-IrHU.an tin. awV W>. J217»-W-t.. L— «pK «om — 41—

Figure 1-5.-Material History Card-Electrical (NavShips 527A).

RESISTANCE TEST RECORD (NavShips 531).-The Resistance Test Record (fig. 1-6) is provided with each piece of electrical equipment to record periodic insulation-resistance readings. The card is Inserted in the material history binder adjacent to the applicable equipment history card.

Current Ship's Maintenance Project

The CURRENT SHIP'S MAINTENANCE PROJECT (CSMP) provides a current record of maintenance, modifications, and repairs to be accomplished by the ship's force, shipyard, or tender during availabilities. The following three basic cards comprise the CSMP:

1. Repair Record, NavShips 529 (blue).

Figure 1-6.-Resistance Test Record (NavShips 531).

2. Alteration Record, NavShips 530 (pink).
3. Record of Field Changes, NavShips 537 (white). As a repair is required, or an alteration or field change

is authorized, the applicable card is filled in and placed in the material history binder adjacent to the appropriate history card. As the binder is examined, the distinctive colors of the CSMP cards readily indicate outstanding work. When the work has been completed and notations to this effect entered on the material history card, all the cards except the Record of Field Changes are removed from the binder and placed in a u completed work" file.

Repair Record Cards and Alteration Record Cards are retained for a period of two years after the work is completed and entries made in the material history. After the two-year period, these cards may be destroyed at the discretion of the commanding officer. When a ship is decommissioned or placed out of service, the active cards are retained onboard the ship.

Electronic Equipment History Cards, NavShips 536, and Record of Field Changes Cards remain with the equipment referred to on the cards. If the equipment is transferred, these cards are transferred with it.

Legal Records

The Engineer's Bell Book and the Engineering Log are official legal records. They can be used in any military

or civilian court as final proof of any action taken on or by the ship, and as evidence for or against any officer or enlisted man of the ship's crew who may be brought before the court or board.

The Engineer's Bell Book (NavShips 116) is a record of all bells, signals, and orders, and of the time they are received regarding the movement of the ship's propellers. The entries are generally made by the throttleman. However, when entering or leaving port, or during any maneuvering activity the entries should be made by an assistant. This procedure permits the throttleman to give full attention to the signals.

The engineering officer of the watch, before going off duty, must sign the Bell Book in the line following the last entry for his watch and the next officer of the watch must continue the record immediately thereafter. In machinery spaces where an engineering officer of the watch is not stationed, this record must be signed by the senior petty officer of the watch. Alterations or erasures are not permitted. An incorrect entry must be corrected by drawing a single line through it and making the correct entry on the following line. Such deleted entries must be initialed by the engineering officer of the watch, senior petty officer of the watch, or O.O.D. (in the case of ships and craft equipped with controllable reversible pitch propellers) as appropriate.

The records for each throttle control station for each day must begin with a new sheet, and the day's records for all stations must be clipped together and filed as a unit.

The Engineering Log (NavShips 117) is a midnight-to-midnight daily record of the ship's engineering department. It is a complete daily record, by watches, of important events and data pertaining to the engineering department and the operation of the ship's propulsion plant. The log must show the average hourly speed in revolutions and knots, the total engine miles steamed for the day, and all major speed changes; draft and displacement; fuel, water, and lubricating oil on hand, received and expended; the engines, boilers, and principal auxiliaries in use, and all changes therein. All injuries to personnel occurring within the department and casualties to material assigned to the department, and such other

matters as may be specified by competent authority, are entered into the engineering log.

The original entries in the log, neatly prepared in pencil or ink, is the legal record. The remarks must be prepared and signed by the engineering officer of the watch or day before leaving his station or being relieved. Any errors must be overlined and initialed by the person preparing the original entries.

The engineer officer must verify the accuracy and completeness of the entries and sign the log daily. The commanding officer must sign the log on the last calendar day of each month, and on the date of relinquishing command.

The Engineer's Bell Book and the Engineering Log must be preserved as a permanent record onboard for a three year period unless they are requested by a naval court or board, or the Navy Department. In such a case, a copy (preferably photostatic) of such sheets that are sent away from the ship are prepared and certified as a true copy by the engineer officer for the ship's files. At the end of the three-year period these records may be destroyed. When a ship is stricken, if either record is less than three years old, it should be forwarded to the nearest Naval Records Management Center.

The electrical log (fig. 1-7) is a complete daily record (from midnight to midnight) of the operating conditions of the ship's service electric plant. The log sheet must be kept clean and neat. Any corrections or changes to entries for a watch must be made by the man that signs the log for that watch. However, corrections or additions must not be made after the log sheet has been signed by the engineer officer without his permission or direction. The station logs are turned in to the log room every morning for the engineer officer's signature and for filing.

Maintenance Records

In addition to the records mentioned, Electrician's Mates are required to maintain a number of maintenance records at their assigned duty stations. These records include Daily Ground Test Sheets, Storage Battery Tray Record (NavShips 151), Storage Battery Charging (Discharging) Record, Ships Memorandum Work Request, and

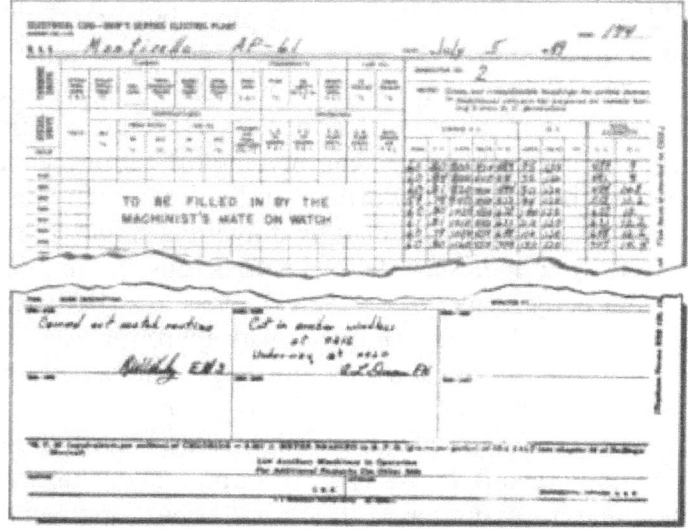

Figure 1-7.-Electrical Log-Ship's Service Electric Plan (NovShips 3649).

others. The maintenance records are of two types-official NavShips forms prepared by the Bureau of Ships and ship's forms prepared by the engineering department of the individual ship.

DAILY GROUND TEST SHEET.-Ground tests are made daily, and the readings are recorded on the Daily Ground Test Sheet (fig. 1-8). When the ground tests are completed, compare the results with the previous readings recorded on the test sheet and note any low readings. A sudden drop in insulation resistance must be investigated immediately and the trouble corrected. A comparison of the readings on the dally ground test sheet will show the difference in readings due to the normal deterioration of the cable and machinery insulation and those caused by a sudden ground.

STORAGE BATTERY TRAY RECORD (NavShips 151). —The Storage Battery Tray Record (fig. 1-9) is a complete history of each lead-acid battery aboard the ship. This record, which is kept up-to-date by the battery electrician, lists the (1) battery number, (2) nameplate data, and (3) record of service, repairs, charges, and test

Figure 1-8.-Daily Ground Test Sheet.

discharges. The information contained in this record shows the true condition of a battery and often indicates trouble in advance of battery failure.

SHIP'S MEMORANDUM WORK REQUEST.-The Ship's Memorandum Work Request (fig. 1-10) is an interdepartmental form used by any department that requires work to be performed by another department. This memorandum ensures the proper routing of work requests between heads of departments. The work requests, or job orders, list the work to be done and include an estimate of the man hours and material required to complete the job. Space is provided also for the shop that performs the work to list the material used, the manhours expended, the date of completion, and the approximate cost. Similar forms are filled in when a ship requires work to be done by a shipyard or tender.

EQUIPAGE CUSTODY RECORD (NavSandA 306A).-Tools and portable test equipment are equipage for which

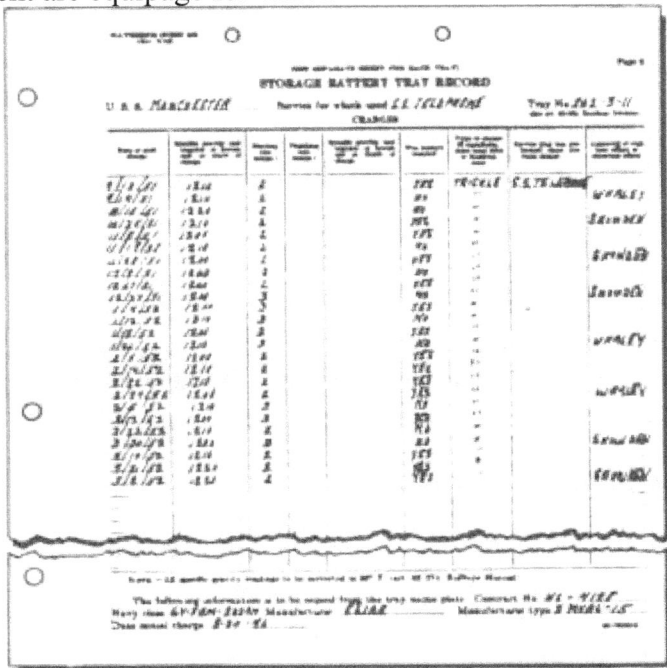

Figure 1-9.-Storage Battery Tray Record (NavShips 151).

custody signatures are required. This equipage is listed on the Equipage Custody Record (fig. 1-11), which is signed by the engineer officer when he receives it from the supply officer. The E division material officer signs a duplicate set of custody record cards when he receives the tools and test equipment from the engineer officer. The material officer, in turn, issues this equipment (when required) to Electrician's Mates who then are responsible for it. Any loss of accountable equipment must be reported to the E division material officer immediately so that a survey report can be made.

PUBLICATIONS

Publications pertinent to the operation and maintenance of electrical equipment and systems are kept in the log

: n roa. I

U.S. S MANCHESTER (CUV torn REQUEST (SOP'S FORCE)

Da.._7 J'V

r*» jlll j-i e^i offU« r
TO btm Ot(X> P.. Olj> . first U. <
1. It I • [iiunMd Ih.t In. r«ll**tM| Hlh b« Ki^l.'t'J
1 i •* i Aiift art i■ (• w
BttU* T*Uphon« SyttM - M JT
l^atiaai Itount U>-7
U3-52
•an la ha aana (tfcraiaft *ft*l«k. alMaalaaa. Pimm. alt., afcara aa.li.aklt.)
Cut off old 20 aLra connection box aotmUd on gun •hieId. Uanuf actur*
braekata to fit n«n nllch box and »ld In placa. Dlaornloni of braekata to b« tayan froa

work.

lark la ba Inopartaal / i
P.I..II, Qj \'7d 1 U..—1 Dafa..a4 <XX) '
•fcl.raal la aW*alr Otflca tart la* ~~
ft*, » »>a. 0.1.. ruaj «2a" 1
tin
Dal. .it atartad j jjRf 19 S2
( ) Ouwmta. (ho. ( ) D.,. ,,,,, raamUlad 10 kU 1952
(XX) Bupfmri RHe» ( ) fall Uckar
( ) Uact Ileal torkahaa ( ) faint Lackar
ftiinatu
e
IfcM*'t^t/'t fttcord

Figure 1-10.-Ship's Memorandum Work Request.

room, which is the office of the engineering department. Refer to the applicable publications before working on any equipment or circuits with which you are not thoroughly familiar.

Publications of primary interest to Electrician's Mates are (1) manufacturer's instruction books. (2) Bureau of Skips Manual, (3) Bureau of Ships Bulletin of Information, (4) Bureau of Ships Technical

DRILL, WUCTf.lt, Frotoblt, llfbt-dutjr, tjpt-A

Figure 1-11.-Equipage Custody Record (NovSandA 306A).

Bulletins, (5) Ship Information Book, volume 3, and (6) ship's plans.

Instruction Books

Manufacturers' Instruction books contain technical Information and Instructions for the operation and maintenance of specific apparatus. This Information usually Includes a general description of the equipment, principles of operation, Installation instructions, operating data, maintenance procedures, and safety precautions.

In most ships, instruction books are issued to responsible personnel who must sign custody receipts. This procedure is necessary because the number of copies allotted to each ship is small and the replacement of missing copies Is very costly.

Bureau of Ships Manual

The Bureau of Skips Manual is the most Important Bureau of Ships publication. This manual contains the administrative and engineering instructions for the use of the engineering department. It describes methods of conducting tests, procedures for making repairs, and many helpful maintenance suggestions. It Is the official authority on operating procedures.

As an Electrician's Mate, you should be familiar with the following chapters of the Bureau of Skips Manual.

Bureau of Ships Bulletin of Information

The Bureau of Ships Bulletin of Information contains data concerning the maintenance and operation of naval vessels. This information includes analysis of casualties, research, developments, and reports concerning tests on material, equipment and apparatus.

Bureau of Ship's Technical Bulletins

Bureau of Ships Technical Bulletins and similar publications are for the dissemination of information concerning the (1) design and construction of ship's machinery and equipment, (2) technical developments, and (3) accomplishments in the field of research.

Ship Information Book , Volume 3

The General Specifications for Ships of the United States Navy require that the contractor furnish copies of the Ship Information Book, which consists of several volumes, to each ship built in accordance with these specifications. Power and lighting, Volume 3 of the Ship Information Book serves a particular ship, and replaces the Record of Electrical Installations and Electrically Operated Auxiliaries, previously required. Volume 3 consists of two parts. Part 1 contains a general description, and design Information of each electrical auxiliary and equipment including the interior communications systems. Part 2 contains the manufacturers' and shipbuilder's test data. This part is particularly valuable as a basis for comparison of test results obtained after the equipment has operated over a period of time.

Damage Control Books

Damage Control Books are issued by the Bureau of Ships to the larger ships of the fleet, and contain information and instructions concerning the ship's damage control systems. A detailed description of the Damage Control Book is contained in Chapter 88, Section I of the Bureau of Ships Manual. Requests for additional copies should be directed to the Bureau of Ships.

Ship's Plans

A complete file of standard plans and blueprints is available in the log room for reference and study. At the time a vessel is delivered, the shipbuilder furnishes a set of blueprints to the commanding officer via the supervisor of shipbuilding. These blueprints are in accordance with a list of working plans corrected to show the equipment, as installed. Two copies of the ship's plan index are also furnished to the ship. This index lists all plans under the cognizance of the Bureau of Ships, which apply to the vessel concerned.

The plans furnished to the ship include elementary and isometric wiring diagrams of all circuits. Each time an alteration is completed, the yard accomplishing the work

furnishes the ship a copy of the plans that show the alteration. A revised copy of the ship's plan index also is furnished, which shows all plans including the latest alteration numbers that apply to the ship. Copies of the latest alterations of Bureau of Ships standard plans applicable to the equipment installed in the ship must be carefully safeguarded. If these plans are lost, replacement plans might be for a later alteration that is not applicable to the equipment actually installed.

QUIZ

1. What are some of the methods for converting civilians into Navy men and Navy men into competent seamen ?

Z. What two broad general classes of duties do Electrician's Mates perform?

3. Which Navy publication lists the military requirements and the professional

qualifications for all enlisted ratings?

4. The rating of the Electrician's Mate consists of how many general service ratings and how many emergency service ratings?

5. As a man shows proficiency in each practical factor, a record of its completion is entered in what Navy form?

6. Which Navy publication lists an annual bibliography of current Navy training courses and other publications that have been prepared for the use of all enlisted personnel concerned with advancement in rating examinations?

7. How can you obtain copies of the Navy training courses ?

8- Whether you are a successful leader or not will be decidednot by compiled lists of desirable traits,but, for the most part, by what final result?

9- How should study about a particular equipment be arranged in relation to actual work on the equipment?

10. What is a good way to review your knowledge of the material contained in each chapter of a training course?

11. The complement of a combat ship is based on the number of personnel required to perform which three general functions?

12. For wartime organization, each division is divided into how many sections?

13. Each section should be able to perform which two important functions under emergency conditions?

14. Assignments to the various billets prescribed by the Ship's Battle Bill, Ship's Watch Organization, Ship's Organization Bills, and departmental and divisional administrative and watch organizations are published in what division bill?

15. Normally, how many divisions comprise the engineering department of a Navy combat ship?

16. Watches in port or at sea are normally how long?

17. Where aboard ship is stationed the electrical petty officer of the watch?

18. Which list supersedes the machinery index and is a listing of all machinery and equipment, except electronic equipment, installed aboard a naval vessel?

19. Which card provides for recording failures and other information pertaining to electrical equipment?

20. Which record card is provided with each piece of electrical equipment to record periodic insulation-resistance readings?

21. The Repair Record, Alteration Record, and Record of Field Changes cards comprise what project?

22. What are the two operating records normally maintained by the electrical division?

23. Maintenance records that the Electrician's Mates are required to maintain at their duty stations are of what two general types according to where they originate ?

24. A daily study of what record will show the difference between readings due to the normal, slow deterioration of the cable and machinery insulation, and those due to a sudden ground?

25. What interdepartmental form is used by any department that requires work to be performed by another department?

26. Custody signatures for tools and equipment are required on what form?

27. Where aboard ship are the publications kept that are pertinent to the operation and maintenance of electrical equipment and systems?

28. In most ships , instruction books are issued to responsible personnel on what type of receipts?

29. What manual contains the administrative and engineering instructions for the engineering department?

CHAPTER

ELECTRICAL CABLE AND ACCESSORIES

Proper cable installation and maintenance are equally as important to the efficient operation of an electrical system as the proper installation and maintenance of the equipment to which the cable is connected. The Electrician* s Mate must have a thorough knowledge of the various types, sizes, capacities, and uses of shipboard electrical cable, and be capable of selecting, installing, and maintaining cable in such a manner as to ensure its adequacy. The repair of battle damage, accomplishment of ship alterations, and some electrical repairs may require that changes or additions to the ship's cable installation be made by the Electrician's Mate. During shipyard and tender availabilities, he may be required to inspect, test, and approve new cable installations.

The adequacy of an electrical cable installation is not only concerned with current-carrying capacity and insulation strength, but because of the varied service conditions aboard ship, is also related to the ability to withstand heat, cold, dampness, dryness, bending, crushing, vibration, twisting, and shock. No one type of cable has been designed to meet all of these requirements; therefore, a variety of types are employed in a shipboard cable installation. Basically, shipboard electrical cables are classified as nonflexing service cable and repeated flexing service cable.

CABLE DESIGNATIONS

Shipboard electrical cables are identified according to type and size. The type designations consist of letters

that indicate construction or use. The size designations, which follow, consist of numerals that indicate their size.

Type and Size Designations

The type letters consist of the first letters of the words used in describing the cable; for example, DBSP (Double, Boat, Shielded, Plain) indicates a double conductor, small boat shielded, plain (unarmored) cable.

The size designation relates to the conductors that indicate the size; the stranding; and the number, or number of twisted pairs, of conductors. No set rule has been established for application of size designations, but generally the following will apply:

1. When the exact number of conductors in the cable is indicated by a letter (S, D, T, or F) in the type letter designation, the numerals following the letter designation will indicate the approximate cross-sectional area of the conductors expressed in thousands of circular mils. For example,type TSGA-9 indicates a three-conductor, shipboard general-use cable with armor and 9016 circular mil conductors.

2. When the numerals immediately following the type letter designation indicate the size of the conductor and are followed by numerals in parentheses, the latter numerals will indicate the number of strands per conductor. For example, SRI-4(7) is a single-conductor, resin-insulated cable with one 4497-circular mil conductor containing seven strands.

3. When the exact number of conductors in the cable is not indicated by a letter in the type letter designation, the numerals following the letter designation will indicate the number of conductors in the cable. For example, MSCA-7 is a multiple conductor, shipboard-control cable

with armor and seven conductors.

4. When the numerals immediately following the type letter designation indicate the number of conductors in the cable and are followed by numerals in parentheses, these numerals will indicate the size of the conductor. For example, MDGA-19(6) is a multiple conductor, degaussing cable with armor and nineteen 6512-circular mil conductors.

5. When a cable contains two conductors that are twisted together to form a pair, the numerals following the type letter designation will Indicate the number of twisted pairs in the cable. For example, TTHFWA-10 indicates a twisted pair, telephone, heat- and flame -resistant, watertight cable with armor and 10 twistedpairs of conductors.

Nonflexing Service

Nonflexing service cable designed for use aboard ship is intended for permanent installation and is commonly referred to as permanently Installed cable. It can be classified according to its application and is of two types-general use and special use.

GENERAL USE.—Nonflexing service cable intended for use in all portions of electric distribution systems (except in special cases and in d-c propulsion circuits for surface vessels in which the impressed voltage is less than 1000 volts) is distinguished by the letter designation SGA (Shipboard General Use Armored) and is illustrated in figure 2-1. This cable employs silicone rubber and glass as primary insulation, making it heat and flame resistant. The specified number of insulated stranded copper conductors are enclosed in an impervious sheath, braided metal armor, and paint. The cable has a maximum voltage rating of 1000 volts a-c or d-c, and is made watertight by the application of waterproof sealing compound to all interstices (voids) of the conductors and cable core. SGA cables are designed to have a minimum diameter and weight consistent with service requirements in fixed wireways on combatant naval vessels, and supersede the type HFA (Heat and Flame resistant—Armored). Type MSCA is a reduced diameter, watertight, multi-conductor cable for use in interior communications and fire control circuits. This type supersedes the older MHFA.

SGA cables are of four different types, depending on the number of conductors they contain. These are SSGA (single conductor), DSGA (double conductor, fig. 2-1), TSGA(three conductor), and FSGA (four conductor). Type SSGA cable is available in conductor sizes ranging from 2828 through 829,300 circular mils (SSGA-3 through

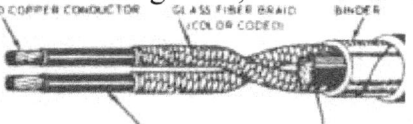

F«T*L,DEC SILICONE RUBBER FILLER

BRAIDED METAL ARMOR AND PAINT

Figure 2-1.-Type DSGA shipboard nonfltxing service cable.

SSGA-800); DSGA and TSGA from 2828 through 413,600 circular mils (DSGA-3 through DSGA-400 and TSGA-3 through TSGA-400); and FSGA from 2828 through 49,080 circular mils (FSGA-3 through FSGA-50).

SPECIAL USE.—There are many shipboard electrical circuits where special requirements of voltage, current, frequency and service must be met in the cable installation and other circuits where general-use, nonflexing service cable may meet the necessary requirements, yet be economically impracticable. For these reasons, there are many different types of nonflexing service cable for specialized use, such as degaussing, telephone, radio, and casualty power. Some of these cables are shown in figure 2-2.

Type MDGA (fig. 2-2, A) is a multiconductor cable used in degaussing circuits. Type SHFP (fig. 2-2, B) is a single conductor cable designed for use in high-voltage (greater than 1000volts) propulsion systems. Type PBJX cable (fig. 2-2, C) consists of one conductor of constantan (red) and one conductor of iron (gray), and is used for pyrometer base leads. Type TTHFWA (fig. 2-2, D) is a multiconductor, twisted-pair cable used for telephone circuits.

Repeated flexing service cable designed for use aboard ship is commonly referred to as being portable because it is principally used as leads to portable electric equipment. It is also of two types—general use and special use.

GENERAL USE.—Repeated flexing service cable, which is designed for use as leads to portable equipment and permanently installed equipment where cables are

Repeated Flexing Service
A TYPE MDGA CABLE

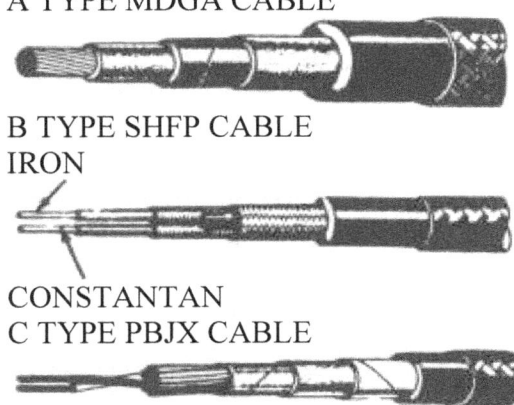

B TYPE SHFP CABLE
IRON

CONSTANTAN
C TYPE PBJX CABLE

D TYPE TTHFWA CABLE
Figure 2-Z-Nonflexing service cable for speciol use.

subjected to repeated bending, twisting, mechanical abrasion, oil, sunlight, or where maximum resistance to moisture is required, is distinguished by the letter designation, HOF (Heat and Oil resistant, Flexible). This cable contains stranded copper conductors that are insulated with butyl rubber, covered with a tape or braid. The designated number of conductors are twisted together, held by a binder, and covered with an impervious sheath (fig. 2-3).

QQ
IMPERVIOUS SHEATH SEPARATOR COTTON TAPE (RUBBER)

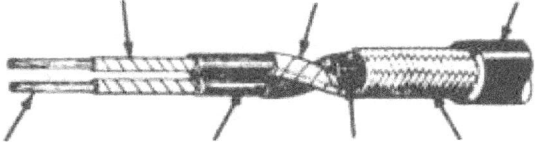

CONDUCTOR SYNTHETIC RUBBER FILLER BINOER
Figure 2-3.-Type DHOF cablt.

Repeated flexing service cable designed for general use is of four different types, depending on the number of conductors.

The types are SHOF (single conductor), DHOF (two conductor), THOF (three conductor), and FHOF (four conductor). Type SHOF cable is available in various conductor sizes ranging from 2594 (SHOF-3) through 812,700 circular mils (SHOF-800) types DHOF and

THOF from 2613 (DHOF-3 and THOF-3) through 413,500circu-lar mils (DHOF-400 and THOF-400); and type FHOF from 2613 (FHOF-3) through 137,800circular mils (FHOF-133).

SPECIAL USE.—There are many different types of repeated flexing service cable designed for special requirements of certain installations, including type TTOP

A TYPE MHFF CABLE

Figure 2*4.-Repeated flexing service cable for special use.

and casualty power cables. Two of these types are shown in figure 2-4. Type MHFF (fig. 2-4, A) is used for control circuits in revolving structures, and type TRF (fig. 2-4, B) is used for arc-welding circuits.

INSTALLATION

Electrical cables installed aboard U. S. Naval vessels must meet certain requirements determined by the Bureau of Ships. These requirements, published in the General Specifications for Ships of the U. S. Navy, are too numerous to cover in detail in this training course; hence, only the more basic ones are included.

The job of installing nonflexing service cable may be performed by the Electrician's Mate whenever necessary to repair damage or to accomplish authorized ship alterations. Before work is begun on a new cable installation, cableway plans should be available. If repairs to a damaged section of installed cable are to be effected, information on the original installation can be obtained from the plans of the ship's electrical system, which are normally on file in the engineering department office (log room) aboard ship. If a ship alteration is to be accomplished, applicable plans not already on board can be obtained from the naval shipyard listed on the authorization for the alteration (SHIP ALT) at the planning yard for the ship.

Cables installed aboard ship must have maximum usable life under service conditions of temperature, moisture, and mechanical abuse. Also, they must have minimum weight for the service intended. The cables must be located to (1) minimize damage from battle action, (2)avoid physical and electrical interference with other equipment and cables, and (3) provide for the maximum dissipation of internally generated heat.

Selecting Cable

Two-conductor cable shall be installed for 2-wire, d-c and single-phase, a-c circuits. Three-conductor cable shall be installed for 3-wire, d-c or 3-phase, a-c circuits. Four-conductor cable shall be installed where two 2-wire lighting circuits are run in the same cable. Four-conductor and multiconductor cable shall be installed for control circuits and communications circuits as necessary.

Ambient temperature, load current, and voltage drop are the principal factors that determine the proper size cable to be used in a particular installation. Ratings and characteristics tables, which specify the maximum current carrying capacities of the various types of shipboard electrical cables at 40° C and 50° C ambient temperatures, are contained in Chapter 40 of the Bureau of Ships Manual. Cables installed in U. S. Naval vessels must comply with the requirements of these tables. A portion of the table for nonflexing service cables is shown in table 1.

To select the proper size cable for a particular installation, it is necessary to know (1) the total connected load current, (2)the demand factor, (3)the resultant load current, and (4) the allowable voltage drop.

The total connected load current for d-c power circuits is determined by adding the sum of the rated current of the connected loads as listed on the identification plates of connected motors and appliances and an additional 100 watts for each receptacle not specifically indicated. For a-c power circuits the connected load current of the connected motors and appliances is added vectorially to obtain the total connected load current.

The demand factor of a circuit is the ratio of the maximum load averaged for a 15-minute period to the total connected load on the cable. If the feeder demand factor for a group of loads cannot be determined, a value of 0.9 maybe assumed. For power systems supplying a single-phase load or for a lighting system branch, submain, and main circuits the demand factor is unity.

The resultant load current is the product of the total connected load current and the demand factor of the circuit. For most installations with which the Electrician's Mate is concerned, the resultant load current and the total connected load current will be the same value. In other words, the demand factor of these circuits will be unity. Cables sizes are usually selected on the basis of maximum current-carrying capacity and allowable voltage drop. However, if the length of the cable installation is short and the total connected load is small, the resultant load current is usually the principal factor that

Table 1.-Ratings and characteristics of cables for nonflexing service (permanent installation) heat and flame resistant

determines the size of the cable to be used, and it is not necessary to calculate the allowable voltage drop. For other cases, however, the allowable voltage drop should be determined.

Determining Allowable Voltage Drop

The voltage drop (difference in voltage between any two points in a circuit) is expressed as a percentage of the rated switchboard (or switchgear group) bus voltage or the transformer nominal voltage. The maximum percent voltage drop allowed for a circuit is specified by the Bureau of Ships and varies according to the intended service of the circuit. For example, General Specifications for Ships specifies that the size of cables forana-c ship's service lighting system shall be such that the voltage drop in the most remote branch, when calculated using a resistivity of 12 ohms per circular mil foot (resistivity of copper at 45° C), will not exceed 6 percent exclusive of transformer regulation.

To obtain a system having minimum weight, the percent voltage drop (VD) allocated to a particular section will be directly proportional to the length of that section. For example, if the total allowable voltage drop for a system having an average over-all length (AOL) of 250 feet is 6 percent, the allowable drop allocated to a particular section having a length of 100 feet and operating at the same voltage will be 100/250 «6 = 1.5 percent.

The length of each section must be based on a common reference voltage. If any section operates at a different voltage, its AOL is multiplied by the ratio of the reference voltage to the voltage of that section. For example, if a particular section has an AOL of 100 feet, an operating voltage of 450 volts, and a common reference voltage is 117 volts, the AOL of the section based on 117 volts will be 100 * 117/450 = 26 feet.

The AOL for any section of a circuit is computed as the arithmetic average of the lengths in feet from the beginning of a section of the circuit involved to the end of the branch circuits connected to the section under consideration. For example, the AOL of a section supplying three branch circuits having lengths of 150 feet, lOOfeet, and 200 feet is

A A
150 + 100 + 200 450

3 = ~3~= 150 feet.

The procedure for determining the allowable percent voltage drop in a circuit is best explained by using the sample ship's service lighting system plan shown in figure 2-5. The information on this illustration is similar to the information found on the plans titled, List of Feeders and Mains for Lighting Systems, furnished for the lighting system of each naval vessel.

The computations for the over-all lengths of the nine branches supplied by feeder F-120 from the branch circuit boxes back to the 117-V Feeder/Distribution Lighting Panel No. 4-149-2 are:

The AOL of feeder F-120 from panel No. 4-149-2 to the end of the system is 2655/9 = 295 feet. The AOL of the other feeders from this panel are combined with that of F-120:

Feeder Over-all Lengths in Feet

F-120 295'
F-122 107'
F-124 262'
F-126 221'
F-128 268'
F-130 219'
F-132 138*
Total length 1510

a e

The average is 1510/7 = 216 feet. The AOL of feeder F-0136 is 216 + 25 = 241 feet. The AOL of feeder F-0436(l) is 241 + [25 * 117/450 = = 247 feet.

The AOL of the other feeders from panel No. 4-148 is combined with that of F-0436(1):

Feeders Over-all Lengths in Feet

F-0436(l) 247' F-438 188' F-440 237' F-442 284'

Total length 956' The AOL is 956/4 = 239 feet.

The AOL of feeder F-0436 is 239 + [100 * 117/450 = 26 = 265 feet from switchboard No. 4 to the end of the system.

The calculations for the percent voltage drops allocated to the various sections supplied by feeder F-0436 are shown in table 2.

The maximum voltage drop allowed (using a resistivity of 12 ohms per circular mil foot) is 8.0 percent in a d-c ship's service lighting system, 15.0 percent in a d-c emergency lighting system, and 12.0 percent (exclusive of transformer regulation) in an a-c emergency lighting system.

The maximum voltage drop allowed from ship's service or emergency switchboard or switchgear group to the terminals of power consuming equipment (using a resistivity of 13 ohms per circular mil foot) is 6.0 percent for a-c systems and 8.0 percent for d-c systems. Under starting conditions, the voltage drop in a-c circuits must not exceed 12.0 percent (exclusive of transformer regulation) except for welders and other appliances and motors where specifically approved by the Bureau of Ships.

In degaussing installations (using ship's service d-c power) the voltage drop must not exceed the value that will permit full design current to flow through the degaussing coil with the rheostat in the all out position and, in the case of a multiconductor cable installation, with approximately a 10-volt drop in the series resistor for compass degaussing compensation.

Table 2.-Percent voltage drops allocated to the various sections
Branch or feeder No.
Length of section
450 v
117 v
AOL
System
Percent voltage drop in
Single section
Input of section to end of system
F-0436
F-0436(l)
F-0.136
F-120
F-120(l)
6-F-120 6-F-120(A)
100' 25'

100 x 117
450
25 * 117 450
25' 180' 110'
= 26'
= 6'
15'
60'
265" 247' 241'
295'
180'
75' 60"
26 265
6 247
•25 241
« 6 = 0.6
5.4 = 0.1
x 5.3 = 0.6
••180
295
***1_10
180
(use 0.4)
« 4.9 = 3.0 (use 2.9)
x 2 = 1.2 (use 1.1)
y| x 0.9 = 0.2
60
60
f£ x 0.7 = 0.7
6 - 0.6 = 5.4
5.4 - 0.1 = 5.3
5.3 - 0.4 = 4.9
4.9 - 2.9 = 2.0
2 - 1.1 = 0.9
0.9 - 0.2 = 0.7

The total dropas viewed from 6-F-120(A) = 0.6 + 0.1 + 0.4 + 2.9 + 1.1 + 0.2 + 0.7 = 6.0 percent. •The cable size required to yield a drop of 0.6 percent in feeder F-0136 is T-300. However, the installed cable sire is based on the transformer rating that requires a T- 35 0 cable to carry the continuous load. The voltage drop resulting from the use of a T-350 cable is 0.4 percent. ••The cable sire that most closely yields a 3-percent drop in feeder F-120 without exceeding this drop is T-50. The voltage drop resulting from the use of T-50 cable will be 2.9 percent. •••The cable size that yields a drop of 1.2 percent in feeder F-120(l) without exceeding this drop is T - JO. The voltage drop resulting from the use of T-30 cable is 1.1 percent.

The formula for determining the percent voltage drop in a 2-wire, d-c circuit supplying ship's service and emergency lighting loads is

$$VD = p \times I \times jLx\ 100 = 2400Jk \text{ percent, cm} \times E \text{ cm} * E$$

where p (greek rho)is the specific resistivity of the conductor in ohms per circular mil foot (12 ohms at 45° C), E the rated voltage at input to circuit, I the resultant load current in one leg of the circuit in amperes, and L the length (in feet) of a single conductor in circuits having equal conductor length in both legs. For example, in a two-wire, d-c lighting circuit 25 feet long, the resultant load current is 2.6 amperes, and the rated voltage at the

fuse box is 120 volts. The maximum allowable percent voltage drop for this circuit is 0.5 percent. Determine if a DSGA-3 cable will satisfy both the current and voltage drop requirements of this circuit.

vn 2400 x 2.6 x 25 « AO

m 2828 x 120 = °46 Per ° ent -

Because this value is less than the maximum allowable value, the DSGA-3 cable will satisfy the required voltage drop. The general ampere rating (maximum) for DSG-3 cable operating at 50° C (table 1) is 12 amperes (d-c); hence, the resultant load current of 2.6 amperes is well below the maximum allowable value, and the DSGA-3 cable will satisfy the current requirement of this circuit.

The voltage drop formula is the same for a 2-wire, d-c ship's service and emergency power system, except that a resistivity of 13 ohms per circular mil foot is used.

For a 3-wire, d-c ship's service or emergency lighting system, the formula for determining the voltage drop is

Trrx P * (k + 0.25 LJ x * 100 1500 x I a x L

^ = cm x E = cm 5 E Percent,

where p is 12 ohms per circular mil foot at 45° C, I* the average of the resultant load currents in the two outside conductors of the 3-wire circuit (assuming a 25 percent current unbalance), and E the voltage to neutral. For example, in a 3-wire, d-c lighting feeder 50feet long with la of 30 amperes and 120 volts between the positive leg and neutral, determine if a size TSGA-23 cable will satisfy a 1.0 percent maximum allowable voltage drop.

1500 x 30 x 50 . q = 22,800 x 120 = °' 8 P ercent -

Thus, TSGA-23 cable will meet the voltage drop requirements of the circuit. From table 1 the general ampere rating (maximum) at 60 cycles is 69 amperes; hence, it is apparent that this cable will also satisfy the current requirements of the circuit.

The voltage drop formula is the same for a 3-wire, d-c ship's service and emergency power system, except that a resistivity of 13 ohms per circular mil foot is used.

The calculations for voltage drop in a-c circuits involve cable reactance and load power factor, in addition to the cable resistance and load current. The resulting vector diagrams with their trigonometric solutions are somewhat lengthy in contrast with the simple formulas for d-c, 2- and 3-wire circuits. However, in the case of relatively short cable runs and loads operating at or near unity power factor, certain simplifying assumptions can be made that will eliminate the need for vector solutions and at the same time give an approximately correct solution. Thus, if the voltage drop in the cable is assumed to be all resistive and in phase with the load voltage, the formula for the percent voltage drop for a single-phase, 2-wire, a-c ship's service and emergency lighting load is the same as that for the corresponding 2-wire, d-c circuit. With similar assumptions the formula for a single-phase, 3-wire, a-c ship's service or emergency lighting system is the same as that for the corresponding 3-wire, d-c circuit.

With similar assumptions the voltage drop, E, for 3-phase circuits may be approximated:

vTp LI — cm

where p is the resistivity in ohms per circular mil foot, L the length of cable (one way) in feet, 1 the load current in amperes per conductor, and cm the cross-sectional area of the conductor in circular mils.

Determine the percent voltage drop in feeder F-0136 by using T-350 cable (fig. 2-5). In this example, p is 12, L is 25 feet, I is 324.8 amperes, the source voltage is Til volts, and cm is 349,800.

/3xp xLx! 1.732 x 12 x 25 x 324.8 n 4 q volts _ = cm ~ 349,800

VP _ = o.41 percent

Wire ways

Before installing new cable, survey the area to see if there are spare cables in existing wireways and spare

R1

stuffing tubes that can be used in the new installation. The cable run must be located so that damage from battle will be minimized, physical and electrical interference with other equipment and cables will be avoided, and maximum dissipation of internally generated heat will occur.

First consideration must be to protect the cable against battle damage. The protection afforded by the ship's structure must be fully used. Do not run cables on the exterior of the deckhouse or similar structures above the main deck, except where necessary because of the location of the equipment served, or because of structural interferences or avoidance of hazardous conditions or locations. Where practicable, route vital cables along the inboard side of beams or other structural members to afford maximum protection against damage by flying splinters or machine-gun strafing. Where exposed runs of vital cables are unavoidable, a method of protection approved by the Bureau of Ships must be used.

Where practicable, avoid installing cable in locations subject to excessive heat, and never install cable adjacent to machinery, piping, or other hot surfaces having an exposed surface temperature greater than 150° F. In general, cables shall not be installed where they may be subjected to excessive moisture. Cable runs that are unavoidably near fire mains and water, steam, oil, or other piping should be provided with drip-proof shields or other barriers to protect against damage from leaks or high-temperature steam.

If any other location is practicable, electric cables should not be run through spaces where hazardous materials are stored or handled (hangar spaces, powder magazines, gasoline stowage compartments, etc.) or through adjacent spaces that are normally open to such spaces. If cables must be run through hazardous spaces, only heat- and flame-resistant armored cable shall be used in such spaces, and through cables shall be of unbroken length within spaces. Cables terminating at equipment in these spaces shall be of unbroken lengths; however, a single cable may be run between lighting fixtures when more than one fixture is installed in such spaces.

▶

* I

*

i

Y

d

:-S

Cables that terminate in, or are routed solely within, highly hazardous spaces, such as gasoline pump rooms, gasoline filter rooms, etc. must be provided with special protection. Protect these cables by installing them in watertight or channel-shaped steel casings of sufficient strength. Air spaces must be provided between protective casings to facilitate heat dissipation and to prevent condensation on the cables. Drainage must be provided for the casings (fig. 2-6).

HORIZONTAL RUNS SINGLE ROM OF CABLES
VERTICAL RUNS SINGLE ROW OF CABLES
TOP PLATE - STEEL, CONTINUOUS _
REMOVABLE PROTECTIVE PLATES STEEL, LENGTH OF SECTIONS 6ET MAX
ANGLE CLIPS STEEL

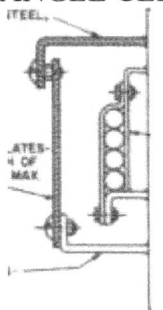

MAX SPACING W ON CENTERS
BULKHEAD OR FRAME
REMOVABLE
PROTECTIVE PLATES STEEL, LENGTH OF
sections err max
ALLOW CLEARANCE AT TOP AND BOTTOM

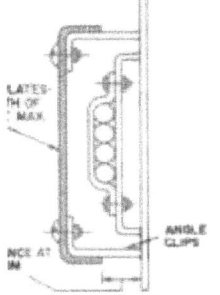

PLAN VIEW
PLAN VIEW

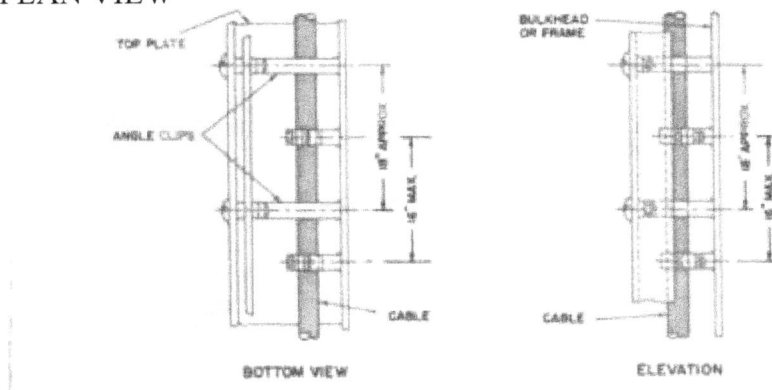

Figure 2-6.-Cable runs with protective casings.

e o

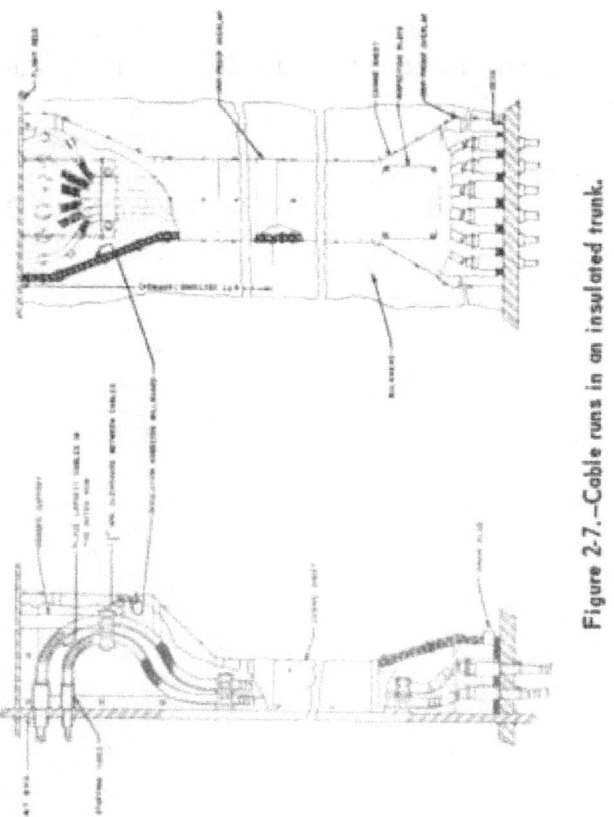

Figure 2-7.—Cable runs in an insulated trunk.

Cables that run through gasoline stowage compartments and spaces containing stowage for other flammable volatile liquids must be installed in an insulated trunk (fig. 2-7). They should be installed so that they are not supported on the trunk itself, and if practicable, bolted access plates should be provided in each stuffing tube.

Cables that run through tanks containing combustible liquids must be provided with concentric pipe enclosures (fig. 2-8).

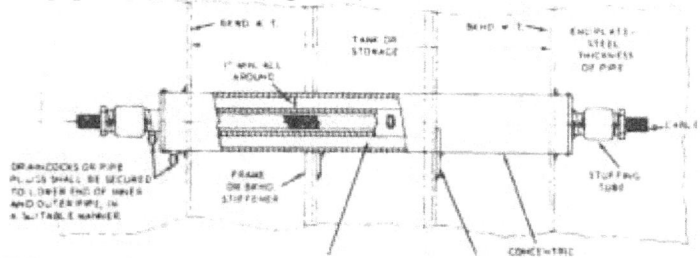

TO DRAINAGE LOCATE »T FRAMES ENCLOSURE
CONDUIT SUPPOBTS-SIE EL n i] SPACEC A Wl Of a FT. I ll APART ANC *EL DEC]| TO INNER PIPE

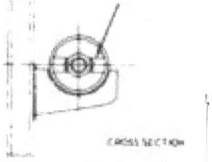

Figure 2-8.-Concentric pipe enclosure cable installation.

Fittings

STUFFING TUBES.-Stuffing tubes (fig. 2-9, A, B, and C) are used to provide for the entry of electric cable into splashproof, spraytight, submersible, and explosion-proof equipment enclosures. Cable clamps, commonly called box connectors of the types shown in figure 2-10, may be used for cable entry into all other types of equipment enclosures, except that top entry into these enclosures shall be through stuffing tubes.

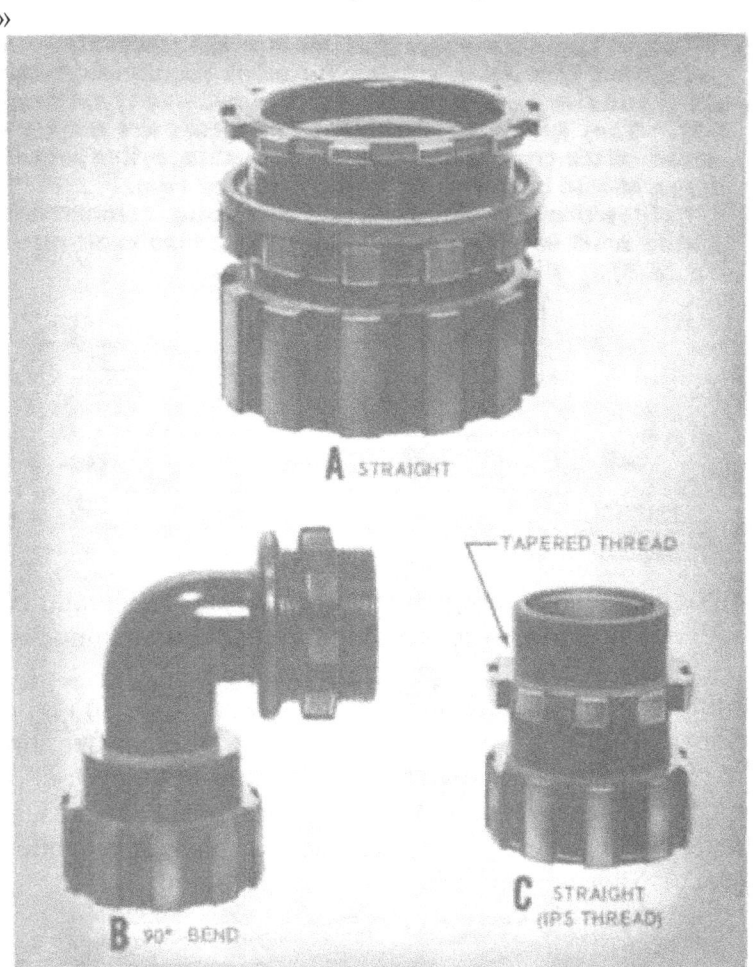

Figure 2-9.-Nylon stuffing tubes.

Below the main deck, staffing tubes are used for cable penetrations of watertight decks, watertight bulkheads (fig. 2-11), and watertight portions of bulkheads that are watertight only to a certain height. Above the main deck, stuffing tubes are used for cable penetrations of (1) watertight or airtight boundaries; (2) bulkheads designed to withstand a waterhead; (3) that portion of bulkheads

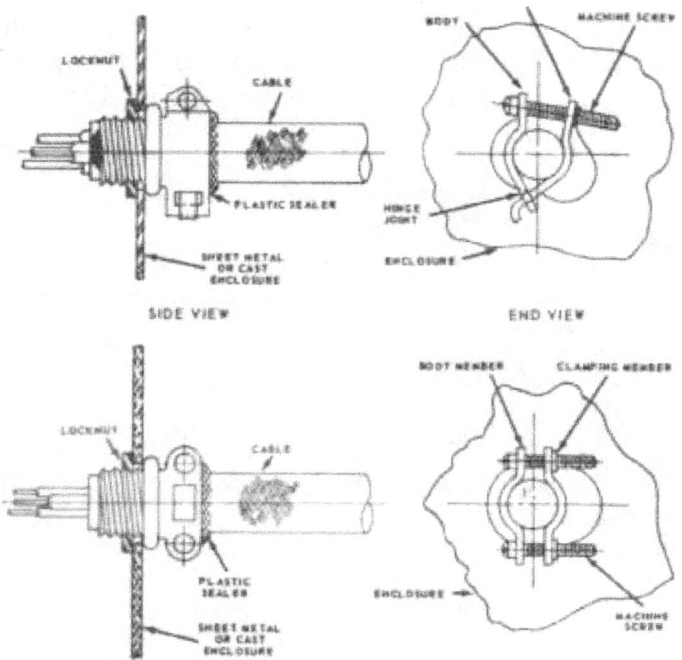

SIDE VIEW END VIEW

Figure 2-10.-Cable clamps.

below the height of the sill or coaming of compartment accesses; (4) flametight or gastight, or watertight bulkheads, decks, or wiring trunks within turrets or gun mounts; and (5) structures subject to sprinkling.

Stuffing tubes are made of nylon, steel, brass, or aluminum alloys. Nylon tubes have practically replaced metal tubes for cable entry to equipment enclosures. Cable penetration of bulkheads and decks is limited, at present, for use above the watertight level of a vessel. The watertight level is the highest expected water level (determined by the Bureau of Ships studies of stability and reserve buoyancy) and is indicated on the applicable ship's plans. The nylon tube is a lightweight, positive-sealing, noncorrosive stuffing tube, which requires only

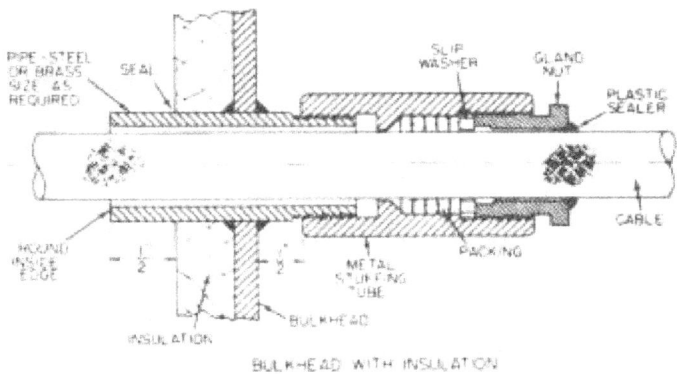

BULKHEAD WITH INSULATION

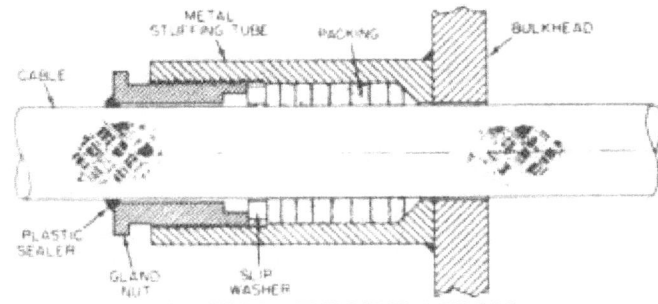

BULKHEAD WITHOUT INSULATION

Figure 2-11.-Stuffing tubes on watertight bulkheads.

minimum maintenance for the preservation of watertight integrity (fig. 2-12). The watertight seal between the entrance to the enclosure and nylon body of the stuffing tube is made with a neoprene "O" ring, which is compressed by a nylon locknut (fig. 2-12, A). A grommet-type, neoprene packing is compressed by a nylon cap to accomplish a watertight seal between the body of the tube and the cable. Two slip washers act as compression washers on the grommet as the nylon cap of the stuffing tube is tightened. Grommets of the same external size, but with different sized holes for the cable, are available.

SHIP'S STRUCTURE

ENCLOSURE

BCOY - TUBE BODY INSERTED FROM INSIDE ENCLOSURE

BOTTOM WASHER

COAT SURFACE INDICATED WITH NEOPRENE CEMENT

WRAP END OF CABLE ARMOR WITH FRICTION TAPE TO A MAX DIAMETER WHICH WILL PASS THROUGH THE SLIP WASHERS

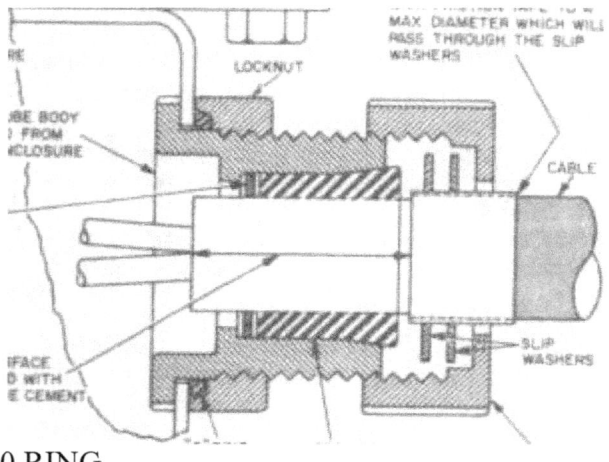

0 RING
GROMMET
CAP
BOOT
BOTTOM WASHER
ENCLOSURE

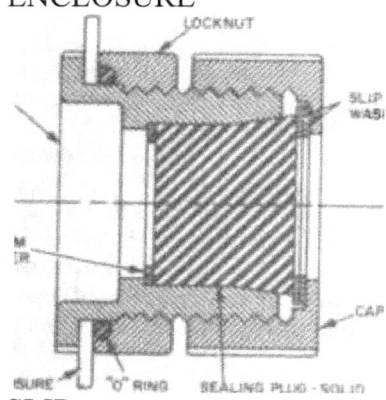

SLIP
WASHERS
CAP
LING PLUG-SOLID
B

Figure 2-12.-Typical nylon stuffing tube installations.

This allows a single-size stuffing tube to be used for a variety of cable sizes, and makes it possible for nine sizes of nylon tubes to replace 23 sizes of aluminum, steel, and brass tubes.

The nylon stuffing tube is available in two parts. The body "O" ring, locknut, and cap comprise the tube; and the rubber grommet, two slip washers, and one bottom washer comprise the packing kit.

A nylon stuffing tube that provides cable entry into an equipment enclosure is applicable to both watertight and nonwatertight enclosures (fig. 2-12, A). Note that the tube body is inserted from inside the enclosure. The end of the cable armor, which will pass through the slip washers, is wrapped with friction tape to a maximum diameter. To ensure a watertight seal, one coat of neo-prene cement is applied to the inner surface of the rubber grommet and to the cable sheath where it will contact the grommet. After the cement is applied, the grommet is immediately slipped onto the cable. The paint must be cleaned from the surfaces of the cable sheath before applying the cement.

Sealing plugs are available for sealing nylon stuffing tubes from which the cables have been removed. The solid plug is inserted in place of the grommet, but the slip washers are left in the tube (fig. 2-12, B).

A grounded installation that provides for cable entry into an enclosure equipped with a nylon stuffing tube is shown in figure 2-13. This type of installation is required only when radio interference tests indicate that additional grounding is necessary within electronic spaces. In this case, the cable armor is flared and trimmed to the outside diameter of the slip washers. One end of the ground strap is inserted through the cap, and one washer is flared and trimmed to the outside diameter of the washers. Contact between the armor and the strap is maintained by pressure of the cap on the slip washers and the rubber grommet.

DECK RISERS.—Where one or two cables pass through a deck in a single group, kickpipes are provided to protect the cables against mechanical damage (fig. 2-14). Steel pipes are used with steel decks, brass pipes with brass decks, and aluminum pipes with aluminum and wooden decks. Kickpipes are of standard pipe size,

Figure 2-13.-Nylon stuffing tube grounded installation."

except that extra strong weight pipe may be used where necessary to overcome welding difficulties. Inside edges on the ends of the pipe and the inside wall of the pipe must be free of burrs to prevent chafing of the cable. Kickpipes including the stuffing tube should be at least 9 inches in height (fig. 2-14, A). Where the height exceeds 12 inches, a brace is necessary to ensure rigid support (fig. 2-14, B). Where the installation of kickpipes is required in nonwatertight decks, a conduit bushing may be used in place of the stuffing tube (fig. 2-14, C).

When three or more cables pass through a deck in a single group, riser boxes must be used to provide protection against mechanical damage. Stuffing tubes are mounted in the top of riser boxes required for topside weatherdeck applications (fig. 2-15).

A riser box Interior Installation is illustrated in figure 2-16. For cable passage through watertight decks inside a vessel the riser box may cover the stuffing tubes if it is fitted with an access plate of expanded metal or

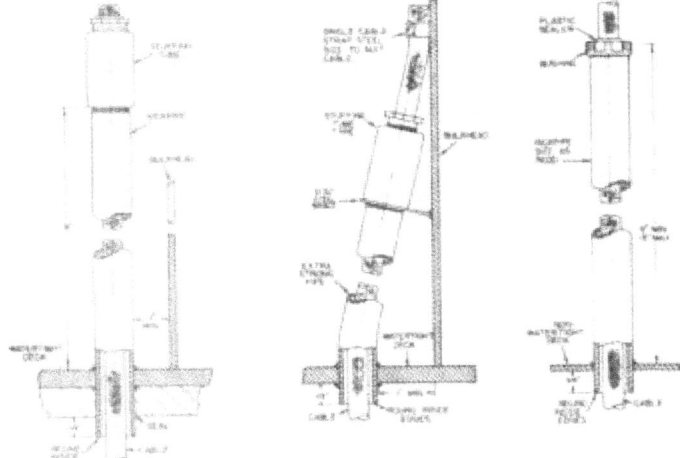

Figure 2-14.-Typical kickpipe installations.

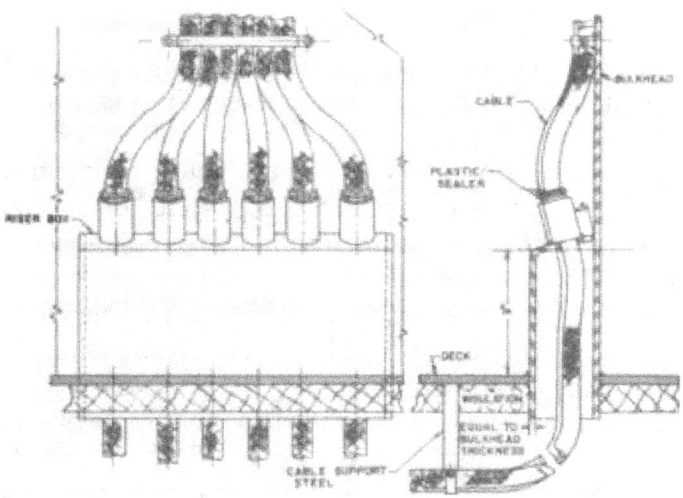

FRONT VIEW CUTAWAY SIDE VIEW

Figure 2-15.-Watertight riser box weatherdeck installations.

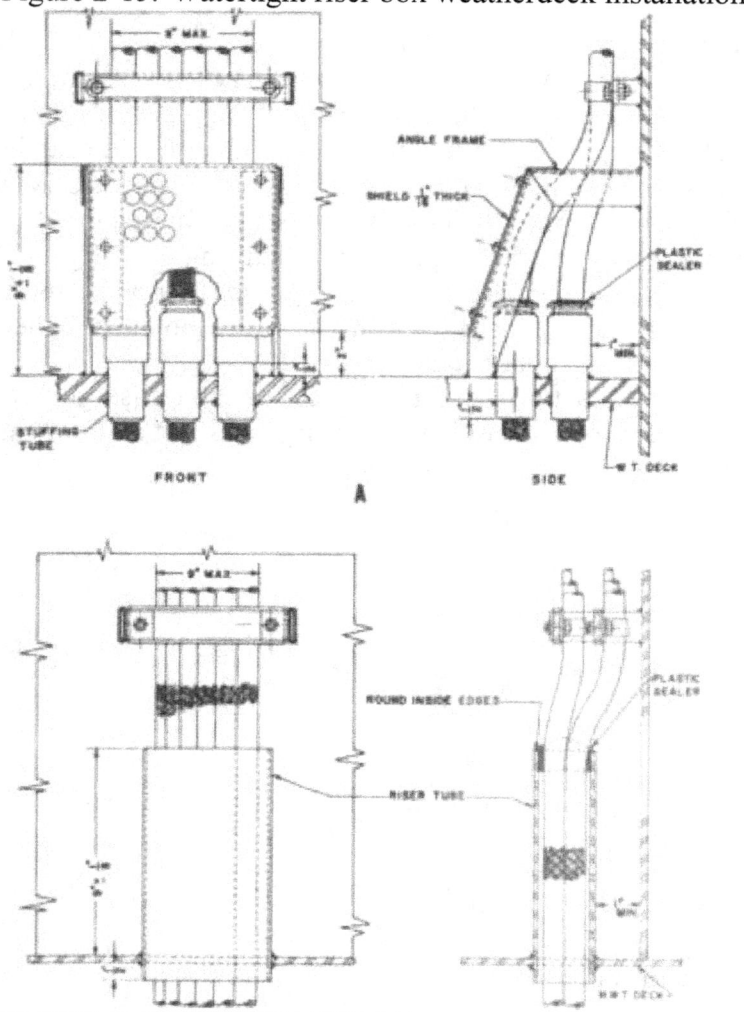

FRONT Q SIDE (CUTAWAY)

Figure 2-16.-Interior riser box installations.

perforated sheet metal (fig. 2-16, A). Stuffing tubes are not required with riser boxes for cable passage through nonwatertight decks (fig. 2-16, B).

a a

The minimum spacing of holes for stuffing tubes and kickpipes is indicated in table 3. Tube sizes indicated are for metal tubes. For nylon tubes, the spacing is the same as for a metal tube of comparable size.

CABLE SUPPORTS.-The single cable strap is the simplest form of cable support. The cable strap is used to secure cables to bulkheads, decks, cable hangers, fixtures, etc. (fig. 2-17). The one-hole cable strap (fig. 2-17, A) may be used for cables not exceeding five-eighths of an inch in diameter. The two-hole strap (fig. 2-17, B) may be used for cables over five-eighths of an inch in diameter.

A more complex cable support is the cable rack, which consists of the cable hanger, cable strap, and hanger support (fig. 2-18). The types of cable straps are (1) contour straps, (2) semicontour straps, and (3) cable bands. The contour and semicontour straps are illustrated in figure 2-18. The cable band is illustrated in figure 2-19. Contour straps do not require the special grouping (by size) of cables that the use of either semicontour straps or cable bands entails. However, contour straps are more difficult to fabricate than semicontour straps, and cable bands are the simplest of the three to install because they do not require preforming.

Cable bands require hangers, which are designed to contain the banding without damaging it. The one-hole hanger (fig. 2-19) is available in sizes to accommodate one or more cables up to 2 inches in diameter. The two-hole hanger (fig. 2-20) ranges in width from 3 inches to a maximum of 15 inches. For all applications, the width of the banding hanger must not be less than the over-all width of the cables that it is intended to support.

Banding material is five-eighths of an inch wide, and may be zinc-plated steel, corrosion resistant steel or aluminum, depending on the requirements of the installation. Apply banding material in the manner shown in figure 2-21. Apply one turn of banding for a single cable less than one inch in diameter (fig. 2-19). Apply two turns of banding for single cables of one inch or more in diameter and for a row of cables (fig. 2-20). Apply three turns of banding for partially loaded hangers where hanger width exceeds the width of a single cable or a single row of cable by more than one-half inch (fig. 2-21).

AC

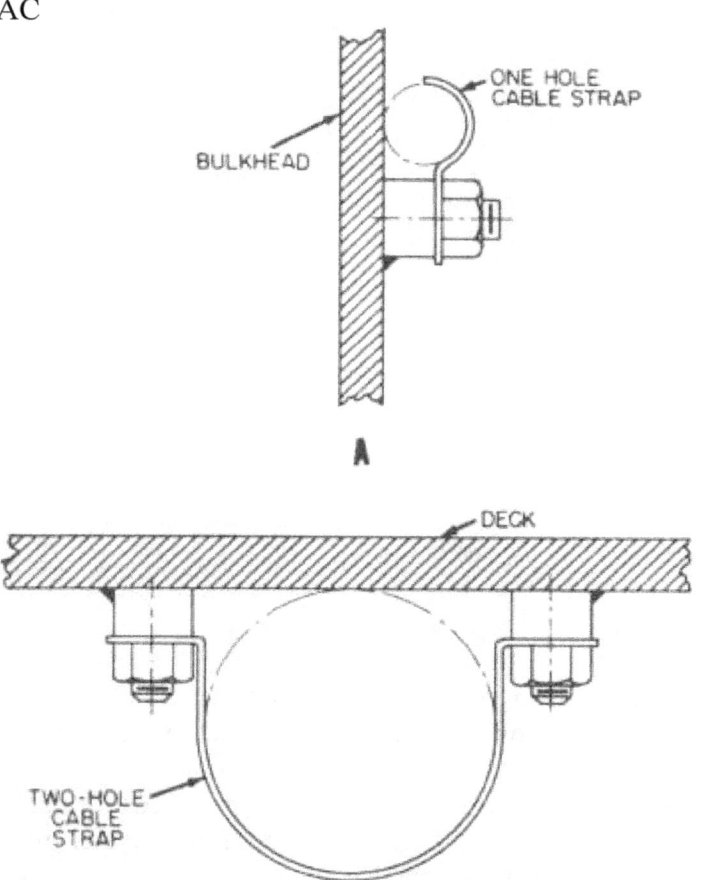

Figure 2-17.—Single cable strap applications.

Add ferrules (fig. 2-20, B) before applying banding material to partially loaded hangers. The application of banding material is illustrated in figure 2-21. A steel band 0.015 inch thick maybe used for cable bands, except when the weight of the cables is supported by the band, and any cable in the row is 2 inches or more in diameter. Then, a steel band 0.020 inch thick must be used. An aluminum band 0.025 inch thick may be used where cables

vssys s /sssysysss s /// /////^

cabie stsap,

Sk#PO*T

H >

. IS" MAI.

CASK hake*

«M.C0«T0UI1 CABl t STRAP

4

3^

ADOHK>.Al MAKERS MAT BE

BUTTED IF DESIRED

SPACES >ASM(,» IS USEO •hen mange R5 ABE feuTTEO AS ShO«"

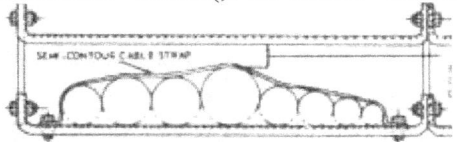

." ulu. CL EARanCE 8fT»Efh TOP Of LARGEST CABl E AND DECK OR BOTTOM
OP HANGEfc

Figure 2-18.-Cables installed in a cable rack.

WATERTIGHT DECK OR BULKHEAD

v////// < m////// /// ////////A

BANDING HANGER

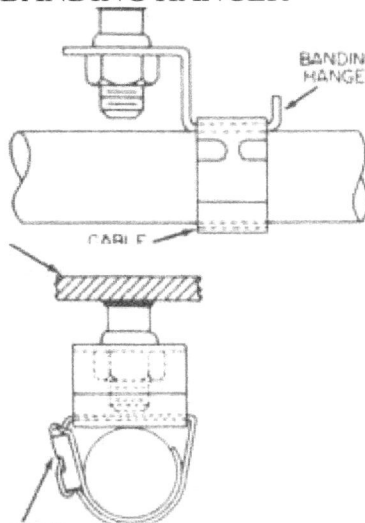

CABLE BAND

BANDING BUCKLE

SIDE VIEW

END VIEW

Figure 2-19.-Single cable band.

are less than 2 Inches in diameter, and cable hangers are on centers no greater than 16
inches. For weatherdeck installations, use corrosion resistant steel band with copper armored

cables; zinc-coated steel with steel armor; and aluminum with aluminum armor.

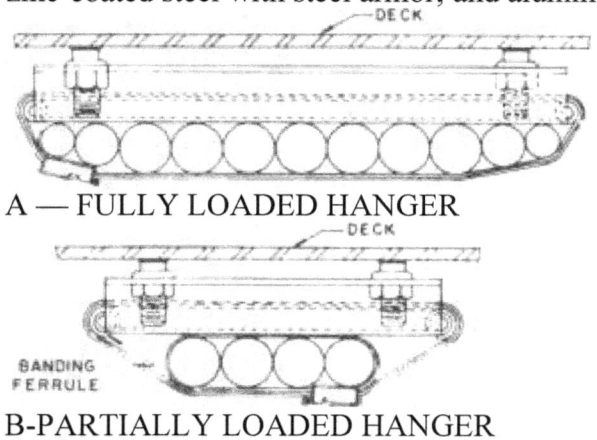

A — FULLY LOADED HANGER

B-PARTIALLY LOADED HANGER
, W.T. DECK
1
C- SIDE VIEW Figure 2-20.-Two-hole banding hanger.

Cables must be supported so that the sag between supports, when practicable, will not exceed one inch. Five rows of cables may be supported from an overhead deck in one cable rack, and two rows of cables may be supported from a bulkhead in one cable rack. As many as 16 rows of cables may be supported in main cableways, in machinery spaces and boiler rooms. Not more than one row of cables shall be installed on a single hanger.

The spacing of simple cable supports, such as those shown in figures 2-17 and 2-19 should not exceed 16 inches

Afl
STEP I
TMHf «0 BANDING MATERIAL THROUGH BUCKLE AND AROUNO CABLES B HANGER

NOLL Of BANDrNS MATENIAL

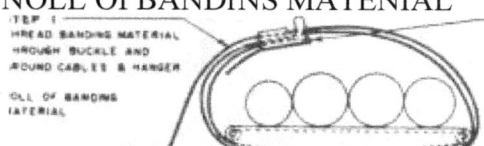

STEP 2
ON TAB CNO Of BUCKLE 'Old unoen a LfNoiM of
8ANDHN0 MATERIAL LONG ENOUGH TO CONTACT AT LEAST ONE CABLE ON A RANT OF THE HANGEN WHEN TlOHTENED
STEP 5
CUT BANDING MATERIAL LONG ENOUGH TO THNEAO THROUGH BAN0IM6 TOOL AFTER TIGHTENING AS MUCH AS POSSIBLE BY HAND. THEN TIGHTEN WITH BANOtNG TOOL UNTIL CABLES ANE F IRMLY HELO. THE TENSION ON THE BANO SHALL NOT EXCEED IIO ROUNDS, EXCERT THAT FOR SOCIO DIELECTRIC COAXIAL CABLES THE TENSION ON THE BAND SHALL NOT EXCEED •O ROUNDS • 10 ROUNDS .

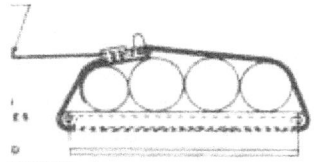

STEP 4
■CNO BANDING MATERIAL
•UFFlClENTLY TO KEEN BAN DWG MATERIAL FROM SLIRRiNG THEN RELEASE TENSION OH BANDING TOO

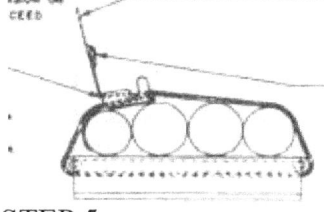

STEP 5
CUT BANDING MATERIAL ll" FROM BEND FOR USE WITH* LONG BUCKLE ON ij "OH ENO WITH SHORT BUCKLE STEP •
FOLD -J"of END OF BANO
IN THE DIRECTION OF BUCKLE
THE
STEP r
St NO ENO OF BANDING MATERIAL BO THAT IT LAYS ON BUCKLE

STEP •
BEND TABS OF BUCKLE SO AS TO PRESS FIRMLY ON BANDING MATERIAL

Figure 2-21.-Applying banding material.

center to center. Spacing of the more complex supports, such as cable racks for main cableways, depends on the requirements of the installation but should not exceed 32-inch centers.

Cable Markings

Metal tags embossed with the cable designation are used to identify all permanently installed shipboard electrical cables. These tags, when properly applied, afford easy identification of cables for purposes of maintenance and replacement.

Two systems of cable marking (the old and the new) may be found aboard naval vessels. The old system used the color of the tag to denote cable classification and the following letters to designate power and lighting cables for the different services:

C—Interior communications D—Degaussing
F—Ship's service lighting and general power
FB—Battle power
G—Fire control
MS—Minesweeping
P—Electric propulsion
R—Radio and radar
RL—Running, anchor, and signal lights S—Sonar

FE—Emergency light and power Other letters and numbers were used with these basic letters to further identify the cable and complete the designation. Typical marking of a power system for successive cables from a distribution switchboard to load would beasfollows: feeder, FB-411; main, l-FB-411; submain, 1-FB-411A; branch, 1-FB-411A1; and subbranch, 1-FB-411-A1A. The feeder number, 411, is indicative of the system voltage. The feeder numbers for a 117- or 120-volt system would range from 100 to 199; for a 220-volt system, from 200 to 299; and for a 450-volt system, from 400 to 499. The exact designation for each cable is given on the electrical wiring plans of each vessel.

The new system for marking power and lighting cable indicates the source and, where practicable, the destination of the cable. Therefore, a thorough knowledge of the system for numbering electrical machinery and equipment aboard ship is necessary to fully understand the new system of cable marking. Except as otherwise specified in this chapter, the general rule for numbering machinery and equipment is used for numbering similar units of electrical machinery or equipment.

The general rule for numbering machinery and equipment states, in effect, that:

1. All similar units in the ship comprise a group, and each group is assigned a separate series of consecutive numbers beginning with 1.

2. Numbering begins in the lowest foremost starboard compartment, and the next compartment selected is to port of the first if it contains similar units; otherwise the next aft on the same level.

3. Proceeding from starboard to port and from forward to aft, continue the numbering procedure until all similar units on the same level have been numbered; then continue on the next upper level, and so on, until all similar units on all levels have been numbered.

4. Within each compartment, the numbering of similar units proceeds from starboard to port, forward to aft, and from a lower to a higher level.

Switchboards or switchgear groups supplied directly from ship's service generators are designated IS, 2S, and so on as necessary to designate all ship's service switchboards. Switchboards supplied directly by emergency generators are designated IE, 2E, and so on, as necessary to designate all emergency switchboards. Switchboards for special frequency (other than the frequency of the ship's service system) a-c generators are designated 1SF, 2SF, and so on, as necessary to designate all special frequency switchboards. Switchboards for coolant-pumping power generators are designated ISC, 2SC, and so on, as necessary to designate all such switchboards. These designations are used only where the coolant-pumping power generators are separate from the ship's service generators.

Sections of a switchgear group other than the generator section are designated by an additional suffix letter starting with the letter, A, and proceeding in alphabetical order from left to right (it is assumed that the viewer is facing the front of the switchgear group). See switchgear group Nos. 4S and 6S (figs. 2-22 and 2-23), respectively.

Some large naval vessels are equipped with zone control, a system of ship's service distribution wherein the ship is divided into areas generally coinciding with fire zones, as dictated by the damage control plan of the ship. Power is distributed within each zone from load center

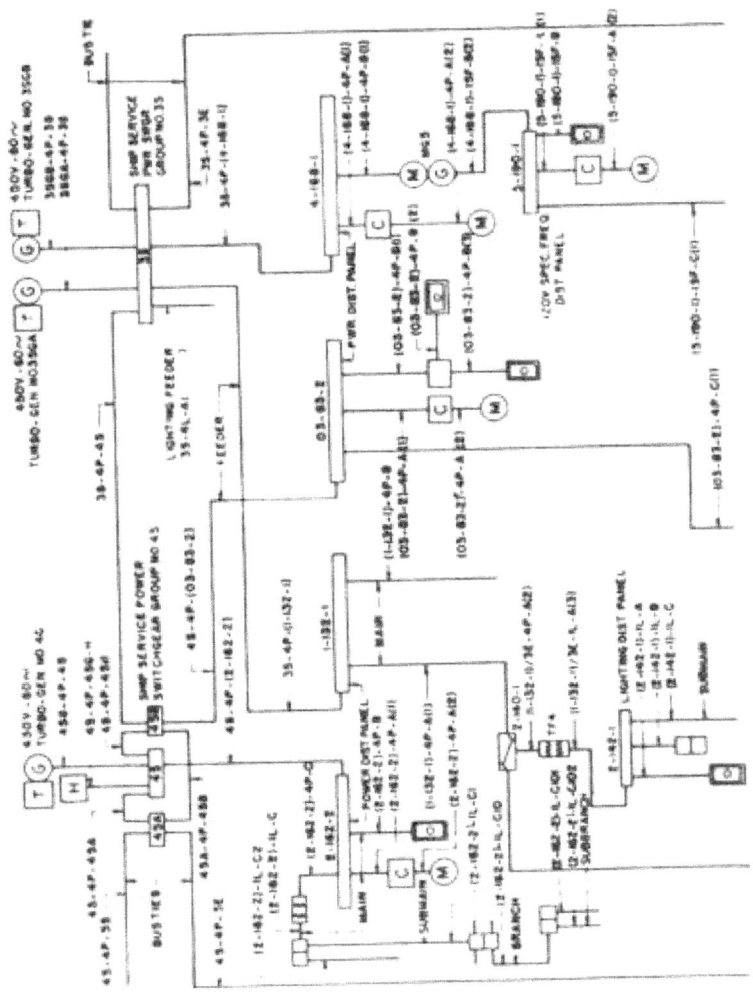

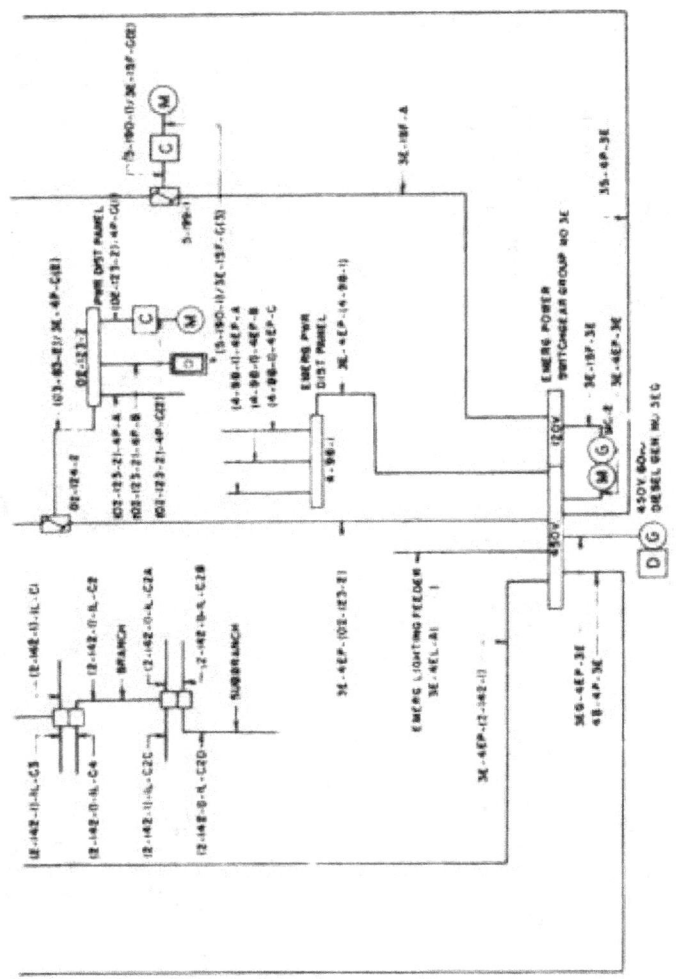

Figure 2.22.—Typical designations for power and lighting system without zone control.

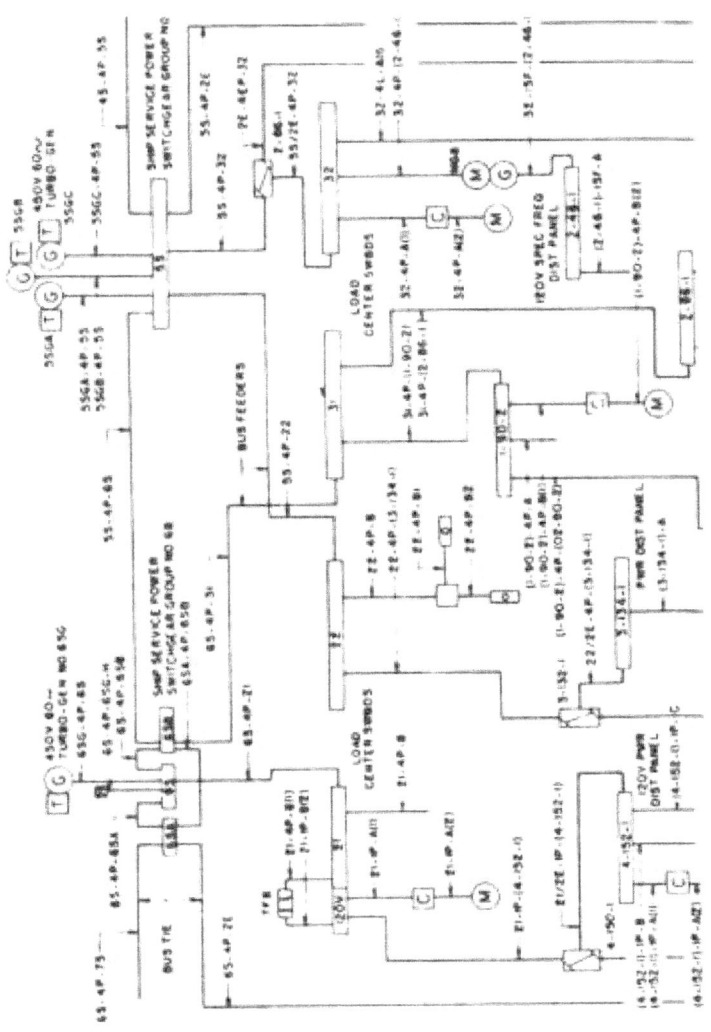

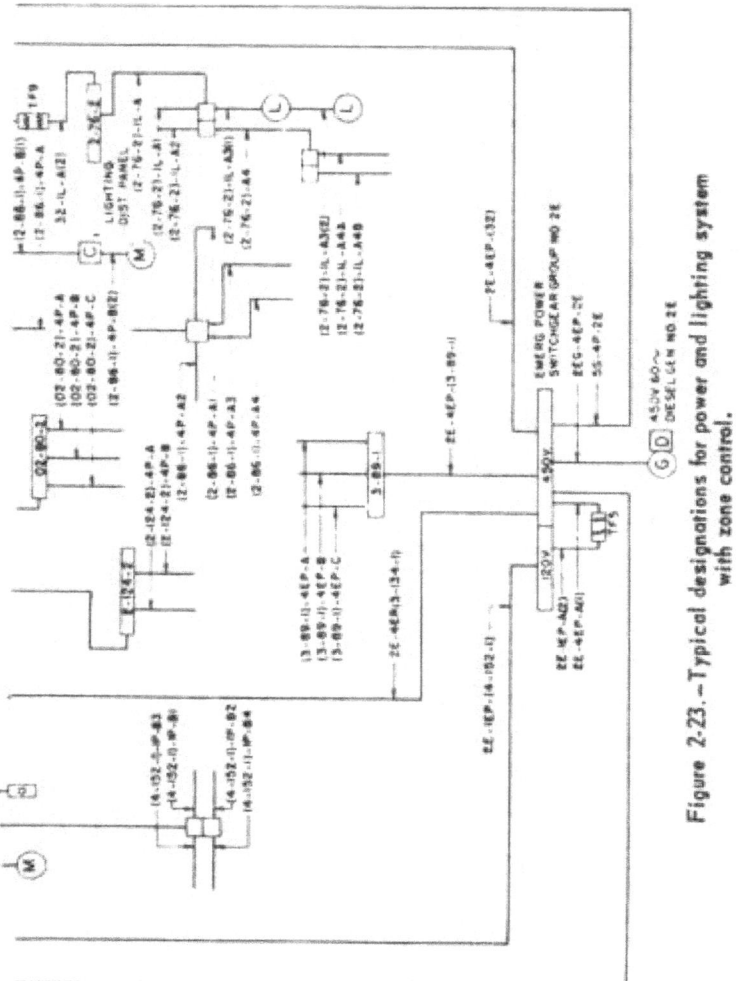

Figure 2-23.—Typical designations for power and lighting system with zone control.

switchboards located within the zone. Load center switchboards and miscellaneous switchboards for ship's service distribution systems on vessels with zone control are given identification numbers, the first digit of which indicates the zone and the second digit the number of the switchboard within that zone, determined in accordance with the general rule for numbering machinery (fig. 2-23).

Load center switchboards and miscellaneous switchboards that are used on other than ship's service systems, all load center switchboards and miscellaneous switchboards for ship's service nonzonal distribution, and all distribution panels and bus transfer equipment are given identification numbers made up of either two or three parts in sequence, and each part is separated by hyphens. The first part indicates the vertical level, at which the unit is normally accessible, by deck or platform number. The second part indicates the longitudinal location of the unit by frame number. For duplicate units on the same deck and frame, the third part indicates the transverse location by the assignment of consecutive odd numbers for centerline and starboard locations and consecutive even numbers for port locations. The numeral one indicates the lowest centerline (or centermost, starboard) component. Consecutive odd numbers are assigned components as they would be observed first as being above, then outboard of the preceding component. Consecutive even numbers similarly indicate components on the port side. For example, a distribution panel with the identification number, 1-142-2, will be located on the main deck at frame 142, and will be the first distribution panel on the port side of the centerline at this frame on the main deck.

Except for the purpose of designating power cables between generators and switchboards, generators (except motor generators) are numbered to correspond to the numbering of the prime mover.

For the purpose of labeling power cables between generators and switchboards, generators are also given another number letter designation. When only one generator supplies a switchboard, the generator will have the same number as the switchboard plus the letter G. Thus, 1SG denotes one ship's service generator that supplies number one ship's service switchboard. When more

than one ship's service generator supplies a switchboard, the first generator determined in accordance with the general rule for numbering machinery will have the letter,

A, immediately following the designation; the second generator that supplies the switchboard will have the letter,

B, and this procedure will be continued for all generators that supply the switchboard and then repeated for succeeding switchboards. Thus, 1SG and 1SGB denote two ship's service generators that supply ship's service switchboard IS. It must be remembered, however, that these numbers are for cable marking purposes only and do not appear on generator information plates.

Motor generators, transformers, and rectifiers are designated by standard symbols, each made up of two capital letters and a numeral. The letters used are MG for motor generators, TF for transformers, and RT for rectifiers. The numeral gives the location of the equipment on the ship, in accordance with the general rule of numbering from forward aft and from starboard to port. MG 1 would designate the motor generator closest to the bow on the starboard side (if there is more than one in the forward location).

The new designation system for power and lighting cables consists of three parts in sequence: source, voltage and service, and, where practicable, destination. These parts are separated by hyphens.

The letters used to designate the different services are:
C—Interior communication
D—Degaussing
G—Fire control
K—Control power
L—Ship's service lighting
N—Navigational lighting
P—Ship's service power
R— Electronics
CP—Casualty power
EL—Emergency lighting
EP—Emergency power
FL-Night flight lights
MC—Coolant pump power
MS—Mine sweeping
PP—Propulsion power

SF—Special frequency power

In the new system, voltages below 100 volts are designated by the actual voltage—for example, 24 for a 24-volt circuit. The number, 1, is used to indicate voltages between 100 and 199; 2 for voltages between 200 and 299; 3 for voltages between 300 and 399; 4 for voltages between 400 and 499; and so on, through 50 for voltages between 5000 and 5999. For a three-wire (120/240), d-c system or a four-wire, 3-phase system, the number used will indicate the higher voltage.

The destination of cable beyond panels and switchboards is not designated except that each circuit alternately receives a letter, a number, a letter, and a number, progressively, every time that it is fused. The destination of power cables to power consuming equipment is not designated, except that each cable to such equipment receives a single-letter alphabetical designation, beginning with the letter, A.

Where two cables of the same power or lighting circuit are connected in a distribution panel or terminal box, the circuit classification will not be changed. However, the cable marking will have a suffix number (inparentheses) indicating the cable section—for example, circuit (4-168-l)-4P-A(l) (fig. 2-22).

For cables that have a normal and an alternate or emergency feeder, the source designation contains (first) the source of the normal feed and (second, separated by a slant line) the designation of the source of the alternate or emergency feed-for example, (5-190-l)/3E-lSF-C(2) in figure 2-22.

Equipment Marking

All distribution panels and bus transfer equipment are provided with cabinet information plates. These plates contain the following Information in the order listed: (1) the name of the space, apparatus, or circuits served; (2) the service (power, lighting, electronics, etc.) and basic location number; and (3) the supply feeder number. For example,

CREW LIVING SPACE, FRAMES 107-149 FIRST PLATFORM LIGHTING PANEL 4-108-2 2S-4L-(4-108-2).

If a panel contains two or more sets of buses and each set Is supplied by a separate feeder, the number of each feeder will be indicated on the identification plate.

Distribution panels are provided with circuit information plates adjacent to the handle of each circuit breaker or switch. These plates contain the following information in the order listed: (1) the circuit number, (2) the name of the apparatus or circuit controlled, (3) the location of apparatus or space served, (4) the circuit load amperes, and (5) the circuit breaker element or fuse rating. Red markers are attached to circuit information plates to indicate vital circuits. Information plates for circuit breakers supplying circle W and circle Z class ventilation systems contain, in addition to the red marker, the class designation of the ventilation system supplied. Information plates without markings are provided for spare circuit breakers mounted in distribution panels. Panel switches controlling circuits that are deenergized during darkened ship operations are marked DARKENED SHIP. The ON and OFF position of these switches are marked LIGHT SHIP and DARKENED SHIP, respectively.

Circuit information plates are provided inside fuse boxes (adjacent to each set of fuses) and indicate the circuit controlled; the phases, or polarity; and the ampere rating of the fuse.

Conductor Identification

Each terminal and connection of rotating a-c and d-c equipment, controllers, and transformers is marked with standard designations. This may be accomplished with synthetic resin tubing or fiber wire markers located as close as practicable to equipment terminals, with fiber tags near the end of each conductor, or by stamping the terminals.

Individual conductors may also be identified by a system of color coding. Color coding of individual conductors in multiconductor cable is in accordance with color coding tables contained in Chapter 40, Bureau of Ships Manual. The color coding of conductors in power and lighting cables is shown in the following table. Neutral polarity, where it exists, is always identified by the white conductor.

Table 4.—Color code for power and lighting cable conductors.

End Seals

Where it is possible for the sheath of shipboard electric cable to act as a conduit in carrying quantities of water from a flooded to an unflooded space beneath the watertight level of a vessel, the cable ends must be sealed to prevent the entrance and discharge of water from cable ends. Because the presently approved method of sealing cable ends requires special equipment not available to shipboard personnel, the installation of cables that require cable end seals should not be attempted by the Electrician's Mate, except in an emergency.

Exercise care when working around cables and switchboards to prevent damage to cable end seals. Make frequent inspections for visual evidence of defects in the seals. If any holes are found in the synthetic tubing used at end seals, make a temporary repair by wrapping the tubing tightly with several layers of synthetic tape (half lapped), with a serving of cord over each end of the tape. Other defects, which are noted, should be repaired as effectively as available materials and equipment permit. All cable end seals to which temporary repairs have been made and all cables that may require end sealing must be tagged and scheduled for permanent repair at a naval shipyard or shore base at the earliest opportunity.

Type SGA and other currently manufactured cables that have a waterproofing impregnant between the strands and in the material between the conductors and the inside of the sheath are not completely watertight. Cable-end sealing is required for these cables.

Cable Connections

CABLE ENDS.—When connecting a newly installed cable to a unit of electrical equipment, the first thing to determine is the proper length of the cable. To do this, form the cable run from the last cable support to the equipment by hand, allowing sufficient slack and radius of bend to permit repairs without renewal of the cable. Carefully estimate where the armor on the cable will have to be cut to fit the stuffing tube (or connector), and mark the location with a piece of friction tape. In addition to serving as a marker, the tape will prevent unraveling and hold the armor in place during cutting operations.

Determine the length of the cable inside the equipment, using the friction tape as a starting point. Whether the conductors go directly to a connection or form a laced cable with breakoffs, carefully estimate the length of the longest conductor, add approximately 2-1/2 times its length, and mark this position with friction tape. The extra cable length will allow for mistakes in attaching terminal lugs and possible rerouting of the conductors Inside the equipment. The length of the cable is now known and it can be cut.

The armor must be removed next. This may be ac -complished by using a cable stripper of the type shown in figure 2-24. Care must be taken not to cut or puncture the cable sheath where the sheath will contact the rubber grommet of the nylon stuffing tube (fig. 2-12). If either a metal stuffing tube or cable connector (fig. 2-14) is used, allow the cable (with armor) to extend at least one-eighth of an inch through the tube.

CUTTING BL AOC

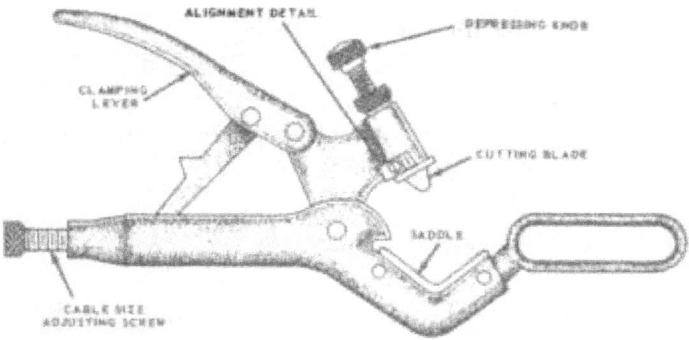

Figure 2-24.-Cable strippers.

Next, remove the impervious sheath, starting a distance of at least 1-1/4 inch (or as necessary to fit the requirements of the nylon stuffing tube) from where the armor terminates. The cable stripper should be used for this job. Do not take a deep cut because the conductor insulation can be easily damaged. Flexing the cable will help separate the sheath after the cut has been made. Clean the paint from the surface of the remaining impervious sheath exposed by the removal of the armor. This paint is conducting. It is applied during manufacture of the cable and passes through the interstices of the armor onto the sheath. Once the sheath has been removed, the cable filler can be trimmed with a pair of diagonal cutters.

The proper method of finishing and protecting cable ends that do not require end sealing is shown in figure 2-25. For cables entering enclosed equipment (connection boxes, outlet boxes, fixtures, etc.) the method shown in figure 2-25, A, should be used. An alternate method CROTCH OF CABLE TO BE GIVEN A COAT OF INSUL ATING VARNISH BEFORE APPLYING TUBING.

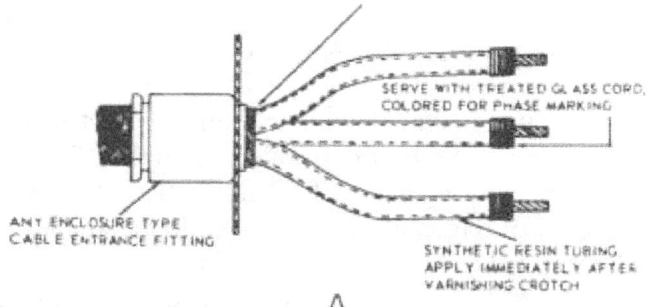

A

AFTER VARNISHING AND BEFORE APPLYING TAPE, PUSH TUBING TIGHTLY INTO CROTCH WHILE VARNISH IS STILL WET.

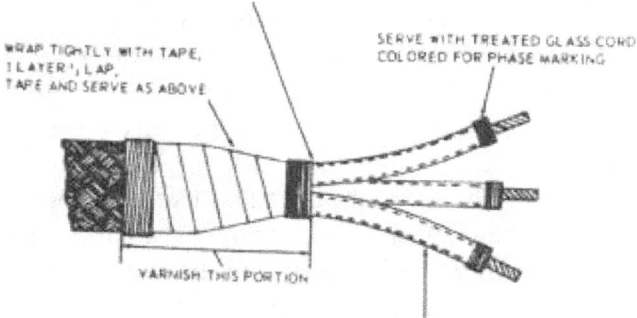

SYNTHETIC RESIN TUBING
B
Figure 2-25.-Protecting cable ends.

(when synthetic resin tubing is not readily obtainable) is to apply a coat of air drying insulation varnish to the insulation of each conductor as well as to the crotch of the cable. The end of the insulation on each conductor is reinforced and served with treated glass cord, colored to indicate proper phase marking. For cables entering open equipment (switchboards,etc.) the method shown in figure 2-25, B, should be used. An alternate method is to wrap each conductor tightly with one layer of synthetic tape (half lapped) and serve with cord as in the tubing method.

Apply tape, air-drying varnish, and serve the crotch and part of the cable for a distance about 2 inches back of the crotch.

CONDUCTOR ENDS.—Wire strippers (fig. 2-26) are used to strip insulation from the conductors. Care must be taken to avoid nicking the conductor stranding while removing the insulation. Side, or diagonal, cutters should not be used for stripping insulation from conductors.

Conductor surfaces must be thoroughly cleaned before terminals are applied. After baring the conductor end for a length equal to the length of the terminal barrel, clean the individual strands thoroughly and twist them tightly together. Solder them to form a neat, solid terminal for fitting either approved clamp-type lugs or solder-type terminals. If the solder-type terminal is used, tin the terminal barrel and clamp it tightly over the prepared conductor (before soldering) to provide a solid mechanical joint. Conductor ends need not be soldered for use with solderless-type terminals applied with a crimping tool. Do not use a side, or diagonal, cutter for crimping solderless-type terminals.

Solderless-type terminals may be used for all lighting, power, interior communications, and fire-control applications, except with equipment provided with solder-type terminals by the manufacturer, and with wiring boxes or equipment in which electrical clearances would be reduced below minimum standards by the use of the solderless-type terminal.

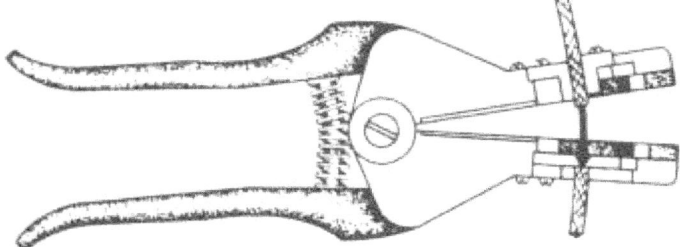

Figure 2-26.-Mechanical wire strippers.

For connection under a screwhead where a standard terminal is not practicable, an alternate method can be used. Bare the conductor for the required distance and thoroughly clean the strands. Then twist the strands tightly together, bend them around a mandrel to form a suitable size loop (or hook where the screw is not removable), and dip the prepared end into solder. Remove the end, shake off the excess solder, and allow it to cool before connecting it.

After the wiring installation has been completed, the insulation resistance of the wiring circuit must be measured with a megger or similar (0-100 megohm, 500 volt d-c) insulation resistance measuring instrument. Do not energize a newly installed, repaired, or modified wiring circuit without first ascertaining (by insulation tests) that the circuit is free of short circuits and grounds.

Small refrigerators, drinking fountains, coffee makers, and bracket fans are plugged into receptacles connected directly to the ship's wiring. To remove stress from the equipment terminal block and its connected wiring, rigidly clamp the cable to the frame of the equipment close to the point where the cable enters the equipment.

LACING CONDUCTORS.—Conductors within equipment must be kept in place in order to present a neat appearance and facilitate tracing of the conductors when alterations or repairs are required. When conductors are properly laced, they support each other and form a neat, single cable.

The most common lacing material is waxed cord. The amount of cord required to single lace a group of conductors is approximately 2-1/2 times the length of the longest conductor in the group. Twice this amount is required If the conductors are to be double laced.

Before lacing, lay the conductors out straight and parallel to each other. Do not twist them together because twisting makes conductor lacing and tracing difficult.

A shuttle on which the cord can be wound will keep the cord from fouling during the lacing operations. A shuttle similar to the one shown in figure 2-27 may easily be fashioned from aluminum, brass, fiber, or plastic scrap. Rough edges of the' material used for the shuttle should be smoothed to prevent injury to the operator and damage to the cord. To fill the shuttle for single lace, measure

J
1"
=! L=i
Figure 2-27. -A lacing shuttle.

the cord, cut it, and wind it on the shuttle. For double lace, proceed as before, except double the length of the cord before winding it on the shuttle, and start the ends on the shuttle in order to leave a loop for starting the lace.

Single lace may be started with a square knot and at least two marling hitches drawn tight. Details of the square knot and the marling hitch are shown in figure 2-28. Do

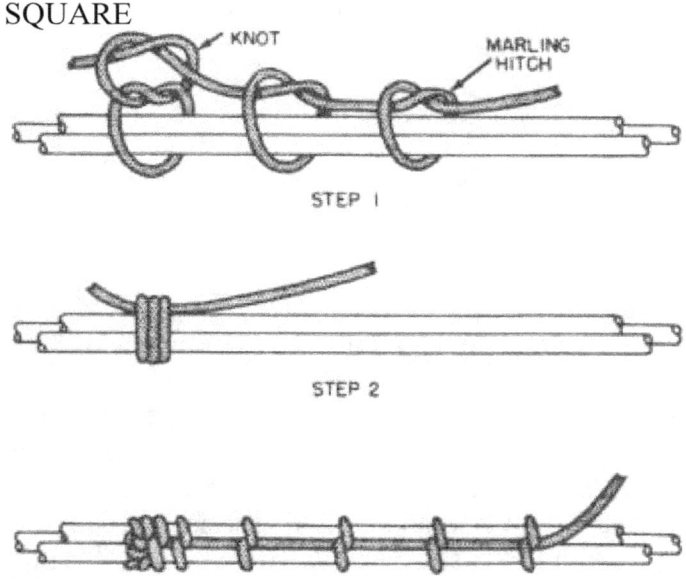

SQUARE

STEP 3

Figure 2-28.-Applying single lace. 86

not confuse the marling hitch with a half hitch. In the marling hitch, the end is passed over and under the strand (step 1, fig. 2-28). After forming the marling hitches, draw them tight against the square knot (step 2, fig. 2-28). The lace consists of a series of marling hitches evenly spaced at one-half to one-inch intervals along the length of the group of conductors, as indicated in step 3.

When dividing conductors to form two or more branches, follow the procedure illustrated in figure 2-29. Bind the conductors with at least six turns between two marling hitches, and continue the lacing along one of the branches (fig. 2-29, A). Start a new lacing along the other branch. To keep the bends in place, form them in the conductors before lacing. Always add an extra marling hitch just prior to a breakout (fig. 2-29, B).

Double lace is applied in a manner similar to single lace, except that it is started with the telephone hitch and is double throughout the length of the lacing (fig. 2-30). Double as well as single lace may be terminated by forming a loop from a separate length of cord and using it to pull the end of the lacing back underneath a serving of approximately eight turns (fig. 2-31).

Lace the spare conductors of a multiconductor cable separately, and secure them to active conductors of the cable with a few telephone hitches. When two or more cables enter an enclosure, each cable group should be laced separately. When groups parallel each other, they should be bound together at intervals with telephone hitches (fig. 2-32).

Conductor ends (3000 cm or larger) should be served with cord to prevent fraying of the insulation (fig. 2-33). When conductor ends are served with glass cord colored for phase marking, the color of the cord should match the color of the conductor insulation.

GROUNDED RECEPTACLES.-Aboard naval vessels, grounded receptacles (fig. 2-34) are used with grounded plugs and portable cables having a ground wire, which grounds the metallic case and exposed metal parts of portable tools or equipment to the ship's structure when the plugs are inserted in the receptacles. The ground wire prevents the occurrence of dangerous potentials between the tool or equipment housing and ship's structure, and thus protects the user from shock.

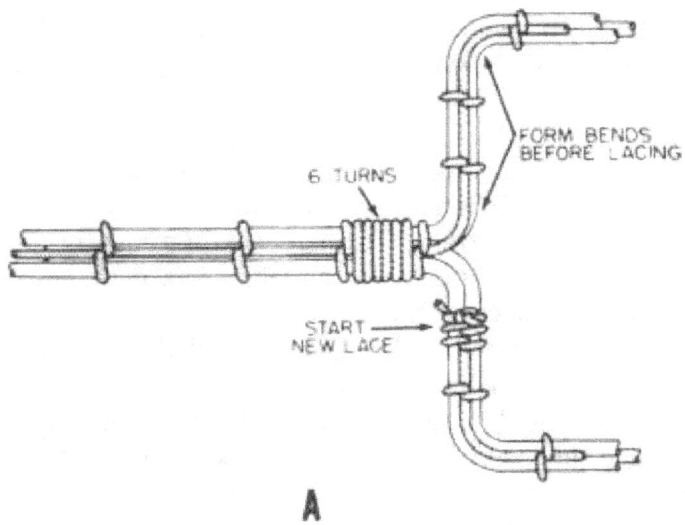

FORM BENDS
BEFORE LACING

6 TURNS

START
NEW LACE

A

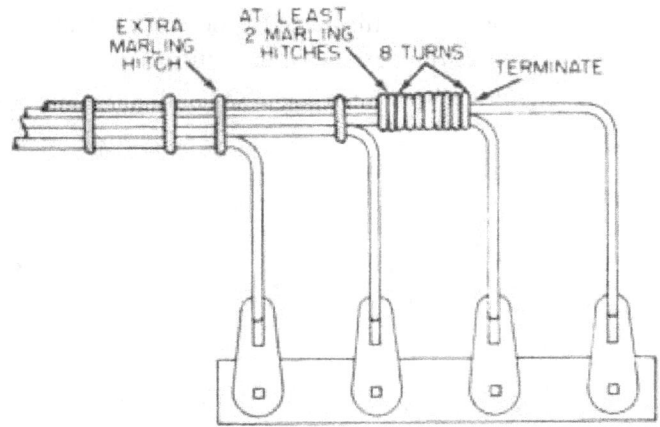

EXTRA
MARLING
HITCH

AT LEAST
2 MARLING
HITCHES

8 TURNS

TERMINATE

B

Figure 2-29.-Lacing branches and breakouts.

WftflsasaistRS«f
n«MMDOooMOOMOMuwM i uuiiM i o »»i wnn ii nr

STEP I

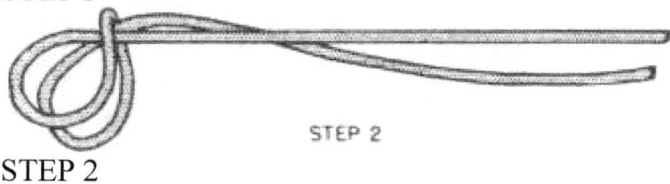

STEP 2

STEP 2

STEP 3

COMPLETED TELEPHONE HITCH

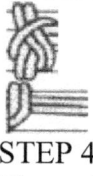

STEP 4

Figure 2-30.-Starting double lace with the telephone hitch.

4 MARLING HITCHES
SERVING OF 8 TURNS
STEP I

STEP 2
Figure 2-31.-The loop method of terminating the lace.
TELEPHONE -HITCH —

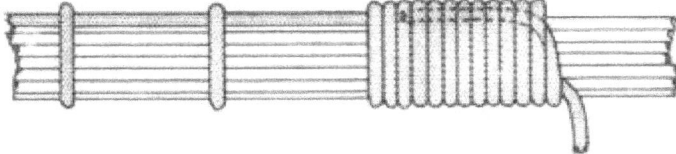

Figure 2-32.-Binding cable groups with the telephone hitch.
STEP I

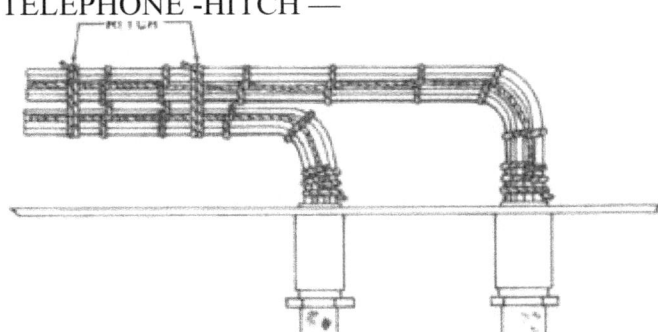

STEP 2
Figure 2-33.-Serving conductor ends.

A- RECEPT1CAL B~ CONNECTOR

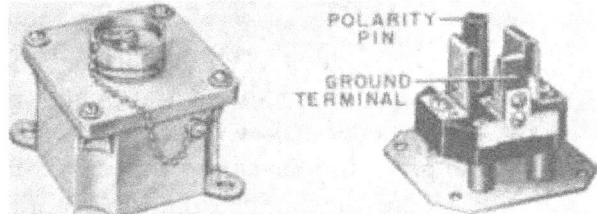

C-COVER D-CAP
Figure 2-34.-Watertight grounded receptacle.

The grounded receptacles most widely used aboard naval vessels have metal enclosures, which are connected internally to the ground terminal of the receptacle. Grounding the enclosures will ground the grounded terminal. Grounded receptacles with plastic enclosures are in use aboard some vessels, and in these receptacles, the grounded terminal is connected to ground through a conductor. The cross-sectional area of the conductor used to connect the grounded terminal to ground must be greater than that of the conductors that supply such a receptacle.

Whenever a grounded receptacle is installed, repaired, or modified in any way, a test must be made to determine whether the connections are correct. To make this test, use a voltmeter or voltage tester to test the voltage between each pair of line terminals (one pair for a single phase receptacle, three pairs for a 3-phase receptacle). Full line voltage should be obtained between each pair. Next, test the voltage between the ground terminal and ground (some metallic part of the ship's structure that is not a part of the receptacle). This voltage should be zero. After the voltage from the ground terminal to ground has been measured and found to be zero, use an ohmmeter or other low-reading (zero to a few ohms), resistance -measuring instrument to measure the resistance from the ground terminal to ground. This resistance should be zero. Unless all of the preceding conditions are satisfied, the receptacle is not correctly connected and constitutes a safety hazard.

To determine the safety of grounded receptacles, inspect them and give the foregoing test at least once a year. A record book should be kept, giving the location of all grounded receptacles and the dates on which they are inspected and tested. The inspection of the receptacle should ascertain that:

1. The phenolic polarity pin in the insert of the receptacle is in place (fig. 2-34, B).

2. The receptacle cover is placed so that the pointed metal polarity tab (brazed to the cover) is superimposed over the phenolic polarity pin within the receptacle.

3. The phenolic plug-guide insert in the aperture of the receptacle cover is firmly in place (fig. 2-34, C). Where the inserts are missing, replace the entire cover. Correct loose inserts by cementing them.

4. The protective cap with rubber gasket is properly attached by chain to the receptacle cover (fig. 2-34, D). Spraytight, watertight, and submersible receptacles must be kept covered when plugs are not inserted.

GROUNDED PLUGS.-The electric cable used with portable electric tools and equipment must be provided with a distinctly marked grounding conductor in addition to the conductors for supplying power. Up to this time (except for a few cases where black was used) red was used for the grounding conductor in three-conductor cables, and green in four-conductor cables. Today, Navy specifications require that green be used for the grounding conductor in cables for all new portable tools and equipment. However, no change should be made in equipment (now on board ship) which has a cable with a grounding conductor color coded in conformity with past practice. Such equipment should be used until cable replacement becomes necessary.

The end of the grounding conductor, which is within the tool (or equipment), should be connected to the metal housing, and the other end should be connected to the grounding blade of a grounded plug (fig. 2-35). Make certain that the ground connection is correctly made. If the grounding conductor is inadvertently connected to a line contact of the plug, a dangerous potential that might prove fatal to the operator will be placed on the equipment casing. To guard against such an occurrence, the connections should be tested after they have been made.

Before using portable electric equipment for the first time, check the plug connections on the equipment for correct wiring. First, connect one lead of an instrument for measuring low resistance (zero to a few ohms) to the metal case of the equipment and the other lead to the ground contact of the plug. Measure the resistance. It should read almost zero. After removing the low-resistance measuring instrument, connect one lead of a megger, or similar high-resistance measuring instrument, to the metal case of the equip- Figure 2-35. — mentand the other lead to one of the line Grounded plug.

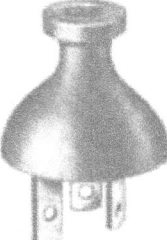

terminals of the plug. Measure the resistance with the switch on the equipment in the ON position and also when it is in the OFF position. In both cases, the readings obtained will be normal insulation resistance (usually well in excess of 1 megohm) if the connections are correctly made. Repeat this procedure with the megger connected to the case of the equipment and the other line contact of the plug (or to each of the other line contacts if there are more than two). In each case the megger should read normal insulation resistance if the connections are properly made.

Two resistance-measuring instruments must be used for the foregoing test because one instrument (unless it has scales for both high and low resistance) will not give reliable results. This test should also be made on all portable equipment that has not previously been tested in this way, even though it may have been used previously without trouble. This test must also be repeated periodically on all portable electrical equipment to ensure that the equipment is in safe condition. Instruct all personnel using the equipment not to attempt to reconnect the cable to the plug if it should ever be pulled out. This connection must always be made and tested by a competent Elec -trician's Mate.

Electrical equipment that is positively and adequately grounded through its mounting (such as bracket fans), or portable lights and other equipment having no exposed metal parts that may become energized may be fitted with a two-conductor repeated flexing service cable and a plug suitable for use with the grounded-type receptacle. However, test to make sure that the natural grounding of the equipment is not blocked off in any way.

Casualty Power Cable

Suitable lengths of portable casualty power cables are stowed throughout naval vessels close to the locations where they may be needed for making temporary connections in the event the installed ship's service and emergency distribution systems are damaged. The casualty power system is completely described in chapter 9 of this training course. Only the portable cable is discussed here. Portable casualty power cables are equipped with

metal tags, which indicate the length of the cable (in feet) and the location of the cable stowage rack (fig. 2-36).

METAL CABLE TAG \'7b CUT FROM SEAMLESS BRASS TUBING OF APPROPRIATE SIZE)

6 APPROX

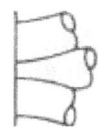

SERVING OF COTTON CORD

Figure 2-36.-Typical portable casualty power cable tags.

Both ends of portable casualty power cable are made up in the same manner as shore power cable illustrated in figure 2-37 except that:

1. The outer sheath of the cable (step 1, fig. 2-37, A) is removed from the cable for approximately 1 foot, or a sufficient distance to permit ready attachment to the casualty power bulkhead, riser, and switchboard, terminals, and portable switches.

2. Ferrules (cable thimbles) of tubular copper material are attached to conductor ends instead of lugs (step 7, fig. 2-37, B).

3. Phase or polarity, as appropriate, of the conductors must be indicated by servings on the conductors. One serving of cotton cord indicates phase A, or positive polarity (black conductor); two separate servings indicate phase B, or neutral polarity (white conductor); and three separate servings indicate phase C, or negative polarity (red conductor). The servings on the conductors afford a means of identifying (by touch) the individual conductors when illumination is insufficient for visual identification.

Portable casualty power cables should be rigged only when required for use or for practice in rigging the casualty power system. At all other times, they should be stowed in the cable rack indicated on the cable tag. When portable casualty power cables are rigged, connections should be made from the load to the supply to avoid handling energized cables.

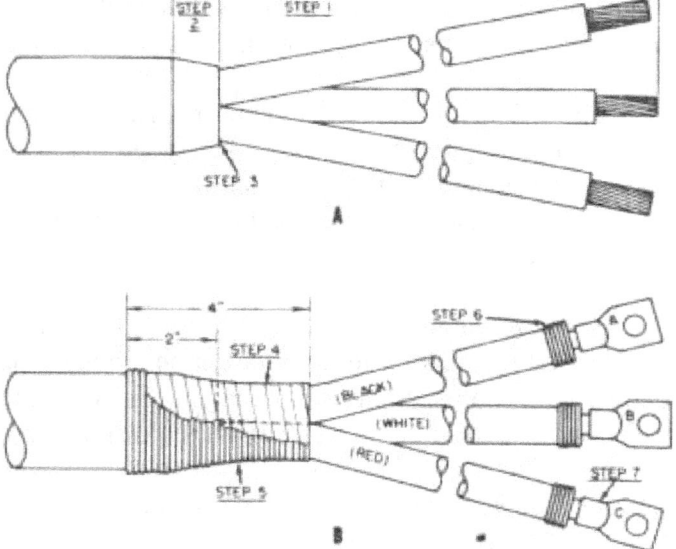

Figure 2-37.-Preparing shore power cable ends.

Shore Power Cable

A shore terminal box is provided at or near a suitable weatherdeck location aboard ship. Portable cables from shore power or from a vessel alongside can be attached to the shore terminal box to supply power to the ship's distribution system. This same connection can be used to supply power from the ship's service generators to a vessel alongside. The shore power system is designed to handle only enough power to operate necessary machinery and provide illumination for habitability and the accomplishment of necessary work. Care must be exercised not to exceed the capacity of the system when it is in use.

Both ends of shore power cables (if not already prepared when received aboard ship) should be made up in the following manner:

96

1. Remove the outer sheath from the cable for approximately 2 feet, or a sufficient distance to permit ready attachment to the shore terminal box (step 1, fig. 2-37, A). Do not damage the insulation of the individual conductors.

2. Taper the end of the outer sheath, starting about 1 inch back from the cut end (step 2, fig. 2-37, A).

3. Apply a coat of air-drying, insulating varnish to the crotch of the cable and allow it to dry (step 3, fig. 2-37, A).

4. Apply two layers of pressure-sensitive synthetic resin tape or friction tape (half-lap), starting about 2 inches back of the crotch and finishing about 2 inches in front of the crotch (step 4, fig. 2-37, B).

5. Reinforce and seize the tape with a serving of treated glass cord, and coat the completed serving with air-drying insulating varnish (step 5, fig. 2-37, B).

6. Apply a serving of treated glass cord, colored for phase marking (step 6, fig. 2-37, B).

7. Attach lugs as furnished with the ship's shore terminal box (step 7, fig. 2-37, B). Stamp the lugs to indicate the phase (A, B, C) or polarity (+, ±, -) of the conductors. The lug markings, the colored glass cord serving, and the color of the conductor should all agree with the phase and polarity code for light and power conductors (table 4).

The terminal connecting block, and terminals of the ship's shore terminal box must be plainly marked to indicate phase sequence, or polarity, as appropriate. For example, the terminal of the black conductor should be stamped (A) to indicate phase A; or stamped (+) to indicate polarity, as appropriate, and so on. The color code of cables should not be implicitly trusted because the conductors (from exposure) may become discolored or lose their original coloring.

When practicable, follow these procedures when connecting shore power cables.

1. Test the ship's shore terminal box for proper marking. Do this by energizing the shore terminal box from the ship's service generator. Check the phase sequence (a-c) markings by testing the terminals with a phase sequence indicator. Check polarity (d-c) markings by testing the terminals with a polarity-indicating voltage tester. An approved type phase-sequence indicator is

shown in figure 2-38. This indicator consists of a miniature, 3-phase induction motor, which indicates phase sequence by the direction in which the rotor (moving element) turns. After it has been determined that the markings are correct, secure the ship's power to the shore terminal box by opening the ship's shore power switch. This switch should be tagged in the manner prescribed for tagging circuits (on which men are working) in the section of this chapter titled "Maintenance."

2. Rig the shore power cable and connect one end of it to the ship's shore terminal box. Make certain that the leads are properly connected in accordance with the terminal markings of the box.

3. Check the terminals on the shore connection with a voltage tester, and after it has been determined that they are deenergized, connect the other end of the shore power cable leads.

4. Energize the cable from the shore power supply, and test either phase sequence, or polarity, (depending on whether a-c or d-c) at the ship's shore terminal box. If the phase sequence, or polarity, matches that of the ship's power, the ship may safely switch to shore power. The procedure for switching to shore power is given in chapter 9.

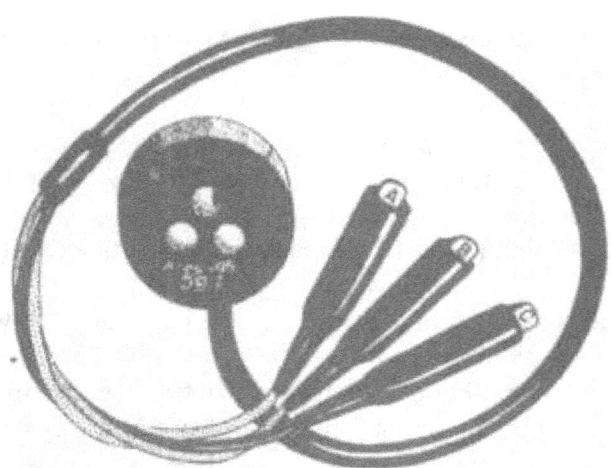

Figure 2-38.-Phase-sequence indicator.

Warning signs indicating the voltage of the cable should be attached to the shore power cable at intervals to warn personnel to keep clear of the cable. All connections must be tightly made to prevent damage to the terminals or leads as a result of overheating and arcing.

The procedures and precautions that follow apply to all portable cables as well as the shore power cables. Spliced portable cables are extremely dangerous and should not be used unless an emergency warrants the great risk involved. Portable cables must be of sufficient length to prevent their being subjected to longitudinal stresses and the need to be pulled taut to make connections. Current-carrying capacity must be ample for the expected power demand. The cable must be checked frequently to ascertain the degree of heating, and any cable that feels uncomfortably warm to the bare hand (placed outside the insulation) should be checked immediately for overloading. Interconnections between lengths of portable cable shall be made only on approved connection block or by other approved fittings that shall be suitably insulated and enclosed to eliminate all possible hazards from fire or shock to personnel.

Always support portable cables above decks, floor plates, and gratings. Never place them where they can be damaged by falling objects, by being walked on, or by contact with sharp corners or projections in the ship's hull or other objects. Where portable cables pass through doorways or hatches, provide steps to prevent cables from being Dinched by a door or hatch cover.

CABLE MAINTENANCE

The primary purpose of electrical cable maintenance is to preserve the insulation resistance. Hence, it is important to know the insulating materials that are used in naval shipboard electrical equipment and the factors that affect insulation resistance.

Insulation

The purpose of insulation on electric cables and equipment is to (1) isolate current-carrying conductors from metallic and structural parts, and (2) to insulate points of unequal potential on conductors from each other. Normally, the resistance of such insulation should be sufficiently high to result in negligible current flow through or over its surface.

Air may be used as an insulator if a solid insulating material is used for support to maintain proper electrical clearance and creepage distances. The CLEARANCE distance is the shortest point to point distance in air between two bare (uninsulated) conductors or between a bare conductor and ground. The CREEPAGE distance is the shortest distance between two bare (uninsulated) conductors, or between a bare conductor and ground, measured along the surface of the solid insulating material. Permissible clearance and creepage distances depend upon the voltage and current involved, the degree of enclosure, and the properties of the insulating material supporting the conductors. Minimum acceptable electrical clearance and creepage distances (in inches) are shown in tables contained in Chapter 40 of the Bureau of Ships Manual.

The electrical insulating materials used innaval shipboard electrical equipment (including cables) are grouped according to their chemical composition; for example, class O, class A, class B, class C, class H, class E,and class T Insulation.

Class O insulation consists of cotton, silk, paper, and similar organic materials when neither impregnated nor immersed in a liquid dielectric. Class O insulation is seldom used by itself in electrical equipment.

Class A insulation consists of (1) cotton, paper, and similar organic materials when they are impregnated or immersed in a liquid dielectric; (2) molded and laminated materials with cellulose filler, phenolic resins and other resins of similar properties; (3) films and sheets of cellulose acetate and other cellulose derivities of similar properties; and (4) varnish (enamel), as applied to conductors.

Class B insulation consists of mica, asbestos, fiber glass, and similar inorganic materials in built-up form with organic binding substances, such as polyvinylacetal or polymide films. A small portion of class A materials may be used for structural purposes only.

Class H insulation consists of (1) mica, asbestos, fiber glass, and similar inorganic materials in built-up

form with binding substances composed of silicone compounds or materials with equivalent properties; and (2) silicone compounds in the rubbery or resinous forms, or materials with equivalent properties. A minute proportion of class A materials may be used only where essential for structural purposes.

Class C insulation consists entirely of mica, glass, quartz, and similar inorganic material. Class C materials, like class O, are seldom used alone in electrical equipment.

Class E insulation is an extended silicone rubber dielectric used in reduced-diameter types of electric cables in sizes 3, 4, and 9. Special care should be exercised in handling the cables to avoid sharp bends and kinks that can damage the silicone rubber insulation.

Class T insulation is a silicone rubber treated glass tape, which is also used in reduced-diameter cables in sizes 14 through 800.

Propulsion generators and motors are usually insulated with class B insulating materials. Ship's service and emergency generators may have either class A, B, or H materials; however, the trend is away from class A. Auxiliary motors are usually class A, although the trend is toward class B and class H materials. Lighting transformers for 60-cycle service are class B insulated, and 400-cycle transformers are class H insulated. Miscellaneous coils for control purposes may be class A, B, or H, but the majority of such coils are class A insulated.

It is important to maintain operating temperatures of electrical equipment within their designed values to avoid premature failure of insulation. Temperatures only slightly in excess of designed values may produce gradual deterioration, which, though not immediately apparent, shortens the life of the insulation. The highest temperature (hot-spot temperature) to which class O insulation may be subjected continuously with normal life expectancy (5 years) is 90° centigrade, class A insulation, 105° centigrade; class B insulation, 130° centigrade; and class H, 200° centigrade. No limit has been established for class C insulation.

The maximum allowable temperature rise and the design ambient temperature for electrical equipment are usually shown on the nameplate, on equipment drawings, and in the technical manual for the equipment. When information is not available from these sources, the maximum permissible temperature rises, shown in Chapter 60, Bureau of Ships Manual, shall be used as a guide for checking or testing windings of rotating electrical machinery.

Insulation Resistance Measurements

The insulation resistance of shipboard electrical cable must be measured periodically with an insulation-resistance-measuring instrument (megger) to determine the condition of the cable. Cable insulation measurements should be made often enough (if operating conditions permit) to ensure that all cables will have an accurate measurement at least once every 3 months. More frequent measurements should be made if a comparison with similar earlier measurements indicates a significant decrease in insulation resistance. In addition, cognizant Bureau Directives or Instructions may require that more frequent cable insulation-resistance measurements be made on some circuits; for example, degaussing coil measurements are required weekly on some degaussing installations.

Factors that affect cable insulation resistance measurements are the length of the cable, type of the cable, temperature of the cable, and the equipment connected in the circuit. Each of these factors must be evaluated to reliably determine the condition of the cable from the measurements obtained.

LENGTH OF CABLE.—The insulation resistance of a length of cable is the resultant of a number of small individual leakage paths or resistances between the conductor and the cable sheath. These leakage paths are distributed along the cable. Hence, the longer the cable, the greater the number of leakage paths and the lower the insulation resistance. For example, if one leakage path exists in each foot of cable, there will be 10 such paths for current to flow between the conductor and the sheath in 10 feet of cable, and the total amount of current flowing in all of them would be 10 times as great as that which would flow if the cable were only 1 foot long. Therefore, to establish a common unit of comparison, cable-insulation

resistance should be expressed in megohms (or ohms) per foot of length. This is determined by multiplying the measured insulation resistance of the cable by its total length in feet.

When measured insulation resistance is converted to insulation resistance per foot, the total length of cable to be used is equal to the length of the cable sheath for single conductor cable and for multiple conductor cable in which each conductor is used in one leg of a circuit. For example, in a TSGA cable with a cable sheath of 100 feet in which the three conductors are phases A, B, and C of a 3-phase power circuit, the total length of the cable is 100 feet, not 300 feet. The reason for this is that each conductor is measured separately. If this cable is connected, either in series or parallel, to a similar cable that has a sheath length of 400 feet, the total length is 500 feet. As another example, 200 feet of type MSCA-7 cable (7-conductor cable) connected to 200 feet of MSCA-24 cable (24-conductor cable) represents a total cable length of 400 feet.

Because degau' ing cable is installed in the form of a loop and the conductors in the multiple conductor cable used are connected in series where the ends of the cable meet to form a single coil with as many turns as there are conductors in the cable, the total length of multicon-ductor degaussing cable is equal to the length of the cable sheath times the number of conductors. For example, the total length of a type MDGA-19 cable (19-conductor cable) with a cable sheath of 500 feet is 9500 feet (500 x 19 = 9500).

TYPE OF CABLE.—Insulation resistance will vary considerably with the nature of the insulating materials employed and the construction of the cable. Therefore, it is possible to determine the condition of a cable by its insulation resistance measurements only when they are considered in relation to the typical characteristics of the particular type of cable. The minimum safe insulation resistance for types SGA, DG, SCA, and HF cables is indicated by the upper curve in figure 2-39. The lower curve is for types TTHFWA and TTHFA telephone cables.

TEMPERATURE OF CABLE.-With nonflexing service cables, the highest permissible operating temperature (85° C at the sheath) and the nature of the insulating

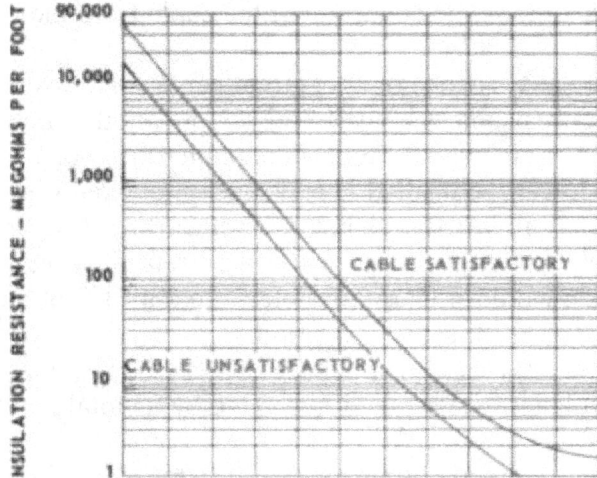

- 10 0 10 20 30 40 50 60 70 90 90 100 C 14 32 50 m 86 104 1 22 140 158 174 1*4 212 f
TEMPERATURE OF SHEATH

Figure 2-39.-Minimum insulation resistance values for nonflexing service cables.

material make it essential that temperature of the cable be considered in conjunction with the insulation resistance measurements. Therefore, fairly accurate estimates or measurements of the temperature of the sheath of the cable must be made to permit proper use of the curves in figure 2-39 to determine (by means of insulation resistance measurements) the condition of the cable. If the results obtained by estimating the temperature indicate that the insulation resistance values are approaching the limiting values on the curve, temperature measurements should be made.

The temperature should be measured by means of thermometers attached to the cable sheath or armor at several points along the length of the cable and these values should be averaged. Place the thermometer bulb in direct contact with the cable armor, scraping away the paint at the point of contact and holding the thermometer in place with pads of felt (or other heat-insulating

material) placed over the bulb and secured with friction tape. The number and location of thermometers used should be such that a representative average of the sheath temperature of the entire cable can be obtained.

EQUIPMENT CONNECTED IN CIRCUTT.-Init ial measurements of insulation resistance of complete circuits should be made to avoid unnecessarily disconnecting apparatus with resultant time, labor, and possible damage to cables by handling. For this purpose, the circuits measured should be considered as beginning at the open switch (or circuit breaker) on the switchboard from which the test potential is supplied and extending through closed switches on distribution panels and boxes, fuse clips (fuses in place), etc. to its extremity, including line terminals of controllers, light fixtures, and switches. When measuring insulation resistance, always record the exact amount of equipment included in the circuit so that accurate comparisons can be made with similar past or future measurements.

MEASURING INSULATION RESISTANCE.-Measurements should be made on each individual leg of d-c circuits and each individual phase lead of 3-phase a-c circuits.

For lighting circuits, the legs or phase leads should include all panel wiring, terminals, connection boxes, fittings, fixtures, and outlets normally connected, but with lights turned off at their switches and with all plugs removed from the outlets (fig. 2-40). If local lighting switches are double pole, the insulation resistance of the local branch circuit will not be measured when the switch is open. In such cases, making an insulation test from one leg or phase lead to ground with the local switches closed will determine whether grounds exist on the circuits and fixtures.

For power circuits, the legs or phase leads should include panel wiring, terminals, connection boxes, fittings, and outlets (plugs removed), motor controller terminals, and other equipment that remains connected when the legs or phase leads are isolated by opening switches (or circuit breakers) at the switchboard, and by leaving motor controller contactors open (fig. 2-41).

For degaussing circuits, measurements should be taken at a degaussing coil connection box; the legs

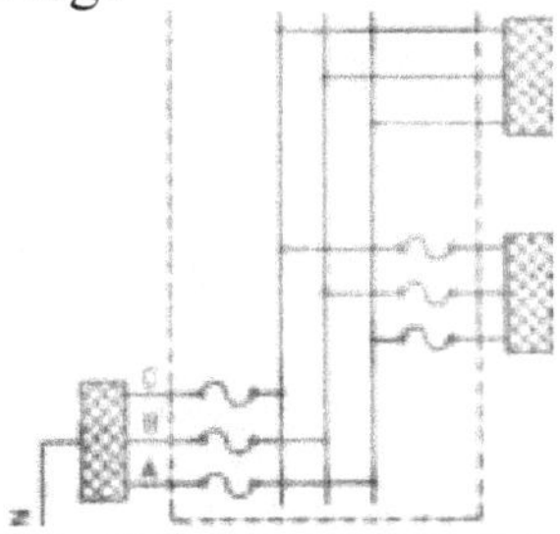

H*- TO OTHER LOADS
1
3
DISTRIBUTION BOX, FUSED
I 1
I C
TO OTHER LOADS
8
DISTRIBUTION BOX, UNFUSED
BRANCH
LOCAL BRANCH CIRCUITS AND FIXTURES
SWITCH OPEN

I BRANCH ' BOX
SWITCH OPEN
LOCAL BRANCH CIRCUITS AND FIXTURES

Figure 2-40.-Measuring insulation resistance of a lighting circuit.

measured should include the coil cables, through boxes, and feeder cables. Disconnect the supply and control equipment by opening the circuit on the coil side of the control equipment. Measure the compass-compensating coil feeder cable with all control equipment disconnected. Additional information on tests of degaussing installations is obtained in BuShzps Manual Chapter 81, and in the degaussing folder furnished with each degaussing installation.

fnn
SWTCH C* C.RCuiT BREAKER OPEN
SVH'TtMBCfO
MOTOR
FEEDER

ADEQUATE LT GROUNDEC 8' CABLE STRAPS
A NO C.IPS
CONTACTORS OPEN
I
controller
S*iTCmES OR C«Cuit BREAKERS.
close c I l
I i
power distribution
PANEL
MAIN
Hi
a

Figure 2-41.-Measuring insulation resistance of a power circuit.

Measurements of these circuits (lighting, power, and degaussing circuits) should be made as follows:

1. Check to see that the cable armor is adequately grounded by measuring between the cable armor and the metal structure of the vessel (step 1, fig. 2-42); normally, grounding has been accomplished by means of cable straps. If a zero reading is not obtained, ground the cable armor.

2. Select one lead to be measured, and connect all the other leads in the cable together and ground them by means of temporary wires (step 2, fig. 2-42).

3. Measure the resistance of the lead being tested to ground (step 3, fig. 2-42). The test voltage should be applied until a constant reading is obtained. Hand-driven generator type instruments (meggers) should be cranked for at least 30 seconds to ensure a steady reading.

4. Repeat steps 2 and 3 as necessary to measure each leg or phase lead to ground (steps 4 and 5, fig. 2-42). When circuits contain permanently connected paths between legs or phases, such as transformers, indicator lights, control relays, etc., measurements need be made only between one lead and ground unless low readings requiring further tests are obtained.

The readings obtained in accordance with the foregoing should be considered satisfactory if they are not less than

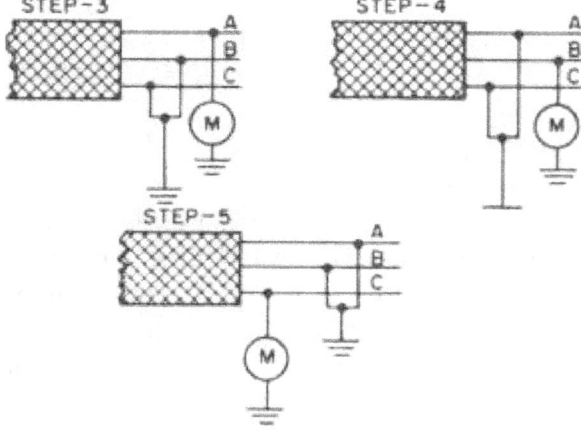

Figure 2-42.—Measuring circuit insulation resistance.

1 megohm for each leg or phase of a lighting or power circuit to ground, or less than 0.2

megohm from any leg, which includes the coil cables, the through boxes, and the feeder cables, of a degaussing circuit to ground. If the readings are lower than these minimum values, they should be considered satisfactory if they are not lower than previous measurements made under the same conditions (same cable with the same equipment connected and approximately the same average sheath temperature), provided the value of the previous measurements has been established as satisfactory by further investigation and satisfactory service operation.

Where the insulation resistance of a circuit is less than the minimum described above, the low resistance may be due to trouble localized in a segment of the circuit. To determine the cause, segregate the circuit into two or more parts by opening switches, circuit breakers, removing fuses, etc. at the feeder distribution panels and boxes, and measure the insulation resistance of each segment. One or more of the segments may indicate abnormally low resistance. The low values on these segments may be due to low insulation resistance at cable junctures or in equipment remaining connected to the segment. Inspect and correct any such faults by cleaning or correc -tive action as necessary. Then, measure the insulation resistance of the circuit and compare the values with readings obtained prior to the inspection and cleaning.

If the value is still low, disconnect all conductors in the questionable segment of the circuit and proceed as follows to ascertain the condition of the cable conductors:

1. Measure or closely estimate the average sheath temperature throughout the length of the cable being tested.

2. Measure the insulation resistance of each conductor in the cable.

3. Convert the measured insulation resistance to megohms per foot, and compare this value with the minimum satisfactory megohms per foot as shown on the graph (fig. 2-39) at the average cable sheath temperature.

4. If the megohms per foot of any conductor in the cable is less than the minimum satisfactory value indicated on the graph, an unsatisfactory condition is indicated and the cable section may require replacement or suitable corrective action.

5. If the megohms per foot of each conductor in the cable is higher than the minimum satisfactory value indicated on the graph, the circuit may be restored to the original condition by connecting the cable; replacing the fuses; and closing the switches, circuit breakers, etc. With the circuit as it was for the original measurements, measure it again. This reading should be recorded on the resistance test record card (NavShips 531) together with the notation as to which equipment remained connected and the cable sheath temperature.

This procedure may be demonstrated using the simplified circuit shown in figure 2-43. Initial measurements,

ABC
ABC
35-4P-(4-168-1) (310') TSGA-100

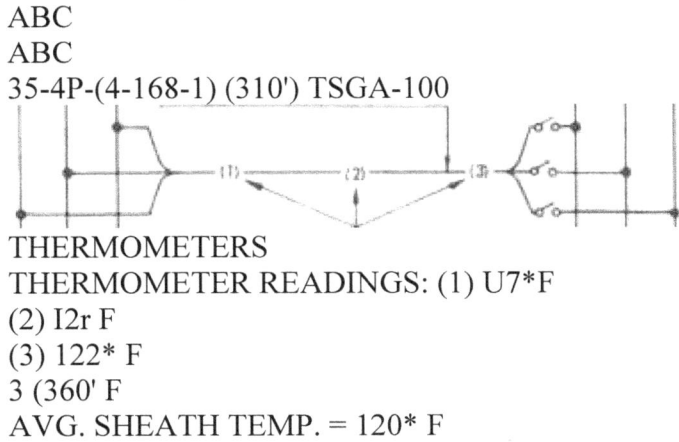

THERMOMETERS
THERMOMETER READINGS: (1) U7*F
(2) I2r F
(3) 122* F
3 (360' F
AVG. SHEATH TEMP. = 120* F

POWER DISTRIBUTION
PANEL 4-168-1 (ALL SW1TCHESOPEN)
Figure 2-43.-Simplified electrical power circuit.

which were made on the complete circuit (all switches closed on panel 4 -168 -1) of which the cable 3S -4P (4-168-1) forms a part, show that the insulation resistance of the complete circuit is low as compared with previously measured values (fig. 2-43). Therefore, further investigation and tests are necessary in order to determine and correct the cause of the low insulation resistance.

The first step is to segregate the circuit into two parts by opening all the switches on panel 4-168-1. Additional measurements are made to determine whether the low insulation resistance is caused by the part of the circuit between the switchboard and the distribution panel. The average sheath temperature, when measured by three thermometers (1, 2, and 3, fig. 2-43), is found to be 120° F, and the measured insulation resistance shows 160,000 ohms for phase A; 140,000 ohms for phase B; and 180,000 ohms for phase C.

The condition of the cable is determined by findingthe insulation resistance per foot of the phase indicating the lowest resistance (phase B). This value is 43.4 megohms per foot (140,000 ohms « 310 ft = 43.4 megohms). The minimum safe insulation resistance of the upper curve (fig. 2-39) at 120° F is 35 megohms per foot. Because

SWITCHBOARD NO. 3S

the lowest measured resistance of 43.4 megohms per foot is above the minimum safe value for type SGA cable, it appears that the condition of the cable is satisfactory.

The next step is to further segregate the circuit by disconnecting the cable 3S-4P (4-168-1) from the distribution panel 4-168-1. An insulation resistance measurement is made on the cable, and this measurement gives a value of 1.5 megohms for the conductor with the lowest insulation resistance. This indicates that the distribution panel is probably contributing to the lowering of the resistance. An examination of the insulation around the bus bars in the panel discloses an accumulation of dust.

After the insulation of the panel is cleaned and the cable reconnected, another measurement is made, which shows a value of 1.2 megohms for the conductor with the lowest insulation resistance. This indicates that the condition has been corrected and no further tests are necessary. The reading of 1.2 megohms is recorded on the Resistance Test Record Card (NavShips 531) for this circuit along with an entry giving the average sheath temperature and the equipment connected in the circuit (entry for 6 June 1960 on fig. 2-44).

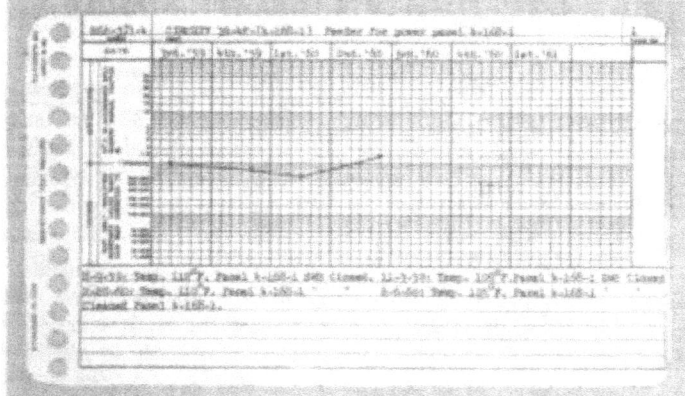

Figure 2-44.-Typical resistance test record card.

It is recommended that a work log be maintained in which the date, insulation resistance, sheath temperature,

equipment connected in the circuit, and other pertinent information (fig. 2-45) are recorded at the time the measurements are being made. This information will prove valuable when comparing measurements, and will provide the data required to be recorded on the resistance test record card (fig. 2-44).

US'". GEAR (CA-IP1)

5/z jjitckboard no. 3 S

DATl!: U JUME I960

Figure 2-45.-Sample resistance test work sheet.

Cable Fittings

STUFFING TUBES.—Periodic inspections should be made to determine the condition of stuffing tubes. Such inspections may be made in conjunction with other tests and inspections. Usually, visual inspection is sufficient to disclose the material condition of the fitting; however, watertightness of stuffing tubes is best determined by air tests. Loose seals occasionally develop due to one or more of several factors, such as loosening of the gland nuts, flowing of the intermediate or soft packing in metal tubes, or "necking" of the cables.

There is only a remote chance of the stuffing tube gland nuts loosening or backing off after the compartment in which they are located has been painted, since the paint oyer the gland nuts will tend to keep them from turning. Loosening of the seal in metal stuffing tubes is more often caused by flow of the soft plastic packing into the interstices of the cable armor, or by depression or "necking" of the cable itself because of excessive packing pressure. Therefore, care must be taken when setting up gland nuts to avoid excessive packing pressure.

It is preferable that no stuffing-tube gland nuts be tightened except those that have obviously loosened or backed off, unless there is positive evidence of leakage through the stuffing tube. When gland nuts must be tightened, an additional turn of softpacking should be inserted if at all practicable.

SUPPORTS.—The bolts and nuts that secure cable hangers, connection boxes, and wiring appliances to bulkheads and other supports may be loosened and lost because of vibration and gun fire. Periodic inspection should be made and all nuts and bolts securely tightened. Such an inspection should be made after firing, especially of those wireways in the vicinity of the guns.

Tests and Inspections

Certain periodic tests and inspections are necessary to ensure the satisfactory operation of shipboard electrical cable installations. The following schedule of tests and inspections should be used so far as practicable and may be supplemented by more frequent tests and inspections as unusual circumstances or conditions warrant.

1. Each watch: Use ground detector voltmeters or ground detector lamps at least once each watch to test for grounds on the circuits that are energized. Record the test and results in the electrical operation log.

2. After firing: Inspect wireways and securely tighten all nuts and bolts.

3. Quarterly: Measure the insulation resistance of all power, lighting, and degaussing cables with an insulation resistance measuring instrument.

QUIZ

1. Basically, shipboard electrical cables are classified according to what two types of service?

2. What does the letter designation, DBSP, mean when used to describe a cable?

3. When the exact number of conductors in the cable is indicated by a letter (S, D, T, or F) in the type letter designation, what do the numerals following the letter designation indicate?

4. When the exact number of conductors in the cable is not indicated by a letter in the type letter designation, what do the numerals following the letter designation indicate ?

5. Nonflexing service cable permanently installed aboard ship is classified how according to use?

6. What is the meaning of the letter designation, SGA, as applied to cable?

7. What are the insulating materials used in SGA cable?

8. What is the principal use of repeated flexing service cable aboard ship?

9. What is the meaning of the letter designation, HOF, as applied to repeated flexing service cable?

10. To select the proper size cable for a particular installation, it is necessary to know what four quantities?

11. What is the meaning of the expression, demand factor, as applied to a circuit?

12. What is the magnitude of the demand factor for power systems supplying a single-phase load or for a lighting system branch, submain, and main circuit?

13. To obtain a system having minimum weight, what is the relation between the percent voltage drop (VD) allocated to a particular section and the length of that section?

14. If a particular section has an average over-all length of 100 feet, an operating voltage of 117 volts, and a common reference voltage of 450 volts, what will be the average over-all length of the section based on 450 volts?

15. A 2-wire, d-c circuit 50 feet long employs DSGA-3 cable to supply a ship's service and emergency lighting load. What is the percent voltage drop if the resultant load current is 3 amperes and the rated voltage at the fuse box is 120 volts?

16. What is the approximate voltage drop between the source and the load for a 3-phase feeder using TSGA-350 cable if the load current per conductor is 300 amperes and the length of the feeder is 30 feet (assume p = 12)?

17. If cable s are run through hazardous spaces, what type of cable must be used?

18. Stuffing tubes are used to provide for the entry of electric cable into what four general types of equipment enclosures?

19. Stuffing tubes are made of what materials?

20. What is a simple form of cable support?

21. How many rows of cables may be supported in one cable rack (a) from an overhead deck and (b) from a bulkhead?

22. In zone control, the ship's service distribution is divided into areas generally coinciding with what zones ?

23. Load center switchboards and miscellaneous switchboards for ship's service distribution systems on vessels with zone control are given two-digit identification numbers, (a) What does the first digit signify? (b) What does the second digit signify?

24. (a) Generators are numbered to correspond to the numbering of what associated unit? (b) Generators are also given another number-letter designation for what purpose ?

25. What are the standard letter designations for (a) motor generators, (b) transformers, and (c) rectifiers?

26. What three general items appear on the cabinet information plates of distribution panels and bus transfer equipment?

27. On a 3-phase system what do the black, white, and red conductors signify?

28. On a 3-wire, d-c system what do the black, white, and red conductors signify?

29. On a 2-wire.d-c system what do the black and white conductors signify?

30. What features of receptacles, plugs, and portable cables with portable tools aboard naval vessels are used to reduce the possibility of electric shock?

31. What color is used to identify the grounding conductor for all new portable tools and equipment?

32. What two items of information are indicated on casualty power tags?

33. What is the primary purpose of electrical cable maintenance ?

34. In general, what materials comprise class B insulation?

35. What dielectric material is used in class E insulation?

36. What class of insulation is usually used in propulsion motors and generators?

37. What classes of insulation are used in ship's service and emergency generators?

38. How is cable insulation resistance related to cable length?

39. For lighting circuit insulation measurements, what precautions should be taken regarding the lights themselves and also associated outlet plugs?

40. For power circuit insulation measurements, what precautions should betaken at the associated switchboard and motor controller?

41. What value of insulation resistance is considered a satisfactory minimum for each leg or phase of a lighting or power circuit to ground?

42. What value of insulation resistance is considered a satisfactory minimum for any leg of a degaussing circuit, including the coil cables, the through boxes, and the feeder cables?

43. Water tightness of stuffing tubes is best determined by what type of tests?

44. As far as practicable, how often should ground detector voltmeters or ground detector lamps be used to test for grounds on circuits that are energized?

45. How often should the insulation resistance of all power lighting and degaussing cables be measured?

CHAPTER

ELECTRIC LIGHTING

PRINCIPLES OF LIGHT

Light is radiant energy of those wavelengths that are capable of affecting the eye to produce vision. Radiant energy is emitted from various sources and varies in wavelength from 10" 12 centimeters for cosmic rays to 10 6 centimeters for radio rays. This range of wavelengths comprises the complete spectrum of electromagnetic waves, which travel through space at the velocity of approximately 186,000 miles per second.

The visible spectrum constitutes a narrow portion of the complete spectrum and includes red, orange, yellow, green, blue, and violet rays (fig. 3-1). The range of wavelengths in the visible spectrum varies from 0.0008 millimeters for red rays to 0.0004 millimeters for violet rays. The spectrum extends beyond these limits into regions that are invisible to the eye. Radiation of greater wavelength beyond the red end is INFRARED and that of lesser wavelength beyond the violet end is ultraviolet.

Propagation

The natural sources of light are the sun, stars, moon, and planets. The artificial sources include the electric -arc lamp, incandescent lamp, and fluorescent lamp. These sources of radiation (except the fluorescent lamp) are bodies at high temperatures, which produce vibration of the particles of matter. The radiated light travels in

BROADCAST X-

INFRARED RAYS
_L_L
SOLAR RAYS
HERTZIAN WAVES
■X RAYS
COSMIC RAYS
I0 6 CM
RAOlO COMM.
ULTRAVIOLET RAYS
VISIBLE SPECTRUM
RED ORANGE
Blue violet
CD
O
8
o
o
8
o
o
8
o
o o * o o *
o o
10"'*CM
GAMMA
RAYS

Figure 3-1.-Spectrum of electromagnetic radiation.

straight lines and behaves like wave motion in which the waves are transverse to the direction of propagation.

Light waves are of different wavelengths; different colors of light correspond to different frequencies of vibration; and electromagnetic waves of all wavelengths travel with the same velocity in free space, but with different velocities in different media. Hence, light waves obey the fundamental equation of wave motion,

~ A A

where V is the velocity of light (3 x 10 10 centimeters per second^J the frequency of vibration (number of waves passing a given point per second), and a the wavelength in centimeters.

Light waves travel in straight lines in a homogeneous medium. When these rays, or lines, of propagation encounter a medium of greater or less density, some of the energy is absorbed and the remainder is either reflected by the boundary between two media or is refracted in passing obliquely from one medium into another, or through a medium of varying density. These phenomena (absorption, reflection, and refraction) usually occur when light impinges on most objects.

Absorption

Absorption is the loss of energy that occurs when light rays strike an object and are dissipated in the form of heat. The amount of absorption depends on the nature of the object, the molecular construction, the wavelength or color of the incident light, and the angle at which the light strikes the object. All objects do not absorb light of different wavelengths in the same proportion. This phenomena accounts for the characteristic color of objects illuminated by either sunlight or artificial white light.

Reflection

Reflection occurs when light rays strike an object and are deflected back from the surface in a different direction. When light falls on a smooth, polished surface, it is reflected in a definite direction and is called REGULAR reflection (fig. 3-2). The light from the source to the polished surface is the INCIDENT RAY. The point at which the incident ray strikes the polished surface is the POINT OF INCIDENCE, P. The ray that leaves the point of incidence is the REFLECTED RAY. The line perpendicular to the polished surface at the point of incidence is the NORMAL TO THE SURFACE, N. The angle between the incident ray and the normal to the surface is the ANGLE OF INCIDENCE, i. The angle between the

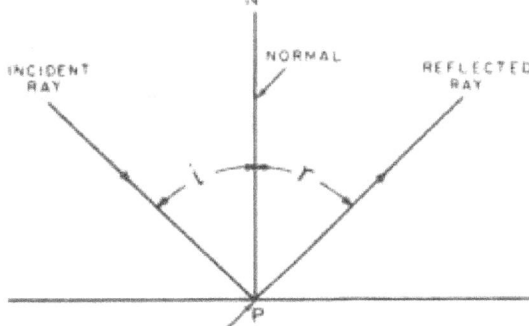

POINT OF INCIDENCE

Figure 3-2.-Reflection of a light ray.

reflected ray and the normal to the surface is the ANGLE OF REFLECTION, r.

The incident ray, reflected ray, and normal to the point of incidence lie in the same plane, and the angle of reflection is equal to the angle of incidence. This condition is true for both plane and curved mirror surfaces. The incident and reflected rays travel with the same velocity because they travel in the same medium.

When light falls on a rough, unpolished surface, it is reflected in all directions and is called diffuse (irregular) reflection. The only essential difference between regular and diffuse reflection is that diffuse reflection takes place from an infinite number of infinitesimal plane surfaces oriented in all directions.

Refraction occurs when light rays pass obliquely from one medium to another of greater or less density and undergo an abrupt change in direction (fig. 3-3). The

Refraction

N

NORMAL

INCIDENT RAY

REFLECTED RAY

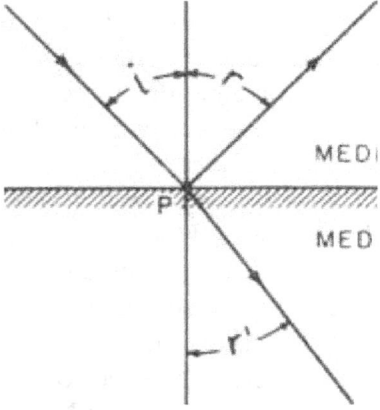

mmmmmmz
\ MEDIUM 2
MEDIUM 1
REFRACTED N' RAY

Figure 3-3.-Refraction of a light ray.

incident ray enters the second medium at the POINT OF INCIDENCE, P. The ray that enters the second medium is the REFRACTED RAY. The line perpendicular to the surface at the point of incidence is the NORMAL TO THE SURFACE, NN\ The angle between the refracted ray and the normal to the surface in the second medium is the ANGLE OF REFRACTION, r'. The magnitude of the angle of refraction depends on the refracting medium and is not necessarily equal to the angle of incidence.

Both reflection and refraction occur in greater or less degrees at the boundary surfaces between most media. The boundary thus serves as reflecting surface for one ray and as a refracting surface for another. The normal to the surface, NN\ and the incident and refracted rays lie in the same plane.

Color

As previously stated, light of different colors travel through space with the same velocity. Hence, in accordance with the fundamental equation of wave motion (equation 3-1), the frequencies are different for different colors and are inversely proportional to their wavelengths as,

$$f = \frac{V}{\lambda},$$

where £ is the frequency, V the velocity, and X the wavelength.

The impression of color depends on the frequency of the light that is reflected or transmitted to the eye. The primary colors of light in the visible spectrum are red, green, and blue. Light of any one of these colors cannot be produced by combining light of any other colors. However, light of any other color than these three can be produced by combining, in the proper proportion, light of any two or all three of the primary colors.

All objects absorb a certain amount of the light that falls on them, but all objects do not absorb the same proportion of light of different wavelengths. If all objects absorbed light in the same proportion, they would appear to have the same color. An object must be observed under white light for it to appear in its true colors. White light

contains (in the proper proportion) all three primary colors.

If an object absorbs all or nearly all of the white light that falls on it, the object appears black in color. If it absorbs all the green and blue portions of white light that falls on it, the object appears red in color because only the red light is reflected to the eye. In other words, only the reflected light is seen. Those components of the visible spectrum absorbed by the object are not seen.

An object appears in its true colors only when the light falling on it contains light of the wavelengths that the object reflects or transmits. Thus, the light falling on a red object must contain red light for it to appear red in color. A red object observed under a light that contains only green and blue rays will appear black in color because the object absorbs the green and blue rays and is capable of reflecting or transmitting only red rays (which are not present).

Candlepower

The luminous intensity, or light-producing power, of a given source of light is a measure of the ability of the source to project light in a certain specified direction. The unit of luminous intensity is the CANDLE, or CANDLEPOWER, I, which is approximately equal to the intensity of ligEt emitted from a seven-eighths-inch sperm (standard) candle burning at the rate of 120 grains per hour. Thus, the candlepower of light is expressed in terms of the standard candle that is considered to emit light equally in all directions. However, this condition is a theoretical concept and does not actually exist, but is necessary to develop the quantities and units used in lighting computations.

Foot-Candle

The intensity of radiation is the light energy radiated or absorbed per unit time per unit area. Light energy is a flow of light waves that always originate from some source, such as the sun, a candle, or an incandescent lamp. The unit of intensity of radiation is the FOOT-CANDLE, E, which is the illumination on a surface that

is 1-foot distant everywhere from a source having a luminous intensity of 1 candlepower.

Inverse Square Law

In figure 3-4 a source of light produces alight intensity of candlepower, I, in the direction of surfaces A, B, and C. The illuminated surfaces are perpendicular to the direction from the light source and at a distance in feet, d, from the source. The illumination in foot-candles, E, falling on surfaces A, B, and C is expressed as an appff-cation of the inverse square law as follows:

E =1 (3-2) ~ d 2

For example, if I is 100 candlepower and d is 1 foot, the illumination on surface A will be

E = - ioo foot-candles. " I 2

If d is 2 feet, the illumination on surface B will be

100

E = = 25 foot-candles.

2 2

If d is 3 feet, the illumination on Surface C will be

100

E = = 11.1 foot-candles.

3 2

This relation results from the fact that the total quantity of light in the cone of light is the same at all distances from the source; whereas, the surface over which it is spread increases as the square of the distance. Hence, the illumination, which is the quantity of light per unit of surface, is inversely proportional to the square of the distance from the source. At any given distance from the source, the illumination is directly proportional to the candlepower.

3 ^^::^Xi,

1 = 100

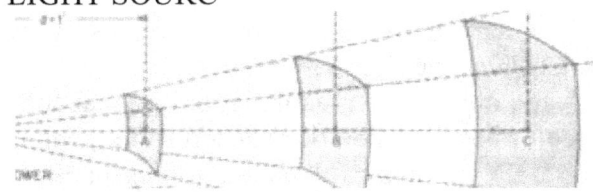

CANOLE POWER
LIGHT SOURC

E'100 FOOT-CANDLES

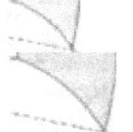

E«25 " FOOT-CANDLES
E » M 1
FOOT-CANOLES

Figure 3-4.-Inverse square law of radiation.

Lumen

Illumination is also expressed in terms of the lumen, which is the unit of luminous flux, or quantity of light. A lumen is that quantity of light flux that must fall on a surface having an area of 1 square foot in order to produce on that surface an illumination of 1 foot-candle.

The number of lumens per square foot is always numerically equal to the illumination in foot-candles, E. Thus, the illumination that is produced by a total flux on an area is

where E is the illumination in foot-candles, * the total flux in lumens, and A the area in square feet. For example,! lumen of flux distributed over an area of 1 square foot will produce 1 foot-candle of illumination over that surface. Similarly, 10 lumens of flux distributed over an area of 5 square feet will produce 2 foot-candles of illumination over the surface. If the distribution of flux over the area is nonuniform, equation (3-3) gives the average value of illumination. In calculating the illumination produced by a lamp in any given direction, the value of the luminous intensity used in the equation must be the value that applies in the appropriate direction.

$$\underline{\mathbf{E}} = \frac{\Phi}{A},$$

(3-3)

The candlepower measured around a light source in a horizontal plane is the HORIZONTAL CANDLEPOWER. The average candlepower in this plane is the MEAN HORIZONTAL CANDLEPOWER. The mean or average value of candlepower measured in all directions about a light source is the MEAN SPHERICAL CANDLEPOWER. Unless otherwise stated, the candlepower rating of a lamp is usually given in mean spherical candlepower.

The total light flux, *, produced by a light source having a mean spherical candlepower, I_ OJ is

*= 12.57 I £ , (3-4)

where 12.57 is the surface area of a sphere of one foot radius, (The area of a sphere is 4trr 2 .)

For example, a light source of 100 candlepower (mean spherical) will produce 12.57 * 100 = 1257 lumens of light flux.

If this light source is considered at the center of a sphere of one foot radius, the illumination on each square foot of surface of the sphere will be E = #/A = 1257/12.57 = 100 lumens per square foot. Similarly, the illumination in foot-candles will be E = _I Q /d 2 = 100/1 2 = 100 foot-candles.

Cosine Law

To find the illumination on a flat surface when the light is inclined to the normal, as illustrated in figure 3-5, the inverse square law (equation 3-2) and the cosine law are applied. The cosine law states that the illumination of a surface that is not normal to the light rays will vary as the cosine of the angle of inclination (d). Thus, the inverse square law and the cosine law are combined and expressed as

I

E = -^ycos 0, (3-5)

where E is the illumination at point P in foot-candles, I the intensity of the light source in candlepower in the direction of P, and d the distance in feet to P (fig. 3-5).

For example, if I is 200 candlepower, d is 10 feet, and 9 is 60°,

E = -^?-cos 60° = 1 foot-candle. 10 2

ILLUMINATED SURFACE

Figure 3-5.-Cosine law of incidence.

LIGHT SOURCES

The sources of electric light used in naval vessel are the (1) incandescent, (2) fluorescent, and (3) glow lamps. The incandescent lamp is the most commonly used type; however, the fluorescent lamp is used to a great extent.

A complete list of Navy type lamps is contained in the Navy Stock List of General Stores, carried aboard all ships. This list includes the electrical characteristics, physical dimensions, applications, ordering designation, and an outline of each Navy type lamp.

Incandescent Lamps

There are over 100 types of incandescent lamps used aboard ship. These types include lamps for general lighting, for many types of signaling, for instrument lighting and indicator lights, for battery-powered flashlights and hand lanterns, for illuminated gun sights, and for a variety of specialized applications.

CONSTRUCTION.—The incandescent lamp consists of a tungsten or carbon filament supported by a glass stem. The glass stem is mounted in a suitable base that provides the necessary electrical connections to the filament.

19R

The filament Is enclosed In a transparent, or translucent glass bulb from which the air has been evacuated. The passage of an electric current through the filament causes it to become incandescent and to emit light. The filament is operated either in a vacuum or a gas-filled bulb to prevent oxidation and consequent failure of the filament, which would occur in air.

All Navy-type, 115- or 120-volt lamps (up to and including the 50-watt sizes) are of the vacuum type, except Navy type TG-24; and all lamps above 50 watts are of the gas-filled type, except Navy type TG-36. The use of an inert gas, which is a mixture of argon and nitrogen gases, retards the evaporation of the filament and allows it to be operated at higher temperatures that result in higher efficiencies. Lamps under 50 watts are of the vacuum type because the effectiveness of the inert gas in increasing the luminous output is less pronounced in the lower-wattage lamps.

The incandescent lamp is further subdivided into tungsten- and carbon-filament types. The tungsten-filament lamps comprise most of those listed in this group.

RATING.—Incandescent lamps are rated in watts, amperes, volts, candlepower, or lumens, depending on their type. Generally, large lamps are rated in volts, watts, and lumens. Miniature lamps are rated in amperes for a given single voltage and in candlepower for a voltage-range rating.

CLASSIFICATION.—Standard incandescent lamps are classified according to the (1) shape of bulb, (2) finish of bulb, and (3) type of base.

The classification of lamps according to the SHAPE OF BULB with the corresponding letter designation is illustrated in figure 3-6. The designation letter, which denotes the shape of the bulb, is followed by a numeral that denotes the diameter of the bulb in eighths of an inch. For example, if the lamp illustrated in figure 3-6, F, is designated PS-52,it is a lamp with apear-shaped bulb and straight sides havinga diameter of 52/8, or 6-1/2 inches.

The lamps commonly used for general lighting aboard ship are 115- or 120-volt, 50-, 100-, and 200-watt rough service types. These lamps (up to and including the 150-watt size) have standard-line shaped bulbs, and those of 200 watts and above have pear shaped bulbs. The other

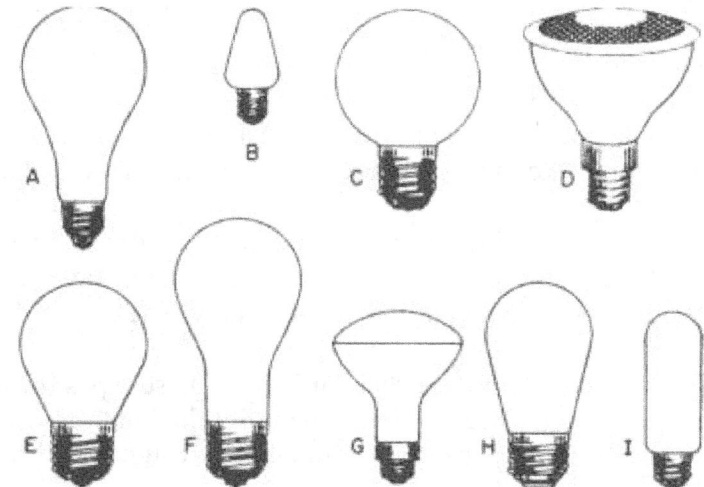

A-STANDARD LINE (A) B- CANDELABRA (C) C- GLOBULAR (G) D-PARABOLIC (PAR) E- PEAR SHAPED (P)

F- PEAR SHAPE WITH
STRAIGHT SIDE (PS) G- REFLECTOR (R) H-STRAIGHT SIDE (S) I-TUBULAR (T)

Figure 3-6.-Classification of lamps according to shape of bulbs.

bulb shapes are designed for lamps that are used for special types of service.

Rough-service (RS) lamps have specially constructed filaments that are supported at several points to increase their ability to withstand vibration and shock. A modification to the RS lamp is the rough-service, high-impact (RSHI) lamp. The RSHI lamp is available only in the 50-watt size because this feature is not successful in larger lamps and is not warranted in smaller lamps. The RSHI lamps are constructed with a resilient element of rubber or copper mesh located between the bulb and the screw base. The cushioning effect of the resilient element makes the RSHI lamps far more shock proof than the standard RS lamps.

The classification of lamps according to the finish of bulb are the clear (CL),the inside frosted (IF), the white bowl (WB), the silvered bowl (SB), and the colored.

The CLEAR lamp consists of a bulb that is made of unclouded or luminous glass, which exposes the filament to view. These lamps are used with reflecting equipment that completely conceals the lamps to avoid glare. Clear lamps can be used with open-bottom reflecting equipment when the units are mounted sufficiently high for the lamps to be out of the line of vision.

The INSIDE FROSTED lamp consists of a glass bulb that has the entire inside surface coated with a frosting, which conceals the filament and diffuses the light emitted from the lamp. These lamps can be used with or without reflecting equipment.

The WHITE BOWL lamp is equipped with a glass bulb that has the lower portion sprayed with a white enamel. The white bowl finish increases the size of the visible light, reduces the brilliancy, and diffuses the light rays to reduce glare. These lamps, which are usually over 120 watts, are used with open-bottom reflecting equipment.

The SILVERED BOWL lamp is provided with a glass globe that has a coating of mirror silver on the lower half, which shields the filament and provides a highly efficient reflecting surface. The upper portion of the bulb is inside frosted to eliminate shadows of the fixture supports. These lamps are used with units that are designed for indirect lighting systems.

The COLORED lamp may consist of a colored-glass bulb. These lamps are used for battle and general lighting, and safety lights.

The classification of lamps according to the type of base is illustrated in figure 3-7. The size of the base is indicated by name, including miniature, candelabra, and intermediate; the type of the base provided with the different sizes is also denoted by name, including screw bayonet, prefocus, and bipin.

The miniature, candelabra, and intermediate types of bases (fig. 3-7, A,B, and C)are used on small-size lamps for detail lighting.

The medium base (fig. 3-7, E), which is the most commonly used type, is used on lamps (up to and including 300 watts) for general lighting.

The admedium base (fig. 3-7, D) is slightly larger in diameter than the medium base and is used on some mercury lamps.

f 0 ^

SCREW

BAYONET FLANGED BIPtN

(StNCLE CONTACT) (SINGLE CONTACT)

A (MINIATURE)

SCREW SCREW WITH BAYONET BAYONET INDEXING

SHORT NUB (SINGLE CONTACT) (DOUBLE CONTACT)

0 ~B" "fl"

BAYONET PREFOCUSING COLLAR PREFOCUSING COLLAR BAYONET SKIRTED

(DOUBLE CONTACT) (SINGLE CONTACT) (DOUBLE CONTACT) (DOUBLE CONTACT)

B (CANDELABRA)

SCREW
C (INTERMEDIATE) D (ADMEDIUM)

SCREW

PREFOCUS (SINGLE CONTACT) SCREW SKIRTED
BIPIN (T-8-F LAMP)

BI POST
E (MEDIUM)
BIPIN (T-I2-F LAMP)

Figure 3-7.-Classification of lamps according to typo of base.

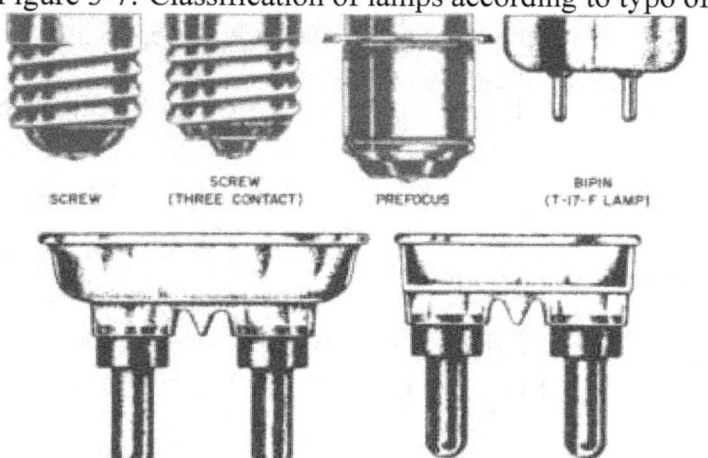

SCREW   SCREW (THREE CONTACT)   PREFOCUS   BIPIN (T-I7-F LAMP)

BIPOST BIPOST (T-24 LAMP) (T-20 LAMP)
F (MOGUL)

Figure 3-7.-Classification of lamps according to type of base-continued.

The mogul base (fig. 3-7, F) is used on lamps rated above 300 watts. The three-control mogul base is used on 3-way lamps.

The medium and mogul prefocused bases are used on lamps provided with concentrated filaments, such as those used for motion picture projection.

The medium bipin base is used on fluorescent lamps described subsequently in this chapter. The medium bi-post base is used on lamps of 500, 750, 1000, 1250 and 1500 watts principally for indirect fixtures. This design allows better heat radiation than can be obtained with the mogul screw base.

The mogul bipost base is used on lamps of 500, 1000, and 1500 watts and above including floodlights.

CHARACTERISTICS.—The average life of standard lamps for general lighting service when operated at rated voltage is 750 hours for some sizes and 1000 hours for others. Every incandescent lamp is designed for operation at a certain specified voltage. The light output, life, and electrical characteristics of a lamp are materially affected when it is operated at other than the design

voltage. Operating a lamp at less than rated voltage will prolong the life of the lamp and decrease the light output. Conversely, operating a lamp at higher than rated voltage will shorten the life and increase the light output. The characteristic curves for gas-filled incandescent lamps (fig. 3-8) show how the performance is affected by operating the lamp at other then the rated voltage. For example, if a 120-volt lamp is operated at 114 volts, the voltage impressed on the lamp is 114/120 * 100, or 95 percent of rated voltage. From the characteristic curves, the lumens emitted from the lamp will be reduced to about 84 percent. Hence, a 5 percent reduction in rated voltage results in a 16 percent reduction in light output. The characteristic curves for vacuum lamps are not exactly the same, but are very similar to those of the gas-filled lamps, and, for all practical purposes can be considered to indicate their performance.

It is apparent from the characteristic curves that lamps should be operated as closely as possible to their rated voltage. Low voltage seriously decreases illumination; whereas, high voltage increases illumination, but an excess voltage of 5 percent will reduce the life of the lamp about one-half.

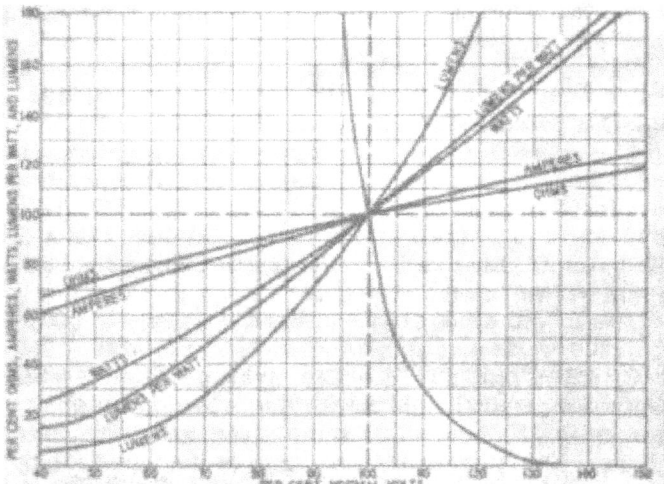

Figure 3-8.-Characteri stic curves for gas-filled incandescent lamps.

Fluorescent Lamps

Fluorescent lamps are used aboard ship for general illumination in offices, messing and berthing spaces, and for detail illumination in plotting rooms, for panels, and for small switchboards.

CONSTRUCTION.—The fluorescent lamp is an electric discharge lamp that consists of an elongated tubular bulb with an oxide-coated filament sealed in each end to comprise two electrodes (fig. 3-9). The bulb contains a drop of mercury and a small amount of argon gas. The inside surface of the bulb is coated with a fluorescent phosphor. The lamp produces invisible, short-wave (ultraviolet) radiation by the discharge through the mercury vapor in the bulb. The phosphor absorbs the invisible radiant energy and reradiates it over a band of wavelengths that are sensitive to the eye.

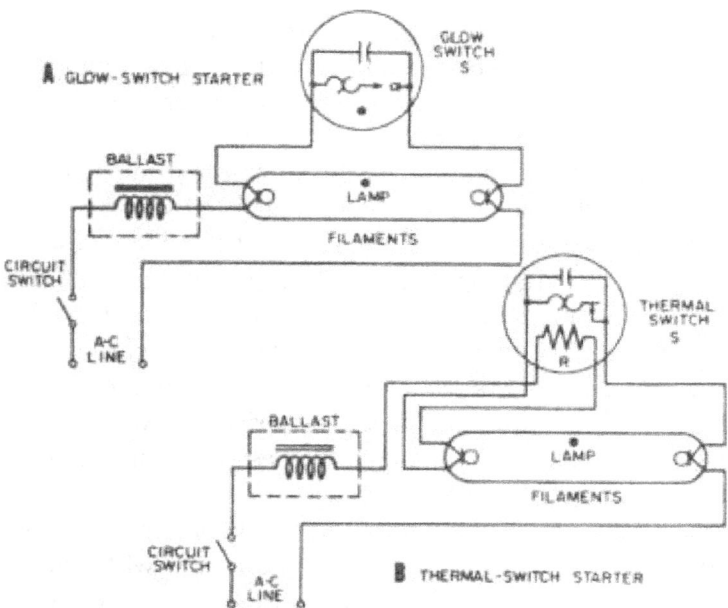

Figure 3-9.-Fluorescent tamp with auxiliary equipment.

AUXILIARY EQUIPMENT. -The fluorescent lamp, like all discharge light sources, requires special auxiliary control equipment for starting and stabilizing the lamp.

This equipment consists of an iron-core choke coil, or ballast, and an automatic starting switch connected in series with the lamp filaments. The starter (starting switch) can be either a glow switch or a thermal switch. A resistor must be connected in series with the ballast in d-c circuits because the ballast alone does not offer sufficient resistance to maintain the arc current steady.

Each lamp must be provided with an individual ballast and starting switch, but the auxiliaries for two lamps are usually enclosed in a single container. The auxiliaries for fluorescent lighting fixtures listed on BuShips Plan 9-S-4953-L are mounted inside the fixture above the reflector. The starting switches (starters) project through the reflector so that they can be replaced readily. The circuit diagram for the fixture appears on the ballast container.

OPERATION.—Fluorescent lamps installed aboard ship are of the hot-cathode, preheat starting type. A fluorescent lamp equipped with a glow-switch starter is illustrated in figure 3-9, A. The glow-switch starter is essentially a glow lamp containing neon or argon gas and two metallic electrodes. One electrode has a fixed contact, and the other electrode is a U-shaped, bimetal strip having a movable contact. These contacts are normally open.

When the circuit switch is closed there is practically no voltage drop across the ballast, and the voltage across the starter, S, is sufficient to produce a glow around the bimetallic strip in the glow lamp. The heat from the glow causes the bimetal strip to distort and touch the fixed electrode. This action shorts out the glow discharge and the bimetal strip starts to cool as the starting circuit of the fluorescent lamp is completed. The starting current flows through the lamp filament in each end of the fluorescent tube, causing the mercury to vaporize. Current does not flow across the lamp between the electrodes at this time because the path is short circuited by the starter and because the gas in the bulb is nonconducting when the electrodes are cold. The preheating of the fluorescent tube continues until the bimetal strip in the starter cools sufficiently to open the starting circuit.

When the starting circuit opens, the decrease of current in the ballast produces an induced voltage across the lamp electrodes. The magnitude of this voltage is sufficient to ionize the mercury vapor and start the lamp. The resulting glow discharge (arc) through the fluorescent lamp produces a large amount of ultraviolet radiation that impinges on the phosphor, causing it to fluoresce and emit a relatively bright light. During normal operation the voltage across the fluorescent lamp is not sufficient to produce a glow in the starter. Hence, the contacts remain open and the starter consumes no energy.

A fluorescent lamp equipped with a thermal-switch starter is illustrated in figure 3-9, B. The thermal-switch starter consists of two normally closed metallic contacts and a series resistance contained in a cylindrical enclosure. One contact is fixed, and the movable contact is mounted on a bimetal strip.

When the circuit switch is closed the starting circuit of the fluorescent lamp is completed (through the series resistance, R) to allow the preheating current to flow through the electrodes. The current through the series resistance produces heat that causes the bimetal strip to bend and open the starting circuit. The accompanying induced voltage produced by the ballast starts the lamp. The normal operating current holds the thermal switch open.

The majority of thermal-switch starters use some energy during normal operation of the lamp. However, this switch ensures more positive starting by providing an adequate preheating period and a higher induced starting voltage.

CHARACTERISTICS.-The failure of a hot-cathode fluorescent lamp usually results from loss of electron-emissive material from the electrodes. This loss proceeds gradually throughout the life of the lamp and is accelerated by frequent starting. The rated average life of the lamp is based on normal burning periods of 3 to 4 hours. Blackening of the ends of the bulb progresses gradually throughout the life of the lamp.

The efficiency of the energy conversion of a fluorescent lamp is very sensitive to changes in temperature of the bulb. The maximum efficiency occurs in the range of 100° F to 120° F, which is the operating temperature that corresponds to an ambient room temperature range of 65° to 85° F. The efficiency decreases slowly as the temperature is increased above normal, but also decreases very rapidly as the temperature is decreased below normal. Hence, the fluorescent lamp is not satisfactory for locations in which it will be subjected to wide variations in temperature. The reduction in efficiency with low ambient room temperature can be minimized by operating the fluorescent lamp in a tubular glass enclosure so that the lamp will operate at more nearly the desired temperature.

Fluorescent lamps are relatively efficient compared with incandescent lamps. For example, a 40-watt fluorescent lamp produces 2400 lumens, or 60 lumens per watt. A 40-watt incandescent lamp produces approximately 400 lumens, or 10 lumens per watt. Thus the fluorescent lamp produces six times as much light per watt as does the comparable incandescent lamp.

Fluorescent lamps should be operated at voltage within ±10 percent of their rated voltage. If the lamps are operated at lower voltages, uncertain starting may result, and if operated at higher voltages, the ballast may overheat. Operation of the lamps at either lower or higher voltages results in decreased lamp life. The characteristic curves for hot-cathode fluorescent lamps (fig. 3-10) show the effect of variations from rated voltage on the condition of lamp operation. Also, the performance of fluorescent lamps depends to a great extent on the characteristics of the ballast, which determines the power delivered to the lamp for a given line voltage.

When fluorescent lamps are operated on a-c circuits, the light output executes cyclic pulsations as the current passes through zero. This reduction in light output produces a flicker that is not usually noticeable at frequencies of 50 and 60 cycles, but may cause unpleasant stroboscopic effects when moving objects are viewed. The cyclic flicker can be minimized by combining two or three lamps in a fixture and operating the lamps on different phases of a 3-phase system. Where only single-phase circuits are available, leading current is supplied to one lamp and lagging current to another so that the light pulsations compensate each other.

The fluorescent lamp is inherently a high power-factor device, but the ballast required to stabilize the arc is a low power-factor device. The voltage drop

ISO 120 IK) 100 90 80

TO

vOLTAOE wa CAUSE OEWEASEO LMHT OUTPUT AND UN-cnriN STARTING

RECOMMENDED OPERATING RANGE

I I M

- TULA MP

FLUORESCENT BALLASTS

EXCESSVE CVER-VCUMC MAY CAUSE INPPHQP. LAMP PtHfOKUAHCt AND OVCftHCATtNO Of

^ jj: £ ^ —

TIT ^iiTfefefcr

109 210 IBS

220

I IB 250 203

lit 120 256 240 209 tit

LINE VOLTAGES

129 290 220

22*

110

w

| 100

s.

3 120

3 no 8

3 10

I H

I!

80

EXCESSIVE UNDER VOUAflC MAT CAUSE OECAEASEO UttHT OUTPU T AND UN-CEP/nUN STARTING

RECOMMENDED OPERATING RANGE

I I I I I I LEAD CIRCUIT

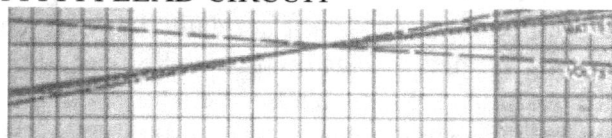

EXCESSIVE OVEP-VCLTAGE MAY CAUSE NFOHOft LAMP PEAPOPMAWLE AND OVOMCATMO OF AUWUAWt

701 109 210 189

110 220 194
119 230 203
119 236 208
120 240 212
123 290 220
2*8 229
UNE VOLTAGES

Figure 3-10.-Characteristic curves for hot-cathode fluorescent lamps.

137

across the ballast is usually equal to the drop across the arc, and the resulting power factor for a single lamp circuit with ballast is about 50 percent. The low power factor can be corrected in a single lamp ballast circuit by a capacitor shunted across the line. This correction is accomplished in a two-lamp circuit by means of a "tulamp" auxiliary that connects a capacitor in series with one of the lamps to displace the lamp currents, and, at the same time, to remove the unpleasant stroboscopic effects when moving objects come into view.

Glow Lamps

Glow lamps are electric discharge light sources, which are used as indicator or pilot lights for various instruments and on control panels. These lamps have relatively low light output, and thus are used to indicate when circuits are energized or to indicate the operation of electrical equipment installed in remote locations.

CONSTRUCTION.-The glow lamp consists of two closely spaced metallic electrodes sealed in a glass bulb that contains an inert gas. The color of the light emitted by the lamp depends on the gas. Neon gas produces an orange-red light, and argon gas produces a blue light. The lamp must be operated in series with a current-limiting device to stabilize the discharge. This current-limiting device consists of a high resistance that is usually contained in the lamp base.

CHARACTERISTICS.-The glow lamp produces light only when the voltage exceeds a certain striking voltage. As the voltage is decreased somewhat below this value, the glow suddenly vanishes. When the lamp is operated on alternating current, light is produced only during a portion of each half cycle, and both electrodes are alternately surrounded with a glow. When the lamp is oper -ated on direct current, light is produced continuously, and only the negative electrode is surrounded with aglow. This characteristic makes it possible to use the glow lamp as an indicator of alternating current and direct current. It has the advantages of small size, ruggedness, long life, negligible current consumption, and can be operated on standard lighting circuits.

A lighting fixture, or unit, is a complete illuminating device that directs, diffuses, or modifies the light from a source to obtain more economical, effective, and safe use of the light. A lighting fixture usually consists of a lamp, globe, reflector, refractor (baffle), housing, and support that is integral with the housing or any combination of these parts (fig. 3-11). A globe alters the characteristics of the light emitted by the lamp. A clear glass globe absorbs a small percentage of the light without appreciably changing the distribution of the light. A diffusing glass

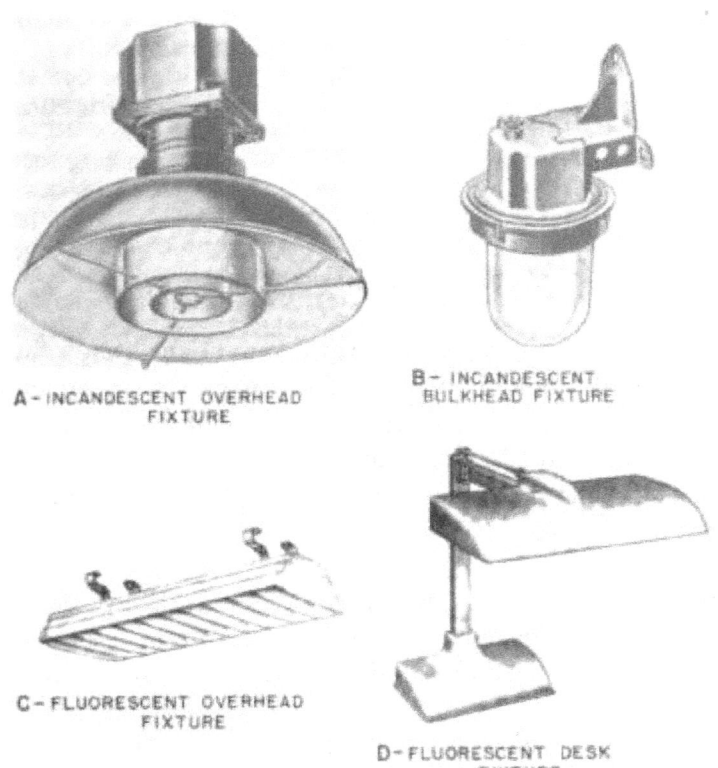

A - INCANDESCENT OVERHEAD FIXTURE

B - INCANDESCENT BULKHEAD FIXTURE

C - FLUORESCENT OVERHEAD FIXTURE

D - FLUORESCENT DESK FIXTURE

Figure 3-11.-Lighting fixtures.

139

globe absorbs a little more light and tends to smooth out variations in the spherical distribution of the light; whereas, a colored-glass or plastic globe absorbs a high percentage of the light emitted by the lamp. A baffle conceals the lamp and reduces glare. A reflector intercepts the light traveling in a direction in which it is not needed and reflects it in a direction in which it will be more useful.

Classification

Lighting fixtures are designated according to the type of enclosure provided, as watertight, nonwatertight, pressure-proof or explosion-proof. They are classified according to use as (1) regular permanent fixtures, (2) regular permanent red-light fixtures, (3) regular portable fixtures, (4) miscellaneous fixtures, (5) navigational lights, and (6) lights for night-flight operations.

REGULAR PERMANENT FIXTURES (incandescent or fluorescent) are permanently installed to provide general illumination and such detail illumination as may be required in specific locations. General illumination is based on the foot-candle intensity required for the performance of normal routine duties. Detail illumination is provided where the general illumination is inadequate for the performance of specific tasks, and includes berth fixtures, desk lamps, and plotting lamps.

REGULAR PERMANENT RED LIGHT FIXTURES (incandescent or fluorescent) are permanently installed to provide low-level, red illumination in berthing areas, in access routes to topside battle and watch stations, and in special compartments and stations. These fixtures are equipped with steam tight inside-acid-etched red globes.

REGULAR PORTABLE FDCTURES (incandescent) are provided for lighting applications that cannot be served by permanently installed fixtures. These units are energized by means of portable cables that are plugged into outlets in the ship's service wiring system and include bedside lamps, desk lamps, and floodlights.

MISCELLANEOUS FIXTURES (incandescent or fluorescent) are provided for detail and special lighting applications that cannot be served by regular permanent or regular portable lighting fixtures. These fixtures

include boom lights, crane lights, gangway lights, portable flood lanterns, hand lanterns, and flashlights.

NAVIGATIONAL LIGHTS (incandescent) include all external lights (running, signal, and anchor), except searchlights, which are used for navigational and signaling purposes between ships to prevent collision, and for communications when underway or at anchor.

LIGHTS FOR NIGHT-FLIGHT OPERATIONS are used to assist pilots (at night) when taking off and landing. These lights also provide visual aid to pilots for locating and identifying the parent aircraft carrier.

Calculations

Interior lighting should provide a sufficient amount of light to supply the desired value of illumination. This condition is accomplished by properly designed lighting fixtures so located as to produce reasonable uniform illumination without glare.

Lighting fixtures are designed to produce (1) direct, (2) semidirect, (3) semi-indirect, and (4) indirect illumination.

The DIRECT fixture consists of a reflecting surface, above the light source, that redirects downward practically all (90 to 100 percent) of the light from the bare lamp.

The SEMIDIRECT fixture has a metal reflector that is open at the bottom and equipped with louvers to conceal the lamps and reduce glare. The unit directs a large portion (60 to 90 percent) of the light downward to the working surface.

The SEMI-INDIRECT fixture consists of a translucent enclosure that is open at the top. The unit transmits some of the light directly to the working surface through the translucent enclosure, and the remainder (60 to 90 percent) of the light is directed to the overhead, which, in turn, reflects it indirectly to the working area.

The INDIRECT fixture is provided with a metal enclosure that is open at the top. The unit transmits 90 to 100 percent of the light to the overhead, which, in turn, reflects the light indirectly to the working area. The majority of lighting installations in naval vessels are of the direct and semidirect types.

CANDLEPOWER DISTRIBUTION GRAPHS.-As previously stated, the candlepower distribution of lamps and fixtures is not uniform. Thus, in calculating the illumination that they produce in any given direction, it is necessary to use the candlepower in that direction. Candlepower distribution curves that indicate the candlepower of the source in all directions are used for this purpose. These distribution curves or graphs are furnished by manufacturers of lighting equipment and are supplied for various types of lamps alone or for lamps when used in conjunction with lighting fixtures.

The distribution graphs of three typical Navy lighting fixtures are illustrated in figure 3-12. A distribution graph is obtained by measuring the candlepower (with a photometer) directly downward from the source. This value is indicated on the vertical to a given scale of a circular chart and represents the candlepower directly below the light (0°). Similarly, the candlepowers around the light are measured at angles of 10° above the vertical. These values are indicated on the chart and represent the candlepowers to scale at the angles above the vertical of 10°, 20°, 30°, and so on around the light. These points are joined by a continuous line that represents the complete distribution graph of the light source. Thus, the candlepower in any direction is the length of the line drawn in that direction from the source to the curve.

The principal methods used in calculating illumination are the (1) point-by-point and (2) lumen methods.

POINT-BY-POINT METHOD.-The illumination at any point on a surface can be calculated by using equation (3-5) in conjunction with the candlepower distribution curve of the specific lamp or lighting fixture. When several fixtures contribute illumination to a given point, the candlepower distribution must be calculated for each fixture, and the actual illumination at any point on a surface will be the sum of the foot-candles contributed by all the fixtures, plus the light that is reflected to the point by the surrounding surfaces.

For example, find the illumination produced on a horizontal surface by two 50-watt, wide-angle, distributing types of overhead fixtures if one unit is located 4 feet away in a direction 30° from the vertical and the other is 7 feet away in a direction 60° from the vertical. The

Figure 3-12.-Candlepower distribution graphs.

first fixture delivers approximately 90 candlepower in the direction 30° from the vertical (curve B of figure 3-12). The illumination produced by the first fixture from equation (3-5) is

I] = ^-(cos 30° = 0.866) = 4.86 foot-candles.

Similarly, the second fixture delivers approximately 60 candlepower in the direction 60° from the vertical

(curve B of figure 3-12), and the illumination produced is

60

E 2 = (cos 60° = 0.5) = 0.612 foot-candles. Then the illumination contributed by both lamps is E = 4.86 + 0.612 = 5.47 foot-candles.

The point-by-point method is useful in many shipboard lighting calculations for computing the illumination in spaces that are very irregular in shape and where there are many overhead obstructions. Also, this method is readily applicable for calculating detail lighting where only a single light source is required.

LUMEN METHOD.—The lumen method gives the average illumination on the horizontal for any system of general lighting where the light sources are spaced sufficiently close to each other to produce approximately uniform illumination.

This method is more practicable for calculating the illumination in larger spaces that require a greater number of lighting fixtures. In this method, a maintenance factor and a coefficient factor must be considered because the lumens required of the light source are calculated from the useful lumens desired on the working surface.

The MAINTENANCE FACTOR is applied to compensate for the difference in the average illumination obtained initially (when the lamps and units are new) and the resulting average illumination maintained in service. A maintenance factor of 70 percent shall be applied in computations, unless otherwise indicated on the standard plan of the lighting fixture being installed.

The UTILIZATION FACTOR, or coefficient of utilization, is the over-all efficiency of the lighting installation and is the ratio of the useful lumens that reach the working area to the total lumens produced by the lighting fixture. Utilization factor tables for the specified types of lighting fixtures are available in the Ship's Design Data Sheet or in the applicable manufacturer's instruction books.

For example, find the size of lamps required to produce an illumination (minimum) of 5

foot-candles in a compartment 20 feet wide by 40 feet long by 8 feet high
with the bulkheads and overhead painted a light color. The lighting units to be installed are the Symbol No. 90 Standard Plan, BuShlps No. 9000-S6401 -73842, which is an incandescent fixture using a type TR-7 lamp. The mounting height, determined by subtracting the fixture height (9 inches) from the compartment height (8 feet) is about 7 feet. The maximum permissible spacing between light sources, obtained from the appropriate tables in the lighting data sheets, is about 7-1/2 feet for this mounting height.

The lighting fixtures for general illumination should be located symmetrically, where practicable, throughout the area to be illuminated. If the number and location of the lighting outlets are not already determined, divide the compartment into eight 10-foot squares and locate a lighting
unit at the center of each r «oft. -j
square (fig. 3-13). This spac- ~J ing may not be satisfactory because it is larger than the £ maximum allowable spacing of ~ 7-1/2 feet determined previ- -i ously for this installation.

Compute the area per outlet Figure 3-13.-De$ign of by substituting in the formula. compartment lighting.

deck area of compartment Area per outlet = num b er of outlets in compartment
20 feet * 40 feet 8 outlets
= 100 square feet.

Compute the lamp lumens per outlet by substituting in the following formula
Lamp lumens per outlet
foot-candles * area per outlet
utilization factor * maintenace factor
5 foot-candles * 100 square feet 0.38 * 0.70
= 1851 lumens,
J"
where the utilization and maintenance factors are 38 and 70 percent, respectively.

Select the lamp that will produce 1851 lumens from table 5, listing the lumen output of the various sizes of incandescent lamps in the lighting data sheets. Assuming the lighting circuit operates at 120 volts, a Navy type TR-14, 120 volt, 150 watt lamp is selected. The 150-watt

Table 5.-Lumen output of the various sizes of incandescent lamps

lamp is used because the lumen output of the next smaller lamp (100 watt) is below the required value and also because the units are spaced 10 feet apart instead of 7-1/2 feet, as previously determined by the mounting height. To determine the foot-candle illumination that will be produced by the selected 150-watt lamp, compute the foot-candles by substituting in the formula
lamp lumens * utilization factor
. „ x maintenance factor
foot-candles =
area per outlet
2050 lumens * 0.38 xQ.70 100 sq feet
= 5.46 foot-candles.

The illumination of 5.46 foot-candles is a satisfactory average for this installation because it exceeds the 5-foot-candle minimum requirement for the illumination of this compartment, as stated initially in this example.

NAVIGATIONAL LIGHTS

The navigational lights installed on naval vessels must be in accordance with Regulations for Preventing Collisions at Sea (Public Law 178, 82nd Congress, chapter 495) or as allowed by an existing waver or a waver to be issued covering the vessel being built. These lights consist of (1) running lights, (2) signal lights, and (3) anchor lights.

Running Lights

The running lights are similar to those used on merchant ships and include (1) masthead light, (2) range light, (3) portside light, (4) starboard side light, (5) stern light (white), and (5) upper and lower towing lights, some of which are illustrated in figure 3-14.

A - MASTHEAD LIGHT

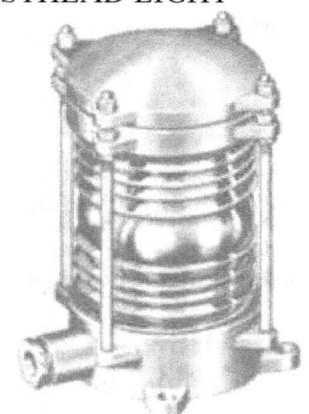

B - SIDE LIGHT

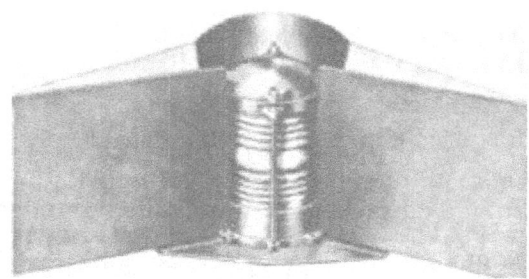

C- STERN LIGHT (WHITE) Figure 3-14.-Running lights. 14ft

The MASTHEAD LIGHT (white) for ships is a 20-point (225°) light (fig. 3-14, A) located on the foremast or in the forward part of the vessel. It is a spraytight fixture provided with a 50-watt, 2-filament lamp and equipped with an external shield to show an unbroken light over an arc of the horizon of 20 points—that is, from right ahead to 2 points abaft the beam on either side.

The RANGE LIGHT (white) for ships is also a 20-point (225°) light located on the mainmast or the fore part of the vessel, but at least 15 feet higher and 15 feet aft of the masthead light.

The PORT and STARBOARD SIDE LIGHTS for ships are 10-point, 112-1/2° lights (fig. 3-14, B) located on the respective sides of the vessel, showing red to port and green to starboard. The fixtures are spraytight, each provided with a 100-watt, 2-filament lamp and equipped with an external shield arranged to throw the light from right ahead, to 2 points abaft,the beam on the respective sides.

The STERN LIGHT (white) for ships is a 12-point (135°) light (fig. 3-14, C)located on the stern of the vessel. It is a watertight fixture provided with a 50-watt, 2-filament lamp and equipped with an external shield to show an unbroken light over an arc of the horizon of 12 points of the compass—that is, from dead astern to 6 points on each side of the ship.

The UPPER and LOWER TOWING LIGHTS (white) for ships not normally engaged in towing operations are 20-point (225°) lights similar to the previously described masthead and range lights. They are portable fixtures, each equipped with a type THOF-3 cable and plug connector for energizing the lights from the nearest lighting receptacle connector. When these lights are used, they are located vertically (6 feet apart) in the fore part of the vessel.

The SUPPLY, CONTROL, and TELLTALE PANEL for the running lights is a nonwatertight, sheet-steel cabinet designed for bulkhead mounting (fig. 3-15). This panel is located in the pilothouse and provides an audible and visible signal when the primary filament burns out in any one of the five running (masthead, stern, range, port side, and starboard side) lights, and, at the same time, automatically switches to the secondary filament so that

Figure 3-15.-Supply, control, and telltale panel.

the defective light will remain in service. A switch is provided for each of the towing (upper and lower) lights for manually switching to the secondary filament when the primary filament burns out. A master control switch with indicator light is also located on the running light supply, control, and telltale panel.

The components required for each of the five running lights consist of (l)an ON-OFF switch, (2) a blue dial indicator with two indicator lamps in parallel, (3) an annunciator, (4) a series relay, and (5) a reset switch. The

circuit also includes a buzzer in parallel with all five running lights (fig. 3-16).

MASTHEAD LIGHT

r

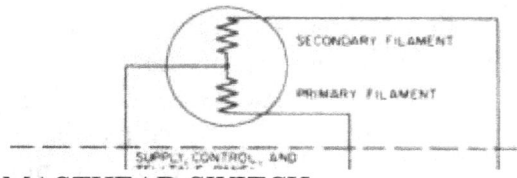

MASTHEAD SWITCH

OFF

SUPPLY, CONTROL, AND TELLTALE PANEL (MASTHEAO LIGHT ILLUMINATED)

s

TELLTALE LAMPS

SERIES RELAY

on

BUZZER

©

OPERATED

POSITION

ANNUNCIATOR

Lt£3

RESET SWITCH (NORMAL POSITION)

Figure 3-16.-Running light control, schematic diagram.

The RESET SWITCH has a normal (unmarked) position and a reset (marked) position. The switch is normally in the unmarked position.

The ANNUNCIATOR is of the drop, or flag, type with the markings, RESET and OUT, embossed on the flag. When the annunciator is restored (deenergized), the flag drops to indicate RESET on the blue dial indicator, and when the annunciator is operated (energized), an electromagnet holds the flag in the normal position to indicate OUT on the dial. However, the indicator lights are not lighted until trouble develops in the primary circuit of the running light.

The SERIES RELAY, the operating coil of which is connected in series with the primary filament, has two sets of contacts. One set of contacts is in the secondary filament circuit and the other set is in the circuit containing the annunciator, buzzer, and indicator lamps. Both sets of contacts are open when the relay is operated, and both sets are closed when the relay is restored.

When the master control switch and the masthead ON-OFF switch are in the ON positions, and the reset switch is in the NORMAL (unmarked) position, the circuit is completed through the operating coil of the series relay and the primary filament. This action holds open the relay contacts to the secondary filament circuit and to the annunciator, buzzer, and indicator light circuit.

If an open occurs in the primary filament circuit (due to a burned-out filament, blown fuse, or broken lead), no current will flow through the operating coil of the series relay. This action causes the relay to restore and to close both sets of contacts. One set of contacts completes a circuit to the secondary filament to keep the (masthead) running light in service. The other set of contacts completes a circuit to the indicator (telltale) lights, buzzer, and annunciator. This action illuminates the blue dial, sounds the buzzer, and operates the annunciator that holds the flag in the OUT position on the indicator.

When a trouble indication occurs, the reset switch is placed in the RESET position to complete another circuit to the indicator lights and to open the circuit to the buzzer and annunciator. This action keeps the blue dial illuminated, silences the buzzer, and restores the annunciator, which allows the flag to drop and indicate RESET on the indicator. This visible warning indication continues until the trouble in the primary filament circuit is corrected.

When the burned-out lamp is replaced, the series relay energizes and opens the circuit to the secondary filament and also opens a circuit to the telltale lamps. However, these lamps remain energized through the reset switch, which is in the RESET position. (The buzzer and annunciator circuit was previously opened when the reset switch was placed in the RESET position. This action also completed another circuit to the indicator lights to illuminate the blue dial.) Hence, to return the circuit to its proper condition, the reset switch must be

returned to the NORMAL (unmarked) position, which also opens the circuit to the indicator lights.

A convenient method of testing the primary filament circuit for proper operation of the series relay, buzzer, annunciator,and indicator lights is to remove the fuse in the primary filament circuit and note the action of the components.

A 0.5-microfarad capacitor is connected across the secondary filament contacts of the series relay for the purpose of arc suppression. If the capacitor should become open, arcing will occur at the contacts and if it should become shorted, the secondary filament will be lighted simultaneously with the primary filament. Replace the capacitor to correct the trouble.

When burned-out lamps are replaced in running lights, the same wattage lamp must be used because the relays are designed to operate in series with a lamp of a specific size. If a smaller lamp is used, the relay may fail to operate; whereas, a larger lamp may cause the relay to burn out.

Signal Lights

The signal lights installed on combatant ships usually include (1) aircraft warning lights, (2) blinker lights, (3) breakdown and man overboard lights, (4) steering light, (5) stern light (blue), (6) wake light, and (7) speed lights (fig. 3-17). The supply switches for these lights located on the signal and anchor light supply and control panel (in the pilothouse) are individual ON-OFF rotary snap switches.

The AIRCRAFT WARNING LIGHTS (red) for ships are 32-point (360°) lights (fig. 3-17, A) installed at the truck of each mast that extends more than 25 feet above the highest point in the superstructure. Two aircraft warning lights are installed if the light cannot be located so that it is visible from any location throughout 360° of azimuth. However, a separate aircraft warning light is not required if a 32-point red light is installed at the truck of a mast for another purpose. The fixtures are spraytight and equipped with multiple sockets provided with 15-watt, 1-filament lamps (fig. 3-17, A).

A-AIRCRAFT WARNING LIGHT
B-BLINKER LIGHT

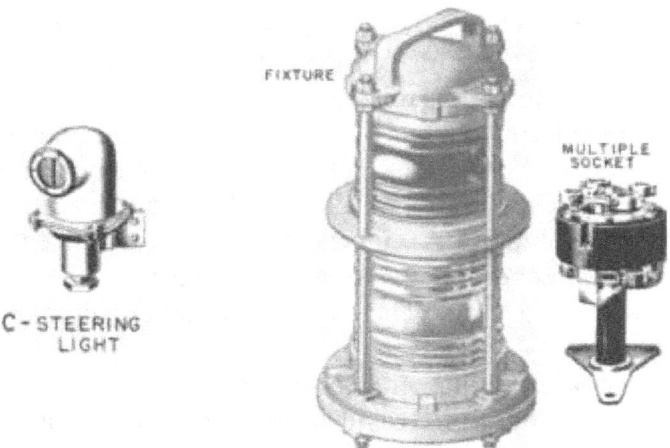

C - STEERING LIGHT

Figur* 3-17. —Signal lights.

The BLINKER LIGHTS (white) for ships are 32-point (360°) lights (fig. 3-17, B) located, one port and one starboard, outboard on the signal yardarm. The fixtures are spraytight, each provided with a 15-watt, 1-filament lamp and fitted with a screen at the base to prevent glare or reflection that may interfere with the navigation of the ship. These lights are operated from signal keys located on each side of the signal bridge. .

The BREAKDOWN and MAN-OVERBOARD LIGHTS (red) for ships are 32-point (360°) lights located 6 feet apart (vertically) and mounted on brackets that extend aft of, and to port of, the mast or structure. This arrangement permits visibility, as far as practicable, throughout 360° of azimuth. The fixtures are spray tight and equipped with 15-watt, 1-filament lamps. When these lights are used as a man-overboard signal, they are pulsed by a rotary snaps witch (fitted with a crank handle) on the signal and anchor light supply and control panel.

The STEERING LIGHT (white) for ships (fig. 3-17, C) is installed on the jack staff or other spar or structure and must be visible to the helmsman in the pilothouse. The light is installed on the centerline if the pilothouse is on the centerline. If the pilothouse is not on the centerline, a vertical plane through the light and helmsman's station in the pilothouse must be parallel to the keel line. The fixture is spraytight and includes a disk screen having a 3/64" x 1" slot (opening) through which light is emitted from a 2-candlepower lamp. A suitable bracket is provided for mounting the light on a jack staff (one-half inch in diameter).

The STERN LIGHT (blue) for ships is a 12-point (135°) light similar to the previously described white stern light (fig. 3-14, C). The light is installed near the stern on ships that are engaged in convoy operations and mounted to show an unbroken arc of light from dead astern to six points on each side of the ship.

The WAKE LIGHT (white) for ships is installed on the flagstaff or after part of the ship to illuminate the wake, and mounted so that no part of the ship is illuminated. The fixture is spray tight and of tubular construction. One end of the fixture is fitted with an internal screen, having a 1-inch-diameter hole provided with a lens (2-5/16" diameter x 3/8" thick) through which light is emitted from a 100-watt, 2-filament lamp. A suitable mounting bracket is included for adjusting the position of the light. Thus, the wake light puts a "target" in the ship's wake.

The SPEED LIGHTS for ships are combination red (top) and white (bottom), 32-point (360°) lights (fig. 3-17, E). They are located at the truck (top) of the mainmast, except when the height of the foremast is such as to

interfere with their visibility; in this case, they are located at the truck of the foremast. Two speed lights are installed if their light cannot be located so that they are visible throughout

360° of azimuth.

Speed lights are provided to indicate (by means of a coded signal) the speed of the vessel to ships in formation. In other words, they indicate the order being transmitted over the engine order system. The white light indicates ahead speeds, and the red light indicates stopping and backing. The fixture is spraytight and equipped with a multiple socket (fig. 3-17, E) provided with nine 15-watt, 1-filament lamps. Six lamps are used in the top of the socket for the red light, and three in the bottom for the white light. Each light is energized from separate circuits.

The controller for the speed lights is located in the pilothouse. The controller components include a (1) signal selector switch with dial, (2) white indicator light, (3) circuit control switch with dial, (4) red indicator light, (5) hand pulse key, and (6) pulsator unit located near the controller. A schematic wiring diagram of the controller components is illustrated in figure 3-18. The pulsator unit consists of a single-phase, 117-volt motor that drives a contact cam through gearing at a constant speed of one revolution every six seconds.

The speed light controller is energized through the supply switch on the signal and anchor light control and supply panel. Automatic operation of the speed light system is accomplished by placing the circuit control switch in the MOTOR PULSE position and the signal (speed) selector switch to the desired position. This action establishes connections to the motor-driven pulsator to provide the following signals.

Signal Selector Switch
Dial Markings Pulsations
Standard speed ahead Steady white light (motor off)
One-third speed ahead One white flash in 6 seconds
Two-third speed ahead Two white flashes in 6 seconds
Full speed ahead Four white flashes in 6 seconds
Flank speed ahead Five white flashes in 6 seconds
Hand pulse key ahead Manually controlled (code same
as above)
Signal Selector Switch Dial Markings
Pulsations
Stop
Slow speed back Full speed back Hand pulse key back
Steady red light (motor off) One red flash in 6 seconds Two red flashes in 6 seconds Manually controlled, code same
as above

If the pulsator unit should fail, manual operation of the speed light system is accomplished by placing the circuit control switch in the HAND PULSE (red or white) position. This action connects the hand pulse key in the speed light circuit so that the signals can be transmitted manually, using the same code as that for the automatic pulsator.

The speed light is used as an aircraft warning light to provide a steady red light by placing the signal selector switch in the stop position and the circuit control switch in the AIRCRAFT WARNING position.

The d-c speed light system is identical to the a-c system except that a resistance unit replaces the transformer to reduce the voltage applied to the dial lights, and a shunt motor replaces the a-c motor in the pulsator unit.

The forward and after anchor lights (white) for ships are 32-point (360°) lights. The forward anchor light is located at the top of the jackstaff or the fore part of the vessel, and the after anchor light Is at the top of the flagstaff. The fixtures are splashproof, each provided with a 50-watt, 1-filament lamp. The anchor lights are energized through individual ON-OFF rotary snap switches on the signal and anchor light supply and control panel in the pilothouse.

Night-flight operation lights are installed on aircraft carriers to assist pilots during take-off and landing at night. They also provide visual aid to pilots for locating

Anchor Lights
LIGHTING EQUIPMENT
Night-Flight Operation Lights

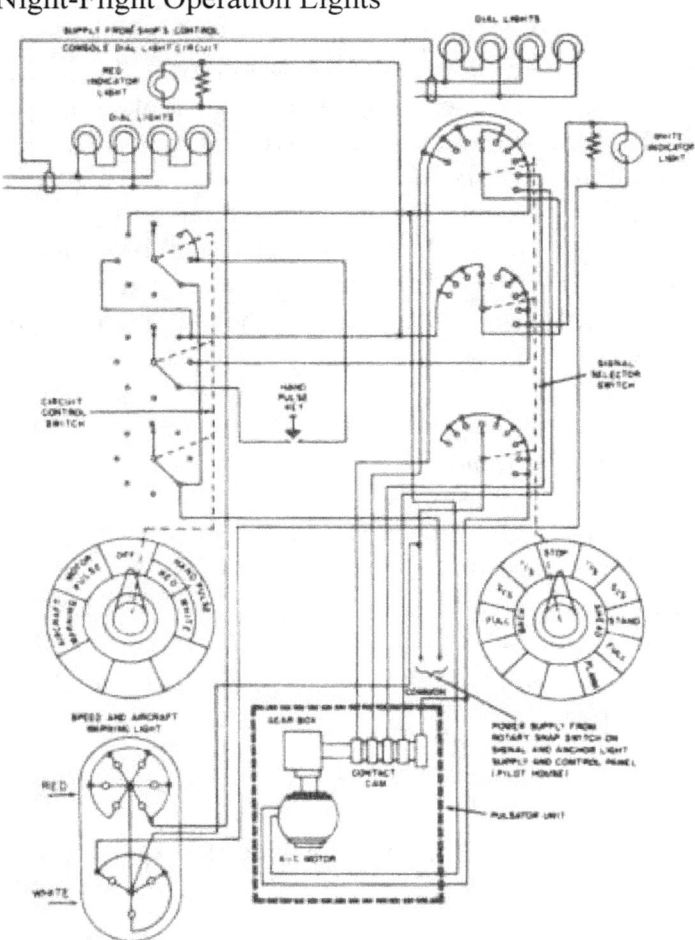

Figure 3-18.—Speed light control, schematic diagram.

and identifying the parent carrier, and for rendezvousing in the vicinity of the carrier. These lights are of a confidential nature and are therefore not listed in this training course.

Darkened-Ship Equipment

Darkened ship is a security condition designed to prevent the exposure of light, which could reveal the location of the vessel. Darkened-ship condition is achieved by means of (1) light traps that prevent the escape of light from illuminated spaces or (2) door switches that automatically disconnect the lights when the doors are opened.

LIGHT TRAP.—A light trap is an arrangement of screens placed inside access doors or hatches to prevent the escape of direct or reflected light from within (fig. 3-19). The inside surfaces of the screens are painted flat black so that they will reflect a minimum of light falling on them. Light traps that are used to prevent the escape of white light should have at least two black, light-absorbing surfaces interposed between the light source and the outboard openings. Light traps are preferred to door switches in locations where (1) egress and ingress are frequent; (2) interruption of light would cause work stoppage in large areas; (3) light might be exposed from a series of hatches, one above the other on successive deck levels; and (4) many small compartments and passages are joined by numerous inside and outside doors that would complicate a door-switch installation.

DOOR-SWITCH.—A door switch is mounted on the break side of a door jamb (inside the compartment) and operated by a stud welded to the door. When the door is opened, the switch is automatically opened at the same time.

Door switches are connected in a variety of ways to suit the arrangement of the compartment concerned. In a single compartment where there are several doors or hatches, door switches are connected in series so that when any one door or hatch is opened, all the lights in the compartment will be extinguished. When an inner compartment is located so that its light is visible from the deck when the doors to both the inner and outer compartments are open, the lights in the inner compartment

T
/ / / / 1/
. / OPEN
/doorway
\ /
^> N_
OUTBOARD DOORWAY
PAINT BLACK TO MERE
OBSERVER
Figure 3-19.-Light trap.

are usually controlled by the door switch of the outer compartment. However, If this arrangement results in an excessive number of light interruptions in the inner compartment, a separate door switch connected in parallel with the outer door switch should be provided on the inner door to control the lights in the inner compartment. Thus, the lights in the inner compartment will be extinguished only when the inner and outer doors are open.

All door switch installations are provided with lock-in devices or short-circuiting switches to change the settings of the door switches, as required from lighted ship to darkened ship and vice versa. Each standard door switch is furnished with a mechanical lock-in device for use when only one door switch is installed. When two or more door switches are connected in series, a single, separately mounted short-circuiting switch is installed in an accessible location to avoid the possibility of overlooking any of the door switches when the change over is made.

When a single door switch at an outer door is connected in parallel with door switches at inner doors, only the door switch at the outer door is provided with a lock-in device, and the lock-in devices are removed from the other outer doors. The location of the control switch is indicated by a plate mounted adjacent to each door switch.

The control switch is marked CAUTION—DOOR SWITCH CONTROL. The portion of the short-circuiting switch that connects the door switches in the circuit is marked DARKENED SHIP, and the portion that disconnects the door switches from the circuit is marked LIGHTED SHIP. Personnel should become familiar with the location of the short-circuiting switch in all compartments and the number of doors that it controls.

Luminous Material

Luminous markers that glow in the dark without external activation are used aboard ship to designate personnel and to delineate objects on weather decks and other areas that are not illuminated during darkened-ship condition. These markers are of the (1) personnel and (2) deck types and contain radium sulphate, which is a radioactive compound normally used on watch and instrument dials.

PERSONNEL MARKER.—The personnel marker is designed primarily to designate personnel as to position, rank, station, or casualty. The marker consists of a radioactive luminous button about 1 inch in diameter. The button is enclosed in a clear plastic case, which is held in a sheet steel bezel (grooved rim). The back of the bezel is provided with a spring pocket clip so that it can be attached to the clothing.

Under normal conditions, the personnel marker is plainly visible at 10 feet, perceptible at 50 feet, almost visible at 100 feet, and invisible at 200 feet. The marker is of rugged construction and can be exposed to rain and sun with very little effect on its efficiency.

DECK MARKER.—The deck marker is similar to the personnel button, except a thin steel disk, or washer, is mounted on the back instead of the spring clip. The steel disk, welded to a bulkhead, deck ladder, hatch, or other object that requires delineation, facilitates the movement of personnel or the handling of gear. (When the button is damaged, it can be removed and replaced with a new one.) If the ship is at sea the marker should be thrown over the side; if in port, it should be disposed of by the physical health personnel.

The deck marker has the same visibility characteristics as the personnel marker. The application includes the delineation of bulkheads, baffles, and coamings of doors and ladders in darkened sections of light traps; bulkheads, door warnings, ladders, and obstructions in darkened passages; bomb and torpedo shaft openings on the weather decks; and walkways, ladders, hatches, and obstacles on the weather decks.

Special Lights

Special lights, in addition to the previously described lighting equipment and installations, are provided aboard ship for various uses. These lights include (1) flashlights, (2) floodlights, (3) hand lanterns, and (4) flood lanterns (fig. 3-20).

FLASHLIGHT.—The general purpose flashlight is supplied in the type, I (fig. 3-20, A), having a prefocused (concentrated) beam and the type, P-O, having a nonfoc using (spread) beam. Both types consist of a watertight plastic tubular case containing two JAN-type, BA-30 batteries. The type I flashlight uses a Navy-type TB-17 lamp, and the type, P-O, uses a Navy-type, TB-8 lamp.

The general-purpose flashlight is used in many routine applications where a small portable light is advantageous. It is valuable as an individual emergency source if the ship's service and emergency lighting systems should fail, and it can be used in an emergency as a signaling light for transmitting messages in Morse code over short distances.

FLOODLIGHT.—The floodlight consists of a splash-proof, parabolic -shaped reflector housing equipped with a rounded glass lens (fig. 3-20, B). The housing that provides an enclosure for the lamp is trunnioned on a yoke, which, in turn, is mounted on a swivel base. The light can be elevated and depressed through an arc of 224* and rotated in train through 360°. The light is secured in train and elevation by a clamp provided on the yoke and on the base. Floodlights are installed on weather decks at suitable locations to provide sufficient illumination for the operation of cranes and hoists, and the handling of boats.

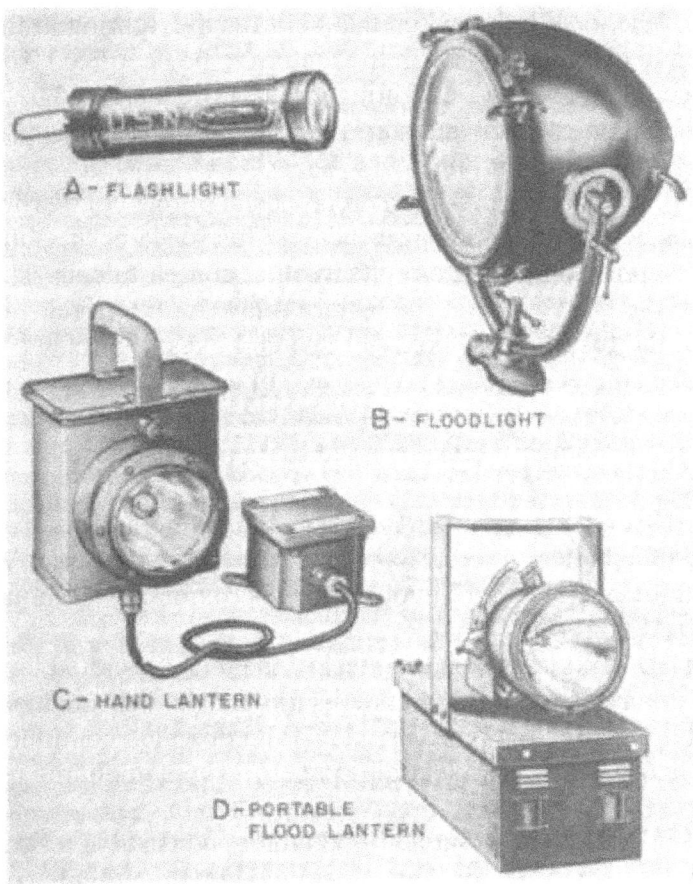

Figure 3-20.-Special lights.

HAND LANTERN. — Type K-10A (steel) and K-lOB (brass) hand lanterns consist of a watertight rectangular case containing two J AN-type, BA-23 batteries (fig. 3-20, C). It includes a plexiglass lens (4-5/16 inches in diameter) with reflector, and uses a Navy-type, TB-5 lamp. The lantern is operated by an external lever that actuates an Interior-mounted toggle switch. A rigid handle secured to the top of the case provides convenient transportation of the hand lantern.

Type K-10A (steel) and K-lOB (brass) hand lanterns are used to ensure against total darkness in the event of failure of the ship's lighting systems. These lanterns can be relay-controlled or manually controlled.

The RELAY-OPERATED hand lantern is provided with a detachable plug and cord for connection to the relay that automatically energizes this lantern when the lighting circuit fails. Relay-controlled hand lanterns are assigned to spaces in which it is necessary to maintain practically continuous illumination. These spaces include essential watch stations, control rooms, machinery spaces, battle dressing stations, and at accesses, companionways, and escape passageways leading from compartments (where personnel are stationed or quartered) to the weather deck. These lanterns must not be installed in any spaces provided with door switches that control the lighting, or in magazines or powder-handling spaces (except those handling spaces in which only fixed or semifixed ammunition is handled), or in any location in which explosion-proof equipment is required. Hand lanterns must not be removed from the compartments in which they are installed unless the compartments are to be abandoned permanently.

The lantern relay is connected in the lighting circuit (in the space in which the lantern is installed) on the power supply side of the local light switch that controls the lighting in the space concerned. Thus, the relay operates and causes the lamp in the lantern to be energized from its batteries only when power failure occurs, not when the lighting circuit is deenergized by the light switch. If the space is supplied with both emergency and ship's service lighting, the hand lantern relay is connected to the emergency lighting circuit only.

The lantern relay is fused so that a short circuit in the relay leads of one compartment will be cleared through low-capacity fuses before the fault causes heavier fuses nearer the source of power to flow and cut off the power supply to lighting circuits in other compartments. The fuses that protect the branch circuits are ample protection for the lantern relay. A lantern relay can be connected directly to the load side of the fuses in fuse boxes or switch boxes. If a relay cannot be connected to a branch circuit, it can be connected to the source side of a fuse box or other point on a submain. If the submaln

supplies lighting to more than one compartment, separate fuses must be installed in the relay circuit.

The lamp used in the type K-10A (steel) and K-10B (brass) hand lantern is rated at 2.4 volts, but is operated at 3 volts (when the batteries are new) to increase the light output. When the batteries are fresh, the lantern can be operated continuously for approximately 10 hours before the light output ceases to be useful.

The MANUALLY OPERATED hand lantern is identical to the relay-operated lantern, except it is not connected to a relay. Manually controlled lanterns are installed as an emergency source of illumination in spaces that are manned only occasionally and in which continuous illumination is not essential. These lanterns are also in certain spaces to supplement the relay-operated lanterns.

PORTABLE FLOOD LANTERNS.—The portable flood lantern consists of a Navy-type SB-1 sealed beam lamp enclosed in a built-in reflector housing equipped with a toggle switch (fig. 3-20, D). The reflector housing is adjustably mounted on a drip proof, acid-resistant case provided with two windows in each end.

The case contains four Navy-type 2V-SBP-25 AH storage cells. Each cell contains a channeled section in which a green, white, and red ball denotes the state of charge of the cell when viewed through the window. When a cell is fully charged, all three indicator balls float at the surface of the electrolyte. The green ball sinks when approximately 10 percent of the cell capacity has been discharged; the white ball sinks when the cell is 50 percent discharged; and the red ball sinks when the cell is 90 percent discharged.

The lamp is rated at 6 volts but is operated at 8 volts to increase the light output. When operated with fully charged batteries, the lamp produces over 200,000 candlepower in the center of the beam. The lantern can be operated for about three hours without recharging the batteries. During this period the light output gradually decreases to about one-half the intensity produced at full charge.

Portable flood lanterns are often referred to as damage control lanterns because they are used by damage control personnel to furnish high intensity illumination for emergency repair work at night or to illuminate inaccessible locations below deck.

The lighting distribution system in naval vessels is designed for satisfactory illumination, optimum operational economy, maximum continuity of service, and minimum vulnerability to mechanical and battle damage. The lighting distribution system comprises the ship's service lighting system and emergency lighting system. These systems consist of feeders from the ship's service or emergency power switchboards, switchgear groups, or load centers to distribution panels or feeder distribution fuse boxes, located at central distribution points from which power is distributed to the local lighting circuits.

Normally, all lighting circuits are supplied from the ship's service lighting system, which is energized from the ships' service generators. 11 the ship's service power supply fails, certain designated circuits on the ship's service lighting system can be supplied from the emergency lighting system, which is energized from the emergency generators through automatic bus transfer equipment.

Lighting distribution systems are either a-c or d-c systems, depending on the power distribution system installed in the ship. The majority of power distribution systems in surface vessels are 450-volt, 3-phase, 60-cycle, 3-wire ungrounded systems.

A-C Lighting System

In ships having a-c ship's service power the ship's service lighting feeders are either 450-volt or 120-volt, 3-phase, 60-cycle circuits. The lighting circuits are 120-volt, 3-phase, 60-cycle, 3-wire circuits supplied from the power distribution system through automatic bus transfer equipment to 450/120 volt transformer banks (fig. 3-21). Each transformer bank consists of three single-phase, delta-delta connected transformers. If one transformer in a bank fails, the two remaining transformers can be operated in open delta to obtain about 58 percent of the initial bank capacity.

In large ships the transformer banks are installed near the lighting distribution panels that are located some distance from the generator and distribution switchboards or switchgear groups. In smaller ships the transformer

1 CO
POWER DISTRIBUTION SWBD.
SHIPS SERVICE LIGHTING FEEDER
3». 430V, tfv
POWER DISTRIBUTION PANEL
31. 450V, SffU
-LIGHTING MAIN
YvV ■VW
9, 130V, 60^
LIGHTING DIST. PANEL
or
LU
o
o
5
—I
g
UJ
S
5
it
31, 120V, «ff\<
<g>—<8>

LIGHTING BRANCH

1», 130V, t£\

<g>—<8>

EMERGENCY POWER SWITCHBOARD

Figure 3-21.-Lighting distribution system, schematic diagram.

banks are located near the generator and distribution switchboard or switchgear groups, and energize the busses that supply the lighting circuits.

The lighting distribution system feeders, mains, and submains are 3-phase circuits. The branches are single-phase circuits and are so connected that the loads on the 3-phase circuits are as nearly balanced as possible.

The emergency lighting system consists of 120-volt, 3-phase, 60-cycle, 3-wire circuits supplied from the emergency generator and distribution switchboards. The circuits to the automatic bus transfer equipment, lighting mains from the emergency lighting feeders, feeder junction fuse boxes, and feeder distribution fuse boxes are 3-phase circuits. If an emergency power system is not installed, an alternate supply from a separate ship's service generator and distribution switchboard or switch-gear group must be provided for the services selected in accordance with the basic principles applying to an emergency system.

Automatic bus transfer equipment is installed at load centers, distribution panels, or loads that are supplied by both normal and alternate and/or emergency feeders. This equipment is used to select either the normal or alternate source of the ship's service power, or to obtain power from the emergency distribution system if an emergency feeder is also provided.

D-C Lighting System

In ships having d-c, 240/120-volt, 3-wlre, ship's service power the ship's service lighting feeders are 240/120-volt, 3-wire circuits,and the branches are 120-volt, 2-wire circuits. The lighting load is arranged to provide satisfactory load balance under all conditions.

The emergency lighting feeders are 240/120-volt, 3-wire, d-c circuits, and the branches are 120-volt, 2-wire, d-c circuits.

MAINTENANCE

The intensity of illumination produced by a lighting installation begins to depreciate at the time the system is placed in operation. This depreciation in the lighting system is due to the decrease in efficiency of the lamp (with use) and to the decrease in reflecting efficiency of the surrounding bulkheads and overhead, resulting from the natural deterioration of the surfaces with age.

The deterioration of an incandescent lamp is caused by blackening of the inside of the bulb because of the evaporation of the tungsten filament. The light output of a tungsten filament lamp near the end of its normal life is about 80 percent of the initial output. The depreciation of a fluorescent lamp is caused by blackening at the ends of the tube due to the exhaustion of the active material on the electrodes.

The depreciation of the light caused by the accumulation of dirt, dust, and film on the lamps and fixtures greatly reduces the efficiency of a lighting system. The actual loss of light from this cause will depend on the extent to which oil fumes, dust, and dirt are present in the surrounding atmosphere, and on the frequency with which the fixtures are cleaned.

In the design of lighting installations, a maintenance factor is used to compensate for the difference in the average illumination obtained initially when the lamps and fixtures are new and the resulting average illumination maintained in service. The amount of depreciation will vary from 25 to 45 percent, corresponding to a maintenance factor of 55 to 75 percent and depends on the type of fixtures, the system of illumination employed, the material of the reflecting surfaces (bulkheads and overhead), the local conditions of dust and dirt, and of the frequency of cleaning the fixtures and repainting the bulkheads and overhead. As previously stated, a maintenance factor of 70 percent is used unless otherwise shown on the standard plan of the installed fixtures.

Lighting Fixtures

The lighting system should be maintained at its maximum efficiency because artificial light has an important bearing on the effectiveness of operation of a naval vessel. All lighting fixtures should be adequately cleaned at regular intervals of about two to six weeks to prevent a waste of energy and low intensity of illumination. The frequency for the cleaning periods depends on the degree of pre valence of dirt, dust, and oil fumes in the surrounding atmosphere, and on the type of fixtures installed in the spaces.

CLEANING.—The schedules for cleaning the lighting fixtures should be developed to suit the conditions and requirements aboard individual vessels. Lamps and fixtures should be inspected at frequent intervals and

cleaned when dirt, oil film, or salt incrustation are first detected. The bulkheads and overhead should be kept clean also as an aid in maintaining the deficiency of the lighting system.

When a fixture requires cleaning, turn off the light and remove the glassware (if any), the lamp, and, if practicable, the reflector (if any). Wash the glassware, lamp, and reflector with soap and water. Avoid the use of strong alkaline and acid detergents when washing aluminum reflectors. Rince the washed parts with clean, fresh water to which a few drops of ammonia are added to remove the soap film. Dry the parts with a soft cloth and replace them in the fixture.

REPLACING LAMPS.-The type J-41E watertight fixture (fig. 3-22) consists of a flanged globe that is secured to the base of the fixture by a securing ring. The inside surface of the flange fits against a molded-rubber gasket (one fourth of an inch square) fitted in a circular groove in the bottom of the base. The outside surface of the flange is provided with a centering, molded-rubber gasket. Abrass slip ring is inserted between the centering gasket and the securing ring. The purpose of the slip ring is that of a compression washer when the securing ring is tightened on the base.

To replace a burned-out lamp in this fixture, unscrew the securing ring with a spanner wrench, remove the globe, and replace the burned-out lamp with a new one. Inspect the one-fourth of an inch gasket in the base, and the centering gasket on the outside of the flange and replace with new gaskets if they are worn or deteriorated. Insert the globe and tighten the securing ring onto the base.

To avoid unnecessary waste, use judgment to determine when an old lamp should be replaced. If blackened, but still operative, lamps that are removed from service should be retained onboard until allowances of new lamps are received. Then the old lamps should be destroyed or turned over to shore activities, depending on the circumstances.

The majority of the difficulties encountered with fluorescent lights are caused by either wornout or defective starters, or by damaged or expended lamps. Hence, when abnormal operation of a fluorescent fixture is observed,

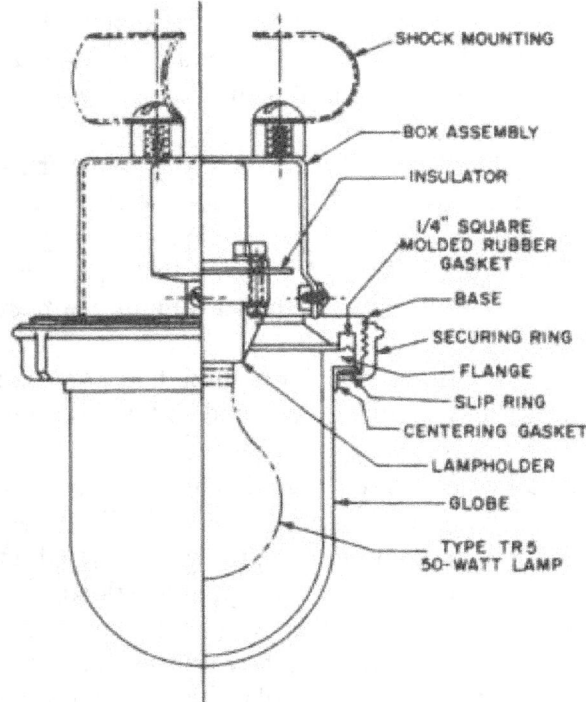

Figor* 3-22.-Type J-41E watertight fixture.

the difficulty can usually be remedied by replacing either the starter or the lamp or both.

Hand Lanterns

To ensure satisfactory operation of the types K-10A (steel) and K-10B (brass) hand lanterns at all times the batteries should be checked at least once every three months. The batteries are checked by connecting a 1.7-ohm resistor across the terminals of each dry cell and reading the cell voltage at the end of one minute. Any cell that has a reading of below 1.2 volts should be discarded and replaced with a fresh cell.

Lanterns that are located in spaces in which the normal temperature is consistently above 90° F should be

checked more often. For example, in boiler rooms it may be necessary to replace the batteries each week to ensure adequate service from the lanterns.

In addition to checking the batteries, the operation of the relays in the types K-10A (steel) and K-10B (brass) relay-operated hand lanterns should be checked at least once each week. The relay is checked by deenergizlng the lighting circuit to which the relay is connected. When the circuit is deenergized, the relay should operate and automatically turn on the lantern.

Portable Flood Lanterns

As previously stated, the degree of charge in a portable flood lantern is determined by observing the charge indicator balls through the plastic windows in the battery case. The batteries should be charged as soon as possible after the green ball (10 percent discharged) has sunk to the bottom. The lanterns should be checked at least once a week to determine if the green indicator balls are floating. If they are not floating, the battery should be charged at a rate of 1-1/4 to 2 amperes until all indicator balls are floating at the indicator line. If the battery is completely discharged, it will require from 20 to 25 hours to recharge it. After the charging voltage has remained constant at 10 volts for one hour, discontinue the charging.

When necessary, add pure water to keep the electrolyte level at the indicator line marked on the front of the cell. Do not add enough water to bring electrolyte level above the line because over filling nullifies the nonspill feature of the battery and may cause the electrolyte to spurt out through the vent tube. However, if the electrolyte level is not at the indicator line, the charge indicator balls will not indicate correctly the state of charge of the battery

Flashlights

Flashlights are checked by switching them on for 3 minutes. If there is a noticeable deer ease in light intensity during the interval, the batteries should be replaced with fresh cells. All flashlights should be given this test every 6 weeks.

SAFETY PRECAUTIONS

When a lighting system is installed, the illumination is a fixed rather than a controllable factor. This is especially true if the work cannot be located or positioned to improve the seeing task. It is natural to move the work closer to the light source when seeing is difficult, and when this can be done, improvement in visibility usually results. When the seeing task is prolonged to any appreciable length of time, improvement in visibility is not the only factor to be considered. It is also necessary to take into account the quality of the lighting as affected by glare and the background illumination.

Glare

Glare is light that hinders, rather than aids vision. It may come from an exposed light source in the field of view (direct glare) or from some shiny surface that reflects a bright source of light toward the eye (reflected glare). In either case, glare is objectionable because it is unpleasant and can cause eyestrain, headache, and fatigue.

Many navy lighting fixtures provide some protection against direct glare from the usual viewing angles, but in many locations it is impossible to mount fixtures so that no part of a bare lamp is in the field of view. Other fixtures have no shielding at all. In a naval vessel it is more important for the lighting fixtures to be simple and rugged, and to provide the essential illumination than to produce entirely adequate and comfortable illumination.

The best protection against the harmful effects of glare aboard ship is the realization by all hands that prolonged exposure to glare should be avoided. Usually, it is possible to position the work and place yourself so that you will not face any of the exposed lamps or uncomfortably bright, reflected-glare spots. If this is not practicable, shield your eyes with a visor.

Always use the proper size lamp in every fixture. For example, glare is invariably produced if a 100-watt lamp is used in a fixture that is designed for a 50-watt lamp. Install shielded fixtures in areas where extended critical seeing tasks are required. Never use exposed lamps or unshielded globes in such spaces. Replace burned out lamps immediately.

If a lamp shatters in its socket, turn off the switch that controls the fixture before attempting to remove the lamp base. Install" and remove fuses with fuse pullers, not with your bare hands. Wiring and rewiring fixtures should be accomplished only by qualified Electrician's Mates and only after authorization has been obtained from the proper authority. When repainting compartments, do not spray paint on the reflecting surfaces of lighting fixtures. Paint will materially reduce the reflection factor of the reflecting surfaces and decrease the illumination.

Background Illumination

The quality of the lighting is affected by the ratio of illumination on the working plane to the illumination on the surroundings. The eyes do not remain constantly fixed on any task until it is completed. Instead, they frequently (often involuntarily) glance from the work for a few seconds. If a great difference exists between the brightness of the work area, the pupils of the eyes (which automatically adjust to the amount of light entering the eyes) must make an adjustment every time the eyes glance away from, and back to, the work. The muscles of the eyes ordinarily will not be overworked if the ratio of brightness of the work to brightness of the surroundings does not exceed 10 to 1. However, if the ratio of brightness is appreciably greater than 10 to 1, eye fatigue will result if the work is continued for a prolonged period. Avoid extreme brightness contrasts between work and background. When lighting fixtures are installed beyond the work area that would illuminate the background, turn on these lights.

Dark Adaptation

When a person leaves a brightly lighted area and enters a darkened (not completely dark) area, he immediately experiences an impairment of vision. At first he sees little or nothing but as his eyes become adapted to the darkness, objects become discernible, and gradually more and more of the objects are visible, depending on the

level of illumination. For most practical purposes, dark adaptation can be considered complete after approximately 25 to 30 minutes. The time required to reach a given state of dark adaptation depends on the intensity of the previous stimulation. The higher the intensity, the longer is the period required for dark adaptation.

When dark adaptation must be maintained, the eyes should be guarded against exposure to white light. If it is necessary to enter an area illuminated by white light, wear red goggles. If the red goggles are not available, close one eye because dark adaptation in one eye is independent of that in the other.

Darkened-Ship Condition

When darkened-ship condition is ordered, check every door switch installation aboard ship to determine that all lock-in devices or short-circuiting switches are set at the DARKENED SHIP position.

Inspect the light traps to determine that they are free of all obstructions. A light colored object of any appreciable size placed in a light trap might be sufficiently illuminated by the interior lighting to be visible beyond the safe limit. Note the positions of the hand lanterns when entering a compartment so that you can find them without delay when they are needed.

Luminous Materials

The luminous materials used in the personnel and deck markers contain a small amount of radioactive substances. No danger to personnel exists under normal usage, but excessive exposure to the radiations emitted is harmful. To keep the exposure below the tolerance limits, observe the following precautions.

1. A person should wear no more than five buttons simultaneously for five hours a day.

2. A person should carry a box of 24 buttons for no more than two hours in a single day.

3. The minimum safe working distance for an 8-hour day from 5 boxes (24 buttons in a box) is about 1-1/2 feet, from 10 boxes is 2 feet, from 20 boxes is 3 feet, and from 50 boxes is 4 feet. Hence, the intensity of radiation varies inversely with the square of the distance.

J"*

The most serious danger associated with radioactive material is from the swallowing or inhaling of the material. This danger is completely absent with respect to the markers as long as they remain intact. For this reason the markers should not be tampered with in any way that will expose the luminous compounds. Broken units should be disposed of immediately.

To avoid fogging, luminous materials should not be stored within a radius of 50 feet of undeveloped photograph film.

QUIZ

1. What is the relation between the angle of incidence and the angle of reflection in figure 3-2?

2. What is the relation between the angle of refraction and the angle of incidence in figure 3-3?

3. What are the three primary colors in the visible spectrum?

4. If an object absorbs all of the green and blue portions of light that fall upon it when it is illuminated with white light, what color will it appear to have to an observer?

5. What is a foot-candle?

6. (a) If a source of 100 candlepower is 2 feet from a flat surface, what is the illumination in foot-candles on the surface ?

(b) If the source is 5 feet from the surface, what is the illumination on the surface?

7. If the illumination on a surface is 8 foot-candles, what is the illumination in lumens per square foot?

8. How many lumens of light flux are contained in a light source of 1 candlepower?

9. If the light source in figure 3-5 has 500 candlepower, the distance, d, is 10 feet, and 0 is 60°, find the illumination in foot-candles at Pointy.

10. Whatare three general sources of electric light used in naval vessels?

11. Standard incandescent lamps are classifiedaccording to what three features?

12. The mogul base is used on incandescent lamps above what rating?

13. What is the average life in hours of standard incandescent lamps for general lighting?

14. How many lumens per watt are obtained from (a) a 40-watt fluorescent lamp, and (b) a 40-watt incandescent lamp ?

15. What current-limiting component is usually included in the base of a glow lamp?

16. Lighting fixtures are classified according to which four types of enclosures?

17. Whatare the two principal methods used in calculating illumination?

18. Find the illumination in foot-candles produced at a point on a horizontal surface by two 50-watt, wide-angle, distributing type fixtures mounted on the over head if each unit is located 4 feet away from the point and in a direction 30 degrees from the vertical.

19. Find the average illumination in foot-candles that will be produced by Navy type TR-14 tungsten filament, 120-volt, 200-watt, 3200-lumen lamps having a utilization factor of 0.38 and a maintenance factor of 0.7 if the area served per outlet is 1 50 square feet.

20. What three classes of navigational lights are employed by naval vessels?

21. (a) If the primary filament circuit of a masthead running light should open (fig. 3-16), what three actions at the telltale panel will indicate this condition?

(b) When the burned-out lamp is replaced, what action must be taken at the control panel to open the secondary filament and extinguish the indicator lights?

(c) What is a convenient method of testing the primary filament circuit for proper operation of the series relay, buzzer, annunciator, and indicator lights?

22. The signal lights installed on combatant ships usually include which seven applications?

23. What are the two methods used to prevent the exposure of light from interior spaces during darkened-ship condition?

24. What is the function of the relay in the relay operated hand lantern?

25. How frequently should the batteries in hand lanterns be checked if the normal temperature is consistently above 90° ?

SEARCHLIGHTS

Naval searchlights are used to project a narrow beam of light for the illumination of distant objects and for visual signaling. This equipment includes an intense and concentrated source of light, a reflector that collects light from the source (to direct it in a narrow beam), and a signaling shutter (to interrupt the beam of light).

Prior to World War n, searchlights were used extensively for navigational and fire control purposes. However, with the advent of radar the use of searchlights for target illumination has become obsolete, except in certain special instances. Searchlights have always been used for visual signaling, but this use has decreased somewhat with the installation of VHF and UHF short-range voice communications equipment aboard ship.

This chapter describes the types of searchlights now used in the Navy. It is intended to acquaint Electrician's Mates with the proper operation, care, and maintenance of this equipment.

Searchlights are classified according to size and light source. The three general classes are the (1) 24-inch, (2) 12-inch, and (3) 8-inch searchlights. The 24-inch searchlight is the carbon-arc type; the 12-inch light is either Incandescent or inert-gas types; and the 8-inch light is the sealed-beam, inert-gas type.

24-INCH CARBON-ARC SEARCHLIGHT

The 24-inch carbon-arc searchlight (fig. 4-1) is used primarily for signaling, and secondarily for navigational

purposes. Itconsistsof a (1) stationary pedestal, (2) turntable with arms, (3) drum with iris and signaling shutters, and (4) carbon-arc lamp. The turntable is supported on the pedestal and can rotate continuously in train, and the drum, which provides a housing for the lamp is trunnioned on the turntable arms and can be elevated or depressed through angles of 110° and 30°, respectively.

SIGNALING SHUTTER OPERATING HANDLE
TURNTABLE. WITH ARMS
TRAIN STOWING LOCK -
COLLECTOR RINGS _ IN PEDESTAL
VENTILATING FAN HOUSING
FRAME
HANDLE

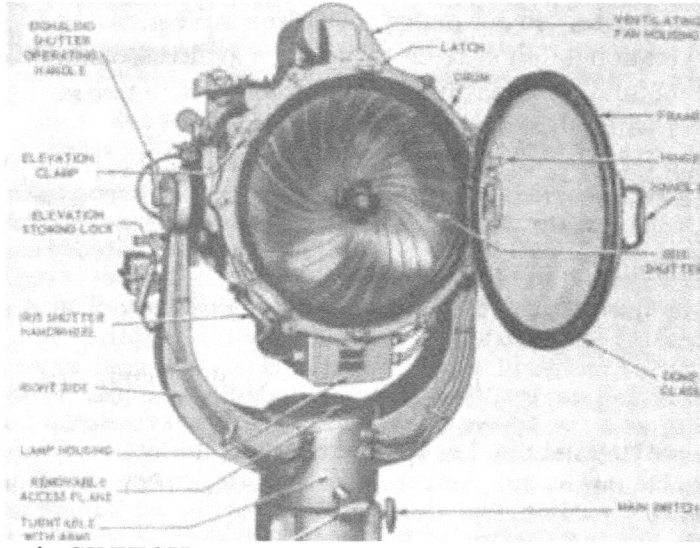

main SWITCH
STATONARY PEDESTAL
Figure 4-1 A.-24-inch carbon arc searchlight. (Front view)

Pedestal

The pedestal is secured to the searchlight platform, or base, and provides a mounting for the turntable and main power switch. The turntable shaft is supported on two large ball-bearing assemblies inside the pedestal (not shown) to allow the turntable and arms to rotate continuously in train.

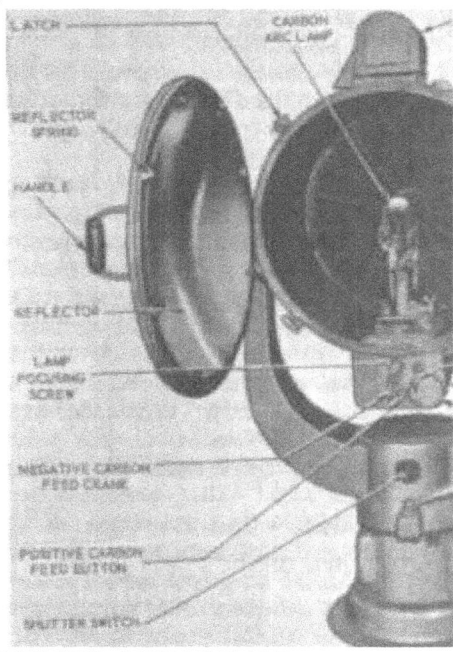

VENTILATING FAN

TELESCOPE
MOUNTING
ARC-IMAGE SCREEN
SIGNAL ING SHUTTER
POSITIVE CAROON FEED BUTTON
SHUTTER SWITCH
SIGNALING f""^ SHUTTER I\ OPERATING I) HANDLE
J
SHUTTER OPERATING UNIT
ARM
SIGNAL KEY
V \
POSITIVE CARBON FEED CRANK
■ LAMP HOUSING
ARC-CURRENT ADJUSTING KNOB COVER

Figure 4-IB.-24-inch carbon arc searchlight. (Rear view)

The MAIN POWER SWITCH is mounted inside the pedestal and is operated by a handle that protrudes from the switch cover. Terminal tubes through the pedestal provide entrances to the searchlight for the d-c supply cable and the remote signaling key circuit (fig. 4-1, A).

Turntable and Arms

The turntable and arms, which support the drum, provide a mounting for the shutter switch, train clamp, elevation clamp, and elevation stowing lock. Three collector rings mounted on the turntable shaft (not shown) supply power to the lamp and the remote signaling key circuit.

SHUTTER SWITCH.-The shutter switch is an indicating snap-action rotary switch located on the turntable below the (removable) access plate to the terminal board inside the turntable. This switch must be in the ON

position to operate the signaling shutter mechanism by means of the local or remote signal keys.

TRAIN CLAMP.—The train clamp Is mounted on the turntable below the shutter switch (fig. 4-1, B). The searchlight can be secured In any position of train by turning the clamp clockwise.

TRAIN STOWING LOCK.-The train stowing lock is located on the turntable opposite the train clamp (fig. 4-1, A). It consists of a spring-loaded pin that engages one of four slots located 90° apart in the top of the pedestal. The searchlight is locked by dropping the pin into any one of these slots. To unlock the searchlight, pull out the pin and turn it clockwise.

ELEVATION CLAMP.—The elevation clamp is mounted on the upper part of the r'^ht turntable arm (fig. 4-1, A). The searchlight can be secured in any position of elevation (between stops) by turning the clamp clockwise.

ELEVATION STOWING LOCK.—The elevation stowing lock is located on the right turntable arm near the elevation clamp (fig. 4-1, A). It consists of a spring-loaded pin (similar to the train stowing lock) for locking the searchlight in either the horizontal (normal) or vertical positions.

Drum

The drum, or barrel, is essentially a sheet-steel shell that provides an enclosure for the lamp. It Is equipped with trunnions that are supported by bearings mounted on the turntable arms to allow the drum to be elevated or depressed through angles of 110° and 30° respectively. Handles are provided at the front and rear of the barrel for swinging the searchlight in train and in elevation to direct the beam of light. The barrel Includes the (1) lamp housing, (2) front door and dome glass, (3) iris shutter, (4) sector-vane (signaling) shutter, (5) rear door and reflector, and (6) ventilating system.

LAMP HOUSING.—The lamp housing is a rectangular section welded on the bottom of the barrel to provide an enclosure for the lamp base. The lamp base, or box, is equipped with four mounting lugs that fit on rails to support the lamp box inside the lamp housing.

FRONT DOOR AND DOME GLASS.-The front door is mounted on hinges welded to the left side of the barrel (fig. 4-1, A). It can be latched in the fully open position and is secured in the (normally) closed position by means of six latches.

The dome glass is mounted in the front door by means of a retaining ring to close the front end of the drum and protect the light source. The glass is heat resistant; its surface is precision ground to a curve (dome) that results in only a slight increase in beam spread, and a shape that is highly resistant to shock. However, the latest 24-inch searchlights are equipped with a flat, tempered glass that is considerably stronger and less susceptible to breakage from gunfire than is the dome glass.

IRIS SHUTTER.—The iris shutter is mounted as a removable assembly inside the drum directly behind the front door (fig. 4-1, A). It is provided to maintain the drum substantially light tight when it is desired to shut off the beam of light without extinguishing the arc. It can be used also to vary the size of the drum opening and hence the amount of light in the searchlight beam. The iris shutter must not be used as a signaling shutter.

SIGNALING SHUTTER.—The signaling shutter is a radial-vane shutter mounted inside the drum behind the iris shutter (fig. 4-1, B). It is spring actuated to close and is manually opened by an operating handle on the forward right-hand side of the drum. It can be automatically opened by local and remote signaling keys in the shutter operating circuit. If the shutter switch is in the ON position, the signaling shutter will open when the key is closed. The shutter can be latched in the OPEN position when the light is used for illumination.

REAR DOOR AND REFLECTOR.-The rear door is hinged to the left side of the barrel (fig. 4-1, B). Similar to the front door, it can be either latched in the fully OPEN position or secured in the (normally) CLOSED position by means of six latches.

The reflector is mounted in a recess of the rear door by means of spring clips and stops. It has a parabolic reflecting surface with the concave side turned toward the light source. The light rays that pass from the light source to the reflector surface are reflected back in a beam essentially parallel to the axis of the reflector.

The light source must be small and at, or close to, the focal point of the reflector to obtain a sharp beam in which all of the reflected light rays are parallel to each other and to the reflector axis.

The materials used in searchlight reflectors are Stellite, Hastelloy, aluminum with a special Alzac finish, chromium-plated steel, and rhodium-plated copper. Stellite is the most satisfactory reflector material and is used in all 24-inch searchlights. It has a reflectivity of about 65 percent, is relatively unaffected by arc fumes, and is highly resistant to corrosion by salt water.

ARC-IMAGE SCREEN.—The arc-image screen is located on the upper right-hand side toward the rear of the drum (fig. 4-1, B). It is a ground-glass screen upon which a lens system projects an image of the arc. A line is inscribed on the screen to indicate the correct position of the end of the positive carbon when it is at the focal point of the reflector.

The arc-image screen is equipped with a hinged cover. A mirror, mounted on the inside of the cover, can be adjusted so that the operator at the rear of the searchlight can observe the arc image in the mirror.

PEEP SIGHT.—The peep sight (not shown) is located below the arc-image screen on the center line of the drum directly opposite the arc. It contains a dense-colored glass light filter through which can be observed the arc and heads.

The peep sight is provided with a cover that is pivoted on the drum. The cover is closed when the peep sight is not in use to protect the glass and to prevent the emission of light.

VENTILATING SYSTEM.-The ventilating system is necessary to remove the gases produced by the arc and to prevent deterioration of the parts exposed to the intense heat of the arc. The lamp must be extinguished if the ventilating system does not operate satisfactorily.

The ventilating system is mounted on top of the drum and consists of a d-c motor that drives an exhaust fan. The motor is connected across the arc terminal studs and is controlled by the main power switch. It is rated at one-sixth horsepower and operates on 125 volts when the lamp is starting, and it operates on 65 to 70 volts when the lamp is in continuous operation.

The air enters baffled openings in the bottom of the lamp housing and at the sides of the drum. It passes over the inside of the dome glass, the front and rear of the reflector, and directly upward through the center, along the inner surface of the drum, and up through the head-supporting column. The air that is drawn up through the head-supporting column is exhausted through a small vent in the top of the obturator on the positive head. All of the air is drawn through the fan at the top of the drum and is exhausted downward over the outer surface on the left side of the drum.

Carbon-Arc Lamp

The carbon-arc lamp utilizes a high intensity d-c arc between special cored carbon electrodes. It is designed for operation with an arc-ballast resistor (located below deck) supplied from the ship's 120-volt, d-c power. The arc current is adjusted for 75 to 80 amperes with an arc voltage of 65 to 70 volts. The arc-ballast resistor is connected in series with the arc to limit the starting current, stabilize the arc, and absorb the difference between the line voltage and arc voltage.

The function of the lamp mechanism is to hold and and control the carbons to produce a source of light (always) at the focus of the reflector. The lamp consists of the positive head, negative head, and lamp operating mechanism (fig. 4-2). The positive and negative heads are secured to the head-supporting column mounted on the lamp base, or box, that contains the mechanism for automatically operating the lamp. The automatic features of rotating and feeding the positive carbon, feeding the negative carbon, starting the arc, and providing ventilation are accomplished by the lamp (feed) motor and associated equipment contained in the lamp base.

CARBON ARC—As previously stated, the lamp utilizes a high-intensity, d-c arc between special carbon electrodes. The carbons, consisting of hard-carbon shells with cores of special materials, are mounted in geared heads. The current is conducted to the carbons through spring-backed, metal brush contacts that press against the carbons. The head assemblies are insulated so that only the brush contacts can conduct current to the carbons.

MioTivt miao rounvt xco

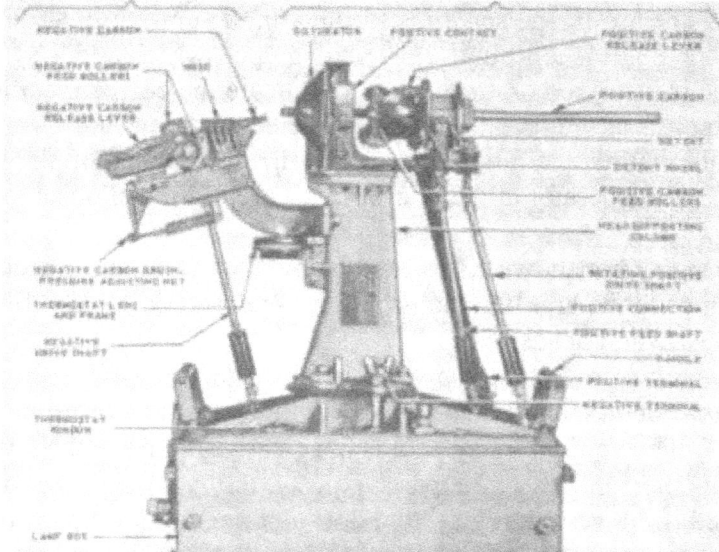

Figure 4-2.—Carbon-arc lamp.

The positive carbon is aligned with the optical axis of the reflector, and the negative carbon is inclined at an angle of 16° to the positive carbon. In operation, the negative carbon burns away to a taper and is fed forward to maintain substantially constant voltage across, or current through, the arc. On the other hand, the core of the positive carbon vaporizes and forms a hollow crater that contains an intensely luminous ball of gas. The positive carbon is rotated continuously about its axis to prevent unsymmetrical burning away of the carbon around the crater and the consequent spilling out of the luminous ball, which is the major source of light. The positive carbon is also fed forward to maintain the crater at the focal point of the reflector, as indicated by the line inscribed on the arc-image screen.

The appearance of the high-intensity arc viewed in the arc-image screen (when burning under different conditions) is illustrated in figure 4-3. When the arc is burning

properly, the luminous ball is in the crater, and a TAIL FLAME extends about 6 inches above the positive-carbon tip in a vertical plane through the axis of the positive carbon. If the tail flame shoots off to one side or the other, Instead of toward the top of the drum, the alignment of the heads should be checked with the alignment gages, and corrected. If the arc current is too low, the appearance of the luminous ball is not very pronounced, and if the arc current is too high, the luminous ball will appear to extend toward the negative carbon. Excessive arc current is also indicated by a thin, black tongue of soot appearing in the center of the tail flame.

NORMAL APPEARANCE NEGATIVE CARBON
OF ARC OUT OP LINE

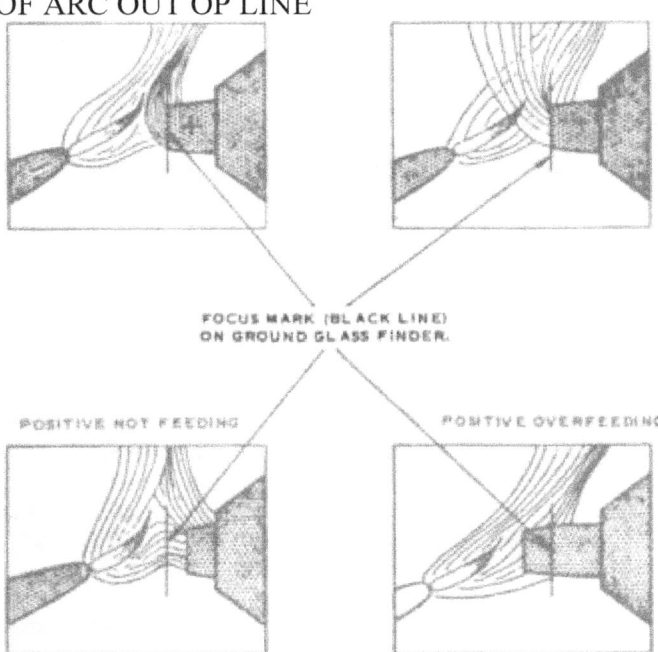

FOCUS MARK (BLACK LINE)
ON GROUND GLASS FINDER.

POSITIVE NOT FEEDING   POSITIVE OVERFEEDING

Figure 4-3.-Appearance of high-intensity arc.

The lamp can be focused by the LAMP-FOCUSING SCREW at the rear of the lamp box. Turning this screw moves the lamp forward or backward on the mounting lugs that slide on rails in the lamp housing.

Both the positive and negative carbons are slowly consumed and must be fed gradually toward the arc. The electrode sizes and burning time for a set, or trim, of carbons can be obtained from the manufacturers' instruction book, furnished with the equipment, or from Chapter 66 of the Bureau of Ships Manual.

POSITIVE CARBON DRIVE.-In addition to continuously rotating the positive carbon to prevent unsymmet-rical burning away of the crater, the positive carbon must be fed forward as fast as it burns away to maintain the luminous ball of gas at the focus of the reflector. The positive carbon is carried in the center of the positive head, which is rotated continuously by the lamp (feed) motor through direct gearing (fig. 4-4). The positive carbon is fed by two serrated rollers that are driven by the positive-head gearing.

Continuous rotation of the positive carbon is accomplished by the bevel ring gear, A, that is driven by the rotating positive drive shaft. Gear A is keyed to a hollow shaft, one end of which provides an eccentric mounting for the shaft of pinion D. Pinion D meshes with gear C, which is keyed to detent wheel B and rotates with it. Detent wheel B and gear C are mounted on, and friction coupled through a ball-bearing sleeve to, the hollow shaft attached to gear A. As gear A turns the hollow shaft, detent wheel B and gear C turn, and pinion D (with the associated positive-head gearing) revolves with it. Thus, the positive carbon is rotated, but does not feed (move along its axis) because pinion D does not turn on its own axis.

Intermittent feeding of the positive carbon is accomplished when the positive feed shaft operates the detent to prevent detent wheel B and gear C from rotating. As gear A turns the hollow shaft, pinion D now turns on its own shaft as it is carried around the circumference of gear C by the eccentric mounting on the hollow shaft. Pinion D turns the worm gear and positive-head gearing that drive the serrated rollers. Thus, the positive carbon is intermittently fed and rotated simultaneously.

The positive-carbon feed can be operated either automatically by a thermostat system, or manually by holding IN the positive-carbon feed button at the rear of the lamp housing.

lAfl

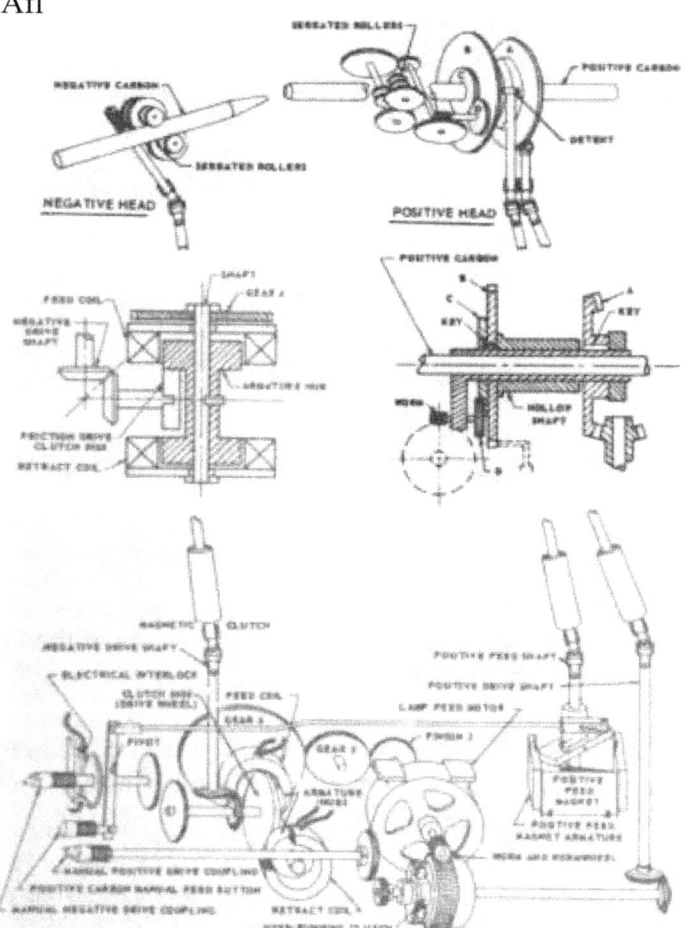

Figure 4-4.-Schematic diagram of carbon-arc lamp.

The THERMOSTAT SYSTEM consists of a thermostat switch in the circuit of an electromagnet (fig. 4-5). The thermostat switch has two bimetallic strips attached to an insulated block. These strips bend with heat and are assembled so that bending, due to changes in the ambient temperature, is the same in both strips. A contact is mounted on the free end of each strip, which

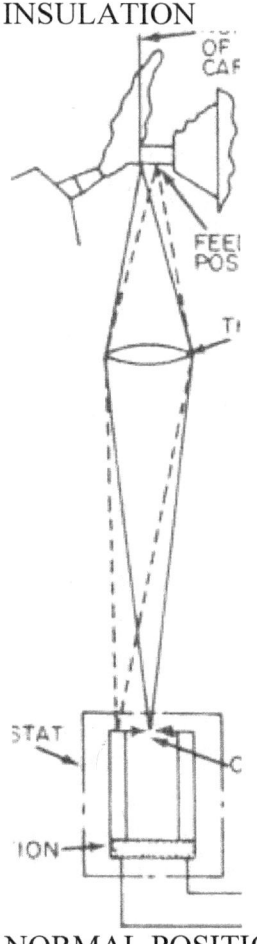

THERMOSTAT
INSULATION

NORMAL POSITION OF POSITIVE CARBON
FEEDING POSITION
THERMOSTAT LENS
ONTACTS MAGNET COIL

Figure 4-5.-Thermostat system of carbon-arc lamp.

closes when the rear strip is heated more than the front strip.

A lens, mounted on the head-supporting column, collects light rays from the arc and projects them through a window in the top plate of the lamp box where they converge on the thermostat. The projection of the positive carbon is of the correct length (nine-sixteenths inch) when

the converging rays strike the space between the thermostat strips. In this position, the thermostat contacts are open and the positive carbon is rotated by the positive head but is not fed into the arc.

As the positive carbon burns away, the concentrated spot of light moves out of the space between the strips and strikes the rear strip, causing it to bend and close the contacts. The contacts complete the circuit to the positive-feed magnet (electromagnet), which operates the positive-feed shaft to prevent the rotation of the detent wheel, B, on the positive head (fig. 4-4). This action causes the pinion, D, to drive the serrated rollers through the positive-head gearing to feed the positive carbon toward the arc. When the carbon is fed to the proper position and the light source is again centered on the space between the thermostat strips, the rear strip cools and opens the contacts to deenergize the circuit of the electromagnet and stop the forward feeding of the positive carbon. The positive-carbon feed is nonreversing and thus cannot pull the positive carbon back.

NEGATIVE CARBON DRIVE.-The negative carbon is fed forward and backward as it burns away to control the arc length and thus maintain the proper current through the arc, or the proper voltage across it. The desired normal value of arc current or voltage is set by adjusting the tension of a spring that opposes the pull of an electromagnet in the negative-feed control circuit.

If the negative-carbon drive is current controlled, the electromagnet (current-regulator coil) is connected in series with the arc. The arc current is fixed by the setting of the negative-feed control and is independent of the line voltage and the resistance connected in series with the arc, provided they do not vary too widely from their normal values. Conversely, if the negative-carbon drive is voltage controlled, the electromagnetic (voltage-regulator coil) is connected across the arc. The arc voltage depends on the line voltage and the setting of the arc-ballast resistor, which is adjusted to the value that provides the most efficient operation.

Although the principle of operation is similar for both the current and voltage control of the negative-carbon feed, only the current control is explained in this chapter.

The negative carbon is carried in the nonrotating negative head and is fed forward or backward by two serrated rollers that are driven by the lamp (feed) motor through gearing (fig. 4-4). The direction of motion is controlled by a feed coil and a retract coil in the negative-drive mechanism.

The negative-drive mechanism contains a shaft that is geared to the feed motor and rotates continuously (fig. 4-4). The shaft is supported on needle bearings in the frames of the FEED COIL and RETRACT COIL. An armature is pinned to the shaft through a slot and is free to slide a limited distance along the shaft and it rotates with the shaft. The clutch disk is located between two hubs on opposite ends of the armature. When neither coil is energized, the hubs rotate free of the clutch disk.

When the feed coil is energized, the armature is pulled back so that an armature hub bears against the rim of the clutch disk. This action transmits motion (caused by friction between the rim of the clutch disk and the armature hub) to the serrated rollers to drive the negative carbon toward the positive carbon. When the retract coil is energized, the armature is pulled in the opposite direction, causing a reversal of the direction of rotation of the clutch disk. This action reverses the direction of rotation of the serrated rollers to drive the negative carbon away from the positive carbon.

CURRENT REGULATOR.-The series-current-regulator coil provides the automatic control of the negative-carbon feed. It consists of an electromagnet connected in series with the arc (fig. 4-6). The current through the arc is changed by adjusting the tension of the armature spring that opposes the pull of the electromagnet. The spring tension is adjusted by means of the ARC-CURRENT ADJUSTING KNOB at the rear of the lamp housing. Turning the knob clockwise increases the spring tension and the arc current. Conversely, turning the knob counterclockwise decreases the spring tension and the arc current.

The arc current flows through the series coil and as the carbon burns away, the length of the arc is increased, thereby increasing the resistance and decreasing the current flow. This action actuates the armature, which

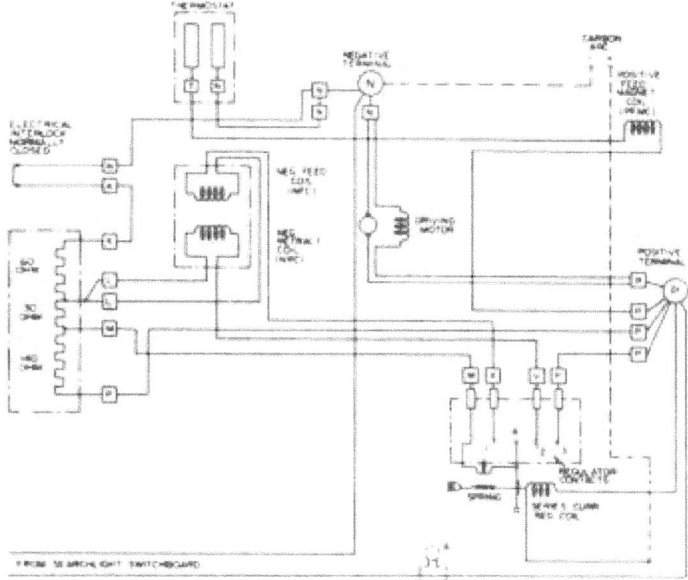

Figure 4-6.—Schematic wiring diagram of lamp control circuit.

carries the movable contact, A, to complete the circuit to the magnetic clutch that drives the negative carbon.

When the lamp is operating at the proper current, the movable armature contact, A, is held midway between contacts 1 and 2 by the balanced pull of the armature spring and the current coil.

When the arc current is low, the pull of the spring is greater than that of the current coil and pulls the movable contact, A, against contact 1. This action places voltage from taps L and M of the resistor across the negative feed coil of the magnetic clutch.

When the arc current is high, the pull of the current coil exceeds that of the spring and pulls the movable contact, A, against contact 2. This action places voltage from taps L and M of the resistor across the retract coil of the magnetic clutch.

When the lamp is started, the negative carbon is in contact with the positive carbon, the starting current is high, and the voltage across the electrodes is low. In

this case the pull of the current coil greatly exceeds the pull of the spring, and regulator contact A Is pulled against contacts 2 and 3. This action short circuits the 140-ohm resistor, thereby compensating for the reduction in voltage across the carbon electrodes and allowing approximately normal current to flow through the negative retract coil. Retracting the negative carbon forms the arc.

LAMP CIRCUITS.-The ship's 120-volt, d-c power is supplied to the searchlight through the searchlight switchboard. The 2-conductor supply cable from the switchboard enters the pedestal through a terminal tube and is connected to a rotary, snap-action, ON-OFF, main-line power switch (fig. 4-7). The 2-conductor, remote-signal cable also enters the pedestal through a terminal tube. The supply and signal circuits are fed through three collector rings, as indicated in the figure.

The negative and R (remote-signal) terminals on the turntable terminal board are connected to the shutter operating unit through the shutter switch. The signaling keys are connected across the incoming signal circuit and if the shutter switch is in the ON position, the signaling shutter will open when the key is closed.

The SHUTTER OPERATING MECHANISM consists of a limited-rotation, d-c motor that is connected mechanically through gears to the sector vanes of the shutter. The armature of the shutter motor is supplied with 65 to 70 volts through the signaling keys. The field of the shutter motor is energized when the shutter switch and the power switch are in the ON positions.

Operation

AUTOMATIC CONTROL.-The arc is started automatically by the negative-feed mechanism. When the power switch is turned ON, the ventilating fan motor and the lamp (feed) motor immediately start and continue to run as long as the circuit to the carbon-arc lamp is energized.

The carbon electrodes are not touching each other; the current through the series-current-regulator coil is zero (fig. 4-6), and the spring holds the movable armature contact, A, against contact 1. The negative feed coil (NFC)

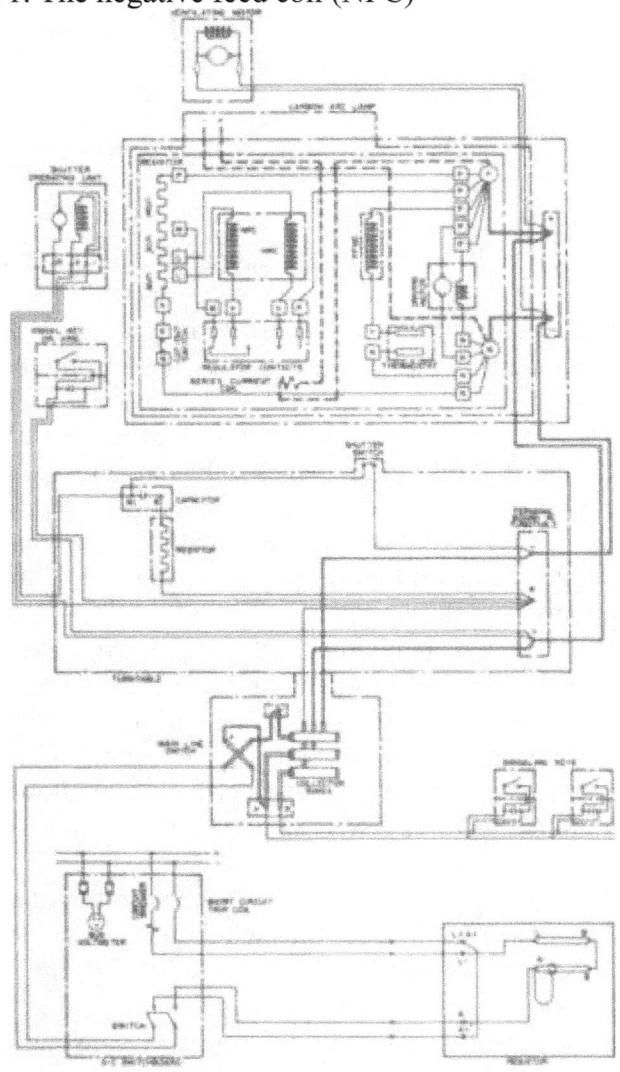

Figure 4-7.—Wiring diagram of 24-inch carbon-arc searchlight.

in the magnetic clutch is energized, and the negative carbon moves toward the positive carbon. When the negative carbon strikes the positive carbon, the arc current increases to between 125 and 200 amperes. The increase in current through the series-current-regulator coil overcomes the spring tension and pulls the movable contact, A, against contacts 2 and 3. This action compensates for the reduction in voltage across the carbon electrodes (previously described) and energizes the retract coil (NRC) in the magnetic clutch to retract the negative carbon. As the negative carbon is drawn away from the positive carbon, the arc is drawn out and the arc current decreases to a stable value (75 to 80 amperes), determined by the setting of the arc-current adjusting knob and the arc-ballast variable resistor. The arc voltage may drop to 40 or 50 volts during the arc-stabilization period and then gradually build up until the arc is stabilized.

During normal operation the series-current-regulator coil maintains substantially constant current through the arc. The movable armature contact, A, normally operates between contacts 1 and 2 because the variations In arc current are usually small. Contact 3 becomes operative only when the arc current Is very high, as when starting the arc, as described previously.

The thermostat will operate the positive-carbon feed when the arc has been held sufficiently long to bum the positive-carbon projection to less than nine-sixteenths inch beyond the obturator. If the positive-carbon tip is slightly ahead or back of the focal line on the arc-image screen after the arc has been established, the automatic feed will quickly bring It Into the proper position. However, If the positive-carbon tip Is more than one-eighth Inch to the right of the focal line, turn the power switch to the OFF position and reset the positive carbon to avoid the possibility of burning the obturator.

MANUAL CONTROL.—The lamp can be manually operated by the controls located at the rear of the lamp box (fig. 4-1, B). In addition to the lamp-focusing screw and arc-current adjusting knob, the controls include the positive-carbon crank, positive-carbon feed button, and negative-carbon crank. These controls are spring loaded and are normally held out of engagement with the motor-drive gearing.

The POSITIVE-CARBON CRANK, when pushed IN and rotated clockwise, rotates the positive head and positive carbon. Rotation of the crank automatically disengages the positive head from the motor-driven nonreversible worm by means of the overrunning clutch, which engages only when the lamp motor is driving the mechanism.

The POSITIVE-CARBON FEED BUTTON, when pushed IN, causes the detent to engage the detent wheel and feed the positive carbon forward. The positive carbon can be fed only when the positive head is being rotated, and it cannot be retracted.

The NEGATIVE-CARBON CRANK, when pushed IN and rotated, feeds the negative carbon when turned clockwise and retracts the negative carbon when turned counterclockwise.

The negative-drive mechanism is provided with an electrical interlock (figs. 4-4 and 4-6) that opens the circuit to the magnetic clutch when the manual control is pushed IN so that the automatic negative-carbon control cannot be engaged at the same time.

Complete manual operation of the lamp can be accomplished provided the ventilating system operates satisfactorily when the power switch is turned to the ON position. To start the arc, push IN the positive-carbon crank (fig. 4-1, B) and rotate it slowly clockwise; at the same time, push IN the negative-carbon crank and rotate it clockwise until the negative carbon strikes the positive carbon. Then, immediately reverse the rotation of the negative-carbon crank to draw out the arc and reduce the current. Excessive current will cause the arc to sputter and give off black soot.

Continuously rotate the positive-carbon crank and note the position of the positive-carbon tip on the arc-image screen. The positive carbon can be fed into the arc by holding IN the positive-carbon feed button as the positive-carbon crank is rotated. Thus, the tip of the positive carbon is maintained at the focal line on the arc-image screen. The positive-carbon crank must be continuously rotated irrespective of the feeding.

To extinguish the arc, turn the power switch to the OFF position.

Preventive Maintenance

Electrician's Mates must be competent to operate and maintain the several types of searchlights used in the Navy. Repairs and adjustments to searchlights aboard ship are accomplished only by Electrician's Mates who are specifically assigned to such work by the Electrical or Engineer Officer.

Signalmen and other personnel who frequently use searchlights must be carefully instructed on how to operate them. When operating carbon-arc searchlights, personnel must be impressed with the importance of renewing carbons before the positive carbon has burned so short that it cannot be gripped by the feed rollers and fed toward the arc. If the positive carbon is permitted to burn after this condition occurs, the arc burns closer to the positive nose (obturator) and will finally cause it to melt.

Operators must frequently observe the arc of an operating searchlight through the peep sight. They should renew the carbons when the positive carbon is no longer fed forward or when the positive-carbon tip projects less than one-half inch beyond the obturator. An alternate procedure to determine when to recarbon is to look into the drum through the front door, taking care to keep out of the searchlight beam. Replace the carbons when the trailing end of the positive carbon (end not in contact with the arc) disappears into the positive head.

RENEWING CARBONS.-An Electrician's Mate must remove the old carbons and replace them with a new pair. Carbons are always renewed by pairs to ensure optimum performance. Normally, the front and rear doors must be kept tightly closed during the operation of the searchlight. The doors should not be opened for any purpose until the drum is placed in the horizontal position and the searchlight is locked both in train and elevation. Before changing carbons or attempting any work on a searchlight, turn the power switch to the OFF position. Open the rear door until the latch engages. Make sure that the components are cool before touching with the bare hands. Lift the negative-carbon release lever upward and remove the negative carbon by pushing it out through the rear of the negative head. Lift the positive-carbon release lever and

remove the stub by pulling it out over the negative head. If the stub is hot, use pliers to remove it.

The positive and negative heads should be reamed before each recarboning to remove the material deposited by the preceding carbons. When using the reamers, the carbon-feed rollers and contacts should be released to prevent the cutting edges of the reamers from damaging these parts. The positive-nose reamer should be inserted from the obturator end, and the negative-head reamer should be inserted from the rear, or release-lever end. Reamers should not be used while the heads are hot unless it is absolutely necessary. However, if it is necessary to ream while the heads are hot, the reamer should be kept turning continuously to prevent its seizure by cooling and contraction of the metal parts.

When the reaming is completed, run the new set of carbons through the heads to be certain that they do not jam. Crooked carbons or those having long cracks, or loose cores should be discarded. Rough spots and blisters can be removed with sandpaper. Adjust the protrusion of the positive carbon from the nose cap, and adjust the arc length in accordance with the values specified for the equipment in the manufacturers' instruction book. The carbons must be centered in their heads and must not touch the nose caps.

After the carbons are renewed, wipe the reflector, front-door glass, thermostat lens, and thermostat window with a clean, dry cheesecloth to remove anycarbon deposit. Turn the power switch to the ON position and operate the searchlight for about 5 minutes to form the crater in the positive carbon and establish a normal arc. The first 5 minutes is the critical period of operation after a carbon renewal. During this interval, closely observe the arc and all lamp parts for any malfunctioning.

CLEANING.—After every extended run of a carbon-arc searchlight, and at laast once each week for searchlights not in use, the interior of the drum and the lamp mechanism should be brushed and wiped thoroughly. Use a small, clean paint brush to collect the carbon refuse from the lamp parts and thermostat lens. Remove fragments of carbon and the accumulations of white deposit from the burning carbons with a soft, dry cloth and blow out the interior of the drum with compressed air. The

proper functioning of the rods, shafts, gears, contacts, and other moving parts of the mechanism depend on cleanliness. Inspect these moving parts to ensure freedom from jams that are invariably caused by dust deposits forming a gum on excessively lubricated parts.

Use the reamers provided for the positive and negative nose caps to clean out any obstruction that may have formed during the extended operation of the lamp.

Clean the contacts of the lamp heads that conduct current to the positive and negative carbons with No. "00" sandpaper wrapped once around a carbon stub. Test the contacts to make certain they are free to move up and down in their slides. Remove all grease, oil, dust, and moisture from electrical insulation and contact surfaces.

Clean the thermostat lens and window, arc-image screen, and exposed faces of the lens system that project the arc image. The cleaning procedure for these parts is the same as that for glass reflectors and front-door glasses, subsequently described.

The effectiveness of a searchlight depends to a great extent on the cleanliness of the reflector and front-door glass. After each extended period of operation of a carbon-arc searchlight, and at least once a week for searchlights not in use, the reflector and front-door glass should be cleaned. When the reflector has cooled, remove all dust with compressed air or a soft, clean brush, and apply a suitable cleaning paste.

Glass and Alzac-finished aluminum reflectors and front-door glasses can be cleaned with a paste prepared by mixing 3 ounces of precipitated chalk with one-half pint of denatured alcohol. Apply the paste with clean, absorbent cotton that has been thoroughly soaked in denatured alcohol, and rub carefully over the entire surface to be cleaned. Both sides of the front-door glasses must be cleaned.

Stellite, Haste lloy, Chromium-plated steel, rhodium-plated copper reflectors can be cleaned by applying any standard Navy bright-work polish to the reflecting surface with absorbent cotton or a soft cloth. Before the deposit of the polish has become dry, lightly rub, or wipe, the surface to be cleaned with a clean, dry, cotton cloth or lens paper to produce a high polish. Move the cloth or lens paper radially outward from the center to the rim of the surface to be cleaned. Do not wipe with a rotary motion.

The surface of a Stellite reflector (after long use, and particularly when exposed to salt spray) may become coated In spots with a white deposit that cannot be removed with bright-work polish. If this condition occurs, prepare a warm 5-percent solution of nitric acid (1 part by volume of concentrated acid and 19 parts by volume of water) and apply with absorbent cotton to the affected spots. Wash thoroughly with hot water, and dry with absorbent cotton. Nitric acid should not be used to clean Stellite reflectors, except when repeated routine cleaning methods fail to remove the white spots.

The reflectors of incandescent searchlights do not require cleaning as frequently as do the carbon-arc types. They should be cleaned after extended periods of use if an inspection of the surface indicates the need for it.

After cleaning and always when the searchlight is not in use, the front door and all other openings into the drum except those provided for normal ventilation must be closed and tightly locked and the searchlight locked in its securing position.

LUBRICATION.—The trunnion and turntable bearings should be greased in accordance with the applicable manufacturers' instruction book as often as an inspection shows this to be necessary.

Lubrication of the lamp mechanism should be limited to the parts specified in the manufacturers' instructions unless they are superseded by Bureau of Ship's orders. The moving parts of the positive and negative heads should be lubricated once a week with a mixture of kerosene and flake graphite. The lubrication should be applied sparingly to avoid any excess from falling on the reflector. Ordinary oil and grease cannot be used for this purpose because of the high temperatures at which the heads operate.

The vanes, gears, and links of the iris shutter and the gears and bearings of the signaling shutter should be lubricated once a month with a kerosene and graphite mixture. The manufacturers' instructions for lubricating shutters and shutter mechanisms should be followed carefully because improper or excessive lubrication will cause the reflector and front-door glass to cloud when the lamp is operated.

PAINTING.—The exterior parts of the drum, trunnion arms, and base of all searchlights must be painted to

preserve the metal. Paint must not be applied to bearing surfaces; working members of any part; or to any bolt, locking nut, or part that must be removed for access to the interior. Nameplates, oil cups, and oil holes must be kept free of paint.

The interior surfaces of the drum, including the signaling shutter blades, must be painted a dull black to minimize stray light.

When painting aluminum alloy parts of searchlights, follow the painting instructions issued by the Bureau of Ships.

Corrective Maintenance

The majority of searchlight repairs are performed by the ship's force. However, reflectors are not repaired aboard ship. If a reflector deteriorates enough to appreciably diminish the candlepower of the light beam, the reflector must be surveyed in accordance with existing instructions and replaced at a naval shipyard.

REMOVING AND REPLACING THE REFLECTOR. -When it is necessary to replace a reflector, engage the train and elevation stowing locks of the searchlight. Remove the rear-door hinge pins and lay the door on blocks with the reflector side up. Remove the eight bolts (fig. 4-1, B), each of which is provided with a clip and vertical stop, and lift the reflector from the door.

Install the new reflector in the door and replace the eight bolts. Position the inside edge of the reflector tightly against the vertical stops on the two bottom points of suspension, and tighten the eight bolts. A shim holder is provided on each of the four bottom stops to prevent the reflector from shifting. Replace the hinge pins and close the door. Start the arc lamp and direct the beam perpendicular to a screen or any other suitable plane surface from 10 to 30 feet distant.

Observe the shape of the beam projection on the screen. If the projection is round, the reflector is properly positioned and requires no further adjustment. If it is not round, adjust the point that corresponds with the portion of the beam that is either inside or outside of the projected circle. This adjustment is accomplished by means of a locking screw and a pressure screw located on the outside of the reflector door and opposite the eight securing bolts.

These screws are countersunk In the rear door and protected by cover screws.

If a portion of the contour of the beam projection is Inside the remainder of the circle, remove the cover screws and loosen the locking screw corresponding to this point and release the pressure screw. Conversely, if a portion of the contour of the beam projection is outside the remainder of the circle, increase the pressure screw corresponding to this point by tightening it. Continue these adjustments until the beam is round, then tighten all eight of the locking screws and replace all of the cover screws.

ADJUSTMENTS.—Proper ALIGNMENT OF THE HEADS is necessary for satisfactory operation of the lamp. The alignment should be checked after the lamp has been removed from, and replaced into, the drum and after the heads have been reassembled. The alignment (fig. 4-8) is checked according to the following procedures:

1. Release the positive-carbon release lever.
2. Insert the positive-head gage with the pin snug against the obturator.
3. Secure the positive-carbon release lever.
4. Rotate the positive-carbon crank (fig. 4-1, B) to bring the gage pin to the vertical (judged by the eye) so that the indication "top" appears upright.

5. Release the negative-carbon release lever.

6. Insert the negative-head gage with the point forward.

7. Secure the negative-carbon release lever (as shown in the figure).

8. Rotate the negative-carbon crank (fig. 4-1, B) until gages touch.

9. Loosen screws and shift negative head to bring negative-head gage point to alignment with positive-head gage punch mark.

10. Secure screws.

The ALIGNMENT OF THE OBTURATOR is accomplished by loosening the two screws that secure it to the positive head, adjusting it so that the positive carbon is approximately centered in the opening of the obturator, and tightening the screws. The obturator must not touch the positive carbon.

Figuro 4-8.-Alignment of hoods.

The ADJUSTMENT OF THE ARC CURRENT is accomplished while the lamp is in operation by means of the arc-current adjusting knob (fig. 4-1, B) at the rear of the lamp box. The nominal arc current is set from 75 to 80 amperes by adjusting the knob and using a portable ammeter, A (fig. 4-6) in series with either leg of the d-c power supply for checking the current readings.

When the arc-current adjusting knob is changed at the searchlight, the arc-ballast resistor must also be adjusted to hold the arc voltage between 65 and 70 volts. Turning the arc-current adjusting knob varies the spring tension on the armature of the series-current-regulator coil (fig. 4-6). To increase the arc current, turn the arc-current adjusting knob clockwise, and to decrease the arc current, turn the knob counterclockwise.

All arc-voltage measurements referred to are taken at the searchlight. If readings of the arc voltage are taken at a distance from the searchlight, the voltage drop in the cables must be considered to obtain accurate adjustments.

The ADJUSTMENT OF THE NEGATIVE CONTACT PRESSURE is accomplished by setting the spring pressure on the negative-carbon brush contact by means of the spring adjustment nut (fig. 4-2). If the spring adjustment is too tight, the negative carbon will be gripped excessively and may stick during operation of the arc. Conversely, if the pressure is too loose, a good contact surface cannot be maintained between the negative carbon and the brush contact. The adjustment should be set to obtain the maximum negative-carbon, brush-contact pressure possible without causing the negative carbon to stick during operation of the lamp.

When the proper adjustment of the lamp is attained, a pull of approximately 8 to 10 ounces parallel to the axis of the negative carbon is necessary to pull the carbon from the head with the feed rollers completely disengaged from the carbon. The positive carbon requires no adjustment.

The ADJUSTMENT OF THE THERMOSTAT LENS should be checked when the reflector is replaced and at any time the feed mechanism should fail to maintain the tip of the positive carbon at the proper distance beyond the obturator. The holes for the screws that attach the thermostat-lens frame to the head support (fig. 4-2) are slotted so that the lens can be moved toward, or away from, the support column. This adjustment is made at the factory so that the beam of light from the lens is centered in the space between the thermostat strips when the positive carbon projects nine-sixteenths inch beyond the obturator. Moving the thermostat-lens frame toward the support column decreases the positive-carbon projection; whereas, moving the lens frame away from the column increases the positive-carbon projection.

To adjust the position of the thermostat lens, start the arc and open the power switch as soon as the forward feed of the positive carbon stops. Measure the projection of the tip of the positive carbon beyond the obturator. Restart the arc and operate the lamp until the positive-carbon feed engages. Immediately stop the arc and again measure the projection of the tip of the positive carbon beyond the obturator. If the average of the two projections is equal to the value specified for the searchlight, the adjustment of the thermostat lens is correct. If the projection is not correct, move the thermostat lens until the proper value of the carbon projection is obtained.

The ADJUSTMENT OF THE THERMOSTAT CONTACTS, which are enclosed in the lamp box, is not

normally required. The clearance between the contacts should be from one-sixty fourth to one-thirty second inch (fig. 4-5). Definitely ascertain that the openings between the contacts are incorrect before making any adjustments. The proper spacing can be accomplished by bending one of the strips.

The ADJUSTMENT OF THE SERIES CURRENT REGULATOR ARMATURE is accomplished by adjusting the armature contacts so that the clearances between contacts A and 1, and A and 2, are approximately 0.006 inch (fig. 4-6). The clearance between contacts 2 and 3 should be approximately 0.008 inch.

The arc-image screen focal line can be adjusted to divide evenly the position of the positive carbon tip by means of four screws that secure the arc-image screen assembly to the drum.

The ADJUSTMENT OF THE ARC IMAGE SCREEN is correct when the focal line is at the mean of travel of the positive-carbon tip with the positive-carbon projection adjusted to nine-sixteenths inch beyond the obturator and with the lamp focused properly, as subsequently described. During the normal cycle of positive-carbon feeding and burning off, the positive-carbon tip will move along the positive-carbon axis. This movement, when reviewed on the arc-image screen, should not extend more than 0.05 inch either way from the focal line.

The FOCAL POINT OF THE REFLECTOR is the point at which the light source must be located to obtain a concentrated beam of parallel light rays. In a carbon-arc searchlight, proper focus of the arc occurs when the image of the tip of the positive carbon appears on the focal line of the arc-image screen. The lamp is adjusted to this position by the lamp-focusing screw (fig. 4-1, B) at the rear of the lamp box. This focusing screw is threaded into the lamp housing of the drum and connected to the lamp by a focusing latch attached to the lamp base. Removing the focusing latch from the guides frees the lamp from the focusing screw. The lamp-focusing screw has a lock collar that must be backed off when turning the screw, and set up when the proper position of the lamp has been obtained. Turning this screw clockwise moves the lamp away from the reflector, and turning it counterclockwise moves the lamp toward the reflector. If the tip of the

positive carbon does not fluctuate about the focal line on the arc-image screen, focus the lamp by turning the lamp-focusing screw until this condition is obtained.

The focal adjustment can be checked by observing the beam at night. If the lamp is focused properly the beam will appear to consist of parallel rays. If the lamp is too near the reflector the beam will spread, and if the lamp is too far from the reflector the beam will converge into an hour-glass shape. The appearance of the beam should be judged from a distance.

TROUBLES.—The conditions that result in operating difficulties of carbon-arc searchlights are considered to facilitate their detection and correction.

If the ARC CURRENT IS HIGH OR LOW and the arc voltage is normal, set the supply voltage to the proper value and adjust the arc-current adjusting knob to obtain the proper current. This adjustment of the arc current changes the arc voltage, thus it is necessary to reset the arc-ballast resistor to obtain the proper voltage (previously described).

If the ARC VOLTAGE IS HIGH OR LOW and the arc current is normal, set the supply voltage to the proper value and reset the arc-ballast resistor to obtain the proper voltage.

The ARC-BALLAST RESISTOR is provided with a clamp-type adjustable connection to permit variation of the amount of resistance in the circuit (fig. 4-7). Proper adjustment of the arc-ballast resistor is necessary for satisfactory operation of the carbon-arc lamp. The adjustable connection on the resistor permits adjustment of the arc voltage of the searchlight. When the arc-ballast resistor adjustable connection is changed, the effect of the change on the arc must be noted immediately.

The arc-ballast resistor must be used only to compensate for permanent changes in the ship's 120-volt, d-c power to the searchlight. It must not be used for making adjustments for transient conditions in the supply voltage.

If the POSITIVE HEAD FAILS TO ROTATE, push IN the manual positive drive coupling, and turn it clockwise (fig. 4-1, B). If the coupling cannot be turned and the head does not rotate, inspect the positive-head gearing for mechanical interference or foreign material in the gears. If the head still does not rotate, remove the lamp from the drum and apply 65-volt test leads to the lamp terminals (without the carbons). Determine that the lamp motor drives the worm gear, and also determine that the motor will operate with no load.

If the POSITIVE CARBON DOES NOT FEED, hold IN the positive-carbon, manual feed button (fig. 4-1, B). If the carbon feeds with the button held IN, determine that the positive-feed shaft and its detent are free to operate and are free from mechanical interference with the positive terminal. Observe through the front door to determine that the thermostat lens focuses the arc on the center of the glass thermostat window on top of the lamp box when the positive-carbon tip is on the focal line of the arc-image screen. If the arc is not focused on the center of the thermostat window, turn the power switch to the OFF position, lock the drum in elevation, open the rear door, and adjust the position of the thermostat lens. Be certain that the thermostat window is clean.

If the arc is focused on the center of the thermostat window and the carbon still does not feed, remove the lamp from the drum and determine that the armature of the positive-carbon feed magnet is mechanically free, the thermostat contacts have the proper clearance, and the contact surfaces are clean. Apply 65-volt test leads to the lamp terminals without the carbons, and short circuit the thermostat by connecting a lead across the thermostat terminals. Determine that the positive-carbon feed magnet operates. If it does not operate, the magnet is probably open circuited and should be replaced.

If the positive carbon does not feed when the feed button is held IN, observe through the front-door dome glass and determine if the rotation of the positive-head detent wheel has been arrested by the positive carbon feed shaft. If the detent wheel is not turning, determine that the feed rollers are gripping the positive carbon. If the rollers are digging into the carbon, ream out the positive head. Inspect the rollers and replace them if the teeth appear to be worn. Determine that the positive-carbon brush contact is free to move up and down in its slide.

If the POSITIVE CARBON FEEDS CONTINUOUSLY, determine that the armature of the positive-carbon feed magnet is mechanically free; that the thermostat contacts have the proper clearance and the contact surfaces are

clean; and that the positive-carbon brush contact has freedom of movement.

If the POSITIVE CARBON TIP IS HELD AT A POINT AWAY FROM THE FOCAL LINE on the arc-image screen, determine that the positive-carbon projection is correct. The correct positive-carbon projection is obtained by adjusting properly the position of the thermostat lens and the focus of the lamp.

If the MOVEMENT OF THE POSITIVE-CARBON TIP ON THE ARC IMAGE SCREEN IS GREATER THAN ONE-SIXTEENTH INCH FROM EITHER SIDE OF THE FOCAL LINE, remove the lamp from the drum and inspect the thermostat contacts for the proper clearance.

If the NEGATIVE CARBON DOES NOT FEED AT THE START, determine that the negative carbon release lever is in the secured position and then determine if the negative-carbon drive shaft turns. If the shaft turns determine that the rollers are gripping the negative carbon. If the rollers are digging into the carbon, ream out the negative head and decrease the pressure on the negative-carbon brush contact by means of the negative-carbon, brush-pressure adjusting nut (fig. 4-2). Inspect the drive rollers and replace them if the teeth appear to be worn.

Inspect the negative-carbon brush contact for freedom of movement. If the shaft is not turning, remove the lamp from the drum and check the series-current-regulator contacts. Check the resistance of the feed and retract coils of the magnetic clutch. The resistance of these coils should be 57 ohms with an allowable variation of ±4 ohms. Check the electrical interlock (cutout switch) on the negative manual feed and determine that it is closed.

If the NEGATIVE CARBON FEEDS SLUGGISHLY, examine the serrated rollers and determine that they grip the carbon firmly. Remove the carbon and examine the contacts and nose, and ream out any obstructions. Replace the negative carbon if it is damp because a damp carbon will cause sluggish feeding.

If the NEGATIVE CARBON DOES NOT RETRACT AFTER STRIKING THE POSITIVE CARBON, turn the power switch to the OFF position and check the circuit breaker on the searchlight switchboard (fig. 4-7) to determine if it has opened, and set the trip mechanism to a higher value (not exceeding 300 amperes). Reclose the

circuit breaker and check the alignment of the heads. Ream the negative head and check the negative-drive mechanism to determine that it is free from mechanical binding.

Remove the lamp from the drum and check the spacing of the series-current-regulator contacts. Check the resistance of the feed and retract coils in the magnetic clutch and replace them if they are open circuited, or if their resistance is not within ±4 ohms of 57 ohms each. Check the resistor that feeds the magnetic clutch coils (fig. 4-6) to determine that proper contact is made and that the resistor is not open circuited. Renew the carbons and set the arc-current adjusting knob, replace the lamp in the drum, and recheck the current setting.

If the ARC BREAKS REPEATEDLY, check the arc voltage and current at the time the arc breaks. The current is usually low when the arc breaks repeatedly, and the arc-current adjusting knob should be turned clockwise until rated current flows. Adjust the arc-ballast resistor to maintain 65 to 70 volts while adjusting the arc current.

If the arc continues to break, remove the lamp from the drum and check the alignment of the heads. Ream the negative head if any obstructions are present. Check the resistances of the feed and retract coils of the magnetic clutch.

If the ARC CURRENT VARIES MORE THAN 5 AMPERES when the lamp is in normal operation, check the contacts on the series-current regulator coil. Adjust the contacts for proper clearances and determine that the contacts are clean. Check the supply voltage and install new carbons if they have become damp.

THE ARC LENGTH IS TOO SHORT (when viewedfrom the arc-image screen is less than five-eighths inch). The length of the arc can be used as a rough check on the current flow. Measure the arc along a straight line from the center of the carbon tips. This is a rough check and no adjustment should be made unless the arc length is off one-eighth inch or more. The arc length can be increased by changing the tap connections to the arc-ballast resistor to decrease the resistance in the circuit.

THE ARC LENGTH IS TOO LONG (when viewed from the arc image screen is more than five-eighths inch). If the arc length is one-eighth inch or more off when measured between the center of the carbon tips, the arc length can be decreased by changing the connection to the arc-ballast resistor to increase the resistance of the circuit.

Safety Precautions

The general instructions and operating precautions must be adhered to by Electrician's Mates, Signalmen, and other personnel when operating or maintaining searchlights.

The assigned Electrician's Mate or lamp operator must be certain that the carbons are properly placed, are firmly gripped by the feed rollers, and make good electrical contact with the current contacts before starting the lamp mechanism of a carbon-arc searchlight. Do not operate the lamp if the ventilating system does not function properly. Do not turn the power switch to the ON position when the carbons are touching, except in the older searchlights that are provided with a hinged negative head for striking the arc.

Frequently observe, on the arc-image screen of an operating lamp, to determine that the positive carbon is properly focused to avoid damage to either head (caused by failure of the carbons to feed properly). A continual or frequent adjustment of the arc focus by means of the lamp-focusing screw indicates a failure of the positive-carbon feed to function properly or other maladjustment that must be corrected.

Turn off the arc and recarbon when the feed rollers fail to engage, and feeding has stopped, or when the positive carbon projects less than one-half inch beyond its nose. Never look directly at the arc. Always view it through the peep sight or arc-image screen.

Under normal conditions the drum must be kept tightly closed during operation of the arc.

Lock the searchlight securely in train and elevation and be certain that the drum is in the horizontal position before opening the front or rear door of a 24-inch, carbon-arc searchlight.

Be certain that the power supply to the shutter motor is deenergized before opening the front door or performing any work inside the drum of a searchlight provided with an electrically operated shutter.

Before attempting to train or elevate a searchlight, be certain that the stowing brakes and locks are disengaged and that the searchlight is free to move.

Be certain that the power switch is in the OFF position before changing carbons. Always recarbon with a pair of carbons and do not use broken or bent carbons. Be certain that the positive carbon does not touch its nose cap and keep the nose cap free from dirt and deposits of carbon dust. Keep the thermostat lens and window clean.

Do not operate the signaling key on the 24-inch searchlight when the signaling shutter is latched open and do not keep a signaling shutter open by holding down on the signaling key.

Keep all electrical contact surfaces clean and free from oil, grease, dust, and moisture. Do not use oil or grease on the positive or negative heads.

When reaming the heads, release the contacts and feed rollers, and do not allow these parts to come in contact with the reamer.

The carbons must be kept sealed in their containers until required. They must be kept dry and protected from any jarring that can cause them to crack or to loosen the cores.

Searchlights that are not in use must be locked in their securing positions by means of the brakes and stowing locks. All openings into the drum must be tightly closed unless otherwise ordered.

In case of threatening or inclement weather at sea, canvas covers must be applied and lashed down on all searchlights, except for those designated as ready lights. During heavy weather at sea, searchlights must be secured with the front-door glasses turned inboard.

12-INCH INCANDESCENT SEARCHLIGHT

The 12-inch incandescent searchlight is used primarily for signaling purposes and secondarily for illumination. The general construction and functions of the principal components are essentially the same as those of the 24-inch carbon-arc searchlight with the exception of the light source.

Construction

The searchlight (fig. 4-9) comprises the (1) mounting bracket, (2) yoke, (3) drum, and (4) lamp (not shown). The MOUNTING BRACKET permits the searchlight to be secured to a vertical pipe or to a flat vertical surface. The YOKE is swivel mounted on the bracket to allow the searchlight to be rotated continuously in train. The steel DRUM provides a housing for the lamp and is trunnion mounted on the yoke to allow it to be elevated and depressed. Clamps are provided for locking the searchlight in any position of train and elevation.

Figure 4-9.-12-inch incandescent searchlight.

The SIGNALING SHUTTER is a Venetian-blind shutter mounted inside the drum behind the frontdoor. It is held in the CLOSED position by two springs and is manually opened by a lever on either side of the drum. The parabolic metal reflector is mounted on the inside of the rear door.

The LAMP is usually a 1000-watt, 117-volt incandescent lamp having special concentrated filaments that reduce the area of the light beam. The lamp is mounted In a mogul bipost socket. The socket is located in front of the reflector and can be adjusted only slightly. The replacement of the lamp is accomplished through the rear door of the searchlight.

The light source must be at the focus of the reflector for minimum beam spread and maximum intensity. Some types of 12-inch incandescent searchlights are provided with focusing adjustment screws. Other types can be adjusted by loosening the screws that hold the lamp-socket support plate in position. Move the entire socket assembly toward or away from the reflector until the beam has a minimum diameter at a distance of 100 feet or more from the light, and retighten the screws. When checking the diameter of the beam, the rear door must be tightly clamped shut.

A screen hood is provided for attachment to the front door to limit the candle power of the beam, to cut down its range, and to reduce stray light, which causes secondary illumination around the main beam. The hood also provides for the use of colored filters.

Short-Arc, Mercury-Xenon Lamp

The Bureau of Ships has developed a mercury-xenon arc lamp that can be installed in the standard Navy 12-inch incandescent signaling searchlights (fig. 4-10). The mercury-xenon arc lamp, when used in these searchlights, produces a maximum beam intensity about 10 times as great as that attained by the incandescent lamp.

The lamp is a 1000-watt short-arc, mercury-xenon lamp requiring 45 amperes to start and 18 amperes to operate. It is supplied from the ship's 117-volt, 60-cycle, single-phase power. The lamp consists of two tungsten electrodes spaced about one-quarter inch apart inside a 2-inch diameter quartz bulb. The bulb contains a small quantity of liquid mercury and xenon gas at a pressure of 3 to 5 atmospheres when the lamp is cold (an atmosphere at sea level is equivalent to 14.7 psi). After the arc is started and the lamp attains a stable operating temperature, the internal pressure increases to about 20 atmospheres. The lamp does not produce full-light output until this pressure is reached and all of the mercury has been vaporized. The lamp differs from the incandescent lamp in that it requires a high voltage, r-f current for starting and a ballast for operating it at rated output.

LAMP ADJUSTER ASSEMBLY.—The lamp adjuster is secured to the top of the searchlight drum and extends

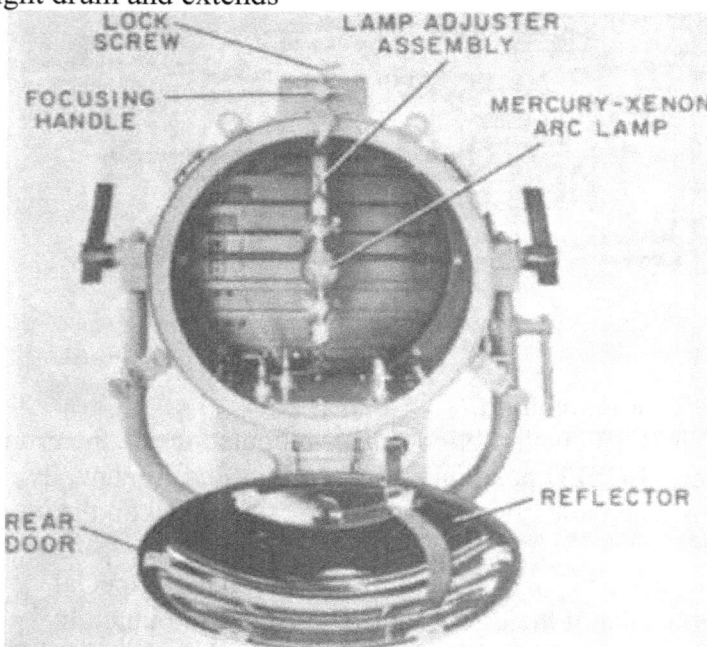

Figure 4-10.-12-inch mercury-xenon searchlight, (rear view showing lamp).

inside the drum to provide a mounting for the mercury-arc lamp (fig. 4-11). The assembly affords longitudinal traverse (toward or away from the reflector), horizontal (from side to side), and vertical adjustments of the lamp.

BALLAST UNIT.—The ballast unit, located out of the weather near the searchlight, consists of a transformer and five resistors enclosed in a vented steel box. The unit provides the necessary ballast voltage in series with the lamp to maintain constant current for normal operation.

STARTER UNIT.—The starter unit, enclosed in a watertight box, is mounted in the lamp housing. It consists of the necessary components for star ting and operating the lamp. The electrical connections from the ballast unit to the starter unit are provided by a cable equipped with a watertight plug.

LOCK SCREW
AXIAL
SIDE PLATE
CLAMP
HORIZONTAL SCREWS
FOCUSING HANDLE
WITNESS MARKS

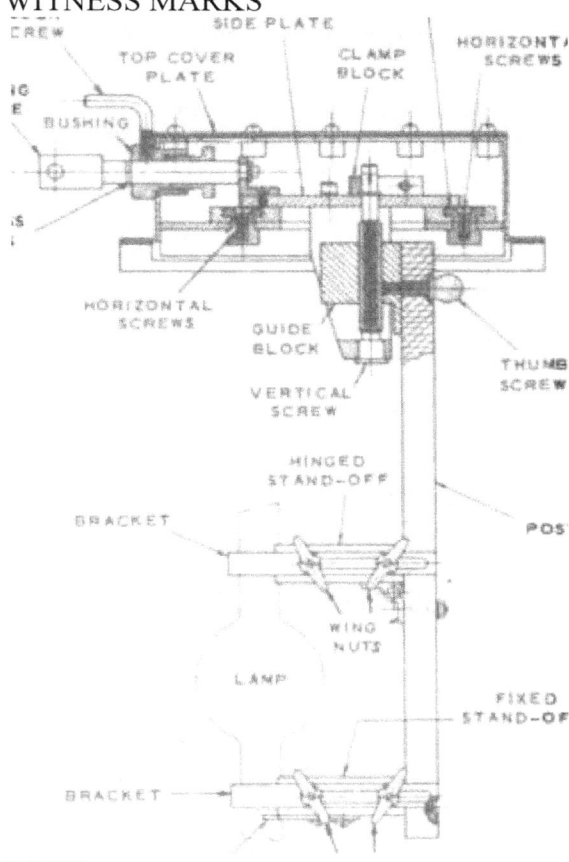

POST
FIXED STAND-OFF
BRACKET
STOP
WING NUTS

Figure 4-11.-12-inch mercury-xenon lamp adjuster assembly.

Operation

The electrical circuit of the 12-inch mercury-xenon arc searchlight comprises the (1) starting circuit and (2) operating circuit. The ship's single-phase, 117-volt, 60-cycle power is supplied to the ballast unit through an OFF and ON rotary power switch (fig. 4-12).

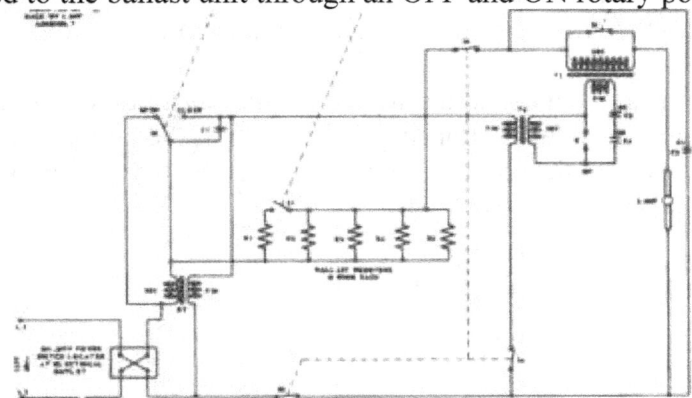

Figure 4-12.-Wiring diagram of 12-inch marcury-xanon searchlight.

STARTING CIRCUIT.-When the lamp is extinguished, whether hot or cold, the internal resistance between electrodes is so high that the arc will not restart by applying normal line voltage. Hence, a special high-voltage r-f circuit is Incorporated to provide Instant starting, irrespective of whether the lamp is cold or hot. This circuit is capable of producing a voltage from 40,000 to 60,000 volts, which is sufficient to start the arc when the lamp is hot and at maximum internal pressure. After the arc is started, the lamp will operate on normal voltage of 60 to 70 volts.

The starting circuit includes the step-up gas-tube sign transformer, T2 (fig. 4-12), the secondary of which supplies high voltage to an r-f circuit containing a spark gap, G, capacitors C3 and C4, and the primary of the pulse transformer, Tl. When gap G breaks down, r-f oscillations build up in C3, C4, and the primary of Tl. These oscillations are fed to the secondary of Tl where high voltage is developed in the circuit comprising the lamp unit and C2 (for starting the arc). Transformer T2 has a special core with a magnetic shunt that prevents the spark gap, G, from effectively short circuiting the primary of T2.

When the arc is struck in the lamp, the starting current is limited to about 45 amperes by the reactance of the

217

secondary of transformer BT In series with the lamp and the parallel combination of the five 10-ohm ballast resistors. The starting current in each resistor is 45/5 = 9 amperes, and the voltage drop across the parallel combination is 9 x 10 = 90 v. The voltage across the lamp during the warm-up period is correspondingly reduced to about 25 volts.

OPERATING CIRCUITS.-After the arc has been started, SI opens and disconnects Rl from the ballast resistor circuit, thereby reducing the load on the circuit (fig. 4-12). The normal operating current of 18 amperes divides equally through the remaining four ballast resistors. The voltage drop across the ballast resistors for this operating current is 18/4 * 10 = 45 v. The lamp voltage is correspondingly increased. Switch S2 opens and disconnects the primary of T2 and shorts the secondary of BT to further increase the lamp voltage to its normal operating value.

To start the arc, turn the power switch to the ON position and rotate the spring-loaded starting knob on the back of the lamp assembly in a counterclockwise direction. This action closes switches SI and S2 and opens switch S5. When switch SI closes, resistor Rl is connected in parallel with resistors R2, R3, R4, and R5 to provide maximum energy for starting the lamp. When switch S2 closes, the primary of transformer T2 is energized. The voltage on the T2 primary is boosted 32 volts above the line supply by the autotransformer action of booster transformer BT. The secondary of transformer T2, in turn, energizes the r-f circuit to superimpose a high-voltage pulse across the lamp electrodes, causing the arc to start.

Release the starting knob Immediately after the arc is started to allow the spring-loaded switch to return to its original position. This action opens SI andS2 and closes S5. It is essential that the starting knob be released as soon as the arc strikes, otherwise the lamp will be overloaded. If the lamp is operated under this condition for a very brief period it will become overheated and fail violently. When the starting knob is released, it rotates clockwise and opens SI to remove Rl from the line (previously described) and opens S2 to deenergize the starting circuit (T2 primary, as previously described). When

switch S5 closes, the secondary of the pulse transformer, T2, is short circuited to remove it from the circuit. Micro switches S3, S4, and S6 are manually operated as a single unit to provide for deenergizing the starter-unit circuits independently of the power switch as an additional safety feature when servicing the lamp circuits.

To extinguish the lamp, turn the power switch at the ballast unit to the OFF position.

Preventive Maintenance

The 12-inch short-arc, mercury-xenon searchlight should be kept clean and the trunnion and yoke bearings lubricated with grease applied with an alemite gun.

The reflector and cover glass should be cleaned weekly. The procedures for cleaning these parts are similar to those for cleaning glass and Alzac-finished aluminum reflectors and front-door glasses described for the 24-inch searchlight.

The shutter assembly should be Inspected periodically for tightness of screws and fastenings, for ease of operation, and for light tightness. If the return action of the shutter assembly becomes sluggish, lubricate the shutter bearings with graphite mixed with kerosene. Use the graphite lubricant sparingly and do not allow it to contact the quartz envelope of the mercury-xenon lamp. If the shutter action remains sluggish after lubrication, replace the shutter springs.

The searchlight is properly focused when shipped. The service lamp is packed in the repair parts box with three additional spare lamps. When assembling or disassembling the lamp, use the face mask and the gloves provided with the equipment. The lamp is filled, under considerable pressure, with xenon gas and may fail violently if it is dropped or struck. The lamp must not be handled without enclosing it in the protective metal enclosure provided with equipment.

To replace the short-arc, mercury-xenon lamp,position it in the lamp adjuster assembly with the collar resting on the stop (fig. 4-11). Press the brackets tightly against the metal collars on the lamp and tighten the wing nuts. Connect the lamp leads to the terminal posts on top of the starter assembly and be certain that the leads have a

clearance of 1 inch from the nearest metal parts. Connect the short, bottom lead on the lamp to the terminal marked, HV.

Remove finger marks or grease from the quartz arc tube with alcohol or other grease-free solvent, and dry carefully with a clean cloth. Operation of the lamp with finger marks or grease on the surface will cause deterioration of the quartz envelope. Inspect all connections for tightness and, using the face guard and gloves, remove the metal safety shield from the lamp and place it in the repair parts box. Close the door assembly and secure it with the three thumb screws. Plug the power cable from the ballast into the receptacle on the starter box assembly.

To check the focus of the lamp, secure the drum in a level position with the shutter locked OPEN. Turn the power switch ON and direct the beam against a perpendicular, flat surface about 100 feet distant. The spot should be a concentrated image of the arc projected along the axis of the drum.

If the lamp is out of focus, the LONGITUDINAL TRAVERSE ADJUSTMENT is accomplished by loosening the lock screw (fig. 4-11) and sliding the plunger, or focusing handle, in or out to move the lamp toward, or away from, the reflector. This adjustment results in increasing or decreasing the diameter of the projected spot. The focusing handle is provided with two witness marks. When the first witness mark is aligned with the edge of the bushing at the top of the box, the lamp will be positioned at the focal point of the reflector and will produce a concentrated light beam of minimum divergence. When the second witness mark is aligned with this bushing, the lamp will produce a beam with a 4-degree spread. The focus should be adjusted for minimum divergence when the light is used at maximum ranges, and a 4-degree spread for close range operation.

It is seldom necessary to make the horizontal or vertical adjustments of the lamp. However, if these adjustments are required, it is necessary to remove the top cover plate.

The HORIZONTAL ADJUSTMENT is accomplished by loosening the four horizontal screws (fig. 4-10) and sliding the mounting plate to the right or left of center as required, and then retightening the horizontal securing screw.

The VERTICAL ADJUSTMENT is accomplished by loosening the socket-head cap screw and turning the vertical screw (fig. 4-10) to elevate or depress the beam, and then retightening the socket-head cap screw, and replacing the cover, making certain that the gasket is in place.

Corrective Maintenance

If the short-arc, mercury-xenon lamp should fail to light, check the wiring to be certain that the lamp leads clear any metal parts by at least 1 inch. Inspect the safety switches S3, S4, and S6 (fig. 4-12) to determine that they are closed. Secure the shutter in the OPEN position, turn the power switch ON and rotate the starting-switch knob counterclockwise; then release it. Observe, through the cover glass, the electrodes of the lamp for signs of arcing. The arc should take place just before the starting switch knob reaches the end of its travel. If arcing is observed and the lamp fails to light, replace the lamp with a new one.

If the lamp still falls to light, check the high-voltage leads from transformer Tl to the HV terminal for high-voltage leaks. Check the circuit to determine that switch SI closes to parallel resistor Rl with resistors, R2, R3, R4, and R5.

Safety Precautions

The short-arc, mercury-xenon lamp must be operated with the shutter assembly secured to the drum. To avoid deterioration of the quartz arc tube, remove finger prints or grease before operating the lamp.

The lamp contains gas under considerable pressure and if dropped or struck it may burst violently. The lamp must not be handled without enclosing it in the protective metal enclosure. This precaution applies to both new and burned-out lamps. Lamps that reach the end of their useful life are potentially dangerous and must be disposed of in the sea. Always use face mask and gloves when assembling or disassembling the lamp to protect the eyes and skin from flying material should the lamp explode.

Before attempting to work on the assembly or to replace the lamp, be certain that the safety switches S3, S4, and S6 (fig. 4-12) are open and that the plug on the starter unit is disconnected. The safety switches must not be bridged or depressed when personnel are working on the assembly or replacing the lamp.

8-INCH, 60-CYCLE SEALED-BEAM SEARCHLIGHT

The 8-inch signaling searchlight utilizes a compact-arc, xenon lamp. The lamp is mounted in a sealed-beam enclosure so that the reflector and arc tube are always clean and in focus. It is designed to withstand high vibratory shock and extreme humidity conditions, and will operate equally well in hot and cold climates. The control unit for this searchlight may be furnished for operation with either a 60-cycle or a 400-cycle power supply. The 8-inch searchlight equipped with a 60-cycle control unit is described in this chapter.

Construction

The searchlight (fig. 4-13) consists of the (1) base, (2) yoke, (3) housing, and (4)lamp. The BASE is equipped with a rail clamp for securing the searchlight to the rail. The YOKE is swivel mounted on the base to allow it to be trained through 360°. The HOUSING provides an enclosure for the lamp and is composed of a front and a rear section. The front section comprises the shutter housing, and the rear section comprises the backshell housing. The two sections are held together by a quick-release clamp ring that permits easy replacement of the lamp. The entire housing is mounted on brackets attached to the shutter housing and supported by the yoke to allow the searchlight to be elevated or depressed. Clamps are provided for securing the searchlight in train and elevation.

The SHUTTER HOUSING contains the Venetian blind shutter, which is held closely by springs and manually opened by a lever on either side of the housing. The front of the shutter housing is sealed by the cover glass and a gasket. The rear of the shutter housing is enclosed by a gasket and adapter assembly. The adapter assembly provides a locating seat for the lamp and incorporates a

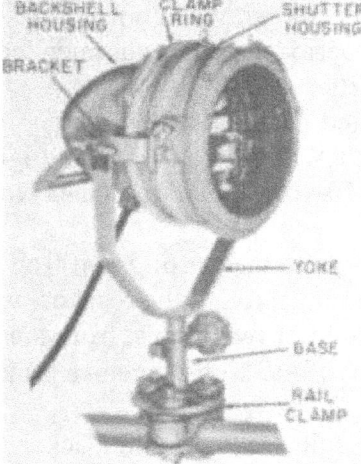

Figure 4-13.-8-inch 60-cycle sealed-beamed searchlight.

hook and key arrangement that aligns the backshell housing and retains it in position while attaching the clamp ring to hold the two sections together.

Three filter assemblies (red, green, and yellow) are provided and can be readily snapped in place over the face of the searchlight. The shutter vanes can be locked in the OPEN position when the searchlight is used as a spot light.

The BACKSHELL HOUSING provides an enclosure for the spun-brass backshell assembly (fig. 4-14). The back-shell (fig. 4-14, A) encloses the hermetically sealed starting unit (fig. 4-14, B) and the xenon lamp (fig. 4-14, C). The front of the backshell is shaped to fit the lamp and to retain it in position against the lamp contacts on the hermetically sealed starting unit. A handle mounted on the rear of the backshell housing is provided to control the direction of the searchlight beam.

Compact-Arc Xenon Lamp

The compact-arc, sealed-beam xenon lamp consists of two electrodes inside a quartz bulb. The bulb is filled

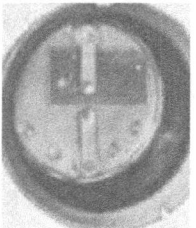

A BACKSHELL
C GASKETEO LAMP

B STARTING UNIT

Figure 4-14.—Back shell assembly of xenon compact-arc lamp.

with xenon gas at several atmospheres pressure. This lamp, similar to the short-arc mercury-xenon lamp, requires a high-voltage, r-f current for starting, and a ballast for operating it at rated output. The xenon lamp has approximately 90 percent light output at start and attains full intensity within a few seconds. It can be extinguished and restarted when hot or cold. However, frequent starting or intermittent operation for short periods is not recommended because the life of the lamp is reduced with the number of starts.

CONTROL UNIT.—The control unit (fig. 4-15), located near the searchlight, consists of a transformer, T, resistor, R, and four relays (RT, RC, RP,and RK) mounted in a spraytight steel case. The transformer is of the high reactance type having a 120-volt primary and a 42-volt secondary. When the arc is started, the transformer

functions as an autotransformer to boost the input voltage to the starting unit from 120 volts to about 165 volts. This voltage is applied to the lamp circuit through the resistor. After approximately 7 seconds, the lamp is transferred by the operation of two relays, from the 165-volt autotransformer resistance circuit to the 42-volt secondary of the high-reactance transformer, and the lamp operates at this voltage.

STARTING UNIT.—The hermetically sealed starting unit, located in the backshell, provides the high voltage, r-f current necessary to ionize the xenon gas and establish an arc in the lamp. It consists of a high-reactance step-up transformer, Tl; spark gap, G; capacitor, C3; r-f transformer, T2; and two by-pass capacitors CI and C2 to prevent r-f current from feeding back into the line.

Operation

The ship's single-phase, 120-volt, 60-cycle power is supplied to the control unit through an OFF and ON power switch, and the control unit is connected to the searchlight through a cable equipped with a receptacle and plug assembly. All arc-type lamps require a ballast circuit to control and limit the current after starting the lamp. The ballast for this searchlight is resistive for lamp starting and reactive for lamp operation after approximately 7 seconds.

When the power switch is operated to the ON position, transformer T functions as an autotransformer to boost the input voltage that is applied through resistor R to the primary of the high-reactance step-up transformer, Tl, in the starting lamp circuit. This transformer is similar to the gas-tube sign transformer, previously described in the 12-inch mercury xenon arc searchlight. The secondary of transformer Tl energizes the r-f circuit containing a capacitor, spark gap, and r-f transformer. The high-frequency current is produced in the oscillating circuit comprising the spark gap, G, capacitor C3, and the primary of T2. The r-f voltage is increased to the required value by the step-up action of r-f transformer T2. The high-frequency high voltage is superimposed across the electrodes. The high-frequency current ionizes a path in the gas through which the initial ballast voltage follows the discharge to establish the arc.

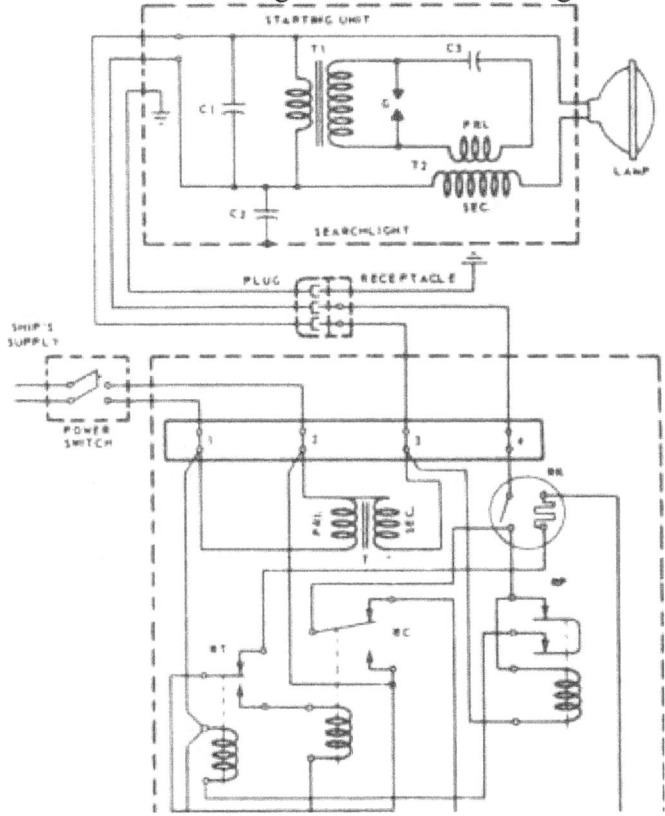

CONTML UKIT

Figure 4-15.—Wiring diagram of 8-inch 60-cycle sealed-beam searchlight.

As soon as the arc is established, the relatively large starting current through ballast resistor R reduces the voltage across the starting unit below the value required to maintain a spark across G, and the r-f voltage is reduced to zero. At the same time, the reduced voltage causes the contacts of relay RP to close and energize the

operating coil of the time-delay relay, RT. Relay RP is connected across the arc lamp and is designed to open its contacts when the voltage across its coil exceeds 35 volts. The contacts of relay RP remain closed when the lamp is operating normally at an arc voltage of about 20 volts.

The time-delay relay, RT, operates in approximately 7 seconds (after the power switch is closed and the arc is started) to apply line voltage to the control relay, RC. The make-before-break contacts of this relay transfer the lamp from the resistive circuit to the 42-volt secondary of transformer T to provide a reactive ballast for the lamp. The operation of relay RT also removes voltage from the heater unit of thermal delay relay RK.

If the arc becomes unstable and is extinguished, relay RP operates and deenergizes relays

RT and RC, thereby transferring the lamp circuit to the resistive ballast to automatically restart the lamp and also applying line voltage across the heater unit of thermal delay relay RK. This action is accomplished within a few seconds and without noticeable flicker of the lamp. If the lamp fails to start or restart, the thermal time-delay relay, RK, will operate in about 25 seconds to disconnect the output circuit to the searchlight, thereby preventing possible damage to the starting unit by the prolonged use of the high-voltage, r-f starting circuit.

Preventive Maintenance

The 8-inch compact-arc, xenon, sealed-beam searchlight should be kept clean and painted. The electrical contacts should be cleaned with a soft cloth. Do not use abrasives to clean these contacts. The front face of the lamp and the cover glass should be kept clean with water or glass cleaner.

To remove the compact-arc, xenon, sealed-beam lamp from the housing for cleaning or replacement, tip the rear end of the searchlight up to its highest position and lock it in place. Release the clamp-ring toggle and remove the clamp ring (fig. 4-13). Remove the backshell assembly by raising it up to disengage it from the hook and tab. Pull the gasketed lamp out of the shutter adapter assembly by gripping the lamp gasket on its periphery and lifting it out to disengage the gasket lugs from the notches in the adapter assembly.

To replace the lamp In the lamp gasket, be certain that the TOP marked on the lamp and on the gasket are aligned and that the lugs of the lamp are firmly seated in the recesses provided in the gasket. To replace the gasketed lamp in the housing, be certain that the lugs are set into the notches in the adapter assembly located inside, and at the rear of, the shutter housing. Set the backshell assembly over the shutter assembly, engaging the shutter hook into the slot of the backshell. Using the hook as a hinge, carefully swing the lower part of the backshell down to the shutter assembly, engaging the shutter tab into the notch in the rolled edge of the backshell. Be careful to swing the backshell down in a straight line to make direct engagement and to ensure proper positioning of the lamp contacts on the terminals of the lamp. Replace the clamp ring, being certain to have the hinge pin set into the notches of the adapter and backshell assemblies.

Corrective Maintenance

If the compact-arc, xenon lamp should fail to light in 15 to 30 seconds, the control unit will automatically shut off the power to the lamp, and starting cannot be repeated until the power switch is turned OFF for about 2 minutes to allow the terminal delay relay to reset. Lock the shutters open, turn the power switch ON, and look for irregular arcing between the lamp electrodes or behind the lamp when viewed through the unplated reflector glass. If there is arcing at either place, the lamp has failed and should be replaced.

If the lamp should fail to light, Inspect the cable and plug connections to check the power to the lamp housing. If the cable connections are satisfactory, remove the plug from the receptacle (fig. 4-14) and with the power switch turned ON, check the voltage across terminals 3 and 4.

Normal operation of the control unit is indicated by voltage readings across terminals 3 and 4 of 155-160 volts at start and then reducing to 40-45 volts for normal operation.

If a zero voltage reading is obtained across terminals 3 and 4, the contacts of the thermal time-delay relay, RK, are probably open or resistor R may be open circuited.

If the voltage across terminals 3 and 4 remains at 155-160 volts and does not reduce to 40-45 volts, the

time-delay relay, RT, or control relay, RC, are probably Inoperative.

If the arc of the lamp is unstable and goes out and will not restart automatically, relay RP is probably inoperative. However, the voltage obtained across terminals 3 and 4 will be 155-160 volts and then reduce to 40-45 volts. The lamp can be relighted in an emergency by turning the power switch OFF and then ON again.

If the control unit output voltage is normal and no arcing is visible in the lamp, the starting unit is defective and must be replaced.

If the lamp falls to light and the voltage across terminals 3 and 4 does not shut off automatically, check the continuity of the heater unit (black leads) of the thermal delay relay, RK. The unit is probably burned out.

In conducting these tests, remember that the control automatically shuts off the power to the lamp housing if the lamp fails to light. The power switch must be turned OFF, and the thermal time-delay relay, RK, allowed to cool until its contacts reclose. This action must occur before again turning ON the power switch.

QUIZ

1. What are the two general uses of naval searchlights?

2. What are the four principal components of a carbon arc searchlight?

3. What are the two principal functions of the iris shutter?

4. What is the function of the line inscribed on the ground-glass arc-image screen?

5. What are two purposes of the arc light ventilating system?

6. What is the magnitude of the arc current and voltage for normal operation?

7. Which carbon (positive or negative) forms a hollow crater that contains an intensely luminous ball of gas ?

8. Which carbon (positive or negative) is rotated continuously about its axis as it is fed forward?

9. What method is used to actuate the thermostatic system of the carbon-arc lamp?

10. Which component in the lamp control circuit (fig. 4-6) provides automatic control of the negative carbon feed?

11. What type of power is supplied to the 24-inch carbon-arc searchlight through the searchlight switchboard?

12. What action should be taken when the positive carbon tip is observed to project less than one-half inch beyond the positive nose (obturator)?

13. Why should the positive and negative heads be reamed before each recarboning operation?

14. Why should the searchlight be operated for about 5 minutes after renewing the carbons?

15. How often should the interior of the drum and the lamp mechanism of the carbon-arc searchlight be cleaned?

16. How is the arc current of the 24-inch carbon-arc searchlight increased (fig. 4-6)?

17. What is the effect on the positive carbon projection of moving the thermostat lens frame of the 24-inch carbon-arc searchlight toward the support column?

18. How should the image of the tip of the positive carbon appear in relation to the focal line on the arc-image screen when the carbon-arc searchlight is in proper focus?

19. What should be the approximate resistance of the feed and retract coils of the magnetic clutch associated with the negative carbon feed mechanism?

20. What is the relative magnitude of the arc current when the arc breaks repeatedly?

21. What precautions should be taken in addition to determining that the circuits are deenergized before opening the front or rear door of a 24-inch carbon-arc searchlight?

22. What is the primary purpose of the 12-inch incandescent searchlight?

23. What is the power rating of the incandescent lamp used in the 12-inch incandescent searchlight?

24. What is the approximate operating pressure of the 1000-watt mercury-xenon lamp that was designed to replace the 1000-watt incandescent lamp in the 12-inch searchlight?

25. What are the essential differences in the mercury-xenon lamp with respect to (a) starting and (b) running as compared to the 1000-watt incandescent lamp?

26. What is the magnitude of the voltage required to (a) start the mercury xenon lamp and (b) operate the lamp?

27. What is the function of booster transformer, BT (fig. 4-12)?

28. If the mercury-xenon lamp is allowed to operate with the starting knob (fig. 4-12) turned fully counterclockwise for a very brief period after the arc is struck, what will happen?

29. When assembling or disassembling the mercury-xenon lamp, what precautions must be taken?

30. When the mercury-xenon searchlight is properly focused for long-range operation, what is the relative divergence of the beam?

31. (a) Because the mercury-xenon lamp contains gas under considerable pressure, what serious action may occur if it is dropped or struck? (b) Lamps that have reached the end of their useful life are potentially dangerous and must be disposed of in what manner?

32. For what primary purpose is the 8-inch sealed beam searchlight used?

33. (a) What is the purpose of the high voltage, r-f cur rent developed in the 8-inch sealed beam starting unit? (b) What is the purpose of the ballast resistor in the control unit?

34. What magnitude of voltage across terminals 3-4 (fig. 4-15) of the control units of the 8-inch sealed beam searchlight should be developed (a) at start and (b) then for normal operation?

QO 1

CHAPTER

5

MAINTENANCE OF MOTORS AND GENERATORS

INTRODUCTION

This chapter describes the proper procedures for the maintenance of motors and generators. Additional information on this subject is contained in Chapter 60 of the Bureau of Ships Manual.

Proper maintenance is equally as important as proper installation and operation in ensuring long, satisfactory service of electric motors and generators. The EM should remember that the most important factor in the maintenance of electric motors and generators is to keep the equipment clean and free of oil, water, dirt, and other foreign particles. Oil vapor, metallic dust, and moisture are present in the air inside a ship. These elements are especially prevalent in the air of machinery spaces where major electric motors and generators are located. All machines require ventilation to dissipate the heat generated by their operation, and, since most motors and generators utilize the air within the compartments in which they are located for this purpose, it is apparent that a considerable amount of these destructive elements will accumulate on, and inside, these machines within a relatively short time. Motors and generators must be cleaned frequently, both internally and externally, and particular care must be taken to keep all air ducts clean.

CLEANING MOTORS AND GENERATORS

The four acceptable methods of cleaning motors and generators are wiping, use of suction, use of compressed air, and use of a solvent.

Wiping with a clean, lint-free, dry rag (such as cheese cloth) is effective for removing loose dust or foreign particles from accessible parts of a machine. When wiping, do not neglect such parts as the end winding, mica cone extension at the commutator, slip-ring insulation, connecting leads, etc.

The use of suction is preferred to the use of compressed air for removing abrasive dust and particles from inaccessible parts of a machine because it lessens the possibility of damage to insulation. If a vacuum cleaner is not available for this purpose, a flexible tube attached to the suction side of a portable blower will serve as a satisfactory substitute. Always exhaust the blower to a suitable sump or overboard when used for this purpose. Grit, iron dust, and copper particles should be removed only by suction methods whenever possible.

Compressed air must be clean and dry when used for cleaning electrical equipment. Air pressure up to 30 pounds per square inch may be used on motors or generators of 50 hp or 50 kw respectively, or less. Pressures up to 75 pounds per square inch may be used to blow out machines that are over 50 hp or 50 kw. A throttling valve should be used on air lines that carry higher pressure than is suitable for blowing out a machine. Before the air blast is turned on the machine, any accumulation of water in the air pipe or hose must be thoroughly blown out, and both ends of the machine must be opened to allow a path of escape for the air and dust.

The use of solvents for cleaning electrical equipment should be avoided whenever possible. However, their use is necessary for removing grease and pasty substances consisting of oil and carbon or dirt. Alcohol will injure most types of insulating varnishes and should not be used for cleaning electrical equipment. Solvents containing highly volatile gasoline or benzine must not be used on board ship for cleaning purposes under any circumstances.

GENERAL MAINTENANCE

When performing any operation on electric motors or generators that produces dust, grit, or shavings, use every precaution to protect all windings and vent spaces from foreign particles. Stationary coils should be protected by a guard, and the armature should be fitted with a canvas, bound on the commutator and armature surfaces in a manner to prevent the entry of dust, grit, etc., into the interior of the machine. Vent spaces under the commutator should be stuffed with rags. After the work is completed, all exposed surfaces must be cleaned and the rags, guards, etc., removed.

Small pieces of iron, bolts, and tools must be kept away from running motors and generators. During soldering operations, do not allow drops of solder to get into the windings. Excess solder, which may later break off, should be removed from the soldered joints. Bolts and mechanical fastenings on both the stationary and rotating members should be checked at regular intervals to make sure that they are tight. However, commutator clamping bolts on d-c machines must not be disturbed, because any interference with them may result in misalignment of the segments and make it necessary to turn or grind the commutator to restore it to service. When an inspection of a rotor reveals loose keys or other fastenings, a further check should be made for evidence of damage due to such looseness.

When a generator is to be inoperative for an appreciable length of time, such as may be necessary for an overhaul of the prime mover, the brushes should be lifted from the collector rings or commutator to prevent an electrolytic action between the brushes and rings or segments. The collector rings and commutator may be covered with grade A paper to prevent corrosion.

At frequent intervals, all electrical connections should be inspected to make sure they are tight. Locknuts, lock-washers, or other similar means should be used to lock connections, which tend to become loose because of vibrations. When electrical connections are opened, all oil and dirt should be cleaned from the contact surfaces before they are reconnected. Contact surfaces of uncoated copper hould be sandpapered and cleaned immediately before

joining. However, sandpaper must not be used on contact surfaces that are silver-plated. They should be cleaned with silver polish. Steel bolts and nuts that are used for making electrical connections should be zinc or cadmium-plated. Make certain that exposed electrical connections are adequately insulated to protect against moisture,and injury to personnel.

When it becomes necessary to disassemble and reassemble a motor or generator, the EM should follow the procedure outlined in the manufacturer's instruction book, exercising care to prevent damage to any part of the machine. The machine rotors (a-c revolving field or d-c armature) should be supported, while being moved or when stationary, by slings or blocking under the shaft or by a padded cradle or thickly folded canvas under the core laminations. To lift the rotor, rope slings (separated by a spreader to prevent the slings coming in contact with the a-c rotor or d-c armature coils) should be placed under the shaft, clear of the journals. If construction of the shaft provides no room for a sling except around the journals, they must be protected with heavy paper or canvas before applying the sling. When the whole unit (stator and rotor) is to be lifted by lifting the stator, the bottom of the air gap must be tightly shimmed unless both ends of the shaft are supported in bearings. It is easily possible, by rough handling or careless use of bars or hooks, to do more damage to a machine during disassembly and assembly than it will receive in years of normal service.

Excessive vibration in electric motors and generators may be caused by an unbalanced rotating part (d-c armature, a-c revolving field, cooling fan, etc.). Therefore, when it is necessary to repair a rotating part in such a manner as to remove or add weight at any point around its periphery, the part must be tested for balance. An unbalanced condition can be corrected by dynamically balancing the part. In dynamic balancing, the test for unbalance is made, with the part turning at normal speed, by determining the exact lengthwise location where whip is maximum, and by spotting the exact radial point at which excessive weight appears. Balancing is effected by removing weight at this point or by the addition of counterweight at a point directly opposite the point of excessive weight. Dynamic balancing requires special

equipment operated by specially trained personnel. Repair ships and shipyards are equipped to perform this task.

SLEEVE BEARINGS

Bearings are of two types: SLIDING and ROLLING. The three types of sliding bearings are the: right line bearing, in which the motion is parallel to the elements of the sliding surface; journal bearing, in which two machine parts rotate relatively to each other; and thrust bearings, in which any force acting in the direction of the shaft axis is taken up. The type of sliding bearing employed in electric motors and generators is the journal bearing commonly called the SLEEVE bearing. These bearings are used in large equipment, such as turbine-driven ship's service generators and propulsion generators and motors. The bearings may be made of bronze, babbitt, or steel-backed babbitt.

Preventive maintenance of sleeve bearings requires periodic inspections of bearing wear, and lubrication.

Wear

Some generators and motors equipped with sleeve bearings are also provided with a gage for measuring bearing wear. Bearing wear on sleeve-bearing machinery not provided with a bearing gage can be obtained by measuring the air gap.

Air gaps should be measured at each end of the machine and can be measured with an air-gap feeler (a machinist's tapered feeler gage) of sufficient length to reach into the air gap without removing the end brackets of the machine. Before making the measurements, clean the varnish from a spot on a pole or tooth of the rotor. Also clean a spot at the same relative position on each field pole of a d-c machine and at least three and preferably four or more spots spaced at equal intervals around the circumference on the stator of a-c machines. Take the air-gap measurements between the cleaned spot on the rotor and the cleaned spots on the field poles or stator, turning the rotor to bring the cleaned spots opposite each other.

On machines with only one end readily accessible, two alternate methods of measuring air gaps are allowed. If possible, take measurements at the more accessible end in accordance with the aforementioned procedure and two measurements approximately 90° apart at the less accessible end. If these two measurements do not differ greatly from the mean of the measurements at the other end, the air gap is reasonably uniform at both ends.

If the above procedure is not possible, average the measurements taken from the more accessible end of the machine in order to obtain the average air gap, or obtain the average air gap from records of preceding measurements. The average air gap is one-half the difference between the inside diameter of the stator or field pole arrangement and the outside diameter of the rotor or armature; this measurement is not affected by bearing wear. Then procure a long, narrow strip of steel of a thickness twenty percent less than the average air gap for average air gaps smaller than 0.040 inch, and ten percent less for average air gaps larger than 0.040 inch. Test the air gap by trying to push the steel strip all the way from one end of the air gap to the other. Make the test at four points 90° apart, or if this is not possible, at not less than three points spaced approximately equal distances around the circumference of the machine. Care must be taken not to damage insulation when inserting the steel strip.

Uniformity of air gaps at all points is more important than to have the gap conform exactly to specified limits. If the air gap at any of the places tested differs from the average by more than twenty percent for average air gaps smaller than 0.040 inch, or by more than ten percent for average air gaps larger than 0.040 inch, the bearings should be realigned, repaired, or replaced. Within these limits, an inequality of air gaps will not usually cause unsatisfactory operation; however, if trouble is experienced, such as vibration or excessive noise which cannot be traced to any other cause than unequal air gaps, it is advisable to realign, repair, or replace the bearings as necessary in order to equalize the air gaps.

Measurements of bearing wear are taken and recorded periodically in accordance with the section of this chapter dealing with periodic tests and inspections.

Lubrication

The purpose of lubrication is to reduce friction and wear. In sleeve bearings complete lubrication exists when the surfaces of the bearing are separated by a fluid film providing the lowest friction factor. Because the oil film must be continuously replenished to compensate for end losses, these bearings are either ring-oiled or supplied with oil under pressure. A cross section of a typical ring-oiled bearing is shown in figure 5-1. Every precaution must be taken to ensure that the oil and bearings are kept clean and free from water or foreign particles.

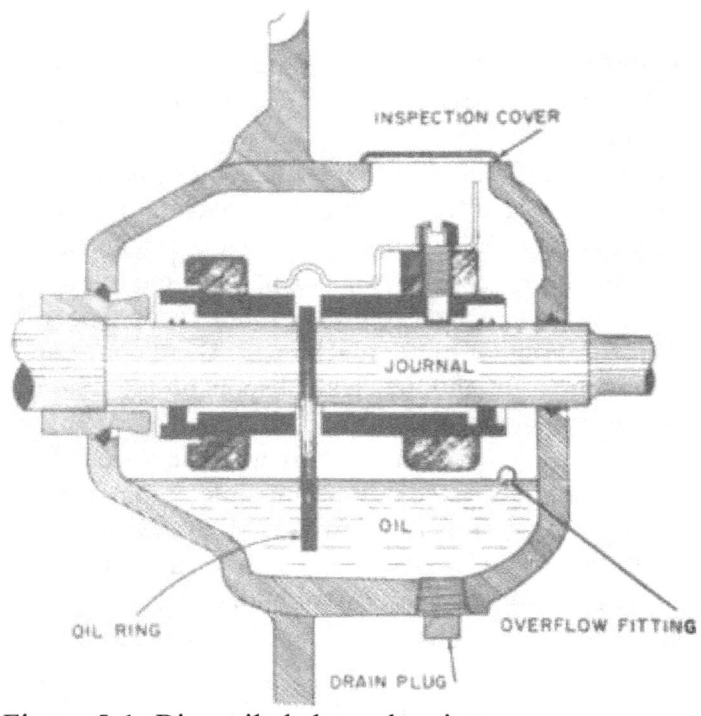

Figure 5-1.-Ring-oiled sleeve bearing.

Sleeve bearings may be equipped with an overflow gage or an oil-filler gage for determining the proper oil level to be maintained. Bearings having an overflow gage are filled until the oil is approximately one-sixteenth inch from the top of the gage; while the machine equipped with an oil-filler gage is filled to the manufacturer's oil level mark, or, in the absence of a mark, the gage should be between two-thirds and three-quarters full at all times. The gage glass and piping must be clean so that the glass will not give false indications of the oil level. If the bearing is not equipped with an overflow gage or an oil-filler gage, it is filled to a level such that the oil ring dips into the oil at a depth of about half the diameter of the shaft. Oil must not be added while the machine is running because this may allow oil mist or spray to escape from the bearing housing and settle on the machine windings.

Frequent inspections of running machines are necessary to ensure that the bearings and bearing housings are in good condition and that oil sight gages and piping connections are tight so that there will be no loss of oil. Much of the oil that leaks out of a bearing will be drawn into the machine by the cooling air and sprayed onto the windings, causing oil-soaked dirt to collect. If this condition is allowed to exist, it will result in insulation failure. When oil leakage is suspected, it can be detected by carefully cleaning and chalking the shaft outside the bearing housing, the outside of the bearing housing, and the parts of the rotor or armature adjacent to the housing. After a short run, discoloration of the chalk will occur if the machine is throwing oil. However, this test will not be dependable unless the parts are perfectly clean before being chalked. Leakage in this area will indicate that the labyrinth seal is ineffective and requires correction.

Some bearing housings are provided with vents, which are another possible source of leakage. These vents require regular inspections to make sure they are not stopped up. They must terminate in still air of atmospheric pressure where no current of air over the vent will suck oil out of the bearing housing or oil vapor out of the vent into the machine. If such oil leakage occurs, it is generally due to overfilling the bearing or trying to fill the bearing through the vent.

Bearing oil should be renewed semiannually. The used oil is drained off by removing the oil drain plug, and the bearing is flushed with clean oil until the drained oil flows clean.

Oil Rings and Bearing Surfaces

An opening Is provided In the top of the bearing for checking the condition of the oil rings and bearing surfaces (fig. 5-1). Periodic Inspections are necessary to make certain that the oil ring Is rotating freely when the machine Is running and Is not sticking due to the motion of the ship. At the same time, the bearing surfaces should be Inspected for any signs of pitting or scoring.

Trouble Analysis

Earliest Indication of sleeve bearing malfunction normally appears as an Increase In the operating temperature of the bearing. This Increase is due to the heat generated In the extremely sensitive oil film separating the rotating Journal from the bearing surface and is governed by the slipping or shearing of the oil molecules. When the bearing is operating satisfactorily, the film temperature remains stable, but when the equilibrium of the oil film is Interrupted, rapid and sharp changes occur in the temperature of the oil film. Oil is supplied under pressure from a central system to sleeve bearings in some large machinery, such as turbine-driven ship's service generators.

Thermometers are inserted in the discharge line from the bearing as a means of visually indicating the temperature of the oil as it leaves the bearing. Thermometer readings are taken hourly on running machinery by operating personnel. However, a large number of bearing casualties have occurred in which no temperature rise was detected in thermometer readings, and, in some cases, discharge oil temperature has actually decreased. Therefore, after checking the temperature at the thermometer, a follow-up check should be made by feeling the bearing housing whenever possible. Operating personnel must thoroughly familiarize themselves with the normal operating temperature of each bearing so that they will be able to recognize any sudden or sharp changes in bearing-oil temperature. If bearing malfunction is indicated, the affected machinery should be secured as soon as possible.

Any unusual noise in operating machinery may also Indicate bearing malfunction. Whenever a strange noise

is heard in the vicinity of operating machinery, a thorough inspection must be made to determine its cause. Excessive vibration will occur in operating machinery with faulty bearings, and inspections should be made at frequent intervals in order to detect its presence as soon as possible.

Corrective maintenance of sleeve bearings is accomplished by personnel of machinery ratings aboard ship or by shipyard personnel. Usually, repairs of this type are of such magnitude as to require lengthy shipyard availability for accomplishment; hence, the importance of practicing proper preventive maintenance of sleeve bearings cannot be overemphasized.

BALL BEARINGS

Rolling, antifriction bearings are of two types: ball and roller bearings. Basically, all rolling bearings consist of two hardened steel rings, hardened steel rollers or balls, and separators. The annular, ring-shaped, ball bearing is the type of rolling bearing used most extensively in the construction of electric motors and generators used in the Navy. This bearing is further divided into three types dependant upon the load it is designed to bear— (1) radial, (2) angular contact, and (3) thrust. Examples of these three bearings are shown in figure 5-2.

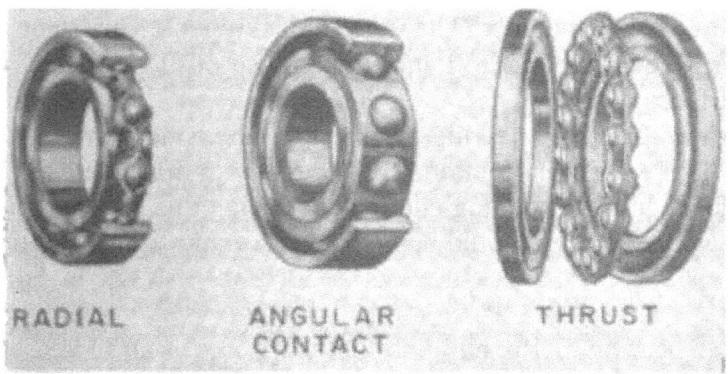

RADIAL   ANGULAR CONTACT   THRUST

Figure 5-Z-Typical ball bearings.

The rotating element of an electric motor or generator may subject a ball bearing to any one or a combination of three loads—radial, thrust, and angular. Radial loads are the result of forces applied to the bearing perpendicular to the shaft; thrust loads are the result of forces applied to the bearing parallel to the shaft; and angular loads are the result of a combination of radial and thrust loads. Because the load carried by the bearings in electric motors and generators is almost entirely due to the weight of the rotating element, it is apparent that the method of mounting the unit is a major factor in determining the type of bearing employed in its construction. In a vertically mounted unit, the thrust bearing would be used; while the radial bearing is common to most horizontal units.

The preventive maintenance of ball bearings requires periodic checks of bearing wear, and adequate lubrication.

Wear

Measuring air gaps to determine bearing wear is not necessary on machines equipped with ball bearings because the construction of the machine is such as to ensure bearing alignment. Ball bearing wear of sufficient magnitude as to be readily detected by air-gap measurements would be more than enough to cause unsatisfactory operation of the machine.

The easiest way of determining the extent of wear in these bearings is to periodically feel the bearing housing while the machine is running to detect any signs of overheating or excessive vibration, and to listen to the bearing for the presence of unusual noise. The Indications thus obtained are comparative, and caution must be exercised in their analysis.

When testing for overheating, the normal running temperature of the bearing must be known before the test can be reliable. Rapid heating of a bearing is Indicative of danger. While a bearing temperature uncomfortable to the hand may be a sign of dangerous overheating, it is not always so. The bearing may be all right if It has taken an hour or more to reach that temperature; whereas, serious trouble can be expected if that same temperature Is reached within the first 10 or 15 minutes of operation.

The test for excessive vibration relies to a great extent on the experience of the person conducting the test.

He should be thoroughly familiar with the normal vibration of the machine in order to be able to correctly detect, identify, and interpret any unusual vibrations. Vibration, like heat and sound, is easily telegraphed, and a thorough search is generally required to locate its source and to determine its cause.

Ball bearings are inherently more noisy in normal operation than sleeve bearings, and this fact must be borne in mind by personnel testing for the presence of abnormal noise in the bearing. A good method for sound testing is to place one end of a screwdriver or steel rod against the bearing housing and the other end against the ear. If a loud, irregular grinding, clicking, or scraping noise is heard, trouble is indicated. As before, the degree of reliance in the results of this test depends on the experience of the person conducting the test.

The one sure method of checking ball bearing wear is also the most difficult. In this test, the bearing caps or other covers provided are removed and the actual condition of the bearing is observed. Each ball bearing should be inspected in this manner at least every two years. The condition of the lubricant in the bearing may be checked at this time; however, no attempt must be made to disassemble double-shielded or sealed-for-life bearings.

Lubrication

Some electric motors and generators are equipped with permanently lubricated ball bearings, which are lubricated by the manufacturer and require no additional lubrication throughout their life. Equipment furnished with this type bearing is not provided with grease fittings or any provision for attaching grease fittings. Name-plates reading DO NOT LUBRICATE should be attached to the housing of all permanently lubricated bearings. A permanently lubricated ball bearing is shown in figure 5-3. Note the absence of grease fittings on the motor.

Ball bearings other than the permanently lubricated type require periodic lubrication with grease or oil. Motors and generators using grease-lubricated bearings are equipped with grease cups attached to the bearing housing to provide a means for adding grease to the bearing.

Figure 5-3.-Motor equipped with permanently lubricated ball bearings.

Grease cups may already be installed on the equipment when it is received aboard ship or they may have been removed and replaced with pipe plugs. In the latter case, the grease cups are delivered with the onboard repair parts or special tools. The parts of a grease-lubricated ball bearing installation are shown in figure 5-4. Whenever feasible, it is recommended that grease cups be attached to electric motors and generators only when grease is being added to the bearing. However, when the grease cup is removed it must be immediately replaced with a suitable pipe plug in order to prevent the entry of foreign particles. This procedure affords a particularly effective means of preventing overgreasing of the bearing by unauthorized personnel.

The frequency with which grease must be added to ball bearings depends on the service of the machine and the tightness of the housing seals, and is determined by the Engineer Officer. Usually, the addition of grease will not be necessary more often than once every 6 months. Overgreasing has been a major cause of bearing failure and must be avoided. In a bearing housing too full of lubricant, the churning of the grease generates heat, causing the grease to separate into oil and abrasive particles. The grease then becomes increasingly sticky and seals the bearing against fresh lubricant until the

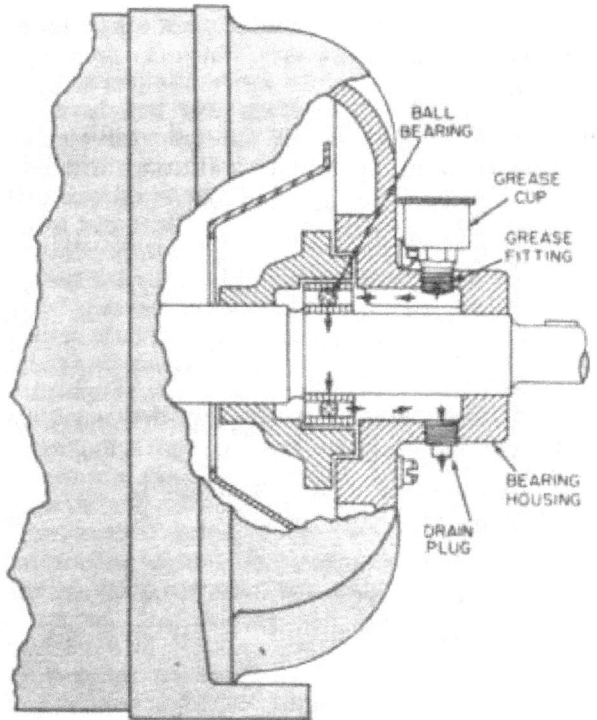

Figurt 5-4.-Grease-lubricated ball bearing.

resulting friction, heat, and wear cause failure of the bearing.

The procedure for adding grease to a ball bearing is to first clean the outside of the grease cup, grease fitting, and drain plug. Then remove the drain plug and clear the drain by probing with a small screwdriver or similar implement. Remove and empty the grease cup, clean it thoroughly, and fill It not more than half full of the proper type grease. Empty and clean out the grease fitting down to the neck, then fill it with clean grease. Replace the grease cup and screw it down as far as it will go. This will protect the machine against overgreasing due to accidental or unauthorized turning of the grease cup if the grease fitting is not replaced with a pipe plug. Run the

machine and let grease run out of the drain hole until drainage stops. This normally takes about 30 minutes. Then, replace the drain plug.

The preferred method for renewing grease Is by disassembling the bearing housing. For bearings with outer bearing caps, first clean all the exterior surfaces and remove the outer bearing cap. Remove the old grease from the accessible portions of the bearing housing and clean these parts thoroughly, taking care not to allow the entry of foreign particles into the housing. Flush out the bearing cap with clean kerosene or diesel fuel oil preheated to about 120° F. Then flush out the cap with a light mineral oil no heavier than S. A.E. 10. Follow this same procedure in flushing out the bearing housing only where It is possible to positively prevent the cleaning liquids from leaking into the machine windings. When the cleaning liquids have thoroughly drained out, pack the housing half full with fresh, clean grease, and assemble the housing.

When conditions do not allow for even partial disassembly of the bearing housing, grease maybe renewed without diassembling the housing if the motor is horizontal, a grease cup is provided for admitting the grease, the housing drain hole is accessible, and the machine is capable of being run continuously while renewing the grease. If all of the above provisions are met, proceed by running the machine after wiping clean all exterior surfaces. When the bearing has warmed up, remove the drain plug and drain piping and clear the drain hole of hardened grease. Remove the grease cup and clear the grease inlet. Clean the grease cup and after filling it with clean, fresh grease, screw it down as far as it will go. Keep the machine running continuously and the drain hole cleared of hardened grease. Stop adding grease when clean grease begins to emerge from the drain hole, and allow the machine to run until the drainage of grease stops. Clean and replace the drain piping and plug. During this operation, every precaution must be taken to prevent the grease from reaching the electric windings of the machine. A probable Indication of excessive leakage inside the machine is the emergence of a quantity of grease around the shaft extension end of the bearing.

A grease gun should not be used to lubricate ball bearings unless there are no other means available. If used,

remove the drain plug and apply just enough pressure to the gun to get the grease into the housing. Grease-gun fittings must be removed from the machine immediately after use, and replaced with a pipe plug.

Some electric motors and generators may be equipped with oil-lubricated ball bearings. Lubrication charts or special instructions are generally furnished for this type of bearing and should be carefully followed by personnel maintaining the equipment. In the absence of other instructions, the oil level inside the bearing housing should be maintained approximately level with the lowest point of the bearing inner ring. This will provide enough oil to lubricate the bearing for a considerable operating period, but not enough to cause churning or overheating.

One common method by which the oil level is maintained in ball bearings is the wick-fed method. In this method, the oil is fed from an oil cup to the inside of the bearing housing through an absorbent wick, which also filters the oil and prevents leakage through the cup in the event momentary pressure is built up within the housing. A typical wick-fed, oil-lubricated ball bearing is shown in figure 5-5.

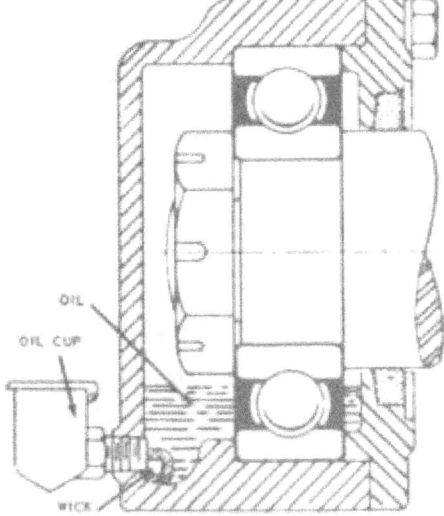

Figure 5-5.-Wick-fed ball bearing.

Corrective Maintenance

A damaged ball bearing may not necessarily be inoperative. However, if the damage continues to the extent that the bearing is no longer suitable for its intended application, it can be considered to have failed. Bearing failure may be indicated by an increased level of noise, vibration, or temperature. Damage of a bearing in service is often a result of a combination of two or more factors—the more prevalent being abuse before and during mounting, improper lubrication, wear from abrasive dirt, corrosion, passage of electric current, and fatigue.

Repair of damaged ball bearings by replacing balls should never be attempted. Cleaning ball bearings should be done only in an emergency and when a suitable replacement is not available. Even when cleaning is carefully done, more dirt may get into the bearing than is removed.

Removal of the bearing from the shaft involves the risk of damage to the shaft, or bearing, or both. A bearing puller applied to the inner race or to a sleeve, which applies pressure to the inner race, should be used to remove the bearing from the shaft. Removal by pulling on the outer race tends to make the balls dent the raceway. Care must be taken to prevent damage to the shaft when removing the bearing.

When installing bearings, use extreme care to prevent dirt or other foreign particles from entering the bearing and bearing housing. A new bearing should not be removed from its original container and wrapping until every preparation has been made to install it. If it is not possible to use a press, a drift pipe of soft steel or malleable iron may be used to mount the ball bearing on the shaft. The inside diameter of the pipe must be large enough to clear the shaft or any locknut threads, and its outer diameter must be no larger than the maximum diameter of the bearing's inner race. Mount the bearing square with the shaft; then, with the pipe fitted squarely against the inner race, lightly tap the pipe with a clean, metal hammer until the inner race is seated tightly against the shaft shoulder.

The thoroughly cleaned bearing housing must be packed half full of clean, new grease and assembled. During reassembly, care must be taken not to omit bearing parts, lubricant seals, grease pipes, plugs, and fittings.

On some generators, the outboard bearing is electrically insulated from the frame to prevent the flow of shaft currents through the bearing. Insulation may be accomplished by means of a shell of insulating material installed between the bearing shell and the bearing housing, or by the use of insulating shims under the pedestal and insulated holding-down bolts, and dowels. Currents caused by the electromotive force generated in the shaft and structural members of a generator, if of sufficient magnitude, will rapidly ruin a bearing. Therefore, make certain that the bearing insulation is not damaged or that conducting paths around the insulation are not inadvertently provided.

BRUSHES

The brushes used in electric motors and generators are one or more plates of carbon bearing against a commutator, or collector ring (slip ring) to provide a passage for electrical current for an external circuit. The brushes are held in position by brush holders mounted on studs or brackets attached to the brush-mounting ring, or yoke. The brush-holder studs or brackets, and brush-mounting ring comprise the brush rigging. The brush rigging is Insulated from, but attached to, the frame of the machine. Flexible leads (pigtails) are used to connect the brushes to the terminals of the external circuit. An adjustable spring is provided to maintain proper pressure of the brush on the commutator in order to effect good commutation. A d-c generator brush holder and brush-rigging assembly are shown In figure 5-6.

Brushes are manufactured In different grades to meet the requirements of the varied types of service. The properties of resistance, ampere-carrying capacity, coefficient of friction, and hardness of the brush are determined by the maximum allowable speed and load of the machine in which it is used. Only the grade of brush recommended by the manufacturer should be used in a machine. The brush grade is shown on the plan of the machine and in the instruction book.

Care

If the correct grade of brushes is used, and the brushes are correctly adjusted and cared for, good commutation

Figure 5-6.-Brush holder and brush-rigging assembly.

will result. Periodic inspections of the brushes and brush rigging are required to ascertain their condition. The brush pigtails must be securely connected at the brushes and terminals. Brushes should move freely in the holders, but must not be loose enough to vibrate. - They should be replaced when they are worn down to half their original length or if chipping has occurred at the corners or edges of the brush. The brush holders and brush rigging should be cleaned before inserting the new brushes. The brush holders should be mounted so that the edges nearest the commutator are the same distance from the commutator (not more than one-eighth inch, nor less than one-sixteenth inch). The leading edges (toes) of all the brushes on each stud must alignment with each other and one commutator segment.

When properly mounted, the brushes will be evenly spaced around the commutator. To check the spacing, a strip of clean paper is wrapped around the commutator and marked where it laps. Then, the paper is removed from the commutator, cut at the lap, and folded or marked into as many equal parts as there are brush studs. Finally, the paper is replaced on the commutator, and the brush holders are adjusted so that the brush toes are at the creases or marks.

The pitting effect on the commutator differs under the positive and negative brushes, making it necessary to stagger the brushes in order to prevent grooving of the commutator, as illustrated in figure 5-7. The positive and negative brushes are staggered in pairs so that the differences in pitting effect are distributed equally over the full brush-contact area of the commutator surface (fig. 5-7, A). In a machine having an odd number of pairs, it is impossible to stagger all the brushes in this manner. In this machine, the brushes are staggered as before; except that the brushes of the odd pairs are staggered separately (fig. 5-7, B).

♦ =3 u u ozn

A

O0D> PAIR!

B

Figure 5-7.-Method of staggering brushes.

As the brushes wear, the brush spring tension must be changed to keep the brush pressure approximately

constant. On some machines the design of the brush holder and spring allow for changing the brush-spring setting. Unless it is stated otherwise in the instruction book for the machine, the proper brush pressure should be between 1-1/2 to 2 pounds per square inch of brush-contact area. The brush pressure is easily measured. Attach a small spring balance to the pigtail end of uie brush, insert one end of a strip of paper between the brush and the commutator, then exert a pull on the spring balance in the direction of the brush-holder axis (fig. 5-8). Note the reading of the spring balance when the pull is barely sufficient to release the paper so that it can be pulled from between the brush and commutator without offering resistance. Then divide this reading by the contact area of the brush to obtain the brush pressure.

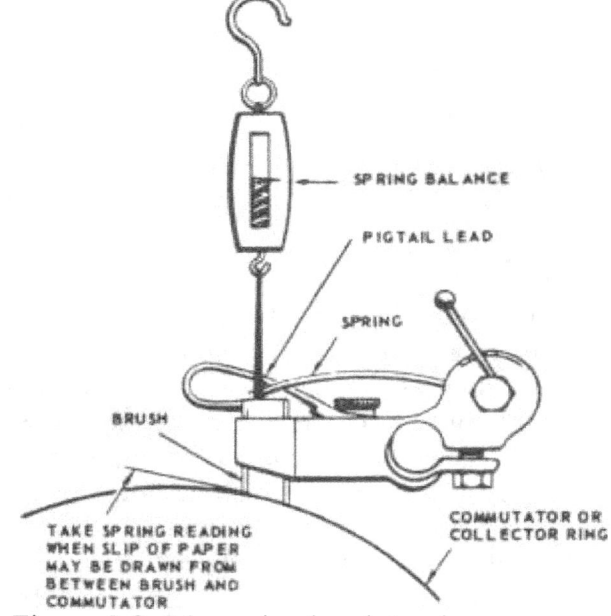

Figure 5-8.-Measuring brush tension.

Setting On Neutral

When a machine is running without load and with only the main-pole field windings excited, the point on the commutator at which minimum voltage is induced between adjacent commutator bars is the no-load neutral point. This is the best operating position of the brushes on most commutating-pole machines. Usually, the brush studs are doweled in the proper position, and the correct setting is indicated on a stationary part of the machine by a chisel mark or an arrow. In some cases, commutation may be improved by shifting the brushes slightly from the marked position.

Two methods of finding the neutral position are the mechanical and reversed rotation. In the mechanical method, the commutator is turned until the two coll sides of the same armature coil are equidistant from the cen-terline of one main-field pole. The position of the commutator bars to which the coil is connected will indicate the approximate mechanical neutral.

Use of the reversed rotation method is possible only where it is practicable to run a machine in either direction of rotation, with rated load applied. This method differs for motors and generators. For motors, the speed of the motor is, at first, accurately measured when the field current becomes constant under full load at line voltage with the motor running in the normal direction. Then, the rotation of the motor is reversed, full load is applied, and the speed is again measured. When the brushes are shifted so that the speed of the motor is the same in both directions, the brushes will be In the neutral position. Generators are run at the same field strength and same speed in both directions, and the brushes are shifted until the full-load terminal voltage is the same for both directions of rotation. To ensure accuracy, a reliable tachometer must be used to measure the speed of the machines for this method.

Fitting

An accurate fit of the brushes must be assured where their surfaces contact the commutator. Sandpaper and a brush seater are the best tools to accomplish a true fit.

All power must be disconnected from the machine, and every precaution must be taken to ensure that the machine will not be inadvertently started before using sandpaper to seat the brushes. The brushes to be fitted are lifted, and a strip of fine sandpaper (No. 1) approximately the width of the commutator, is inserted (sand side up) between the brushes and the commutator. With the sandpaper held tightly against the commutator surface to conform with the curvature and the brushes held down by normal spring pressure, the sandpaper is pulled in the direction of normal rotation of the machine (fig. 5-0). When returning the sandpaper for another pull, the brushes must be lifted. This operation is repeated until the fit of the brush is accurate. Always finish with a finer grade of sandpaper, No. 0. A vacuum is required for removing dust while sanding. After sanding, the commutator and windings must be thoroughly cleaned to remove all carbon dust.

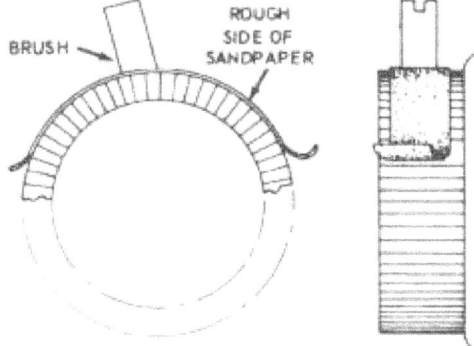

END VIEW SIDE VIEW

Figure 5-9.-Method of sanding brushes.

The brush seater is compounded of a mildly abrasive material loosely bonded, and is formed in the shape of a stick about 5 inches in length. The brush seater is applied to the commutator while the machine is running, and every precaution should be taken to prevent injury to the person applying it. The brush seater is touched lightly, for a

second or two, exactly at the heel of each brush (fig. 5-10). If placed even one-fourth inch away from the heel, only a small part of the abrasive will pass under the brush. Pressure may be applied to the brush by setting the brush spring tension at maximum or by pressing a stick of insulating material against the brush. The dust is removed during the operation, and the machine is thoroughly cleaned afterwards in the same manner as for sanding brushes.

BRUSH BRUSH BRUSH

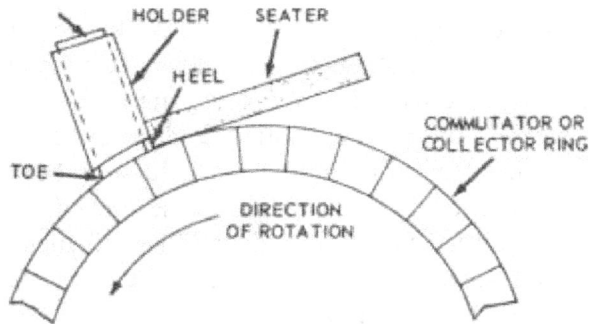

Figure 5-10.—Using the brush seater.

COMMUTATORS AND COLLECTOR RINGS

After being used approximately two weeks, the commutator of a machine should develop a uniform, glazed, dark brown color on the places where the brushes ride. If a nonuniform or bluish colored surface appears, improper commutation conditions are indicated. Periodic inspections and proper cleaning practices will keep commutator and collector-ring troubles at a minimum.

Cleaning

One of the most effective ways of cleaning the commutator or collector rings is to apply a canvas wiper while the machine is running. The wiper can be made by wrapping several layers of closely woven canvas over the end of a strong stick between one-fourth and three-eighths inch thick (fig. 5-11). The canvas may be secured with rivets if they, in turn, are covered with linen tape to

prevent the possibility of their contacting the commutator. When the outer layer of canvas becomes worn or dirty, it is removed to expose a clean layer. The wiper is most effective when used frequently. On ship's service generators, it may be desirable to use the wiper once each watch. When using the wiper, exercise care to keep from fouling moving parts of the machine. The manner of applying the wiper to a commutator is illustrated in figure 5-12.

STICK I 4 TO \'7d 6

INCH THICK RIVET CANVAS

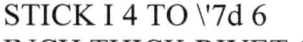

Figure 5-11.-Canvas wiper.

0

e e

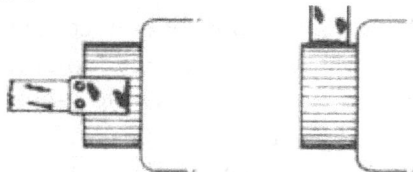

Figure 5-12.-Using the canvas wiper on a commutator.

When machines are secured, a toothbrush can be used to clean out the commutator slots, and clean canvas or lintless cloth may be used for wiping the commutator and adjacent parts. In addition to being cleaned by wiping, the commutator should be periodically cleaned with a vacuum cleaner or blown out with clean, dry air.

A fine grade of sandpaper, No. 00, may be used to clean a commutator that Is only slightly rough, but not out of true. Sandpapering is recommended for reducing high mica, and for finishing a commutator that has been ground or turned. The sandpaper, attached to a wooden block shaped to fit the curvature of the commutator, is moved slowly back and forth across the surface of the

commutator while the machine Is running at moderate speed. Rapid movement or the use of coarse sandpaper will cause scratches. Emery cloth, emery paper, or emery stone should never be used on a commutator or collector ring.

Care of Commutators

Commutators must be true within close limits. For the most satisfactory operation, runout (eccentricity) of the commutator surface (as checked on the radius with an indicator) should not exceed two mils (0.002 inch). Handstoning, grinding with a rigidly supported stone, and turning the commutator are measures that will correct some or all out-of-true conditions of the commutator.

For handstoning, the machine should be running at, or only slightly below, rated speed. For motors, remove all except enough brushes to keep the armature turning at the proper speed. The stone to be used should fit the curvature of the commutator and have a surface substantially larger than the largest flat spot to be removed. Hold the stone in the hand and move it very slowly back and forth, parallel to the axis of the commutator, applying only enough pressure to keep the stone cutting (fig. 5-13). Crowding the stone will roughen the surface. Care must be exercised to avoid electric shock and to prevent jamming the stone between fixed and moving parts of the machine.

Either a nonrotating or a revolving stone can be used when grinding the commutator with a rigidly supported stone (fig. 5-14). Irrespective of which stone is used or whether the grinding is done with the commutator within the machine or in a lathe, extreme care must be taken to align the supports so that the motion of the stone is accurately parallel to the axis of the commutator. Failure to properly align the supports will taper the commutator, and failure to maintain the support rigid will cause the stone to dig into the commutator.

For turning, the armature should be supported in a lathe (fig. 5-15). A cutting tool that is rounded sufficiently so that the cuts will overlap must be used to make the cut. Proper cutting speed is about 100 feet per minute with the feed about 0.010 inch per revolution. Depth of the cut should not exceed 0.010 inch.

Figure 5-13.-Handstoning the commutator.

The windings of the machine must be adequately protected from grit during stoning and grinding operations conducted with the commutator in the machine. Afterwards, the brushes, brush holders, and commutator must be thoroughly cleaned, and a complete insulation test should be made to determine that no grounds or short circuits exist. After the trulng-up operation has been completed, regardless of the method used, always finish with a fine grade of sandpaper, undercut the mica to a depth no greater than one-sixteenth inch, and slightly bevel the edges of the commutator bars.

When the oxide film (dark brown color) has been removed by truing operations, it can be replaced by burnishing the commutator with a hardwood block. After the end grain of the block has been shaped to the curvature of the armature, the block is pressed hard against the surface of the commutator while the machine is running. A commercial burnishing stone may also be used for this purpose. Less pressure is required in applying the stone

Figure 5-14.-Grinding the commutator with a rigidly supported stone.

because friction is greater, and the heat developed is high. Do not raise the commutator temperature above its normal operating level.

Commutator mica that has become carbonized loses its insulating value. It should be scraped out and replaced with sodium silicate or other insulating cement. Poor commutation will develop if the commutator bars are worn down to, or below, the level of the mica. The mica should be undercut to a depth of between three sixty-fourths inch and one-sixteenth inch below the level of the commutator bars. A small motor-driven, circular saw especially designed for this purpose, a slotting file having an angle of 60° between faces, or a hacksaw blade that has been ground to the right thickness and fitted to a handle may be used for undercutting mica. Before using the motor-driven circular saw, install a canvas cover around the armature in a manner to prevent copper dust from

Figure 5-15.-Truing commutator by turning.

becoming embedded In the armature windings. When undercutting has been completed, the edges of the bars should be beveled to a depth about one thirty-second inch below the surface. Finally, all mica, copper dust, and other foreign materials must be cleaned from the slots and commutator.

Care off Collector Rings

Collector rings (slip rings) require the same careful attention as the commutator. Out-of-round conditions of the rings may be corrected in the same manner as for commutators, except for the fact that crocus cloth is used to apply a mirror-like finish following any turning, grinding, or sanding operations.

Pitting can develop because of the electrolytic action on the surface of collector rings caused by current flow. It may occur in only one ring, but will be general over the whole ring area. This condition can be corrected by reversing the polarity of the rings every few days. Reversing the polarity of the d-c field of a 3-phase generator will not affect the phase rotation of the generator.

Field current must not be left on while a machine is secured because it will cause spot pitting and burning of the rings beneath the brushes.

ARMATURES

Preventive maintenance of an armature consists of periodic inspections and tests to determine its condition, and proper cleaning practices to preserve the insulation.

Frequent checks must be made of the condition of the banding wire that holds down the windings of the d-c armature to see that the wires are tight, undamaged, and have not shifted (fig. 5-16). At the same time, the clips securing the wires should be checked to see If solder has loosened. When repairs are required, banding-wire size, material, and the method of original assembly should be duplicated as far as possible. Only pure tin must be used for soldering banding wire.

BANDING WIRES

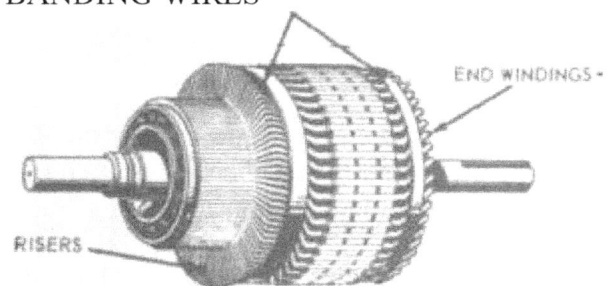

Figure 5-16.-D-c armature.

Periodically, all end windings should be Inspected and cleaned (fig. 5-16). Allow sufficient clearance between the end windings and end brackets or any air deflecting shields to prevent chafing or other damage. In cases where chafing is slight or where shop overhaul is not feasible, air-drying varnish may be applied to the windings by brush after they have been cleaned and dried.

Risers must be Inspected periodically to determine the condition of the solder that secures the windings to the segments (fig. 5-16). All dirt and lint should be removed

by thorough cleaning to ensure that cooling passages will not be clogged. It may be necessary in the case of generators to do this each time the machine is secured. Cleaning is easier when performed while the machine is warm.

Trouble Indications

Some armature troubles may be detected while making inspections of running machines. Heat and the odor of burning insulation may indicate a short-circuited armature coil. In a coil that has some turns shorted, the resistance of one turn of the coil will be very low, and the voltage generated in that turn will cause a high-current flow, resulting in excessive heating, which will cause the insulation to bum. If the armature is readily accessible, the short-circuited coil can be detected immediately after stopping the machine because the shorted coil will be much hotter than the others. In idle machines, a short-circuited coil may be identified by the presence of charred insulation.

An open armature coil in a running machine is indicated by a bright spark, which appears to pass completely around the commutator. When the segment to which the coil is connected passes under the brushes, the brushes momentarily complete the circuit; when the segment leaves the brushes, the circuit is broken, causing a spark to jump the gap. Eventually, it will definitely locate itself by scarring the commutator segment to which one end of the open coil is connected.

When a ground occurs in an armature coil of a running machine, it will cause the ground test lamps on the main switchboard to flicker on and off as the grounded coil segment passes from brush to brush during rotation of the armature. Two grounded coils result in the same effect as a short circuit across a group of coils. Overheating will occur in all of the coils in the group and bum out the winding. Grounded coils in idle machines can be detected by measuring insulation resistance. A megger, or similar insulation measuring device can be connected to the commutator and to the shaft or frame of the machine in order to properly measure the resistance of the insulation of the coils.

Locating Troubles

D-c armature troubles are usually confined to one coil or group of coils, and if not readily apparent, the segments to which they are connected can be located by a bar-to-bar test. In some cases, this test may be conducted with the armature installed in the machine. A low-voltage, d-c source, such as a storage battery, lighting circuit, or welding set, is required for this test. The machine must be disconnected from its normal power supply before the test is made and all except one pair of brushes lifted from the commutator. The voltage is applied across the + and - brush through a resistance, lamp, lamp bank, or rheostat. A low-reading voltmeter or millivoltmeter is necessary for taking measurements.

To locate a ground in a d-c armature coil, one lead of the voltmeter is connected to the shaft, and, with the armature in a fixed position, the other lead is touched to each commutator bar in turn. If there is a ground, two or more bars will indicate practically zero readings. Some of these will be real and others will be phantom grounds (fig. 5-17, A). All such bars should be marked with chalk. The armature is then rotated a few degrees and tested again. The real grounds will remain in the same bars while the phantom ones will shift to other bars (fig. 5-17, B). For example, in figure 5-17, B, the phantom ground has shifted from bar b to bare, while the real ground has remained in bar a. The ground will be in a coil connected to the bar, showing a real ground with the lowest voltage reading.

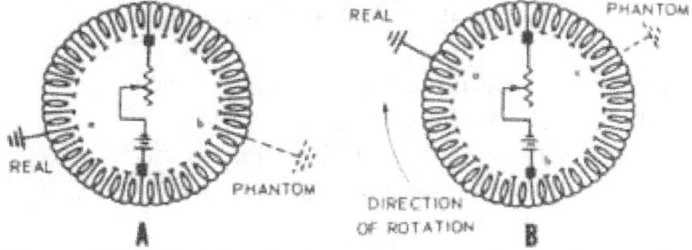

Figure 5-17.-Real and phantom grounds in a bar-to-bar test.

To locate an open or short circuit in a d-c armature, remove all brushes except those of one positive brush holder and an adjacent negative brush holder. Connect the low-voltage potential to these brushes and adjust the current, if need be, so that the readings obtained with the millivoltmeter will be roughly one-third to one-half full scale. The current must not exceed one-fourth that normally carried by one set of brushes. The voltage drop between two adjacent commutator bars is measured with a millivoltmeter. The armature is held in a fixed position, and the meter leads are moved from one pair of adjacent bars to the next until a test has been made of all the pairs of bars included between the brushes (fig. 5-18). The armature is then turned to bring different bars between the brushes, and these bars are tested. This is repeated as necessary to test all around the commutator. In a simplex winding, an open coil is located where the meter reading is a maximum and a shorted coil, where the reading is a minimum.

D-C SUPPLY

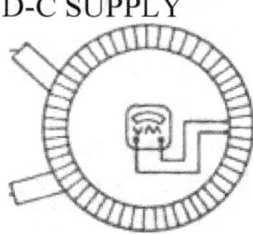

Figure 5-18.-Testing for an open coil.

For most armatures in use aboard naval vessels, the windings will be free from fault if all the voltage readings are a small fraction of the voltage between the brushes and are equal within the limits of measurement. However, in some cases, a duplex winding may be encountered. This type of winding is indicated when the readings are only a small fraction of the voltage between the brushes and follow each other in a regularly repeating pattern, such as O, R, O, R,0,R,andsoon, where R is a reading different from zero. When this happens, a further test must be made by measuring the voltage drop between alternate bars—1 and 3, 2 and 4, 3 and 5, 4 and 6, and so on. If these readings are equal within the limits of measurement, the winding will be free from faults.

When an open circuit is present, the voltmeter reading across one pair of adjacent bars will be approximately equal to the voltage between the brushes, and zero readings will be obtained on several pairs of bars on each side of the pair with the high reading. The open-circuited coil will be connected to one or both of the bars in the pair with the high reading. Should the voltmeter readings taken between adjacent pairs of bars increase or decrease in magnitude and be alternately plus and minus, a duplex winding is indicated. A further test by measuring the voltage drop between alternate bars is then necessary to locate the open circuit. When a reading approximating the voltage between the brushes is thus obtained, the open-circuited coil will be connected to one or both of the bars in the pair with the high reading.

When a short circuit is present, the interpretation of the indication given by readings between adjacent bars or between alternate bars (duplex windings) depends upon whether the armature has a lap or a wave winding. In an armature having a lap winding, a voltmeter reading considerably lower than the others will indicate a short-circuited coil is connected between the pair of bars that shows the low reading. A short-circuited coil in an armature with a wave winding will cause low readings to be obtained on as many pairs of bars as there are pairs of poles, and the short circuit will be in a coil connected to bars in these pairs.

The best method for locating the ends of a faulty coil in a wave-wound armature is to separate the coil from the rest of the winding in the following manner. In a six-pole machine, a short-circuited coil in a wave-wound armature would be indicated at three positions during the test. These positions should be marked with chalk. When the riser connections on these segments are lifted, six coils will be isolated from each other and the rest of the winding. The shorted coil is located by comparing the resistances of the six coils, and it will have less resistance than the others.

Emergency repairs can be effected by cutting out a short-circuited or open-circuited armature coil. This will permit restoration of the machine to service until permanent repairs can be made. However, permanent repairs should be made as soon as possible. The coil is

cut out by disconnecting both ends of the coil and installing a jumper between the two risers from which the coil was disconnected. The coil, itself, is then cut at both the front and rear of the armature to prevent overheating of the damaged coil. A continuity test from one end to the back of the coil will locate the turns of the faulty coil. If a pin or needle is used to puncture the Insulation for this test, insulating varnish can be used to fill the tiny hole in the event the wrong coil is pierced. All conducting surfaces exposed by the change in connections should be insulated, and all loose ends should be tied securely to prevent vibration.

A-C ROTORS

Basically, the rotors in a-c machines are of two types— the cage rotor (fig. 5-19) and the wound rotor. The cage rotor usually consists of heavy copper or aluminum bars fitted into slots in the rotor frame. These bars are connected to short-circuiting end rings by bolts or rivets and are then brazed or welded together (fig. 5-19, A). In some cases, the cage rotor is manufactured by die-casting the rotor bars, end rings, and cooling fans into one piece (fig. 5-19, B). The cage rotor requires less attention than the wound rotor. However, the cage rotor should be kept clean, and the rotor bars must be checked periodically for evidence of loose or fractured bars and localized overheating.

In the wound rotor, the uninsulated bar winding of the cage rotor is replaced with a distributed winding of preformed coils similar to those of a d-c armature. The windings are wye-connected and the ends are brought out to collector rings (fig. 5-20). Wound rotors, like other windings, require periodic inspections, tests, and cleaning. The insulation resistance of the winding may be tested with a megger to determine if grounds are present.

An open circuit in a wound rotor may cause reduced torque accompanied by a growling noise, or failure to start under load. In addition to reduced torque, a short circuit in the rotor windings may cause excessive vibration, sparking at the brushes, and uneven collector ring wear. With the brushes removed from the collector rings, a continuity check of the rotor coils will reveal the

A - WELDED ROTOR BARS AND END RINGS

B - DIE-CAST ROTOR BARS AND END RINGS

Figure 5-19.-Cage rotors.

presence of a faulty coil. Emergency repair of a faulty coil in a wound rotor may be effected in the same manner previously prescribed for cutting out a damaged armature coil.

Some single-phase, a-c motors, such as the split-phase motor, are equipped with a centrifugal switch mounted on the rotor or rotor shaft. This device functions to open the starting winding circuit when the motor has reached almost normal speed. The condition of the device must be checked periodically to determine that the switch

Figure 5-20.-Wound rotor.

contacts are clean and that all moving parts function properly. Stalling while starting or failure to start may indicate a faulty centrifugal switch. If this happens, power to the motor must be secured immediately or the starting winding will soon overheat and burn out.

FIELD COILS

Preventive maintenance of field coils requires periodic Inspections and tests to determine the condition of the coil. Coils should be cleaned periodically to remove any foreign particles, which might have collected on them.

Locating Troubles

The insulation on field coils should be tested periodically to determine its condition. A megger or similar resistance-measuring device may be used for this purpose. If a ground is detected in the field circuits (shunt, series, and Interpole) of a d-c machine, the circuits must be disconnected from each other and tested separately to locate the grounded circuit. Then all the coils in that circuit must be opened and tested separately to locate the grounded coil, which can be repaired or replaced as necessary.

If an open circuit develops in the field windings of an a-c or d-c generator that is carrying load, it will be indicated by immediate loss of load and voltage. An open in the shunt field winding of an operating d-c motor may be indicated by an increase in motor speed, excessive armature current, heavy sparking, or stalling of the motor. When an open occurs in the field circuit of a machine, it must be secured immediately and examined to locate the faulty circuit. The open circuit will usually occur at the connections between the coils and can be detected by visual inspection. An open in the coils generally causes enough damage to permit detection by visual inspection. If the faulty coil is not readily apparent, it can be located by applying a low-voltage source (dry batteries) to the terminals of the field winding and using a low-range voltmeter to measure the difference of potential between the terminals of each coil (fig. 5-21). The open-circuited coil will develop the greatest difference in potential between its terminals.

A short-circuited field coil in a machine develops an unbalanced magnetic pull and causes vibration. If the short circuit is severe, smoke or the odor of burning insulation will be present. In a generator, a shorted field coil is indicated when it becomes necessary to increase field current in order to maintain voltage with the machine running at normal speed. A machine with a shorted field

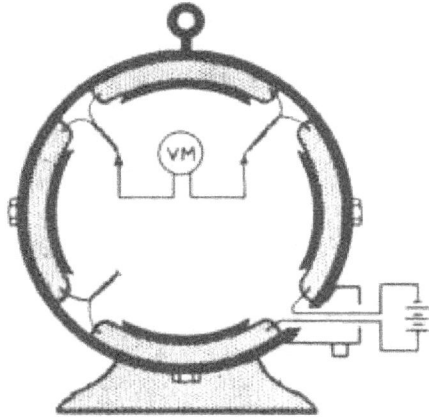

Figure 5-21 .-Testing for an open-circuited field coil.

coil must be secured immediately. The faulty coil can be located by passing normal current through the field circuit and measuring the voltage drop across each coil. The coil indicating the lowest voltage will be the shorted coil.

Replacing Coils

A field coil may be removed from a large machine without removing the armature. However, the armature should be covered with heavy paper or canvas to prevent damaging it while the coil is being removed. To remove the coil, the field windings are disconnected, and then the bolts securing the pole piece to the frame are removed. The coil and pole piece can be slid from the machine intact. Care must be taken not to lose or misplace any of the shims found on the pole piece.

Before installing a new or repaired coil, its polarity must be determined and it should be tested for shorts, opens, and grounds. A small magnetic compass may be used to determine the polarity of a field coil. A small battery is connected to the coil leads, and the compass is held several inches above the coil. If the south pole points toward the center of the coil, the face of the coil nearest the compass will be a north pole. This will indicate that the coil should be placed on a north pole in the same position it was in during the test, and the field current should flow through the coil in the same direction.

To protect the armature, the same precautions that were observed during removal of the coil must be observed when installing it. All of the shims originally removed from the pole piece must be in position when it is replaced. With the coil in position in the machine, it should be temporarily connected to the other coils in the field circuit and a compass and battery again used to check its polarity. For this test, connect the battery to the proper field leads and check the polarity of all the coils with the compass (fig. 5-22). Adjacent poles must be of opposite polarity. If need be, polarity of the new coil can be reversed by reversing its leads. When the polarity is correct, the coil is connected, and the pole-piece bolts are tightened. Air gaps should be measured to ensure uniformity. Before starting the machine, test

it thoroughly to ascertain that no grounds, shorts, or opens exist as a result of the repairs.

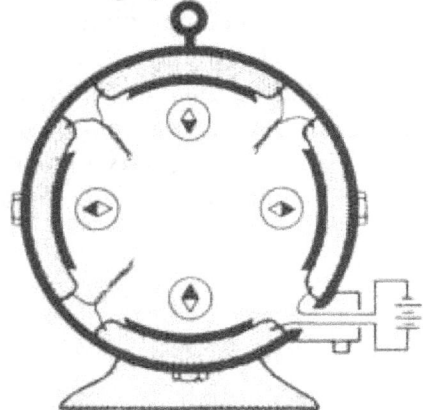

Figure 5-22.-Testing polarity of field coils.

A-C STATOR COILS

A-c stator windings require the same careful attention as other electrical windings. For a machine to function properly, the stator windings must be free from grounds, short circuits, and open circuits. Frequent tests and inspections are necessary to determine the condition of the windings, and they must be kept clean to preserve the insulation.

A short circuit in the stator of an a-c machine will produce smoke, flame, or the odor of charred insulation. The machine must be secured immediately; the faulty coil may be located if the coil ends are felt before they have time to cool. The shorted coil ends will feel perceptibly hotter than those adjacent to it. Emergency repairs of a faulty stator coil may be effected by cutting it out in the same manner as cutting out a faulty armature coil. However, a new coil must be installed as soon as possible.

Open circuits in a-c stator windings can sometimes be found by visual inspection because the open is usually the result of damaged connections where the coils and circuits are connected together. Should visual inspection

fail, resistance measurements between the phase terminals will reveal the presence of open-circuited coils. The coil ends in the faulty phase are tested with an ohmmeter to locate the open-circuited coil. When the open circuit is in an inaccessible location and cannot be reached for repairs, the machine can be repaired for emergency use by cutting out the faulty coil.

Grounds in a-c stator windings can be detected with a megger or similar resistance-measuring instrument. Both ends of each phase are opened and tested to locate the grounded phase. Then, each circuit in the grounded phase is opened and tested to locate the grounded circuit. Finally, the ends of each coil in the grounded circuit are opened and tested until the grounded coll is located.

New or repaired stator coils should be tested (before and after installing) for grounds, shorts, and opens. Polarity of the coil must be checked after installation by employing a low d-c voltage source and a small compass, as in the polarity test for field coils. The compass needle will reverse Itself at each adjacent pole if the stator is properly connected. When the same compass needle indication is obtained from two adjacent poles, a reversed coil is indicated. The coil connections must be changed to correct the polarity.

MOTOR AND GENERATOR AIR COOLERS

Some large electric motors and generators, such as propulsion generators and motors, are equipped with surface-type air coolers. In this system the air is circulated by fans on the rotor in a continuous path through the machine windings and over the water-cooled tubes of the cooler. The cooler is of double-tube construction (one tube inside another) to minimize the possibility of damage due to water leakage. The location of the air cooler in a generator is shown in figure 5-23.

The air and water sides of air cooler tubes must be kept as clean as possible because foreign deposits will decrease heat transfer. When the air side of the tubes requires cleaning, the individual tube bundles may be removed and washed with hot water or cleaned with a steam jet. The water side of cooler tubes must be cleaned in accordance with instructions contained in Bureau of Ships Manual, Chapter 46.

Figure 5-23.-Generator equipped with an air cooler.

When a leak between an inner tube and the tube sheet occurs, water will seep from the cooler head through the leaky joint into a leak-off compartment and out the leakage drain. If a leak occurs in an inner tube, water will seep into slots in the outer tube where it is carried to a leak-off compartment and out the leakage drain. The leakage drain line is equipped to give a visual indication of the presence of water in the line.

When a leaky tube is found, both ends of the tube should be plugged with plugs provided as spare parts or with condenser plugs. When the number of plugged tubes in a cooler section becomes large enough to adversely affect the heat-dissipating capacity of the cooler, the cooler section must be removed and replaced.

Zinc plates or rods are provided in air coolers to protect the cooler tubes from corrosion. Zincs should be inspected once a month and should be kept clean by wire brushing or scraping. They should be replaced before they are one-half consumed. It is necessary to drain the cooler before removing the zincs in order to avoid discharging water on, and possibly into, the machine.

PERIODIC TESTS AND INSPECTIONS

To determine whether or not proper maintenance procedures for motors and generators are being carried out,
a checkoff list of the maintenance schedule should be prepared and maintained. To be effective, the checkoff list must identify the equipment and indicate when and by whom the maintenance was accomplished. The list should be inspected periodically by the Electrical

Officer or Engineer Officer, and, when applicable, entries to the effect that certain maintenance has or has not been performed should be made in the appropriate electrical history, current ships maintenance project, or operating logs.

It may not be possible to carry out a complete maintenance schedule on some shipboard equipment because of its inaccessible location. However, the Bureau of Ship's required maintenance schedule for motors and generators which follows, should be carried out so far as practicable, and if experience or manufacturer's instructions dictate, should be supplemented by additional or more frequent tests and inspections.

DAILY:

1. Check oil level and condition of oil rings in oil-lubricated bearings, and the flow of oil (by sight gage) in force-feed lubricated bearings.

2. Inspect motor and generator surroundings for dripping water, oil, steam, acid, excessive dirt, dust, or chips, and any loose gear that might interfere with ventilation or jam moving parts.

3. Observe running motors for vibration and unusual or excessive noise.

4. Examine each running generator set for cleanliness, vibration, unusual or excessive noise, heating; and condition of brushes, commutators, collector rings, bearings, bolts, and mechanical fastenings.

WEEKLY:

1. Check temperatures of bearings and frames. Estimate by touch, and if temperature appears to be excessive, measure with a thermometer.

2. Inspect for leakage of lubricant from generator and motor bearings, as shown by oil or grease on the shaft extension or lubricant creeping towards the winding.

3. Inspect for leakage of lubricant from the bearings of prime movers that drive generators.

4. Inspect commutators of idle machines for commutator condition.

5. Inspect collector rings of idle machines for evidence of corrosion.

6. Inspect all running motors for unusual sparking.

7. Run each generator at partial or full load for at least 30 minutes once a week, and record in log. If it is not practicable to run each set every week because of naval shipyard work, extensive overhaul, or casualty, an entry shall be made in the log stating the facts.

8. If operating conditions permit, measure the insulation resistance of each ship's service and emergency generator and exciter, and of propulsion generators, motors, and exciters in electric drive ships. Record results of the tests on the appropriate megger cards.

9. Blow out generators that have been in use with clean, dry, compressed air, and wipe with alintless cloth.

MONTHLY:

1. Run each generator continuously for at least 4 hours once a month at full-rated load and voltage. Record in log. If it is not practicable to apply full load, the maximum load possible should be used, and an entry should be made in the log, giving the load used, and the reason why full load was not practicable. If it is not practicable to run each generator for this test every month, an entry should be made in the log stating the facts.

2. Inspect the zincs in motors and generators equipped with air coolers.

3. Remove the drain plugs provided in Navy Class A spraytight, watertight, and submersible motor enclosures to drain off water that may have collected in the enclosures. Be sure to replace the drain plugs immediately after draining the motor enclosures. The draining of

the motor enclosures should be entered in the log or electrical history, as appropriate, together with an entry giving a rough idea of the amount of water drained off.

QUARTERLY:

1. Inspect pulleys, belts, belt guards, mounting-frame bolts, end-shield bolts, and mechanical fastenings for mechanical soundness and tightness.

2. Check clearance between bearings and shaft on machines with sleeve bearings.

3. Check air gaps, if accessible for measurement, on machines with sleeve bearings. Record measurements taken on appropriate electrical history card.

r -t

4. Check end play of motor and generator shafts.

5. Rotate motors through one complete turn and inspect the commutator.

6. Replace worn out brushes.

7. Inspect and tighten brush "pigtails."

8. Check brush alignment parallel to commutator segments.

9. Check distance of brush holders from the commutator.

10. Check brush pressure.

11. Make sure that brush holders are clean and that the brushes move freely in the holders.

12. Blow out and clean motors thoroughly to remove dirt from the commutator, ventilation ducts, and insulation.

13. Measure and record the insulation resistance and temperature of motors.

14. Operate motors at normal load and temperature. SEMIANNUALLY:

1. Drain, flush out, and renew oil in sleeve bearings.

2. Add grease to ball bearings if required. Record the date and the fact that the machine was lubricated.

3. Inspect all gaskets, particularly lubricant seals. Replace worn gaskets and seals.

4. Inspect armature banding and slot wedges.

5. Inspect the connections of armature coils to commutator risers.

6. Inspect and tighten all electrical connections.

7. Inspect commutator clamping ring.

8. Clean out slots in the commutator and undercut mica if necessary.

9. Inspect the ends of cage rotors for evidence of loose or broken bars or localized overheating.

10. Inspect fans for loose or broken blades.

ANNUALLY:

Inspect all windings and Insulation. Clean and repair insulation as necessary.

EVERY TWO YEARS:

Remove bearing caps or other covers as may be provided and observe the condition of ball bearings and lubricant. If no repairs are indicated, flush out the old grease, and replace with fresh, clean grease.

QUIZ

1. What is the most important factor in the maintenance of electric motors and generators?

2. What are four acceptable methods of cleaning motors and generators?

3. Why is the use of suction preferred to the use of compressed air for removing abrasive dust and particles from inaccessible parts of a machine?

4. What precautions should be taken in soldering operations on motors and generators?

5. When a generator is to be inoperative for an appreciable length of time, such as may be necessary for the overhaul of the prime mover, why should the brushes be lifted from the collector rings or commutator?

6. What precautions should betaken regarding all electrical connections ?

7. What material should be used to clean contact surfaces that are silver-plated?

8. When lifting the rotor of an a-c or a d-c machine

(a) what precaution should be taken with respect to the rope slings placed under the shaft and clear of the journals? (b) Why?

9. Sleeve bearings equipped with an overflow gage or an oil-filler gage for determining the proper level of oil to be maintained are filled until the oil is approximately how far from the top of the gage ?

10. If the bearing is not equipped with an overflow gage or an oil-filler gage it is filled to a level such that the oil ring dips into the oil at what relative depth in terms of the diameter of the shaft?

11. How frequently should bearing oil in motors and generators be renewed?

12. (a) What is a good method of testing for the presence of abnormal noise in a bearing? (b) How is trouble indicated?

13. (a) What condition may cause the grease in a bearing to churn and generate heat with the result that the grease separates into oil and abrasive particles ?

(b) How does this action affect lubrication?

14. On some generators, why is the outboard bearing electrically insulated from the frame?

15. When should brushes in motors and generators be replaced ?

16. When a d-c motor or generator is running without load and with only the main-pole field windings excited, what is the point called on the commutator at which minimum voltage is induced between adjacent commutator bars?

17. In determining the no load neutral by the mechanical method, the commutator is turned until the two coil sides of the same armature coil are what relative distance from the centerline of the main-field pole ?

18. What should be the general appearance and color of the commutator of a d-c motor or generator?

19. What abrasives shouldnever be used on a commutator or collector rings?

20. What action may be corrected by reversing the d-c polarity of the slip rings of an a-c generator every few days ?

21. A bright spark, which appears to pass completely around the commutator, is an indication of what type of fault in a d-c machine?

22. When locating grounds in a d-c armature coil (fig. 5-17) the voltmeter indicates what relative magnitude between the commutator bar leading directly to the coil and the ground?

2 3. When shifting the commutator between readings (fig. 5-17), how are the real grounds distinguished from the phantom grounds ?

24. What is the relative magnitude of the meter reading (fig. 5-18) to indicate (a) an open coil and (b) a shorted coil ?

25. What two actions are required in order to cut out a short-circuited or open-circuited armature coil?

26. What two basic types of rotors are found in a-c induction motors?

27. Reduced torque in a form-wound rotor induction motor accompanied by a growling noise or failure to start under load are symptoms of whatkind of trouble ?

28. An increase in motor speed, excessive armature current, heavy sparking, or stalling of the motor are symptoms of what kind of trouble in a d-c motor?

29. In locating an open field coil (fig. 5-21) the voltmeter reading has what relative magnitude when connected across the ends of the open coil?

30. What two devices may be used to determine the polarity of the field coil?

31. What is the purpose of a checkoff list of the maintenance schedule for motors and generators?

32. How often should running motors be observed for vibration and unusual or excessive noise?

33. How often should the temperatures of bearings and frames be checked?

34. How often should the zincs in motors and generators equipped with air coolers be inspected?

35. How often should the end play of motor and generato r shafts be checked ?
An , i

36. How often should sleeve bearings be drained, flushed out, and the oil renewed?

CHAPTER

MAINTENANCE AND REPAIR OF BATTERIES

This chapter further describes the principles of battery operation contained in Basic Electricity, NavPers 10086, and also includes battery maintenance and repair procedures. The EM, if assigned, must give the same careful attention to the operation of a storage battery that he gives to the operation of any other power unit.

DRY BATTERIES

A dry (primary) cell is a unit source of d-c electrical energy derived from chemical action. Dry cells are of the (1) Leclanche, (2) mercury, and (3) low-temperature types. One or more dry cells constitute a dry battery. A dry battery is not completely reversible, and therefore cannot be recharged economically after it has been discharged. Dry batteries are easily transported from place to place and are used as a source of electrical energy to supply circuits that require alow and intermittent current.

Leclanche Cell

The Leclanche cell has a nominal open-circuit voltage of 1.5 volts. The positive electrode is a mixture of manganese dioxide and powdered carbon in contact with a carbon rod. The negative electrode is zinc and usually comprises the container. The electrolyte is a paste consisting of ammonium chloride (sal ammoniac), a small amount of zinc chloride and mercuric chloride, water, and starch. The electrical energy is obtained from reactions between the zinc, sal ammoniac, and manganese

dioxide, as described under Dry (primary) Cells in Basic Electricity, NavPers 10086.

Mercury Cell

The mercury cell has a nominal open-circuit voltage of 1.3 volts. The positive electrode is a mixture of mercuric oxide and carbon. The negative electrode is a zinc plate or compressed zinc powder. The electrolyte is an aqueous solution of potassium hydroxide. The electrical energy is obtained from the reduction of mercuric oxide to metallic mercury and the oxidation of zinc to zinc oxide.

Low-Temperature Cell

The low-temperature cell is composed of the same materials as the Leclanche cell, except that the electrolyte is modified to give better, low-temperature (below zero, Fahrenheit) performance. The positive electrode is made of highly active manganese dioxide; consequently, low-temperature cells do not store as well as normal cells. Their use is not recommended unless they can be shipped, stored, and used at temperatures below 35*° F.

The mercury cell and the low-temperature cell have superior characteristics for special applications, but are not used as extensively as the Leclanche cell because of the greater cost.

Classification

Dry cells are usually cylindrical or flat, and are manufactured in a wide variety of sizes and weights. The classification and requirements for dry batteries used in the Navy are contained in the Military Specification for Dry Batteries, MIL-B-18B. This specification includes other pertinent information concerning military dry batteries.

LABEL.—In addition to specification sheets for individual battery types, all dry batteries are labeled to provide installation and replacement data (fig. 6-1). This data includes the type designation and code number.

The TYPE DESIGNATION comprises (1) the battery component, (2) battery type number, and (3) installation indicator.

MIL BATTERY BA-236/U

1958-PHILA.-51-7 037

J & J MFC. CO. PLAINFIELD, N. J.

Figure 6-1.-Dry battery label.

The battery component denoted by the symbol, BA, identifies a dry battery.

The battery type number, 236, following the component symbol, identifies the basic design of the battery and the kind of cell with which it is assembled. Battery type numbers 1 to 999 inclusive indicate Leclanche cells; 1,000 to 1,999 inclusive indicate mercury cells; and 2,000 to 2,999 inclusive indicate low-temperature cells.

The installation indicator denoted by the symbol, /U, indicates general utility service. This symbol follows battery type numbers of 200 and above.

The ORDER NUMBER, 1958-Phila.-51-7, appears next and is self-explanatory.

The CODE NUMBER, 037, identifies the date of manufacture. The first two digits, 03, indicate the third month (March) and the last digit, 7, indicates the year (1957).

The remainder of the label data includes the manufacturer's name and address.

TERMINAL MARKINGS.-The terminals of a dry battery are indicated with the proper polarity markings, which appear on the top or side of the battery as close as possible to the applicable terminal. The location of the polarity markings depends on the type of terminal, which can include wire leads, Fahnestock clips, binding posts, sockets, and flat caps.

Characteristics

Dry cells are rated in volts and hours for various conditions of operation. In other words, the capacity of a dry cell is determined by the length of time required before the terminal voltage drops to a specified end value when the cell is discharged continuously through a fixed resistance. The time is in hours and the load is in ohms.

The operating conditions that determine the discharge characteristics of adry battery are the (1) initial capacity, (2) delayed capacity, (3) initial arctic capacity, and (4) delayed arctic capacity.

INITIAL CAPACITY.—The initial capacity is the time required at normal temperature for the battery voltage to fall below the specified test end voltage, or to show evidence of electrolyte leakage, or swelling of the case when discharging through a specified load.

DELAYED CAPACITY.-The delayed capacity is the time required at normal temperature for the battery voltage to fall below the specified test-end voltage, or to show evidence of electrolyte leakage, or swelling of the case after the battery has been stored for a period of time.

INITIAL ARCTIC CAPACITY.-The initial arctic capacity is the time required, at 40° F below zero, for the battery voltage to fall below the specified test-end voltage, or to show evidence of electrolyte leakage, or swelling of the case.

DELAYED ARCTIC CAPACITY.-The delayed arctic capacity is the time required, at 40° F below zero, for the battery voltage to fall below the specified test-end voltage, or to show evidence of electrolyte leakage, or swelling of the case after the battery has been stored for 48 hours at this temperature.

An example of the capacity ratings for battery BA-2039/U are:
1. Initial capacity 36 hr
2. Delayed capacity 31 hr
3. Initial arctic capacity 3.6 hr
4. Delayed arctic capacity 3.1 hr

Installation

Applicable installation practices and safety precautions must be followed when installing dry batteries. Fresh batteries of the proper type, size, and shape should be installed initially, or as a replacement for a specific installation. They must be connected in accordance with the instructions furnished with the equipment, and all electrical connections should be clean, tight, and well insulated.

If equipment is idle for two weeks or more, remove the batteries, and, if such batteries are fairly fresh,

r t

return them to store, with a record showing how long they have been used. Remove and scrap batteries that have had long and hard use.

Storage

Dry batteries are perishable and will deteriorate when not in use. To minimize deterioration, they should be stored in refrigerated spaces (10° F to 35° F) that are not dehumidified (dry). If a refrigerated space is not available, they should be stored in the coolest space possible where they will not be subjected to excessive dampness or large variations in temperature.

Batteries should not be shipped or stored in the equipment with which they are to be used because they may become discharged by internal chemical reaction or by leakage currents across normally open contacts. These conditions generate water within the cells, and electrolyte may leak out and corrode the equipment. To ensure reliable operation, install fresh batteries in the equipment as near as possible to the time that it will be operated. To avoid damage from electrolyte leakage, remove dead batteries from the equipment.

SHELF LIFE.-The shelf life of dry batteries is the length of storage time beyond which a group of batteries will contain so many dead cells that the entire group is considered unusable. Shelf life depends not only on the kind and size of a cell but also on the application. Batteries containing the smaller cells do not retain their capacity as well as those containing the larger cells. As to use, the shelf life is less for a battery that requires a high initial capacity and a low internal voltage drop on discharge than for a battery that requires less initial capacity retention and a high internal voltage drop on discharge. Shelf-life tables are issued by the various supply offices. These tables are based on general use requirements and indicate a longer shelf life than is tolerable for more critical applications.

The assignment of a certain shelf life to a specific type of battery signifies that a substantial percentage, but not all, of the batteries in the group will be usable if the end of the shelf life is not reached. Test the usability of an individual battery by trying it out in the equipment

for which it is to be used or by connecting it to an equivalent load and measuring the closed-circuit voltage. If this voltage is above that required for satisfactory equipment operation, the battery is suitable for use. Always survey batteries when they have reached the end of their shelf life.

ISSUE.—No known nondestructive tests will show the amount of capacity remaining in a battery after it has been partially discharged or stored for a considerable length of time. It is usually good practice to issue the oldest batteries first if they are to be used in the close proximity of the warehouse where replacement stock is readily available. However, batteries issued for use aboard ship where premature failure would cause serious inconvenience, should be less than 6 months old.

Safety Precautions

Certain types of batteries generate hydrogen, which is very explosive. When working with these batteries, interrupt the current at a remote point before disconnecting them from the equipment.

When a battery is disconnected from the operating equipment, the wire lead terminals of the battery should be insulated to prevent the possibility of a short circuit, which might generate sufficient heat to cause afire. Also, the discharge caused by a short circuit usually produces excessive water in the cells, thereby causing them to burst and spill corrosive electrolyte onto the equipment.

Multicell dry batteries should not be used after the closed-circuit voltage has dropped below 0.9 volt per cell. If the battery is further discharged, current will be forced through some cells that may be completely discharged. This action will generate hydrogen and oxygen due to the electrolysis of the water, with the consequent danger of a hydrogen explosion.

STORAGE BATTERIES

A wet (secondary) cell is a unit consisting of positive and negative plates, separators, cell covers, and electrolyte, properly assembled in a single jar or in one compartment of a monobloc case (molded in one piece). A

pa*

tray consists of one or more cells assembled in a common container or monobloc case. One cell, one tray of cells, or a number of trays of cells connected in series, parallel, or series-parallel constitute a storage battery. A storage battery is reversible because the active materials consumed during discharge can be restored by passing a charging current through the battery. The storage cells in general use are the nickel-cadmium and lead-acid types.

Nickel-Cadmium Cell

The nickel-cadmium cell is an alkaline storage cell. It is used in aircraft and in the Nike and Corporal guided missiles, and may be installed aboard ship in the foreseeable future. The construction and general arrangement of the cell are similar to that of the lead-acid cell. The negative plate assembly is a cadmium-oxide compound; the positive plate assembly is a nickel-oxide compound; and the electrolyte is a 30 percent solution of potassium hydroxide. Separators between the positive and negative groups prevent internal short circuits. The cell components are assembled in a single plastic jar. The battery case is of fabricated steel. A 6-volt battery contains 5 series-connected cells. When fully charged, the cell voltage ranges from 1.39 to 1.45 volts.

CHARGING.—The effect of the charging current is to change the active material of the negative plate (cadmium-oxide) to metallic cadmium (CdO to Cd). The active material of the positive plate (nickel-oxide) is changed to a higher state of oxidation (NiO to Ni_2O_3). As long as the charging current continues, this action occurs until both materials are completely converted. Toward the end of the charging process and during overcharge, the cell will gas due to the electrolysis of the water in the electrolyte. This action liberates 4 atoms of hydrogen gas ($2H_2$) at the negative plate for every 2 atoms of oxygen gas (O_2) liberated at the positive plate. The amount of gas liberated depends on the charging rate.

The electrolyte does not enter into any chemical reaction with the positive or negative plates. It acts simply as a conductor of current between the plates, and its specific gravity does not vary appreciably with the amount

of charge. The effect of the reactions during charge is a transfer of oxygen from the negative to the positive plates. In this respect the nickel-cadmium storage battery is similar to the lead-acid storage battery. Unlike the lead-acid storage battery the specific gravity of the electrolyte remains constant, except at the end of a charge or on overcharge, when it increases because of the electrolysis of the water. The proper specific gravity can be obtained by adding distilled water.

Nickel-cadmium batteries can be charged by the (1) constant voltage, (2) constant current, and (3) stepped constant-current methods. The most efficient performance is obtained when the charging rate is such that 140 percent of the rated ampere-hour capacity of the cell is delivered to the cell within a 3-hour interval.

DISCHARGING.—When the cell is connected to an external circuit containing a resistance, the chemical action which occurs is the reverse to that of the charging process. The negative plate gradually regains oxygen (Cd to CdO), and the positive plate gradually loses oxygen (Ni 2 0 3 to NiO); that is, the action during discharge effects a transfer of oxygen from the positive to the negative plates. There is no gassing on discharge due to the interchange of oxygen. The discharge process is the conversion of the chemical energy of the plates into electrical energy. The rate at which this conversion occurs is determined by the external resistance, or load, to which the battery is connected. The internal resistance of the cell is extremely low because of the cell construction.

The nickel-cadmium battery is smaller and weighs less than the comparable lead-acid battery. There is little or no local action, and the positive plates do not shed, or flake off, as do the positive plates in the lead-acid cell. The battery life is from 10 to 15 years with constant capacity over the entire period, and satisfactory operation can be obtained over a temperature range from -65° F to 165° F.

Lead-Acid Cell

The lead-acid cell is the storage battery most extensively used in the Navy. As explained in Basic Electricity , NavPers 10086, the positive-plate assembly consists

of lead peroxide; the negative-plate assembly is pure sponge lead; and the electrolyte is a mixture of sulphuric acid and distilled water.

CHARGING.—The effect of the charging current is to change the lead sulphate formed on both the negative and positive plates by the reaction of the electrolyte with the active material, back to the original active form of lead peroxide (PbS0 4 toPb0 2) on the positive plate and sponge lead (PbS0 4 to Pb) on the negative plate. At the same time the sulphate is restored to the electrolyte, and the specific gravity of the electrolyte increases. When all of the sulphate is restored to the electrolyte, the specific gravity is maximum and the cell is fully charged. Toward the end of the charging process, hydrogen (H 2) is liberated at the negative plate, and oxygen gas (0 2) is liberated at the positive plate. This action is due to the electrolysis of the water in the electrolyte.

DISCHARGING.-When the cell is connected to an external circuit containing a suitable load resistance, the action of the electrolyte on the active material of the negative and positive plates causes a current to flow. Thus, the chemical energy stored in the plates is transformed into electrical energy as the cell discharges. The reaction of the electrolyte with the active material forms lead sulphate on both the negative and positive plates. As the discharge continues, the acid content of the electrolyte becomes less as it is used in forming lead sulphate, and the specific gravity of the electrolyte decreases. When so much of the active material in the cell has been converted into lead sulphate that it can no longer produce sufficient current to be of practical value, the cell is discharged.

Classification

The classification and requirements for storage batteries used in the Navy (except for aircraft and automotive vehicles) are contained in the Military Specification for Portable Lead-Acid Storage Batteries, MIL-B-15072A. This specification includes other pertinent information concerning these batteries.

Storage batteries are classified as Class ER, electrolyte-retaining and Class FE, free-electrolyte

batteries. Class ER batteries have separators and plates that absorb and retain within the cell at least 80 percent of the electrolyte, and operate with not more than 20 percent of the electrolyte in a free condition. Class FE batteries operate with at least 50 percent of the electrolyte in a free condition.

NAMEPLATE.—Supplementary to the specification sheets supplied for the individual battery types, all storage batteries are provided with nameplates that contain installation and replacement data (fig. 6-2). The TYPE DESIGNATION contained on the nameplate is in the same form as that previously described for dry batteries. It comprises the battery component, battery type number, and installation indicator.

MIL BATTERY BB-2S4U

(MANUFACTURER'S NAME)

CLASS fV.SBM.W0AM CONTRACT NO. C

MFR TYPE 1 | TYPE OF SEPARATORS I

CAPACITY 100 AM AT 10-MR RATE AT wf * DISCHARGE RATE/10 AMP FOR 10 HR [2*0 AMP FOR S MIN FINAL VOLTAGE 1.7S VOLTS PER CELL

AT 10-HR RATE CHARGE RATE ^1 START

AMP AMP

1 FINISH

MAXIMUM SPECIFIC GRAVITY 1.220 AT W' F HEIGHT OF ELECTROLYTE 1/2 INCH OVER TOP

OF SEPARATORS [1 DATE OF INITIAL CHARGE

|) DATE ELEMENTS RENEWED

|) DATE ELEMENTS RENEWED

Figure 6-2.—Storage battery nameplate.

The battery component denoted by the symbol, BB, identifies a lead-acid storage battery.

The battery type number, 256, following the symbol, identifies the basic design of the battery. Batteries designed for low-temperature operation are identified by increasing the established battery type number by 2,000. For example, a low-temperature battery is identified if the established battery type number, 256, is increased to 2256.

The installation indicator denoted by the symbol, /U, indicates general utility service. Battery type numbers of 200 and above are appended with this symbol.

Prior to the Military Specification (MIL-B-15072A), Navy standard batteries were identified according to voltage, use, and rating. This former type designation, CLASS 6V-SBM-100AH, also appears on the nameplate (fig. 6-2).

The voltage of the fully charged battery denoted by the symbol, 6V, identifies a 6-volt battery.

The use of the battery denoted by the symbol, SBM, identifies a monobloc storage battery. Other uses denoted by symbols, SBP and SBMD, identify portable and monobloc diesel-starting storage batteries, respectively.

The rating of the battery denoted by the symbol, 100AH, identifies a battery that can supply 100 ampere-hours when discharged at the 10-hour rate (10 amperes x 10 hours = 100 ampere-hours).

The remainder of the data contained on the nameplate requires no explanation.

TERMINAL MARKINGS.-The positive and negative terminals of storage batteries are legibly indicated by POS or P and NEG or N, respectively. These polarity markings are raised or depressed characters placed on, or as close as possible to, the applicable terminal. For further identification of the battery polarity, portions of the positive and negative terminals that are not contact surfaces can be painted red and black, respectively.

Characteristics

CAPACITY.—The capacity of a storage battery is the constant current that the battery can supply continuously at a definite rate of discharge before the voltage drops below a specified limiting (final) voltage. This capacity varies with the rate of discharge; that is, the lower the rate of discharge the greater the capacity, and vice versa.

DISCHARGE RATE.-Navy storage batteries arerated at the 10-hour discharge rate, which is the constant current in amperes that the battery (starting with an initial electrolyte temperature of 80° F) can supply continuously for 10 hours before the voltage drops to the low-voltage limit. Other hourly discharge rates are defined similarly.

FINAL VOLTAGE.—The final voltage, or low-voltage limit, is that (set by the manufacturer) beyond which very little useful energy can be obtained from the battery. Generally, the low-voltage limits for batteries at different rates of discharge vary slightly with the size and make of the battery. At the conclusion of a discharge at the 10-hour rate, the closed-circuit voltage will be approximately 1.75 volts per cell, and the specific gravity will be approximately 1.06 0. At the completion of a charge, the closed-circuit voltage at the finishing rate will be approximately 2.4 to 2.6 volts per cell, and the specific gravity will be from 1.210 to 1.220 at 80° F.

Installation

Lead-acid storage batteries are shipped (1) filled and charged and (2) dry. When shipped to naval vessels they are usually in the filled and charged condition. When shipped to replenish stock at naval shipyards or other shore establishments, or on board repair ships or tenders, they are usually in the dry condition.

CHARGED AND WET.-When batteries are shipped (filled and charged), the cells are completely assembled, contain electrolyte, and are sealed ready for use. When batteries are received in this condition, the height of the electrolyte should be examined, and specific gravity and temperature readings should be taken of all cells to ascertain the state of charge.

If there is no evidence of spilling or leakage, and the height of the electrolyte is below the bottom of the filling tube or the level mark on polystyrene cells, sufficient distilled water should be added to bring the level even with the bottom of the filling tube or the level mark. Unless otherwise specified, the electrolyte level should be three-eights of an inch above the separators.

If the electrolyte has been spilled during shipment, it should be replaced with electrolyte of 1.215 specific gravity corrected to 80° F. If the loss of electrolyte is due to a cracked jar or broken case, the element should be removed and immersed in pure distilled water in a rubber or glass container until a new jar or case can be obtained. Wood or metal containers should not be used.

Am

If water or electrolyte is added, the battery should be charged at the finishing rate until specific gravity readings, corrected for temperature, are constant for a period of 5 hours. If the battery is not to be placed in service immediately, an equalizing charge should be given at 1-month intervals, or sooner if the specific gravity has dropped to 1.180, and again just before placing it in service.

UNCHARGED AND DRY.—When batteries are shipped dry, the cells are completely assembled and sealed, but contain no electrolyte, and the plates are in the uncharged condition. When batteries are received in this condition and are not to be placed in service, the vent plugs must be kept tightly in place and the batteries stored in a clean dry space until they are required.

If the battery is to be placed in service, follow the instructions contained on the accompanying tag. These instructions specify the specific gravity of the electrolyte to be used in filling the cells, the time to allow for the electrolyte to soak into the plates before charging, the temperature limit, and the method for conducting the initial charge.

Only authorized personnel at repair activities are permitted to fill a cell or battery with electrolyte. Electrolyte of a specified specific gravity is added until its level is even with the bottom of the filling tube or the level mark on polystyrene jars. The battery will become heated when filled with electrolyte and should be allowed to stand for at least 12 hours to cool before starting the initial charge. The battery should not stand for more than 24 hours (after filling) before starting the charge. During the standing period the level of the electrolyte will drop because it soaks into the plates and separators. The level should be restored to the prescribed height by adding electrolyte of the same specific gravity as used initially before starting the charge.

PLACEMENT IN SERVICE.-Whenbatteries are to be placed in service, the trays are arranged so that the positive terminal of one tray can be connected to the negative terminal of the next tray throughout the battery. If the battery terminals, or posts, are not properly marked, the polarity can be determined by means of a d-c voltmeter across the terminals. If a voltmeter is not readily

available, immerse the terminal leads in salt water, holding them about one-half inch apart. Bubbles will collect on the negative lead. Do not allow the leads to touch each other when conducting this test.

The contact surfaces of battery terminals are usually covered with a thin film of acid, which should be neutralized and removed to prevent corrosion before making the connections. For this purpose, diluted ammonia or a solution of bicarbonate of soda can be used, exercising care not to allow the solution to enter the cells. After the contact surfaces have been neutralized, they should be brightened with a wire brush or fine sandpaper. The contact surfaces are then bolted together and coated with petrolatum or cup grease.

All connections should be checked to make certain that the polarity is correct and that all exposed metal in the connectors is protected with petrolatum or cup grease. The top and sides of the battery should be cleaned with a cloth dampened in a solution of dilute ammonia or bicarbonate of soda, after which the battery should be thoroughly washed with pure water.

BATTERY NUMBERING.—Batteries must be assigned a number to provide a systematic method of identification and to facilitate the maintenance of battery records. This identification is accomplished by assigning a number to each tray. For example, if a ship has a total of 100 trays of portable storage batteries for all purposes, each tray is numbered consecutively from S62-3(1),S62-3(2) and so on, to S62-3(100). In addition, the cells of a tray are numbered consecutively, beginning at the positive end.

The battery number should be stenciled in a conspicuous place on the side of the container for ready identification. When a new tray replaces an old one, give it the number of the old tray, and begin a new log.

PILOT CELLS.—Pilot cells (one from each tray) must be selected at random from which cell readings are taken during routine inspections and tests. The cell readings provide an approximate indication of the battery condition without the necessity of taking readings of all the individual cells.

RECORDS.—A complete history for each storage battery tray in the ship must be maintained on the Storage

Battery Tray Record, NavShips 151. This record lists the tray number, nameplate data, and record of service, repairs, charges, and test discharges. This information shows the true condition of each battery and often indicates trouble in advance of failure. This record must be inspected, approved, and signed each month by the Engineer or Electrical Officer. It must accompany the battery tray when transferred, and when surveyed, it must be delivered to the survey officer. The record can be destroyed when the service of the battery is terminated.

The Storage Battery Charging-Discharging Record must be maintained in addition to the Storage Battery Tray Record, NavShips 151. This record should contain supplementary data of applicable voltage, temperature, and specific gravity readings taken during normal, equalizing, and emergency charges, and test discharges.

Storage

Storage batteries,like dry batteries,deteriorate when not in use. Batteries in a wet condition should not be stored for considerable periods of time. When batteries are to be stored temporarily, they should be placed in a clean, dry space at an average temperature of 75° F and never subjected to temperatures below 50° F or above 100° F. Batteries that have a capacity of 80 percent or more when taken out of service should be maintained in a fully charged condition and issued at the first opportunity.

WET STORAGE.—When placing batteries in wet storage, give them an equalizing charge, and coat (with petrolatum) all exposed terminals to prevent corrosion. The batteries are then placed on racks, located in the battery locker, to allow free circulation of air around the top and sides.

Batteries in wet storage must be inspected at frequently regular intervals, and the electrolyte level maintained above the tops of the separators. Give these batteries an equalizing charge each month, or more often if thespecific gravity falls below 1.180, and maintain a battery tray record during the period that the battery is in wet storage.

ACID. —Carboys containing sulphuric acid must be stored in spaces where freezing cannot occur. The

freezing temperatures of sulphuric-acid electrolyte depend on the specific gravity of the solution and vary over a wide temperature range, as indicated by the specific gravity versus temperature curve shown in figure 6-3. For example, the freezing temperature of sulphuric acid having a specific gravity between 1.614 and 1.667 is indeterminate (drops rapidly) below -40° F. On the other hand, if the specific gravity of concentrated acid is reduced from 1.835 to 1.800, the freezing temperature changes from -29° F to +42° F.

2 •

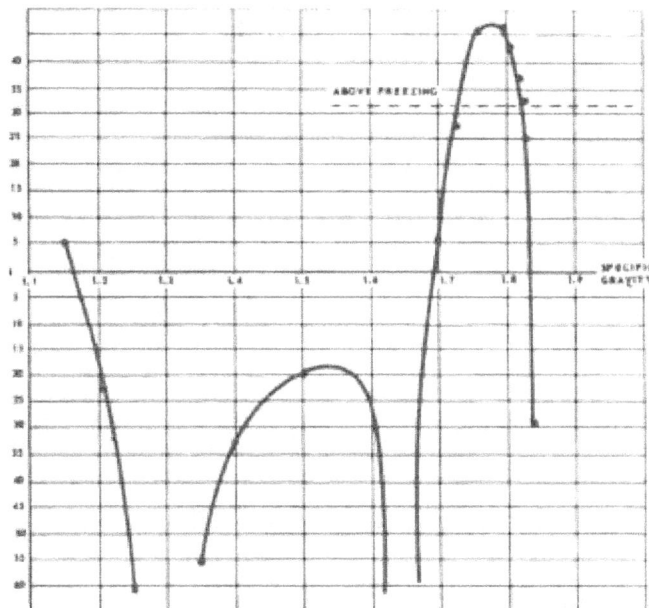

Figure 6-3.-Specific gravity versus temperature curve.

Because sulphuric acid is highly hygroscopic (readily absorbs and retains moisture), the carboys must be kept absolutely airtight. If a small quantity of water is allowed to enter a carboy containing sulphuric acid, it can reduce the specific gravity sufficiently to freeze (in cold weather) the solution, causing the carboy to break, with the consequent danger of acid burns to personnel. At any

temperature, the addition of a small quantity of water to concentrated sulphuric acid can cause an explosion with the sudden evolution of heat.

MAINTENANCE

Lead-acid storage batteries will deteriorate rapidly if they are not maintained properly and kept fully charged. The actual life of these batteries is indeterminate, but when properly cared for they should give a useful life of four or more years, depending on the type and use.

Care

The service for which a battery is used primarily determines the nature of the care and maintenance that will ensure maximum reliability and life.

ENGINE-STARTING BATTERIES.-Batteries that are used to start propulsion engines, motor-boat engines, and ship's service engine-generator sets are subjected to moderately heavy use and may require frequent charging in addition to the charging provided by the engine generator. If the specific gravity has dropped 30 points below the prescribed operating value of specific gravity, give the battery an equalizing charge, taking care that it is not overcharged by the engine generator. To prevent overcharging, set the charging rate of the engine generator to maintain the specific gravity of the electrolyte about 20 points below the prescribed operating specific gravity.

AUXILIARY LIGHTING AND POWER BATTERIES.-Batteries on stand-by service for auxiliary lighting and power are seldom discharged by actual use. A normal battery on open circuit will discharge 50 percent or more in a period of 3 or 4 months. This is a normal condition and is usually called local action or self-discharge. The lead sulphate formed during self-discharge is more difficult to reduce by charging than the lead sulphate formed during a regular discharge. Therefore, stand-by batteries should be given an equalizing charge when the specific gravity drops to 1.180, and at 30-day intervals, irrespective of the specific gravity (if the batteries are on open circuit and not delivering power). If the monthly equalizing

charges fail to maintain the prescribed specific gravity, the batteries should be cycled to restore them to a normal condition.

TELEPHONE, F. C, AND I. C. BATTERIES.-Batteries on stand-by service for dial telephones, fire control, and interior communications are usually floated on a low-voltage supply line from motor-generators through reverse-current circuit breakers. Batteries for automatic telephones are intermittently charged automatically from motor-generators that supply the telephone system.

GUN FIRING BATTERIES.-Batteries used for gunfir-ing and sight-lighting purposes are idle a large part of the time. These batteries should be given an equalizing charge once a month, or when the specific gravity drops to 1.180. These batteries should also be cycled to restore them to a normal healthy condition if the monthly equalizing charges fail to maintain the specific gravity of the electrolyte above 1.180.

Routine Inspections

Routine inspections of batteries are necessary to ascertain good operating conditions. The following schedule of inspections is considered to be the minimum requirements consistent with good engineering practice for batteries used under average operating conditions. However, if experience indicates that these inspections are not sufficient to ensure that the batteries are capable of performing their functions when needed, the frequency of inspections should be increased as necessary. For example, in cold weather operations, it may become necessary to make daily inspections of boat batteries.

DAILY INSPECTIONS.-The ventilation system in battery rooms and battery lockers should be inspected daily to ascertain that all parts of the system are in properly operating condition. Also, battery hydrometers should be cleaned daily to prevent the accumulation of sticky substances that collect on the float inside the barrel, and result in inaccurate readings. The hydrometer parts should be cleaned with an ammonia solution, flushed with clear water, dried, and allowed to air before they are reassembled.

WEEKLY INSPECTIONS.-The specific gravity and temperature readings of pilot cells for all batteries should be taken and recorded each week. The batteries should be watered if the height of the electrolyte in the pilot cells is at the low mark. Pure distilled water can be added to a battery at any time to replace that which has spilled or evaporated. However, it is preferable to add water just prior to placing the battery on charge in order to mix it with the electrolyte. Otherwise, in cold weather, If the water is allowed to remain on top of the electrolyte, it may freeze and crack the monobloc cases or jars. The charging rate of engine battery-charging generators and the voltage at which batteries are being floated should be checked each week to prevent overcharging.

MONTHLY INSPECTIONS.-All batteries, except those that are charged by an associated engine generator or those that are being floated, should be given an equalizing charge each month. A complete set of voltage, temperature, and specific gravity readings should be taken on these batteries at the completion of the equalizing charge. Batteries should be cleaned and inspected for broken or c racked cases. The terminals should be coated with petrolatum, if necessary, to prevent acid from coming in contact with the connections. Sulphate is formed when acid attacks these parts, making it difficult to disconnect leads from the terminals. Also, the sulphate increases the resistance of the terminal connections, with a consequent reduction of the voltage on discharge. Battery connections should be examined and all faulty condition corrected.

QUARTERLY INSPECTIONS.-All batteries that are charged from an associated engine generator or are being floated should be given an equalizing charge each quarter. A complete set of voltage, temperature, and specific gravity readings should be taken on these batteries at the completion of the equalizing charge.

SEMIANNUAL INSPECTIONS.-Each battery should be given a test discharge every 6 months.

Charging

BATTERY CHARGING AND DISCHARGING SWITCHBOARD.—A typical battery charglng-discharglng switchboard installed In a destroyer is illustrated in figure 6-4. This switchboard is located in the battery compartment

and consists of a single dead-front type of panel comprising an upper (hinged) section and a lower (removable) section. The upper panel section contains the necessary meters, indicating lights, and power circuit breaker. The lower panel section contains the required rheostats, ammeter switches, and circuit switches.

Figure 6-4.— Battery charging and discharging switchboard.

This switchboard is provided with two local circuits and two remote circuits for charging and discharging lead-acid storage batteries in trays of three cells each. Each tray comprises a 6-volt battery. The two local circuits (1 and 2) are designed for charging and discharging 50 ampere-hour and 100 ampere-hour batteries. The two remote circuits (3 and 4) are designed for charging and discharging 175 ampere-hour, motor boat and diesel-engine batteries, respectively. The 4 charging and discharging circuits are capable of handling from 1 to 12 batteries of the prescribed capacities in series. Two transfer switches are provided so that the remote circuits can be used also as local circuits.

The 120-volt, d-c power for charging batteries is normally supplied to the switchboard over the battery-charging feeder from the forward d-c generator and distribution switchboard. A typeACB power circuit breaker provides overload and reverse current protection, as indicated by the schematic diagram (fig. 6-5).

When the normal power supply is not available, a limited supply (20 amperes, maximum) can be obtained from the I. C. switchboard over a limited-capacity feeder. This limited power is supplied through a DPDT selector switch, the midpoint of which is connected to the charging circuit through the power circuit breaker on the switchboard. This selector switch is mounted on the base of the power circuit breaker, and its position can be changed only by opening the upper (hinged) section of the panel. A white indicator light (C) indicates when power is available from the I. C. switchboard. The majority of battery charging-discharging switchboards are not designed to supply the limited power from the I. C. switchboards, and therefore are not equipped with the feeder selector switch and white indicator light.

The POWER CIRCUIT BREAKER, located in the center of the upper section, is a manually operated, 2-pole, 250-volt, 100-ampere, type ACB breaker. It is equipped with two overcurrent coils and an undervoltage coil. The over-current coils can be calibrated from 100 to 200 percent of the breaker rating, and the undervoltage coil functions to trip the breaker on low voltage to prevent discharging the batteries. An amber (A) and a blue (B) light located

120 v D C
fROM FWO
VA
FRI
sum
A/w—0^
r*' VW—0
BLUE
■t*—VW—0

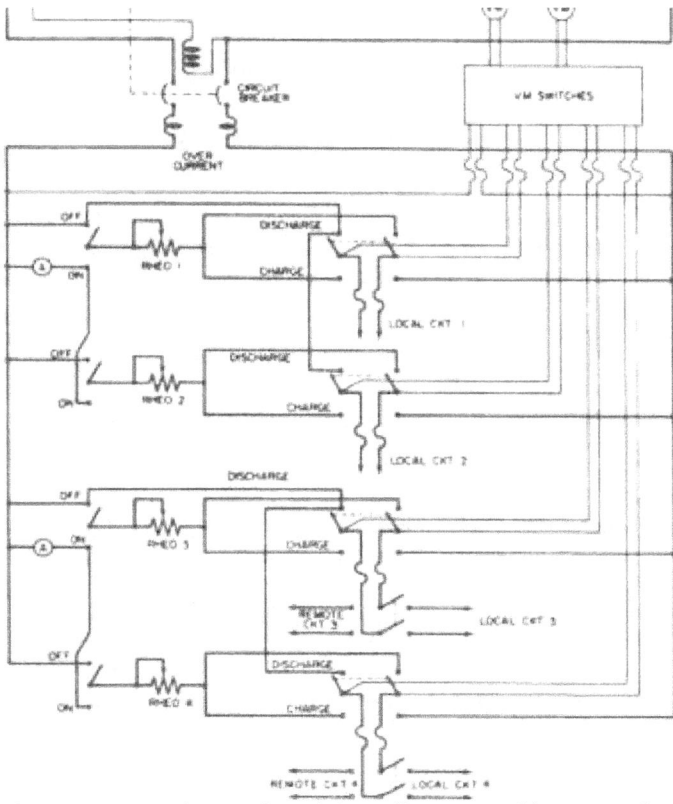

Figure 6-5.-Schemotic wiring diagram of battery charging and discharging switchboard.

directly above the circuit breaker indicate the OPEN and CLOSED positions, respectively, of the breaker.

A DOUBLE-SCALE VOLTMETER having scales of 0 to 15 volts and 0 to 150 volts is located at the top on the left-hand side of the upper section. This voltmeter is provided with a 6-position switch for (1) cutting out the voltmeter, (2) reading the voltage of only one of the four switchboard circuits, or (3) reading the bus voltage. A scale-changing switch is also provided for changing to the low-voltage scale when taking voltage readings on only one or two batteries. The scale-changing switch must remain in the high-scale position, except when actually taking low-voltage readings, because of possible overloading, with consequent damage to the voltmeter.

A SINGLE-SCALE VOLTMETER having a scale of 0 to 3 volts is located adjacent to the double-scale voltmeter. This voltmeter is connected to a jack receptacle centrally located near the top of the lower section. A 15-foot cord having a plug at oneend and battery test clips at the other end is provided for taking voltage readings of single cells in a battery. The test leads of this cord and plug must not be connected to more than one cell at a time because of possible overloading of the voltmeter.

Two AMMETERS having scales of 0 to 20 amperes and 0 to 40 amperes are located at the top on the right-hand side of the upper section. These ammeters are provided for taking current readings of circuits 1 and 2 and 3 and 4, respectively.

A SPDT, spring-loaded AMMETER SWITCH, located directly below the associated rheostat handle on the lower section, is provided for each of the four switchboard circuits. The ammeter switch is used to take readings of the charging and discharging current for the associated circuit when the switch is held in the ON position. Only one ammeter switch of each set of circuits (1 and 2 or 3 and 4) should be closed at any one time because of the possibility of overloading and damaging the ammeter.

A RHEOSTAT, located on the lower section, is provided for each of the four switchboard circuits. These rheostats are used when charging or discharging batteries to cut IN or to cut OUT sufficient resistance in order to obtain the necessary current flow for the number of batteries connected in the associated circuits at the time.

A DPDT CHARGING and DISCHARGING SWITCH for each of the four switchboard circuits and a DPDT local and remote transfer switch for each of the two remote circuits (3 and 4) are located at the bottom of the lower section. The local battery charging and discharging circuits are connected to a terminal block in the rear of the battery switchboard. The remote battery charging and discharging circuits are connected to the common terminal (stud) of the two local and remote DPDT switches.

The charging and discharging rates for the battery switchboard are contained in the manufacturer's instruction book. These rates should not be exceeded because overloading can damage either or both the switchboard circuits and batteries.

When batteries are charged, the voltage of the charging line must exceed the total voltage of the batteries connected in series to be charged. When more than one tray is to be charged in series, the negative terminal of one tray is connected to the positive terminal of the next tray, irrespective of the number of trays connected. After the trays are connected in series addition, the remaining positive terminal on the first tray and the remaining negative terminal on the last tray are connected to the corresponding positive and negative terminals of the charging circuit. If the batteries are of different sizes, the charging rate in amperes must not exceed the maximum charging rate of the lowest rated battery in the series. The maximum charging rates are obtained from the battery nameplates. For example, a series-connected line containing two 6-volt batteries and one 12-volt battery with maximum charging rates of 5, 7, and 10 amperes, respectively, must be charged at a 5-ampere rate.

GASSING.—When a batterynears full charge, the voltage rises to about 2.35 volts per cell and the battery begins to gas freely. When gassing begins near the end of the charge, the charging current must be decreased to the finishing rate to avoid heavy gassing.

Excessive gassing is an indication that the battery is being charged too rapidly, and the rate must be reduced. Otherwise, the gas bubbles will tend to loosen the active material from the plates and eventually cause disintegration of the positive grids. This loosened, active material is precipitated to the bottom of the cell, resulting in loss

of battery capacity. An Internal short circuit will result if this accumulation of active material reaches the bottom of the plates. Also, excessive gassing increases the danger of a hydrogen explosion.

INITIAL CHARGE.—An initial charge is a long, low-rate, forming charge that is given to a new battery (uncharged and dry) to place it in service.

To conduct an initial charge, fill the battery with electrolyte of the proper specific gravity, and replace the vent plugs, making certain that the vent holes are clear. The battery should be allowed to cool before beginning the charge.

The battery is connected to the charging and discharging switchboard, and all the resistance is cut in by means of the rheostat.

The battery switch is closed to the charging position, and the rheostat is adjusted until three-fourths of the finishing rate (obtained from the nameplate) is indicated by the ammeter. The rheostat is further adjusted when necessary to maintain the proper charging rate.

Hourly readings of the voltage, temperature, and specific gravity are taken. If the temperature of the battery reaches 125° F.,the charging rate is either reduced or discontinued until the battery cools. The charge is continued after the battery cools until the voltage and specific gravity have remained constant for a period of at least 5 hours.

The number of ampere-hours for the initial charge should never be less than five times the rated capacity at the 10-hour rate, as specified on the nameplate. If this rate is maintained, the battery will require approximately 84 hours of charging.

After the initial charge is completed, the specific gravity should be adjusted (corrected to 80° F) to that specified on the nameplate, as explained in Basic Electricity, NavPers 10086. The capacity of all Navy standard portable batteries, except Class 2V-SBP-20AH, is based on a fully charged electrolyte between the limits of 1.210 and 1.220 specific gravity at 80° F. Class 2V-SBP-20AH batteries are always operated with electrolyte specific gravity between the limits of 1.270 and 1.285. If the electrolyte has been mixed carefully, very little adjustment should be necessary.

The date of the initial charge must be stamped or scribed legibly in the space provided on the nameplate. The date of the initial charge is important because it indicates the beginning of the useful life of the battery.

NORMAL CHARGE.—A normal charge is a routine charge that is given in accordance with the nameplate data during the ordinary cycle of operation to restore a battery to a fully charged condition.

To conduct a normal charge, ascertain the proper charging rates for starting and finishing the charge from the nameplate data. Add distilled water to each cell to bring the electrolyte to the proper level, and replace the vent plugs. Connect the battery to the switchboard (the positive terminal to the positive side and the negative terminal to the negative side) and cut in all the resistance. The battery switch should be closed to the charging position and the rheostat adjusted until the proper charging rate is indicated by the ammeter. As the battery voltage rises, the rheostat is adjusted to maintain this charging rate.

Hourly readings are taken of the voltage, temperature, and specific gravity of the pilot cells during the charge. The battery temperature should not exceed 125° F. The charging rate is decreased to the finishing rate when the battery begins to gas freely or when the voltage reaches 2.35 volts per cell. The finishing rate should not exceed that specified for the particular battery.

The normal charge is completed when the specific gravity, corrected for temperature, has reached a value within five points (0.005) of that obtained on the preceding equalizing charge.

EQUALIZING CHARGE.—An equalizing charge is an extended normal charge that is given a battery at the finishing rate. It is given periodically to ensure that all the sulphate is driven from the plates and that all the cells are restored to a uniform maximum specific gravity. To conduct an equalizing charge, continue the normal charge at the finishing rate and take 30-minute readings of the temperature and specific gravity.

The equalizing charge is continued until the hydrometer readings, corrected for temperature, show no increase in corrected specific gravity for any cell over a period of 4 hours.

All Navy standard batteries are given an equalizing charge at 30-day intervals, except for engine-starting batteries and batteries normally on floating charge, which are given an equalizing charge at 90-day intervals.

FLOATING CHARGE.—A floating charge is given a battery that is standing idle or used on stand-by service to maintain it in a fully charged condition.

To conduct a floating charge, connect the battery across a power line that has a voltage maintained between the limits of from 2.13 to 2.17 volts per cell in the battery. The charging rate is determined by the battery voltage, not by a definite rate of current. The voltage is adjusted periodically to maintain an average of 2.15 volts per cell. Special equipment is necessary to give a battery a floating charge. If this equipment is not included with the battery-charging panel, a battery is maintained by giving it monthly equalizing charges.

EMERGENCY CHARGE.—An emergency charge is given to recharge a battery in the minimum amount of time. The charging rate is much higher than is normally used for charging. This charge can be accomplished by using the (1) constant-potential and (2) modified constant-potential (multiple-step) methods.

The CONSTANT-POTENTIAL METHOD employs a starting rate sufficiently high to bring the voltage up to about 2.35 volts per cell. If gassing starts, lower the voltage until gassing occurs only slightly, and use this voltage as the charging voltage. Maintain this voltage, with a steady decreasing current, until the finishing rate is reached. The finishing rate is held constant until the charge is completed. The charge can be stopped at any time, but it should be completed if possible.

The MODIFIED CONSTANT-POTENTIAL METHOD is started at about 2.25 to 2.30 volts per cell. The current rate obtained at this starting voltage is maintained until the voltage increases to 2.35 volts per cell. At this time the charging rate is dropped to a lower rate. This lower rate is continued until the voltage again increases to 2.35 volts per cell, and the charging rate is again dropped to a lower rate. The charge is continued in successive steps until the finishing rate is reached. At this time no further reduction in the charging rate is made.

The constant-potential and modified constant-potential methods of conducting an emergency charge cannot be used for charging cells of different sizes.

Tests

STATE OF CHARGE.-The specific gravity of the electrolyte gradually decreases on discharge and gradually increases on charge. This decrease and increase is nearly in direct proportion to the number of ampere-hours taken out of, or put back into, the battery, and is thus a check on the state of charge at any time. As explained in Basic Electricity, NavPers 10086, the state of charge, or the number of ampere-hours, expended by a battery can be determined from the (1) specific gravity when fully charged; (2) specific gravity after the battery has been discharged; and (3) reduction in specific gravity per ampere-hour. The voltage alone is not a reliable indication of the state of charge, except when approaching the low-voltage limit on discharge, and hence should not be used for this purpose.

TEST DISCHARGE.—A test discharge is the best method of determining the capacity of a battery. All Navy standard portable batteries must be given a test discharge at least once every 6 months. Also, a test discharge must be given when the cell voltage and specific gravity fall below specified limits at the completion of a periodic equalizing charge. The cell voltage readings are compared with the values indicated by the cell voltage versus electrolyte temperature curve (fig. 6-6).

The battery should be given a test discharge if any cell has a voltage that is less than that shown on the curve for the corresponding electrolyte temperature, or if the specific gravity of one or more cells cannot be brought up to within a 0.010 of the full charge, irrespective of whether the voltage after the equalizing charge is satisfactory (equal to or greater than the value indicated on the curve). To ensure that reliable data is obtained, a test discharge must be preceded by an equalizing charge.

To conduct a test discharge, connect the battery to the charging and discharging switchboard, and cut in all the resistance by means of the rheostat. The battery switch is closed to the discharging position, and the rheostat is

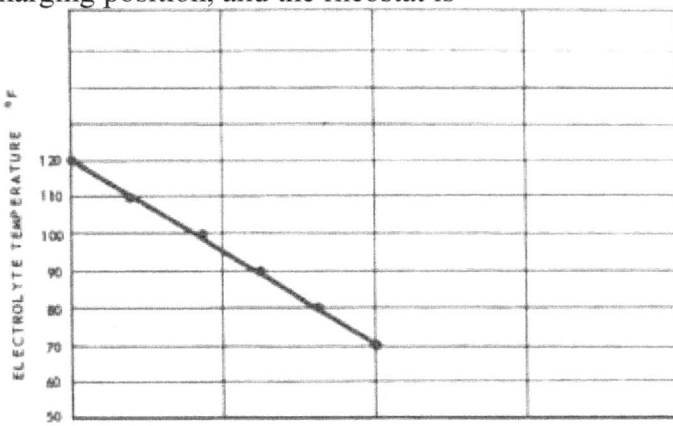

2.32 2.40 2.48 CELL VOLTAGE

Figure 6-6.-Cell voltage versus electrolyte temperature
curve.

adjusted for the proper 10-hour discharge rate. For example, the correct discharge rate for
a 200 ampere-hour battery is 200/10, or 20 amperes. The 10-hour discharge rate is continued
until (1) the total battery voltage drops to a value that is equal to 1.75 times the number of cells
in series, or (2) the voltage of any individual cell drops to 1.65 volts, whichever occurs first.

If a battery has 100-percent capacity, the discharge will continue for a period of 10 hours
before reaching the low-voltage limit, provided the temperature of the electrolyte at the
beginning of the discharge is exactly 80° F. Otherwise, the time duration of the discharge for
100-percent capacity must be corrected for the actual electrolyte temperature existing at the start
of the test.

The corrections, in the time to be applied to the 10-hour rate for 100-percent capacity for
various temperatures of electrolyte, can be obtained from the electrolyte temperature versus time
curve shown in figure 6-7. The corrections, in the time of discharge indicated by the curve, apply
only to the 10-hour rate.

As explained in Basic Electricity, NavPers 10086, the ampere-hour capacity of a storage
battery is calculated om the equation*

where C is the percentage of ampere-hour capacity available, H a the total hours of
discharge, and H, the total hours for 100-percent capacity. If the ampere-hour capacity is 80
percent or more, the battery should remain in service; whereas, if it is less than 80 percent, the
battery may be surveyed.

40

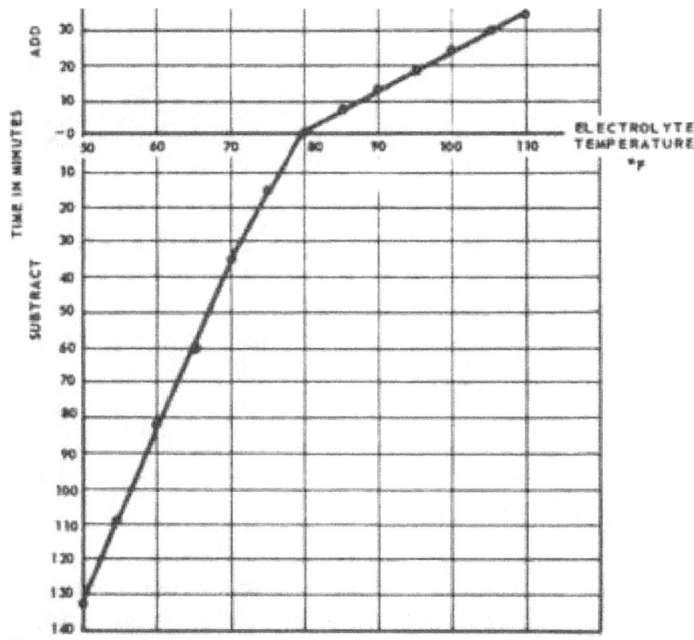

Figure 6-7.-Electrolyte temperature versus time curve.

TROUBLES

The most common trouble encountered in the operation of storage batteries in that of low cells. The specific gravity of low cells is usually below that of the other cells,
both on charge and discharge; hence, the expression, LOW, refers to specific gravity. This condition is caused by poor connections, sulphation, loss of active material, local action, short circuits, loss of electrolyte, hardened negative plates, disintegration of positive grids, broken positive pencils (iron-clad), and impure electrolyte. Low cells can be detected by progressive loss of capacity, low voltage on discharge, low specific gravity, abnormally high or low voltage on charge, and abnormally high temperature.

When trouble is experienced with any cell, it should be examined immediately for low electrolyte, poor connections, and voltage readings should be taken at the cell terminals while the battery is being charged or discharged. If the voltage readings are appreciably higher or lower than that of normal cells, the trouble is probably within the cell and must be Investigated further.

Sulphation

Sulphation is one of the most common causes for the failure of cells to maintain their normal charge. Consequently, low cells are often erroneously attributed to sulphation, with the result that the actual defective condition is not remedied but aggravated by the sulphation treatment.

Sulphated cells are indicated by a falling off of the specific gravity, low voltage on discharge, and loss of capacity. Sulphation is usually evidenced by an inspection of the element, but only after a normal charge, because a discharged plate is always somewhat sulphated. The plates of a sulphated element are hard and gritty, and the material lacks cohesion (fig. 6-8).

A sulphated condition will result in batteries that are given undercharges or partial charges, or if they are not given periodic equalizing charges. In normal battery operation it is difficult to determine when sulphation begins. The only way that sulphation can be detected in its early stages and corrected or prevented is by giving the required equalizing charges and comparing the specific gravity and voltage readings of individual cells.

A battery that is allowed to stand in a partially charged condition will become sulphated when the sulphate

Figure 6-8.-Sulphated element.

deposited on the plates hardens and closes the pores in the material. Batteries should be charged as soon as practicable after discharge, and they should never be allowed to stand in a completely discharged condition for more than 24 hours.

If the level of the electrolyte is permitted to fall below the tops of the plates, the exposed surfaces will harden and become sulphated. Also, a cell in which sulphation exists becomes aggravated with the addition of acid.

In general, sulphation is more likely to occur and will be more difficult to reduce in fully charged cells that have high values of specific gravity. The possibility of sulphation is increased in cells that have a specific gravity of 0.015 or 15 points above the average. High temperatures accelerate sulphation, particularly in idle, partially charged batteries.

Sulphation of a battery, except in extreme cases, may be reduced by a long charge, finished at a low rate. The charge should be started at the highest available rate that can be utilized without excessive gassing, and continued at progressively reduced rates to keep within the allowable

temperature limits before the charging current is reduced to the finishing rate. This procedure breaks down the sulphate that has not been reduced by previous normal charges. The finishing rate is approximately

$$FR = 1,1 \ 10 = 0.055 \ xAH \ 10 \ ,$$

where FR is the finishing rate in amperes, and AH 10 is the ampere-hour capacity at the 10-hour rate. The charge is continued at this rate until the specific gravity and voltage are constant for a period of 5 hours.

To measure the results of this treatment, compare the specific gravity and voltage readings with similar readings of healthy cells in the battery. These readings are also considered in conjunction with the capacity of the battery, as indicated by a test discharge. If the comparison is satisfactory, the battery has been restored to a normal condition, and no further treatment is necessary. Otherwise, the sulphation treatment should be repeated.

Loss of Active Material

The loss of active material, or shedding, can be determined by a visual inspection of the plates and by measuring the amount of sediment in the bottom of the jar or case (fig. 6-9). A normal loss of active material occurs in all batteries due to erosion caused by gassing at the end of a charge. An excessive loss of active material can be caused by too much gassing, overcharging, and expansion due to high temperatures.

Excessive shedding results in a reduction of capacity. It is a sign of wear and is often evident in batteries that have given long service. One of the first indications of shedding is a slightly muddy appearance of the electrolyte caused by the lead peroxide in suspension. There is no remedy for active material that has already left the plate. Additional shedding can be prevented by reducing the rate of charge, particularly at the end of a charge. Some batteries are provided with a mat of spun glass placed against the positive plate to retard the loss of active material after it has become soft and muddy.

Figure 6-9.-Loss of active material.

Local Action

Local action, or internal discharge, is indicated by an excessive drop in the specific gravity of the electrolyte when the cell is on open circuit. This condition is also indicated by excessive gassing when the cell is on discharge. In general, the higher the specific gravity the greater will be the tendency for local action, and the more rapidly the specific gravity will fall when the battery is standing idle.

Internal discharge is caused by impurities and non-homogeneity of the material in the plates or by impurities in the electrolyte. Internal discharge is sometimes caused by sediment in the bottom of the jar or by broken separators, which also cause considerable gassing.

Iron is the most common impurity. If impurities are suspected, the electrolyte should be dumped, and the plates should be rinsed with distilled water. The cells should be refilled with new electrolyte of the same specific gravity as that discarded and given an equalizing charge.

Short Circuits

Short circuits are indicated by abnormal temperature, low specific gravity, low voltage, and reduced gassing on charge. Short circuits can be caused by (1) faulty separators, (2) lead particles or other metal forming a circuit between the positive and negative plates, (3) buckling of the plates, (4) excessive accumulation of sediment in the bottom of the jar, (5) mossing that usually occurs between the tops of the negative plates and the positive straps, and (6) a cracked partition between cells in a monobloc case.

A cell that contains a short circuit will gradually discharge and will become unduly heated on charge. The element should be removed and examined to determine the cause of the short circuit. If mossing or sediment caused the faulty condition, the cell should be removed and the element replaced in the container. The element should be replaced also if the separators are damaged or if the plates are badly worn.

Loss of Electrolyte

Loss of electrolyte is indicated by an excessive drop in the level of electrolyte compared with that of normal cells, and this indicates a crack in the jar or case.

Hardened Negatives

Hardened negatives are the result of the formation of a hard, nonporous material on the negative plates. This condition can be detected by scratching the active material. A healthy plate is easily scratched and shows a bright metallic luster. A hardened plate contains a hard, insoluble coating, which is difficult to remove.

Reversal of Cell Voltage

A low cell, in series with other cells, may reverse its polarity when discharged considerably below its low-voltage limit. If this action occurs before the adjacent cells are near their low-voltage limit, it should be assumed that the condition is caused by faults within the cell.

A reversed cell can be detected by a voltage reading of the cell and it (reversed cell) will cause a rapid fall in the battery voltage. This fall in battery voltage may be as much as 3 or 4 volts because the cell voltage may

change from 1.7 volts in the proper direction to as much as 2.0 volts in the reverse direction in a short time. Except in an emergency, as soon as a reversed cell is detected, the battery must be taken out of service until the cell or tray is cut out.

REPAIRS

Major repairs to batteries, except for cleaning the sediment spaces and the removal of moss from the tops of the plates, must be performed only by authorized personnel at naval shipyards or other shore establishments, or aboard repair ships and tenders. The following instructions for the overhaul and repair of storage batteries primarily concern Electrician's Mates assigned to repair ships and tenders.

Dismantling

A storage battery should not be opened for repair until it has been given an equalizing charge followed by a test discharge to determine the approximate capacity as compared with the nameplate data. A cell should be opened when cycling the battery under ordinary conditions fails to restore the plates, and when a definite indication of defects within the cell is apparent.

REMOVING THE CONNECTORS.-The intercell connectors must (first) be removed to open a battery by (1) using a connector puller or (2) boring into the connectors. If apulleris used, remove the vent plug on the cell to be dismantled and place the puller in a vertical position on the connector. Then gradually force the plunger of the puller until the connector is free from the post. The puller provides a quick and easy method of removing the connectors, but its use necessitates trimming the posts before reassembling the cell.

If a connector puller is not available, the connectors can be removed by using a brace and bit to bore out each end of the connector to loosen the joint between the connector and the terminal post. The bit can be a twist drill or a wood bit and should have the same diameter as that of the post. Before drilling, center the bit accurately on the connector and bore the hole to a depth of about

three-sixteenths of an Inch. The vent plug should be in position (while boring) to prevent lead chips from falling into the cell. After drilling the connector to the proper depth, insert a screwdriver beneath the connector and pry gently, but firmly, on the connector. Repeat this operation on the other side of the post and continue until the connector is free.

The connector can also be removed by gripping it with pliers. Twist it back and forth to break the joint at the bottom of the drilled hole, and then lift it free (fig. 6-10). If the connector does not break loose, use the lead-burning torch to melt the lead slightly at the bottom of the drilled hole, and lift the connector off. Brush the lead chips off the top of the battery with a whisk broom and remove the vent plugs from the cells. Avoid damage to the cell cover and avoid short circuiting the cell by not allowing tools to come in contact with both terminal posts at the same time.

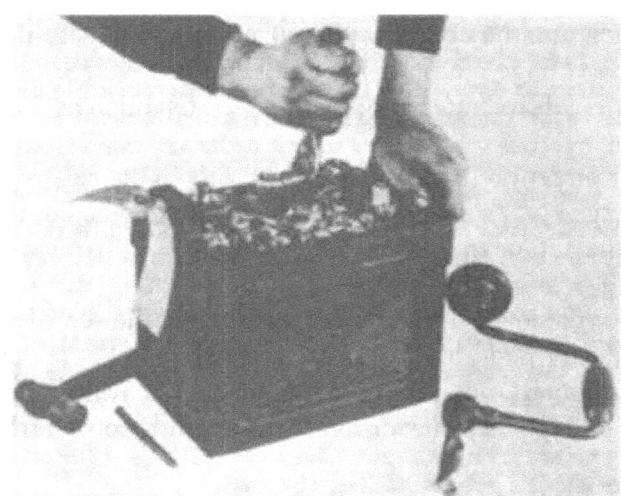

Figure 6-10.-Removing the connectors.

REMOVING THE SEALING COMPOUND.-Remove as much of the sealing compound as possible with a screwdriver. The remaining compound can be removed with heated putty knife. All of the compound should be removed

from the monobloc case to prevent damage to the separators when pulling the element. The compound that remains on the cell cover can be removed after the element is pulled from the case. Cell covers of batteries that have polystyrene parts are cemented to the monobloc case or jar and are not provided with sealing compound.

REMOVING THE ELEMENT.—The elements of batteries assembled in polystyrene cases or jars cannot be removed. The elements of all other batteries are readily removed by placing the battery on the deck and gripping the outside of the case between the feet. Then grasp both terminal posts with heavy pliers and pull upward steadily. When removing the element, do not allow it to strike the edges or bottom of the case and thus damage the separators. After pulling the element, place it on top of the case in order to drain for several minutes and then turn it upside down, with the posts resting on the work bench. Then pour the electrolyte carefully into a glass, lead, or rubber container; wash out the sediment and clean the inside of the monobloc case with fresh water. If the electrolyte is to be used again, allow the sediment to settle because only clear electrolyte should be poured back into the cells. The electrolyte should not be poured out while the element is in the case because the sediment will become lodged between the plates, with the consequent danger of short circuits after reassembly of the cell.

INSPECTING THE ELEMENT.-After the cell has been opened, a thorough inspection is necessary to determine the cause of failure and the amount of repairs required. The element can be readily inspected by slightly spreading the plates while it is in an inverted position. When spreading the plates, avoid excessive strain that can result in a broken plate or plate lug.

If the inspection discloses that either the positive or negative plates require renewal because of loss of active material, broken grids, or excessive sulphation, it Is usually good practice to replace the entire element, including the separators. However, if the positive and negative groups are in good condition and only the separators are cracked or badly worn, then replace the separators.

Rebuilding

Unless the element is to be replaced, it is not usually necessary to detach the cell cover when the element is removed from the case for inspection.

DETACHING THE COVER.-If the element must be renewed or rebuilt, the cell cover is readily detached by removing the sealing nut and lifting or prying the cover vertically. The covers of some batteries are secured to the terminal posts of the element by lead-burning the posts to lead bushings molded into the covers. These covers are removed by boring out the posts, similar to the method previously described for removing the inter-cell connectors.

REPLACING THE ELEMENT.-The monobloc cases of Navy portable batteries are of standard dimensions, and any type of element can be replaced with elements available in standard stock. In an emergency, if it is necessary to rebuild the element, select the proper size plates and set them up in positive and negative groups in a section of the lead-burning rack, which provides the proper distance between the plates (fig. 6-11). The terminal post must be placed properly so that it will fit the cover because the distance between the post holes in the covers is not the same for all batteries. The distance from the center of the post hole to the end of the cover should be measured and allowance should be made for this distance so that when the positive and negative-plate groups are assembled, the ends of the cover will align with the edges of these groups.

After a complete group of plates is set up in the lead-burning rack, the plate lugs are welded to the horizontal strap, which, in turn, is welded to the post strap. Lead-burning is the process of welding these parts together and is accomplished by means of a small oxyacetylene torch. Lead-burning requires skill, especially if some of the parts that are to be welded are thicker than others (the lighter parts melt first and flow away from the heavier parts).

The lead-burning rack is provided with small fences, which are set up around the plate lugs to hold the molten lead during the lead-burning operation. As the horizontal strap and the post strap are brought to the fusing point, a

Figure 6-11.-Setting up plates in a lead-burning rack.

lead-burning strip is inserted under the flame and melted to fill in, and build up, the weld to the proper size. When the torch is removed, the parts cool quickly, allowing the lead to set and form a solid junction.

When the lead-burning process is completed, the positive and negative groups are removed from the lead-burning rack and assembled, as illustrated in figure 6-12. Next, the rubber gaskets and cell cover are installed, and, finally, the nuts are secured that lock the posts to the cell cover.

Figure 6-12.-Assembling the element. 31

The separators are Inserted between the plates, with the grooved sides of the separators always placed adjacent to the positive plates and the flat sides adjacent to the negative plates. They are driven into position with a hammer and a block of wood. A trimmer is used to cut the separators so that they extend about one-sixteenth of an inch beyond the edges of the plates.

SEALING THE COVER.—The completed element, consisting of the positive and negative plate groups, separators, and cover, is then installed in the battery case (fig. 6-13). When all the elements have been placed in the case, the sealing compound is poured into the joints between the cell cover and the case to provide an acid-resistant and acid-tight seal.

Figure 6-13.-Installing the element,

The sealing compound is usually melted in an electrically heated, thermostatically controlled pot, the temperature of which can be adjusted from 350° F to 375° F. The sealing space is half-filled with the melted compound, which is allowed to cool for about a minute. The remaining space is then completely filled with the compound. A gas flame is applied lightly over the surface of the compound to smooth off any irregularities and to ensure complete adhesion. Do not scorch the cover or the container, and be sure that the seal is tight.

If the posts are not of the proper size, it is often necessary to trim them with a hacksaw or trimmer before burning on the connectors. The posts should be trimmed sufficiently to permit the connector to fit close to the lead nuts that hold the cell covers to the posts.

BURNING ON THE CONNECTORS.-Intercell connectors are provided with a hole (in each end) that fits over the terminal post. After the connector is placed in position for burning on, apply the torch flame to the top of the post, using care to keep the flame away from the thin end of the connector at the hole. Continue applying the heat flame to the top of the post until it begins to melt. Then use a circular motion to bring the flame simultaneously to the outer edge of the post and in contact with the inner surface of the hole in the connector. The circular motion is continued until the two parts fuse together. After the connector is fused to the posts, apply additional lead with a lead stick to completely fill in the connection.

It is usually necessary to build up the positive and negative terminal posts of a rebuilt battery. A steel mold (post builder) that fits over the post is used for this purpose. Post builders are furnished in two sizes, one for the positive, and the other for the negative posts. The appropriate post builder is placed in position, and the torch is applied to the top of the post until it is melted; lead is then added from a lead-burning strip until the mold is filled. When the lead has cooled and set, remove the post builder by slowly twisting it. The rebuilt post is tested by gripping it at the top with pliers, and twisting it. If the top breaks off, the lead has not fully fused and the operation must be repeated. When the post is properly rebuilt, it will have a smooth exterior surface, free from

cracks and pits. A wire brush is used to burnish newiy finished lead-burning jobs.

When the battery has been completely reassembled, the cells are filled with electrolyte of the proper specific gravity, corrected for temperature, and given an equalizing charge, as previously described.

Electrolyte

As previously mentioned, the capacity of all Navy standard portable batteries, except Class 2V-SBP-20AH, is based on a fully charged specific gravity of the electrolyte between the limits of 1.210 and 1.220 at 80° F.

INCREASING THE SPECIFIC GRAVITY.—If the specific gravity of any cell has dropped belowthe 1.210-1.220 limit, the affected cells are given a 4-hour equalizing charge. At the end of this charge, determine accurately the specific gravity and correct it for the electrolyte temperature existing at the time. For each point that the actual specific gravity of the cell is below 1.215, the amount of electrolyte to be withdrawn must be approximately equal to 1 percent of the total volume of electrolyte normally contained in the cell. For example, if the specific gravity of the low cell is 1.195, it will be (1.215-1.195) x 1000, or 20 points below normal. Thus, the amount of electrolyte to be withdrawn will be 20 x .01, or 20 percent, of the normal volume of electrolyte in the cell. The withdrawal must never lower the electrolyte level below the tops of the separators.

To ensure thorough mixing, the level of the electrolyte is restored to normal by adding electrolyte of 1.300 specific gravity while the cells are being charged and are gassing.

The affected cells are charged at the finishing rate for a 1-hour period. At the end of this period the specific gravity is accurately determined and corrected for the electrolyte temperature existing at the time.

The foregoing procedure is repeated as many times as necessary to increase the specific gravity of the affected cell within the limits of 1.210 to 1.220 at 80° F.

DECREASING THE SPECIFIC GRAVITY.-If the specific gravity of any cell is above the upper limit of 1.220, the affected cells must be given a 4-hour equalizing charge.

At the conclusion of this period, the specific gravity is accurately determined and corrected for the electrolyte temperature existing at the time. For each two points that the specific gravity of the cell is above 1.215, the amount of electrolyte to be withdrawn must be approximately equal to 1 percent of the total volume of electrolyte normally contained in the cell. For example, if the specif ic gravity of the high cell is 1.230, it will be (1.230-1.215) x 1000, or 15 points, above normal. Thus, the amount of electrolyte to be withdrawn will be 15/2 x .01, or 7.5 percent, of the normal volume of electrolyte in the cell.

To ensure thorough mixing, the level of the electrolyte is restored to normal by adding pure distilled water while the cells are being charged and are gassing.

The affected cells are charged at the finishing rate for a 1-hour period. At the end of this period, the specific gravity is accurately determined and corrected for the electrolyte temperature existing at the time.

The foregoing procedure is repeated as many times as necessary to decrease the specific gravity of the affected cell below the 1.220 limit at 80° F.

SURVEY

A battery should be discarded, according to the Bureau of Ships policy, when it is definitely ascertained that it cannot be made to give more than 80 percent of its rated capacity. Care must be exercised in determining the capacity of a battery because its capacity maybe only temporarily low.

Scrapping Procedure

When an EM surveys a battery he should not scrap it unless he definitely determines that it cannot be made to give further service. However, he should exercise care that time and money are not spent in an effort to keep a worthless battery in service.

Batteries that are sent to shore establishments or repair ships and tenders for survey must be given an equalizing charge, followed by a test discharge, as previously explained. During the equalizing charge, if any

«JOQ

cell has a higher temperature, a lower voltage, or less gassing than the other cells, a short circuit is indicated within this cell. If this condition exists in a battery prior to the normal 4-year useful life expectancy, it should be opened, cleaned, and repaired, if conditions warrant.

On the other hand, if no troubles are indicated by any cells in a battery at the conclusion of the equalizing charge, the battery should be discharged at the 10-hour rate to the low-voltage limit, and the capacity thus obtained should be compared with the rated capacity from the nameplate data, as previously explained. If the battery delivers at least 80 percent of the rated capacity on the test discharge and is otherwise in good condition, it should be recharged and restored to service.

If the battery does not deliver at least 80 percent of the rated capacity on the test discharge, it should be repaired; if this is not possible, it should be scrapped. Normally, a battery must not be scrapped until a recommendation (for scrapping) has been received from authorized survey and repair personnel. In an emergency, if it becomes necessary to scrap a battery without authorized recommendation, the EM should state clearly the reasons in the survey.

When portable storage batteries contained in hard-rubber monobloc cases are surveyed, the elements can be scrapped, but the cases should be cleaned, set aside, and credited in the Appropriation Purchases Account (APA). These cases are sent to repair ships and tenders for replacing damaged cases in reassembled or rebuilt batteries.

Data Required

All requests for survey of portable storage batteries must contain the following information for each tray to be surveyed:

1. Name of manufacturer.
2. Navy type and manufacturers' type.
3. Purchase contract or order number.
4. Date of initial charge.
5. Service for which used.
6. Details of special treatments given to bring the battery (tray) to 80 percent of rated capacity or over.
7. Complete description of the condition of the battery at time of survey.
8. The Storage Battery Tray Record, NavShips 151, must accompany the survey request.

SAFETY PRECAUTIONS

Battery charging stations must be provided with adequate ventilation to remove the explosive mixture of hydrogen and air. As previously explained, when batteries are being charged, hydrogen gas is liberated; if the surrounding air contains from 4 to 8 percent hydrogen, a mixture is formed that will burn, if ignited. If the mixture contains more than 8 percent hydrogen, it will explode. Hence, ample ventilation is necessary to keep the hydrogen concentration below an established safe limit of 3 percent.

When a large number of batteries are concentrated in a single compartment, the ventilation problem becomes serious, and care must be exercised to ensure that the ventilating system is operating at all times. The system should be supplied with fresh air from the exterior of the ship, and the exhaust from the battery compartment should lead overboard. Battery explosions are dangerous and will result in serious injury to personnel, and in damage to equipment.

The ventilating system, in addition to preventing the formation of the explosive mixture, provides a means of keeping the temperature of the battery-charging station down to 95° F or lower, and thus facilitates the charging of batteries (the charging rate is limited to a value that will cause the electrolyte temperature not to exceed 125° F).

After a battery compartment that has been sealed is opened, it must be thoroughly ventilated before light switches are turned on, any other electrical connections made or broken, or work of any kind performed. Flames and sparks must be kept away from the vicinity because storage batteries give off a certain amount of gas at all times.

The ventilating system must be operating properly before starting a charge. The charge must be stopped if ventilation is interrupted; except in an emergency, the charge should not be resumed until the ventilation has been restored.

All batteries must be kept clean and free from the accumulation of acid and dirt; otherwise corrosion will occur and eventually lead to troublesome grounds. Grounds are formed by the collection of dirt and acid on the cell tops and sides of a battery. They will result in dissipation of battery energy and disarrangement of the circuit in which the battery is connected. Also, a ground in the vicinity of a battery can furnish a spark that will ignite an explosive mixture.

Acid that has collected on a battery can be removed with a cloth moistened in a solution of dilute ammonia or bicarbonate of soda. Also the top and sides of a battery should be cleaned after each watering.

A battery should always be charged at the prescribed rates; never at a higher finishing rate than that indicated on the nameplate. When charging more than one battery at the same time, the voltage of the charging line should be known to exceed the total voltage of all the batteries being charged in series. Also, the charging rate in amperes should be known not to exceed the maximum charging rate of the battery having the lowest ampere-hour capacity in the charging line. The charging rate should be lowered as soon as the battery begins to gas or the temperature reaches 125° F. Except in an emergency, a battery should not be discharged below the low-voltage limit. As previously stated, a battery should never be allowed to stand in a completely discharged condition for more than 24 hours.

Only tools with insulated handles should be used while servicing a battery, and care should be exercised not to short circuit the terminals. No repairs to battery connections should be made when current is flowing in the circuit, and batteries should never be connected to, or disconnected from, the charging line without first turning off the charging current. When batteries are used with one terminal grounded, the grounded terminal should be disconnected (first) when removing the battery and connected (last) when replacing the battery. This action will avoid the possibility of grounding the hot terminal and shorting the battery when the ungrounded terminal is disconnected first.

When mixing electrolyte, always pour the acid into the water, not the water into the acid.

The acid must be added slowly to prevent excessive heating, and cautiously to prevent splashing. Stir the solution continually, while mixing, to prevent the acid (which is heavier than water) from flowing to the bottom of the vessel. The solution becomes very hot when the concentrated acid is diluted.

Personnel handling or mixing electrolyte must wear rubber aprons, rubber boots, and rubber gloves to prevent the acid from coming in contact with the skin or clothing. The eyes, in particular, must be protected by wearing goggles.

Sulphuric acid having a specific gravity greater than 1.350 should not be added to a battery. Pure water or pure electrolyte only should be added, and the level of the electrolyte should be maintained above the tops of the separators. Sulphuric acid should be stored in locations where freezing cannot occur.

QUIZ

1. What are three general classes of primary cells ?

2. What is the open-circuit voltage of the Leclanche cell?

3. What is the open-circuit voltage of the mercury cell?

4. In addition to specification sheets for individual battery types, what is done to provide installation and replacement data for each individual dry battery?

5. What does the battery code number indicate?

6. What is the battery rating called that indicates the time required (at normal temperature) for the battery voltage to fall below the specified test-end voltage, or to show evidence of electrolyte leakage, or swelling of the case when discharging through a specified load ?

7. What is the battery rating called that indicates the time required (at normal temperature) for the battery voltage to fall below the specified test-end voltage, or to show evidence of electrolyte leakage, or swelling of the case after the battery has been stored for a period of time?

8. What is the battery rating called that indicates the length of storage time beyond which a group of batteries will contain so many dead cells that the entire group is considered unusable?

9. What should be the maximum age of batteries issued for use aboard ship where premature failure would cause serious inconvenience?

10. What should be the minimum closed-circuit voltage per cell of multicell dry batteries below which the batteries should not be used?

11. What two types of storage batteries are in general use in the Navy ?

12. How many series-connected cells are contained in a 6-volt, nickel-cadmium cell?

13. What is the approximate life of the nickel-cadmium battery ?

14. What is the temperature range over which satisfactory operation can be obtained?

15. What type of battery is indicated by the symbol, BB, which is denoted as the battery component in the type designation contained on the nameplate ?

16. What is the meaning of the rating of a battery denoted by the symbol, 100 AH?

17. At the completion of a charge at the finishing rate, what should be the approximate voltage per cell of a lead-acid storage battery?

18. What is the reason for using diluted ammonia or a solution of bicarbonate of soda on the contact surfaces of lead-acid storage battery terminals before making the connections ?

19. After bolting the contact surfaces together, battery connectors should be coated with what type of material ?

20. A complete history of each storage battery tray in the ship must be maintained on

what form?

21. Why should water never be poured into a solution of concentrated sulphuric acid?

22. A normal lead-acid storage battery on open circuit will discharge 50 percent or more in approximately how long ?

2 3. Batteries used for gun firing and sight-lighting purposes, that are idle a large part of the time, should be given an equalizing charge how often?

24. In cold-weather operations, how often may it be necessary to make inspections of boat batteries ?

25. How often should battery hydrometers be cleaned?

26. How often should the specific gravity and temperature readings of pilot cells for all batteries be taken and recorded?

27. How often should each lead-acid storage battery be given a test discharge?

28. In the battery charging and discharging switchboard installed in a destroyer (figs. 6-4 and 6-5), what type of component is used in each of the four charging and discharging circuits to obtain the necessary current flow for the number of batteries connected in the associated circuits?

29. If a series-connected line contains two 6-volt batteries and one 12-volt battery with maximum charging rates of 5, 7, and 10 amperes, respectively, what should be the charging rate of the circuit?

30. The number of ampere hours for the initial charge should never be less than what amount of the rate specified on the nameplate ?

31. Why is it important that the date of the initial charge be indicated clearly on the nameplate of the battery?

32. The equalizing charge given to a lead-acid storage battery is contained until the hydrometer readings, corrected for temperature, show no increase in corrected specific gravity for any cell over a period of how many hours ?

33. What is the correct test discharge rate for a 200 ampere-hour battery?

34. What is the most common trouble encountered in the operation of storage batteries ?
•JOG

35. Additional shedding of the active material from the positive plate can be prevented by what action with respect to the charging rate?

36. What condition is indicated by abnormal temperature, low specific gravity, low voltage, and reduced gassing on charge ?

37. When a low cell, in series with other cells, is discharged considerably below its low-voltage limit, what may happen to its polarity?

38. A battery should be discarded when it cannot be made to give more than what percent of its rated capacity?

39. Why must battery charging stations be provided with adequate ventilation?

40. When a battery begins to gas or the temperature reaches 125 F, what should be done to the charging rate ?

41. What is the maximum time that a battery should be allowed to stand when it is completely discharged?

42. When mixing electrolyte, is the water poured slowly into the acid or the acid poured slowly into the water?

CHAPTER

PROTECTIVE DEVICES

In complex electrical systems, such as those used aboard ship, protective and control devices must be used in order to carry out normal operations.

This chapter describes how and why protective and control devices are used. Because an important part of your duties is to maintain uninterrupted operation of equipment, you must understand the purpose of each device.

This knowledge will enable you to quickly find the cause of interruption, clear the trouble, and restore operation with a minimum of lost time.

Any trouble, whether due to overload or fault, must be cleared before new protective devices (for example, fuses) are installed. Otherwise the new device will be ruined by the same defect that caused the original breakdown.

Protective devices may differ considerably in appearance and construction. Regardless of their appearance or construction, they are used for one purpose only—to protect life and equipment.

Most protective devices are designed to interrupt the power to a circuit when circuit conditions become dangerous.

Control equipment is designed to control the power to a circuit. Electrical devices may function at timed intervals or operate instantly when certain conditions are reached. Some equipment is manually controlled.

The purpose, construction, and testing of each device will be described individually. When two devices are used on the same circuit, the purpose and operation of each device will be explained. By studying the equipment in this manner, you should be able to visualize power supply systems and the methods used to protect and control entire systems as well as individual branches.

FUSES

The simplest protective device is a fuse. A complete fuse consists of a metal alloy strip or wire and terminals for electrically connecting the fuse into the circuit. All fuses are rated according to the amount of current that is safely carried by the fuse element. Usually, the current rating is in amperes, but some instrument fuses are rated in fractions of an ampere.

The fuse element is designed to melt when the current through it exceeds a predetermined value. Thus, when the circuit is overloaded, or a fault develops, the fuse element melts and opens the circuit that it is protecting. However, all fuse openings are not the result of overload or circuit faults. Aging of the fuse element, poor contact at the fuse holder, and the condition of the surrounding atmosphere will affect the time required for the element to melt.

Delayed-Action Fuses

Some equipment, such as electric motors, requires more current during the starting cycle than for normal running. Thus, a fuse rating that will give running protection might blow during the period when high current is required. Delayed-action fuses have been developed to handle these situations.

A heater element is connected in parallel with the fuse element in order to get the delayed action. During normal operation the heat developed in the fuse link is not great enough to melt the link. The melting, or opening, of the fuse link depends on the transfer of heat to the link from the heater. Therefore, more time is needed to melt the link than would be required if the link were directly heated.

Because the heater and fuse element are in parallel, the opening of the fuse element will cause the total circuit current to flow through the heater. The high current will cause the heater to burn out and completely open the circuit.

Another type of delayed-action fuse has the fuse element and heater connected in series.

Current above that of the rated value for a short time will have no effect on the fuse or heater. However, prolonged overloads cause the heater section to become hot enough to melt the junction between the elements. This action also opens the circuit.

Delayed-action fuses are sometimes called "slow-blow" fuses, and two trade names, Fusestat and Fusetron, are in common use.

Plug Fuses

The plug fuse is constructed so that it can be screwed into a socket mounted on the control panel or distribution center. The fuse link is enclosed in an insulated housing of porcelain or glass. The construction is so arranged that the fuse link is visible through a window of mica or glass. Therefore, an open element may be located by visual examination. The plug fuse is used primarily to protect low-voltage,low-current circuits. The operating ratings range from 0.5 to 30 amperes up to 150 volts.

Although the plug fuse is seldom found aboard naval vessels, it is used extensively in shore installations. When found to be defective, the fuse is discarded and a new fuse installed in its place. The plug fuses are listed in Group 5920, Section 2 of General Stores Catalog.

Usually, the spacing of plug fuse contacts is about 1/2 of an inch—too close for high currents. The spacing between the fuse contacts must be increased when high currents are to be handled. In addition, the physical size of the fuse link must be increased as the current through it increases. This means that greater distance between line contacts is needed.

Cartridge Fuses

The cartridge fuse fills all of the requirements. In operation, the cartridge fuse is exactly the same as the plug fuse. In construction, the fuse link is enclosed in a tube of insulating material with metal ferrules at each end (for contact with the fuse holder). The dimensions of cartridge fuses vary with the current rating. A cartridge

fuse rated at 30 amperes is approximately 5/8 of an inch in diameter and 2 inches long. A cartridge fuse rated at 100 amperes is approximately 7/8 of an inch in diameter and 4 inches long. Fuses of the same amperage rating may vary in length according to voltage. In addition, fuses of one rating for use on a-c may have different dimensions than those designed for d-c use at the same ratings.

Because an arc is produced when excessive current causes the fuse link to open, the fuse tube is filled with materials, such as shredded asbestos or boric acid, which act as an arc quencher. Cartridge fuses are listed in Group 5920, Navy Stock List of the Electronics Supply Office Catalog.

NONRENEWABLE AND RENEWABLE FUSES.-Cartridge fuses are divided into two types—nonrenewable and renewable. They differ in mechanical construction only. The end ferrules, or contacts, of the nonrenewable types are permanently attached. Thus, when this type of fuse becomes open circuited, it must be discarded. Because high-current fuses are expensive, a fuse with a renewable link has been developed. In this type, the end ferrules are threaded so that they can be removed. The open fuse link may then be replaced with a good unit, which has the same current rating as the original. If an arc-quenching substance has been used, it should be replaced before replacing the ferrule.

Blown-Fuse Indicators

It is not always possible to detect a blown fuse by a visual examination. Hence, fuses are often equipped with a device that will provide a visual indication so that a blown-fuse condition can be readily detected (fig. 7-1). These devices consist of the spring-loaded and the neon-lamp types of blown-fuse indicators.

In the spring-loaded type (fig. 7-1, A) when the link opens, it releases a spring that is held under tension. This action exposes an indicator, which makes the visual location of the blown fuse possible.

The neon-lamp type (fig. 7-1, B) is designed to be mounted on the fuse. When the link opens, a neon lamp glows to show a blown fuse.

A" SPRING-LOADED INDICATOR

B - NEON- LAMP INDICATOR

Figure 7-1.-Blown-fuse indicators.

When no indicator is used, it is necessary to test the fuse continuity with a megger, ohmmeter, or voltmeter. Various methods of testing will be described later in this chapter.

Most fuse panels and switchboards are of the enclosed panel type. The term, "dead-front," means that all fuses and bus connections are enclosed in a metal cabinet when the cover is closed. The use of this type of construction reduces the possibility of property damage and danger to personnel. Modern switchboards are of the "dead-front" type.

However, the complete enclosure of the equipment makes it less accessible for test purposes. Therefore, most fuses used on "dead-front" switchboards have indicators that show when a fuse is blown. The fuse holder consists of a molded phenolic base, plug, and cap with a built-in indicator lamp (blown-fuse indicator). The lamp is usually a small neon bulb, which normally is shunted by the fuse element. When the fuse opens, the shunt is removed, causing an increase in the voltage across the neon lamp. The lamp then glows, indicating the open fuse.

TROUBLESHOOTING ELECTRICAL CIRCUITS

The complete electrical system of a ship may consist of a comparatively small number of circuits or, in the case of larger ships,the installation maybe equal to that of a fair sized city.

Regardless of the size of the installation, it consists of a source of voltage (generator or batteries); a means of transferring power from the source to the load (wires or bus bars); and the load, which is made up of all the lights, motors, and other electrical equipment.

The entire electrical load is divided into circuits or branches to reduce the possibility of one part failure interrupting the power for the entire ship. The distribution boxes are illustrated in figure 7-2. The feeder distribution boxes (fig. 7-2, A) and the branch distribution boxes (fig. 7-2, B) contain fuses to protect the various circuits.

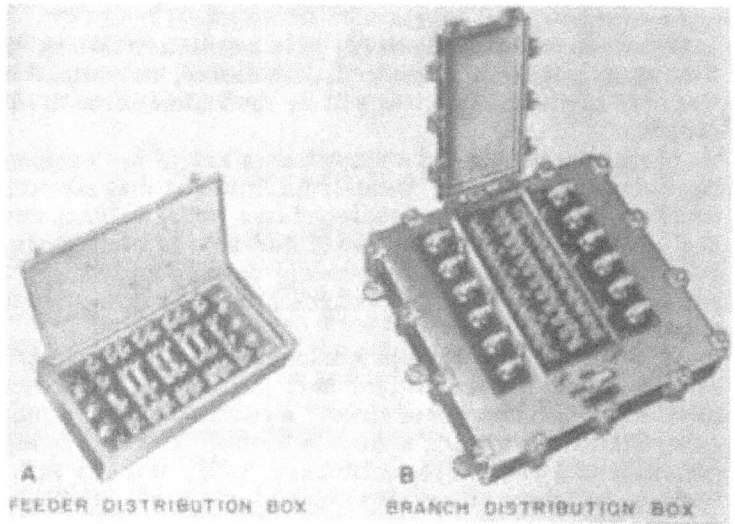

Figure 7-2.—Distribution boxes.

The distribution wiring diagram that shows the connection that might be used in a ship's lighting system is illustrated in figure 7-3. A ship installation might have several feeder distribution boxes, each supplying six or more branch circuits through branch distribution boxes.

I a!

!

I z

L_

L

FEEDER SUPPLY

<? (? e

FEEDER SUPPLY

<? e <?

1

r^i ^2 ^3

X

o

Z

a

1 I

O

o

-FEEDER CABLE

FEEDER DISTRIBUTION BOX

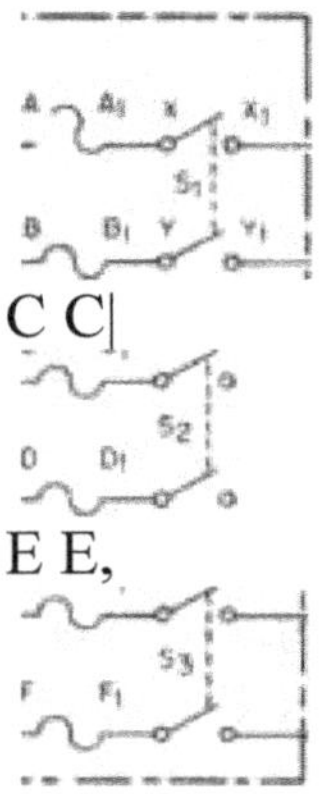

2 2 3

BRANCH NO 1

BRANCH NO 2

L ! !

BRANCH NO 3

BRANCH DISTRIBUTION BOX

Figure 7-3.-Distribution wiring diagram.

Fuses Fl, F2, and F3 protect the main feeder supply from heavy surges such as short circuits or overloads on the feeder cable. Fuses A-Al and B-Bl protect branch 1. If trouble develops or if work is being done on that circuit switch SI may be opened to isolate branch 1. Branches 2 and 3 are protected and isolated in the same manner by their respective fuses and switches.

Branch Circuit Tests

Usually, receptacles for portable equipment and fans are on branch circuits separate from lighting branch circuits. Test procedures are the same for any branch circuit. Therefore, a description will be given of the steps necessary to (1) locate the defective circuit and (2) follow through on that circuit and find the trouble.

VOLTAGE TESTER.—The voltage tester issued to each electrician is of the solenoid type. Make certain that it is in good condition. The indicator is a plunger, which moves in and out of a coil according to the amount of voltage applied to the coil. Because of the construction of the voltage indicator, a slight vibration is felt, and an audible hum is produced when it is connected to a source of alternating voltage. This type of tester is shown in figure 7-4.

Remember that tester leads may break, making it impossible to measure voltage. To check the tester, attempt to measure voltage at several pairs of fuses. If no voltage is indicated, the tester should be carefully checked for open leads. If you find voltage at several points but none at one pair of fuses, you can assume that you have located the fuses that protect the inoperative circuit.

It is easy to fall into the habit of assuming that normal voltage is present when you feel the vibration and hear the hum. However, these indications only mean that SOME voltage is present. It is necessary to actually READ the scale to determine the amount of voltage because certain defects will cause less than normal voltage. Unless this is done, you may assume that normal voltage is present when the actual voltage is less than normal. Less than normal voltage on certain tests (shown later) is a definite clue to the location of faults.

Never use a lamp in a "pigtail" lamp holder as a voltage tester. Lamps designed for use on low voltage (120 v) may explode when connected across a higher voltage (440 v). In addition, a lamp would only indicate the presence of voltage, not the amount of voltage. Learn to use and rely on standard test equipment.

LOCATING THE DEFECTIVE CIRCUIT.-Assume that, for some reason, none of the lights are working in

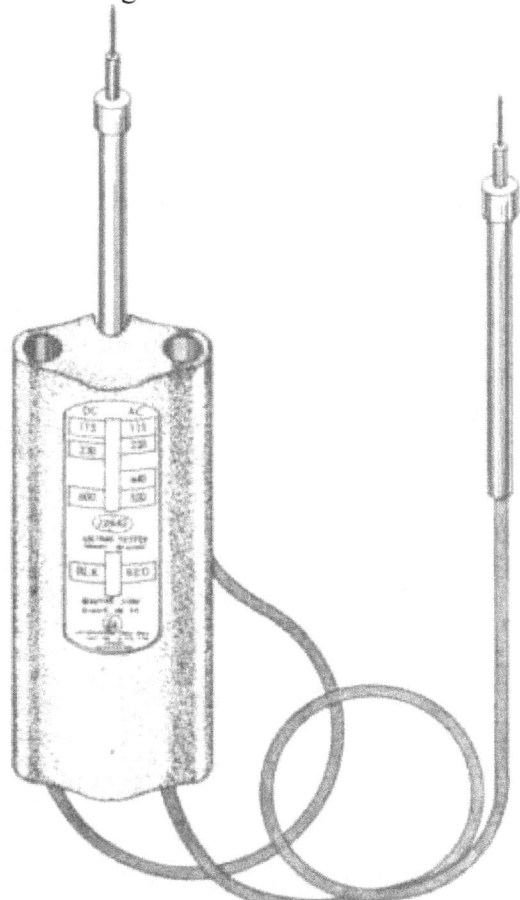

Figure 7-4.-Voltage tester.

a certain compartment. Because several lights are out, it will be safe to conclude that the voltage supply has been interrupted.

To verify this conclusion, you must first locate the distribution box feeding the circuit that is inoperative. Then make sure that the inoperative circuit is not being

supplied with voltage. Unless the circuits are identified in the distribution box, you will first need to measure the voltage at the various circuit terminations.

To pin down the trouble, connect the voltage tester to the load side of each pair of fuses in the branch distribution box. No voltage between these terminals indicates a blown fuse or a failure in the supply to the distribution box. To find the defective fuse, make certain SI is closed, then connect the voltage tester across A-Al, and next across B-Bl (fig. 7-3). The full -phase voltage will appear across an open fuse, provided circuit continuity exists across the branch circuit. However, if there is an open circuit at some other point in the branch circuit, this test is not conclusive. If the load side of a pair of fuses does not have the full-phase voltage across its terminals, place the tester leads on the supply side of the fuses. The full-phase voltage should be present. If the full-phase voltage is not present on the supply side of the fuses, the trouble is in the supply circuit from the feeder distribution box.

Assume that you are testing at terminals A-B (fig. 7-3) and that normal voltage is present.

Move the test lead from A to Al. Normal voltage between Al and B indicates that fuse A-Al is in good condition. To test fuse B-Bl, place the tester leads on A and B, and then move the lead from B to Bl. No voltage between these terminals indicates that fuse B-Bl is open. Full-phase voltage between A and Bl indicates that the fuse is good.

This method of locating blown fuses is preferred to the method in which the voltage tester leads are connected across the suspected fuse terminals because the latter may give a false indication If there is an open circuit at some point In the branch circuit.

TESTING FOR GROUNDS.-If no voltage is supplied to the branch circuit, determine the type of circuit: lighting, portable equipment, receptacles, or other. This is necessary to avoid false interpretations of tests made later.

To test for grounds, find or prepare a spot on the bare metal to avoid the insulating effect of paint or enamel. This spot should be on a bulkhead or overhead to ensure a good ground. To protect yourself while making tests, leave the fuses out of the holder. Then, close SI in order to make the ends of the circuit accessible Inside the cabinet. In addition, 11 you close the switch, its contacts and internal connections are included In the tests.

Connect one megger test lead to the bare spot prepared on the metal and the second test lead to terminal X (fig. 7-3).

A resistance of 1 megohm or more indicates that the lead from terminal X to the end of the circuit is free from grounds. The next part of the test is to move the test lead from terminal X to terminal Y. The second test lead should remain connected to ground. If the resistance from Y to ground is 1 megohm or more, the lead from Y to the end of the circuit may be considered free from grounds.

If both leads are found to be free from grounds, check for overloading. This might happen if a machine tool had been connected to a circuit that was already near maximum loading. The additional load will cause the fuse to blow.

OTHER TROUBLE FINDING HINTS.-Make it apart of your circuit-testing steps to check if any additional loads were connected when the fuse opened. If you find an overload, it should be removed. If no overload is present, go back to the branch distribution box, open SI, and replace fuses A-Al and B-Bl. Then, close SI. If the fuse does not blow, you can assume that the first fuse failure was due to aging or some other transient effect.

However, if the replacement fuse blows, the overload may be due to a short circuit between conductors, which would not show up in a megger test for grounds. A megger is not suitable for checking low resistances, and it will be necessary to use an ohmmeter (lowest range scale) in making further tests. Because an ohmmeter can be damaged by connecting it in a circuit that is carrying current, open Si and remove fuses A-Al and B-Bl before starting the tests. Then close Si so that It will be included in the tests.

The ohmmeter test leads should be connected to terminals Al and Bl. Zero resistance between these terminals indicates a direct short circuit across the line. However, you should be careful in interpreting the results of this test. For example, consider a case where 10 lamps are connected in parallel.

The number of lamps and the individual wattage of the lamps will determine the circuit resistance measured between Al and Bl. The resistance checked in this manner is the "cold" resistance of the lamp filaments. This resistance is not the same value that you would obtain in an Ohm's law calculation, using rated voltage and power because for this condition, the filament is hot, and, for tungsten, this will be a large increase in resistance. For example, the cold resistance of a 40-watt, 120-volt lamp may vary between 40 and 50 ohms; whereas, the hot resistance is of the order of 360 ohms. Ten of these lamps connected in parallel will have a "cold" resistance between 4 and 5 ohms. These low values are difficult to measure unless you use considerable care in setting up the ohmmeter and in reading the indicated value. Unless these precautions are taken, you may easily mistake the normal "cold" resistance of a parallel connected group of lamps for a short circuit.

Assume that your resistance tests have shown a short circuit. How do you locate the trouble?

Although branch No. 1 (fig. 7-3) shows three lamps, an actual circuit may contain more or less than this number. Choose a lamp that appears to be near the center of the affected group and open the housing to expose the leads.

The wire leads are attached to the fixture with screw-type fasteners. The socket shell screws are inserted in metal bars, which serve to connect the lamp socket terminals to the line. Two other screws anchored in the socket insulation serve as feed-through terminals for line wires entering and leaving the outlet box. Therefore, it will be easy to "pair up" the leads for testing. Disconnect each pair of leads from the fixture terminals.

As a safety precaution, when you open a box or outlet, always check for voltage at that location; never assume that the circuit is dead.

It is possible to determine the direction of the fault from the opened outlet box. To do this, connect the ohm-meter to the pair of leads that go back to the distribution box. If the meter indicates a short circuit, the fault is in that direction. If an open circuit is indicated, the ohm-meter leads should be removed and connected to the pair of wires running toward the end of the circuit. Remember

to take lamp resistance into account or remove the lamps from the sockets.

If no short circuits show up when you make this test, the short is in the fixture that has been disconnected. The trouble will naturally disappear when the defective fixture is disconnected.

Assuming that your tests at the opened box have shown a fault toward the end of the line,open another outlet box at a point approximately halfway between the first opened outlet box and the end of the line. The leads in the second opened outlet should be disconnected from the fixture. Each pair of wires should be tested for a short circuit. This test will show whether the trouble is between the two open boxes or between the second open box and the end of the circuit.

If the trouble is found to be between the two opened outlet boxes, open any boxes between them, disconnect the fixture, and make the tests for short circuits at each box and at the end of the line. Bear in mind that the trouble may be in the cable running between the two boxes. In this case, it is usually necessary to pull in new cable. However, some cables have a spare wire, which might be used to save running a new cable.

On circuits for portable equipment, disconnect each item. If a short circuit disappears, the trouble is in some of the portable equipment. In this case, the defective fuse can be replaced and the circuit energized. If the short remains with the portable equipment disconnected, the circuit should be tested, as described for lamp circuits.

Of course, if the trouble is due to defective portable equipment, each item should be carefully checked before it is reconnected to the line. Any defective equipment should be repaired.

To test a circuit for portable devices, it is sometimes helpful to connect an ohmmeter at the distribution box and watch the meter (or have someone watch it for you) as eachportable device is disconnected from the line. If the short circuit disappears when one item is disconnected, assume that this item was the cause of the trouble.

Three-Phase Feeder Distribution Circuits

The tests previously described relate to individual branch circuit troubles. However, branch circuit faults

may not always be confined to an individual branch. Instead, the trouble may be in the feeder distribution system.

For example, assume that you get a report that one-third of the lights in a section or compartment are normal and the remaining two-thirds are dim. Because these lights are on different branch circuits, use the report information to determine possible causes of trouble.

The feeder supply circuit is 3-phase, and the lighting branches are single-phase circuits. The changeover from 3-phase to single-phase is made in the branch distribution box by connecting the various branch circuits to the proper feeders. Figure 7-3, A, illustrates how these connections are made.

Now, assume that a fuse in the feeder distribution box opens. This fuse will open one leg of the 3-phase supply. When this happens, the loads will then form two branch circuits in series across a single-phase branch circuit. It will be easier to visualize this condition by examining figure 7-5, which is a simplification of the figure shown in figure 7-3.

In this circuit, consider that fuse Fl is open. Branch circuit 3 still has normal connections to phase B-C. Therefore, it will receive normal voltage, and all lamps on this circuit will light to normal brilliancy. However, because of the open fuse, Fl, branches 1 and 2 cannot receive normal voltage. They do, however, receive voltage from terminals B and C. The supply path is from terminal B through fuse F2, branch 1 to branch 2, to fuse F3 and terminal C. Therefore, branches 1 and 2 are in series across the terminals that supply branch 3.

If the wattages of the series-connected lamps are equal, the voltages will divide equally across each load, and lights connected in these circuits will light to about half brilliancy. However, any load unbalance will cause unequal voltage division in the branch circuits so that, in an actual case, there may be several degrees of lamp brilliance in lighting circuits connected to the same distribution box.

The same action will be noted if fuse F3 opens, except that branch 1 will have normal voltage, and the voltage of this branch will divide across branches 2 and

Pi

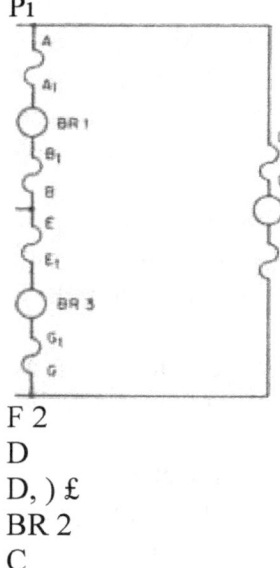

F 2

D

D,) £

BR 2

C

^3
Figure 7-5.-Simplified main and branch wiring.

3 because they will be in series across the supply voltage.

If fuse F2 opens, branch 2 will receive normal voltage, and branches 1 and 3 will be in series across the voltage supply.

Thus, you can see that an open fuse in any leg of the 3-phase supply mains will give an immediate indication as to the cause of the trouble. In this case, the first clue is the dimming of lights on some circuits and full brilliance of lights on others. Suspicions of an open-phase fuse will then be confirmed by voltage measurements.

VOLTAGE TESTS.-At the feeder distribution box (fig. 7-3) determine which fuse is defective. To do this, connect the test leads of the voltage tester across fuse F1. If the fuse is open (fig. 7-5), the full voltage of that phase will be indicated by the tester, provided circuit continuity is maintained through connected loads in branches 1, 2, and 3. If the fuse is in good condition, it will shunt the voltage indicator, and no voltage will be shown.

There is, however, an exception to this test. If an open circuit is at another point in that circuit, no voltage will appear across an open fuse. However, this condition is not likely to exist in tests at the feeder distribution box because of the relatively large number of connected loads compared with the number of loads at the distribution box (fig. 7-5).

OPEN-CIRCUIT TESTS.-If fuse F1 is open and branches 1 and 2 are in good condition and loaded, the full voltage will appear across the open fuse (fig. 7-5). If an overload on branch 1 has caused fuses A-A1 and F1 to open, a test across F1 will show full voltage because of a path through branch 2 and its associated fuses.

If F1 and D-D1 open but A-A1 stays closed, the path will be completed through branch 1 and its fuses.

The only condition that prevents voltage across an open feeder distribution fuse are the simultaneous open circuits in two branch circuits.

Simultaneous open circuits in branches 1 and 3 will prevent voltage from appearing across an open fuse F2. No voltage will appear across an open fuse F3 if branches 2 and 3 open simultaneously.

Klrchoff's law states that, in a series circuit, the full source voltage will appear across an open circuit. No voltage drop will appear across any load resistance in the circuit because no current can flow in an open circuit. In order for a voltage drop to appear across a resistance (load), current must be flowing in the circuit.

If only one branch circuit is interrupted, it is safe to assume that the feeder distribution box fuses are in good condition and that a fuse is blown in the branch distribution box. However, do not overlook the necessity for actually measuring voltage. As mentioned before, do not mistake vibration or hum in the voltage tester as an indication that normal voltage is present.

If two branch circuits are affected but each branch has some voltage, check the feeder distribution box fuses.

In another case, you might find that all lights on branch 1 (fig. 7-3) are out. First, try to decide which parts, if defective, will open the circuit.

If S1 is open, it is obvious that the lights will be out. If the switch is in the OFF position, the trouble is readily apparent, and the cure is to turn the switch ON. However, if the switch is in the ON position but open circuited because of an internal defect, the trouble will not be apparent through a visual examination.

Another cause of circuit failure could be a blown fuse, A-A1 or B-B1. They should be tested as described for branch circuit tests.

If these tests prove that fuses A-Al and B-Bl are in good condition, the open circuit will be between the fuses and the first light on the circuit. If it were at some other point, one or more of the lights on the circuit would be operating.

Check the switch by removing the front of the distribution box so that terminals Xl and Yl are exposed (fig. 7-3). The switch wafer can be turned to the ON position with a screwdriver. The voltage tester leads should be connected to Xl-Yl. No voltage means one or both switch arms are open. To check, connect your voltage tester to X-Y. The phase voltage should be present. Leave one lead on X and move the other lead to Yl. Normal line voltage between these points is an indication that switch arm Y-Yl is in good condition. No voltage means that the switch arm or contacts are defective, and the switch should be replaced. Arm X-Xl is tested by placing one test lead on Y and the other on Xl.

If the switch is in good condition and voltage is present at Xl-Yl, the open is between Xl-Yl and the first light.

The foregoing discussion shows that the effect is a clue to the type of trouble. If only one circuit is out of commission, look for a blown fuse in the branch distribution box. If one circuit is normal and two other circuits have less than full voltage, look for a blown fuse in the feeder distribution box.

Safety Precautions

Figure 7-3 illustrates that all fuses are "hot" when power is being furnished by the main supply. Therefore, use extreme care while working at fuse locations. Always remove and replace fuses with an approved fuse puller. This tool is insulated so that there is less danger from electric shock to the electrician while he is working with the fuses.

Fuses should be removed and replaced only after the circuit has been deenergized. This action prevents drawing an arc when the fuse is removed.

Branch circuits are made safe to work on by opening the switch associated with that circuit. These switches should be tagged, "This circuit was opened for repairs

and shall not be closed except by direct order of

If more than one repair party is engaged in repair work on an electrical circuit, a tag for each party should be placed on the supply switch. After work has been completed, each party should remove its own tag, but no other.

CONTROL AND PROTECTIVE ACCESSORIES

Control and protective accessories refer to devices that control the operation of equipment. In its simplest form, the control applies or removes voltage to a device. In more complex control systems, the initial control device may set in action other devices that control the motor speeds, the compartment temperatures, the depth of water, the aiming and firing of guns, and the direction of guided missiles. In fact, all electrical systems and equipment are controlled in some manner.

Pushbutton Switch

The simplest form of control is a pushbutton switch. No doubt you have seen these used to operate doorbells or similar devices. However, a pushbutton control may take one of several forms.

In construction, pushbuttons may vary from a single -spring contact to several ganged springs and fixed contacts operated by one button. One weatherproof type is made so that the button is covered by a rubber diaphram, and operation is accomplished by pressing the center of the diaphr am.

Pushbutton contacts may be arranged so that the circuit is normally open and then closed when the button is pressed. Another contact arrangement is such that the circuit is normally closed, and opens when the button is pressed. Still another arrangement makes it possible to have one or more circuits normally closed by the button while other circuits connected to the same button are open. Pressing the button opens one circuit or circuits and closes others.

Pushbutton controls are shown on equipment schematics, and figure 7-6 illustrates the symbols used to indicate the buttons and the contact arrangement at each button.

Figure 7-6.-Pushbutton symbols.

Pushbutton control is suitable for momentary application or interruption of current. When a pushbutton is used to apply current for a considerable time, a latching device is incorporated to hold operation (maintain contact) until the latch is released.

Knife Switch

Next in simplicity of operation and construction is the knife switch. Like pushbuttons, knife switches may be very simple devices to turn current ON or OFF, or they may be complex in construction.

In its simplest form, a knife switch has a movable conductive arm (usually copper), which is hinged or pivoted to a fixed support that forms one terminal of the switch. The second switch terminal is fixed in position so that it may be engaged by the moving arm. Usually, this contact is made of copper and is arranged so that it contacts both sides of the switch arm when the switch is closed. The second terminal has a natural spring to provide a tight contact that is self-wiping. That is, each time the switch is opened or closed, the moving arm will be rubbed by the fixed contact, keeping the contact surfaces clean and bright.

Another type of knife switch may have the arms and contacts arranged to perform a number of actions. This type, with one moving arm and one closed position, is a single-pole, single-throw (SPST) switch, and may be used to open or close one conductor only. In many cases, it is necessary to open or close two conductors simultaneously. Switches for this purpose are double-pole, single-throw (DPST) and have two moving arms connected by an insulating bar. This type has two hinged and two fixed contacts.

The number of arms on a switch is determined by the number of conductors it controls. Thus, we may have three-pole, single-throw (3PST) or four pole, single-throw (4PST) switches. In each case, the moving arms will be connected by an insulated bar, for simultaneous operation.

Knife switches may be mounted on base of bakelite, hard rubber, or slate in order to insulate the conducting parts of the switch from other surfaces. However, open switches can be extremely dangerous because the conductors are exposed. Therefore, most switches are enclosed in a metal cabinet. The switch arm or arms are then operated by an external lever, which is mechanically coupled to the insulating bar on the switch arms.

Knife switches may be used to supply power to two different circuits (loads). When they are used in this manner, the moving arms are arranged so that when they are thrown in one direction one set of contacts is engaged and when the arms are thrown in the other direction, a second set of contacts is engaged. These switches may have any number of poles and are identified as single-pole, double-throw (SPDT); double-pole, double-throw (DPDT); triple-pole, double-throw (3PDT); or four-pole, double-throw (4PDT). The schematic symbols for each designation are illustrated in figure 7-7.

o SPST o c< o SPOT

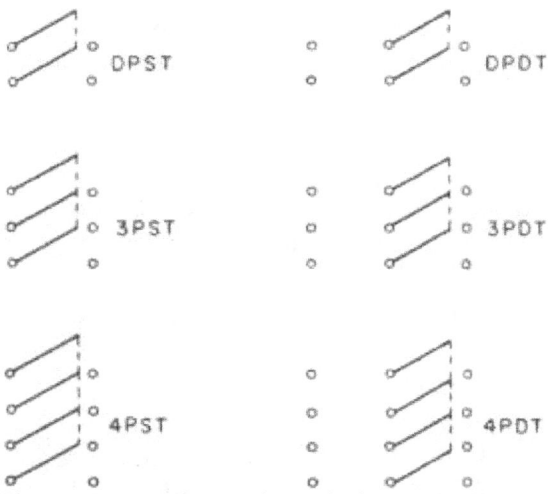

Figure 7-7.-Schematic switch symbols.

Rotary Switch

One type of switch used to open or close one or more conductors is the rotary type. This type is constructed so that the moving arm or arms are pivoted in the center. These arms are insulated from each other and from the pivot post. The contact arrangement is such that a moving arm is connected to two fixed contacts at the same time. Because the circuit wires are connected to the fixed contacts, the arm closes the circuit.

It is possible to construct a rotary switch in a very compact form. However, the amount of current and voltage to be handled by the switch determines the size of the switch arms and contacts as well as the spacing between them. The advantage of a rotary switch is that the number of positions and the number of poles can be increased by arranging the fixed contacts on wafers.

Because there are 360 degrees in a circle, it is possible to have 35 contacts and an OFF position, each

position spaced 10 degrees apart. For applications requiring low voltage and current, one set of contacts can be placed on one side of the wafer and another set of contacts placed on the opposite side of the same wafer. Because wafers may be separated as little as 3/8 of an inch in some applications, it is possible to have an 8-pole, 36-position switch in a space approximately 2 inches in diameter and 1-1/2 inches deep. This kind of rotary switch is used in many electronic applications. However, one type of rotary switch is rated at 500 volts a-cat 200 amperes and may be used on power applications.

Master Switch

The switches encountered in naval equipment will fall into the three general types that have been described. However, special applications change the previously described simplified forms. A master switch (pushbutton, knife, or rotary) starts or stops a sequence of operations. A device that is controlled by a master switch cannot be used until the switch is closed. The actual closing of the switch may be done manually or automatically. In some circuits, the master switch may be followed in sequence by other controls that are designed for a specific purpose. These will be described later.

Master switches are used with the face-plate and cam-type controllers. The principal difference is in the method of making contact between the fixed and moving terminals. A schematic diagram of face-plate internal connections is given in the d-c motor starter circuit of figure 7-8.

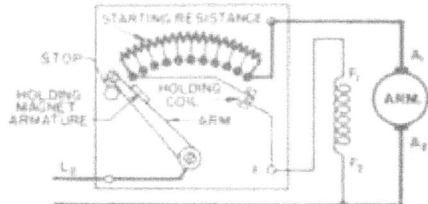

Figure 7-8.-D-c motor starter circuit.

In the face-plate switch, connection is made from one side of the line to a lever that carries a contact brush, or shoe. The brush moves over a series of contacts (fixed), which are connected to resistors. As the lever arm moves across the contacts, the amount of resistance in series with the armature gradually decreases. At the last contact, all series resistance is out of the circuit, and the armature receives full voltage.

When the lever closes the circuit at the first contact, it establishes a circuit to an electromagnet. The current flowing through the magnet winding produces a magnetic flux, which holds the lever at the last contact.

Face-plate switches are usually used as starting switches in d-c motor circuits. They will seldom be found on larger ships that have a-c electrical systems. In the three-terminal type (fig. 7-8) the current through the holding magnet winding is actually the shunt field current of the motor being controlled. Therefore, if the voltage supply for the motor field is interrupted for any reason, the reduction in field current will reduce the strength of the holding magnet, and the lever will be released. Because the lever is spring loaded, it will return to the OFF position. This feature makes it impossible to open a motor field circuit at some point and then close the circuit (with the motor connected to the line). As soon as the holding coil magnet releases the lever, it returns to the OFF position, and the motor cannot be started without going through the entire starting procedure.

Cam-type switches (fig. 7-9) are suitable for use with multispeed, reversing and nonreversing controllers.

Cam-type switches are manually operated by turning a hand wheel or lever located on one side of the switch housing. Turning the hand wheel rotates the shaft on which the operating cams are mounted. These cams engage rollers mounted on the switch contact levers and cause the switch contacts to open or close.

Some cam switches are self-centering. That is, it is necessary for the operator to hold the operating control at the desired position. When the control is released, it returns to the OFF position. In the nonself-centering type, the operating control remains in the position selected until the operator turns it to another position.

A star wheel mechanism provides positive positioning on the nonself-centering type of master switch given in figure 7-10. Parts A and B show how rotation of the cam

Figure 7-9.-Cam-type switch mechanism.

CAM

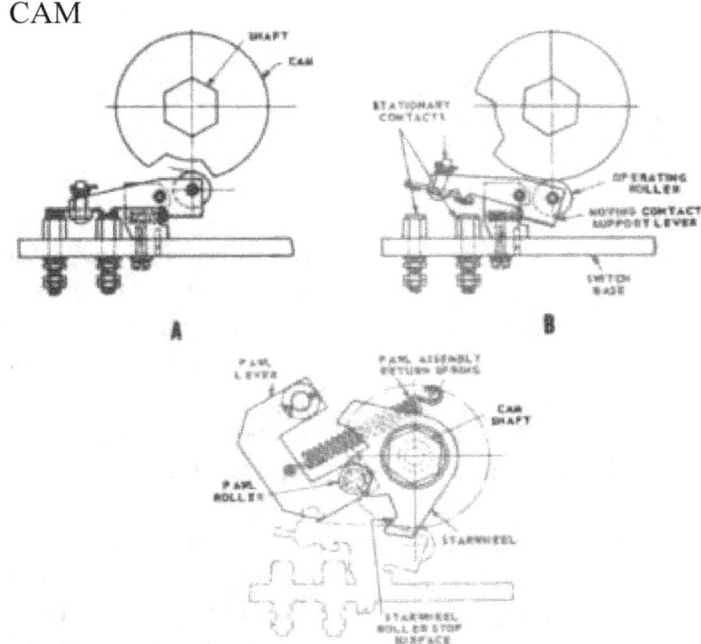

Figure 7-10.-Nonself-centering cam switch.

causes the switch contacts to open and close, and part C illustrates how the star wheel latches the switch in position.

The operating cams are made of phenolic material and are mounted on a hexagonal steel shaft. The operating sequence of the switch contacts is determined by the shape of the cams and their orientation on the shaft. Cam-type master switches may be mounted on a bulkhead, deck, or pedestal. Regardless of the type of mounting or the number of poles or positions, the method of operation is the same.

Limit Switch

In certain applications the ON-OFF switch does not give enough control to ensure safety of equipment or personnel. A limit switch is incorporated in the circuit in order that operating limits are not exceeded.

The limit switch is installed in series with the master switch and the voltage supply. Any action causing the limit switch to operate will open the supply circuit.

One application of limit switches is in equipment that moves over a track. It is possible to apply power so that operation will continue until the carriage hits an obstruction or runs off the end of the track.

If limit switches are installed near the end of travel, an arm or projection, placed on the moving section, will trip a lever on the limit switch. The switch then opens the circuit and stops the travel of the carriage. This type of control is a direct-acting, lever-controlled limit switch. Another type, an intermittent gear drive limit switch, may be coupled to a motor shaft to stop action when a definite number of shaft revolutions is completed.

Float Switch

A float switch (fig. 7-11) controls electrically driven pumps used to fill or empty tanks.

CONSTRUCTION.-In a tank installation (fig. 7-11, A) the deck and overhead flanges are welded to the deck and overhead of the tank. The float guide rod, E, fits into the bottom flange and extends through the top flange. The float guide rod then passes through an opening in the switch operating arm. Collars A and B on the guide rod

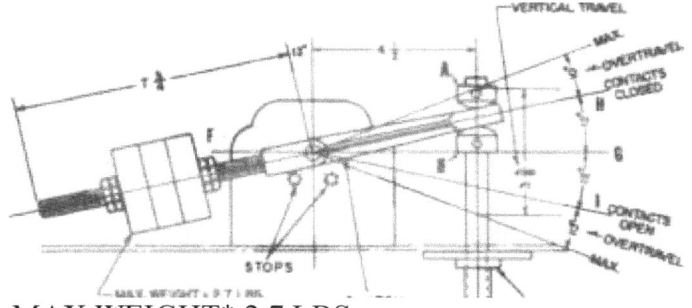

-MAX.WEIGHT* 2 7 LBS.
TC^P^TVIDE COUNTERBALANCE OPERATION
REDUCE WEIGHT IF NECESSARY
t6ftt wei»Tt or operating
SWITCH
operating
ARM
fiNG SWITCH
ALLOWANCE FOR TRAVEL of
operating ROC
5 INCHES

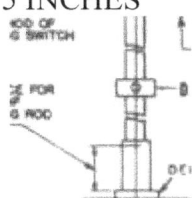

OVERHEAD FLANGE
— E
DECK FLANGE

Figure 7-11 A.-Float switch tank installation.

exert up or down pressure on the operating arm as the float approaches minimum or maximum depth positions. The switch operating arm is fastened to the shaft, which is coupled to the switch contact mechanism (fig. 7-11, B). Collars C and D on the guide rod are held in position by setscrews so that their positions can be changed to set the operating levels to the desired points.

As the float goes up and down, corresponding to the liquid level, it does not move the operating rod, E, until contact is made with either collar. When the float comes in contact with either collar, the external operating arm of the switch is moved and the switch is operated.

Although the switch assembly is of rugged construction, it must be Checked Figure 7-1 IB.-Float switch regularly for proper per- mechanism, formance.

MAINTENANCE. -The switch contacts should be kept clean. Often times this is done by operating the switch by hand. Otherwise, they should be cleaned with fine sandpaper (00 grade) or a fine file from the ignition tool kit. Of course, contacts should never be cleaned while power is applied to the circuit. When the motor control is set for manual operation, the float switch is connected to one side of the line supply.

After cleaning the contacts, apply a thin coat of vaseline to the wiping contacts to retard corrosion. Do not apply vaseline to nonwiping contact surfaces.

USE AS PILOT DEVICE.-Float switches used as pilot devices control the pump operation through other controls. A typical control circuit is shown in figure 7-12.

Switch SI makes it possible to have either manual or automatic operation of the motor-driven device. When this switch is in the manual position (the circuit is closed by the start stop switch, S2) current flows through the master switch magnet, holding coil Ml, which pulls line contacts M closed and applies power to the motor. The motor will then operate until it is stopped manually.

With SI in the automatic position and the start stop switch closed, the motor will not run unless the switch on the pilot device is closed. When the pilot device (in this case, the float switch) closes the circuit, current flows through the arm of SI to the operating coil, Ml, which then pulls the line switch contacts closed and starts

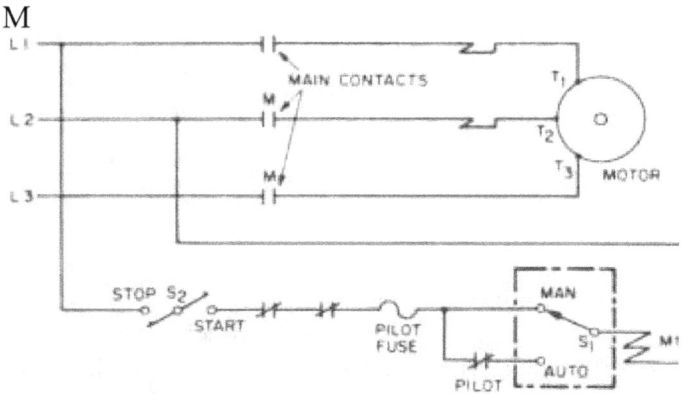

M

Figure 7-12-—Typical pilot device control circuit

the motor. The motor drives a pump that fills the tank. When the liquid in the tank rises to the desired level, the pilot device opens the circuit and removes voltage from the operating coil. This action allows the line contacts to open and stop the motor.

This start-stop action will continue automatically as long as the start-stop switch is in the ON or START position and SI is in the automatic position.

Pre ssure--Temperature Switches

These two controls have been grouped together because the switching mechanism is the same for both controls; the difference is in the operation.

Pressure-controlled switches are operated by changes in pressure in an enclosure such as a tank. Temperature-controlled switches operate from changes in temperature that take place in an enclosure or the air surrounding the temperature-sensing element. Actually, both switches are operated by changes in pressure. The temperature element is arranged so that changes in temperature cause a change in the internal pressure of a sealed gas or air-filled bulb or helix, which is connected to the actuating device by a small tube or pipe. Temperature changes cause a change in the volume of the sealed-in gas, which causes movement of a dlaphram. This movement is transmitted by means of a plunger to the switch arm. The moving contact is on the arm, and a fixed contact maybe arranged so that the switch will open or close on a tern -perature rise.

When the switch is used to control pressure, the temperature element is replaced by a tube that leads to the pressure tank. The pressure inside the tank then operates the switch mechanism.

Pressure or temperature controls may be used as a pilot device (fig. 7-12). The circuit operation is exactly the same regardless of the kind of pilot device used to control the circuit. To maintain more or less constant temperature or pressure, switch contacts are arranged to close when the pressure or temperature drops to a predetermined value and to open when the pressure or temperature rises to the desired value. The reverse action can be obtained by changing the contact positions.

The difference in pressure for contact opening and closing is the differential. The switch mechanism has a built-in differential adjustment so that the differential can be varied over a small range. Once set, the differential remains essentially constant at all pressure settings.

Each switch has a range adjustment that sets the point at which the circuit is closed. Changing the range adjustment raises or lowers both the closing and opening points without changing the differential.

OPERATING ADJUSTMENTS. —To set the operating range of the switch, turn the differential adjustment screw (fig. 7-13) counterclockwise against the stop for minimum differential. Bring the pressure to the value at which the circuit is to be closed. If the switch contacts are open at this pressure, turn the range screw slowly clockwise until the contacts close. If the contacts are closed when the desired pressure is reached, turn the range screw counterclockwise until the contacts open; then turn the screw slowly clockwise until the contacts close. These adjustments set the closing pressure.

The pressure (keep in mind that changes in temperature are converted to changes in pressure) is now raised to the point where the circuit is to be opened. When this point is reached, turn the differential screw slowly

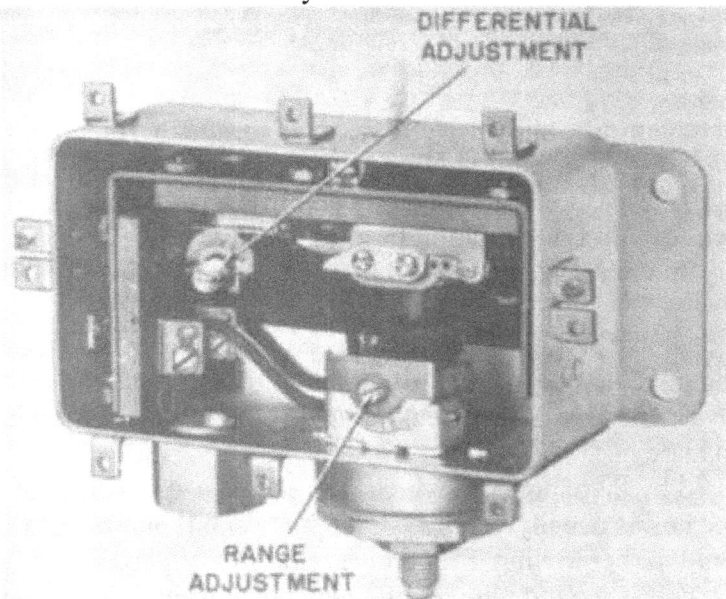

Figure 7-1 3.-Pressure-operated switch.

clockwise until the contacts open. This adjustment sets the opening pressure.

THERMAL UNIT TYPE.-The bulb and helix units can be connected to the switch section (fig. 7-14). The bulb unit (fig. 7-14, A) is normally used when liquid temperatures are to be controlled. However, it may control air or gas temperatures, provided the circulation around it is rapid and the temperature changes at a slow rate.

The helical unit has been specifically designed for air and gas temperature control circuits. To be most effective, the thermal unit must be located at a point of unrestricted circulation so it can "feel" the average temperature of the substance that is to be controlled.

Some switches are stamped, WIDE DIFFERENTIAL, and are adjusted in the same manner described for the regular controls. However, because of slight design changes, it is possible to get wider variation in differential settings.

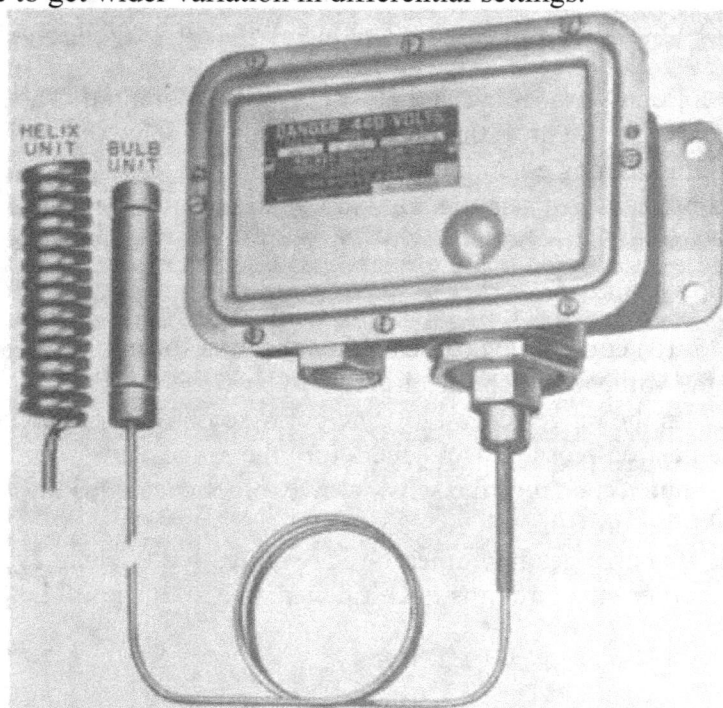

Figure 7-14.-Thermal units.

When adjusting temperature controls, allow several minutes for the thermal unit to reach the temperature of the surrounding air, gas, or liquid before setting the operating adjustments. After adjusting the operating range of pressure or temperature controls, check the operation through at least one complete cycle. If you find variation from the desired operating values, go through the entire procedure again and observe operation through a complete cycle.

D-C Relays and Contactors

A relay is a magnetically operated switch. The operating coil can be connected in series with a supply line to the load or shunted across the line. A contactor, like the relay, is a magnetically operated switch except the main contacts are designed to carry the heavier current of the load device.

The coil design is influenced by the manner in which the relay is used. When the relay is designed for series connection, the coil is usually wound with a fairly small number of turns of large wire because the load current will be flowing through the winding. When the relay is designed for shunt connection, the coil is wound with a large number of turns of small wire, which will increase the resistance and thus lower the current through the coil.

The contacts of relays and contactors may open or close when energized. This means that relays can be used as protective devices, as control devices, or to perform both functions simultaneously. Because of this flexibility, relays and contactors are used in many shipboard applications.

SHUNT.—The shunt-type contactor (connected across the line) operates when line voltage is applied to its^operating coil (7) (fig. 7-15). Usually, the contacts @ are arranged to supply or interrupt an electric circuit. In this

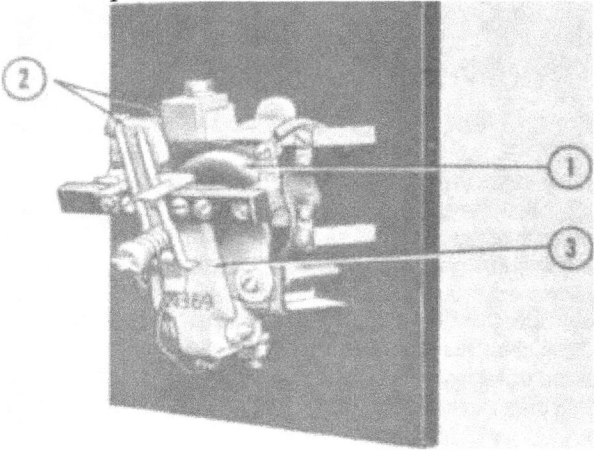

1. Operating ceil
2. Contacts
3. Armature

Figure 7-15.-Shunt-type d-c contactor.

arrangement the contacts are connected in series with the voltage supply to the controlled circuit. The operating coil is connected in shunt with the voltage supply to the controlled circuit. When voltage is applied to the coil, it attracts the armature © which closes the main contacts.

When the voltage supply to the coil is interrupted, the magnetic pull on the armature is removed, and the armature spring pulls it away from the magnet. This action opens the contacts and deener gizes the controlled circuit. The double-pole, shunt-type contactor (fig. 7-16) may be used for miscellaneous power control functions so long as the current rating of its contacts is not exceeded.

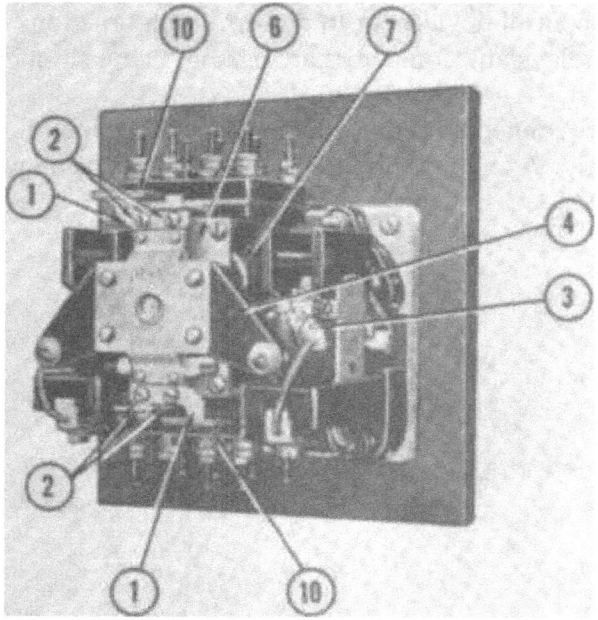

1. Interlock operating arm
2. Interlock arm screws
3. Connector screws
4. Front armature Figure 7-16.-Double-p
6. Coil clamp
7. Operating coil
10. Interlock operating bar
e, shunt-type contactor.

The construction of this relay is such that it can withstand severe mechanical shocks. This feature is obtained by using two movable armatures of equal mass that are connected together by levers so that they move simultaneously but in opposite directions. When this construction is used, any mechanical shock (regardless of direction or intensity) will tend to move both armatures in the same direction. The resulting forces will neutralize each other through the lever mechanism and prevent false operation.

Figure 7-17 shows two parts of the relay assembly of the double-pole, shunt-type relay. Numbers on these parts correspond to numbers on the complete assembly shown in figure 7-16 for easy identification.

To disassemble the unit, first remove the interlock operating arms (l) by taking out the four screws ®. Then take out the connector screws (z) and lift off the front

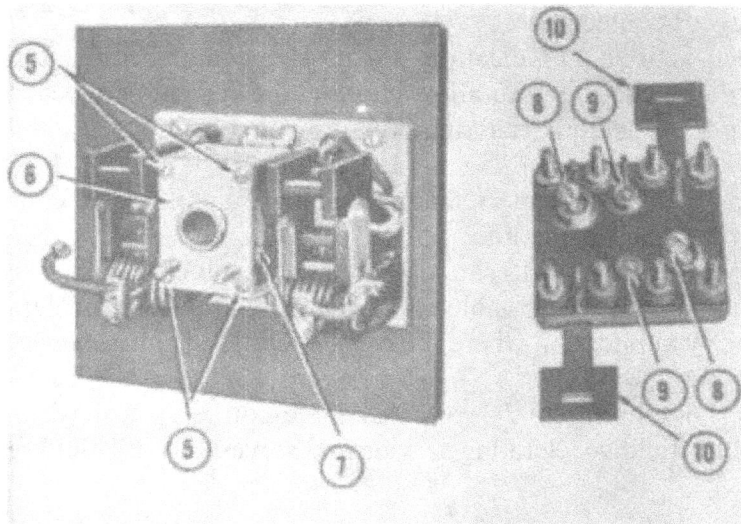

A-FRONT ARMATURE B"ASSEMBLY
AND INTERLOCK
5. Coil clamp screw* 8. Interlock mounting screws
6. Coil clamp 9. Interlock cover screws
7. Operating coil 10. Interlock operating bar
Figure 7-17.-Double-pole, shunt-type contactor components.

armature ®. The four screws (5) holding the coil clamp (6) are removed to expose the coil (7) (fig. 7-17, A). Disconnect the coil leads and lift the coil out of the magnet frame.

The interlock contacts may be inspected by removing the two screws (§) and lifting the interlock assembly (fig. 7-17, B) off the magnet frame. This exposes the two cover screws (9) which, when removed, allow access to the interlock contacts. When the interlock mechanism is replaced, make sure the interlock operating arms (T) (fig. 7-16) properly engage the bakelite operating bars @ (fig. 7-17, B). The interlock assembly looks as though it could be mounted in either of two positions. However, there is a slight difference in the length of the operating bars and therefore the interlock mechanism may be mounted in one position only.

The interlock contacts should never be filed or lubricated. When they become excessively worn, they should be replaced with new contacts.

A detailed view of the contacts is shown in figure 7-18. The main contacts should be removed when the dimension, A, becomes 1-9/16 inches for copper contacts or 1-11/16 inches for carbon contacts. These contacts should never be lubricated. Grease, dust, or copper oxide act as insulators, which increase contact resistance and cause unnecessary heating. Accumulations of dust and grease should be wiped off while the circuit is deenergized. Cop-peroxide can be removed with a fine file. The formation

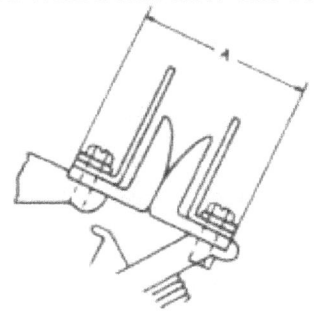

Figure 7-18.-Detailed view of contacts.

of copper oxide on contacts is often an indication of excessive contact temperature. If excessive heating occurs, check the remaining wear allowance and the current being carried by the contacts. If both are satisfactory, weak contact spring pressure may be the fault, and the spring should be replaced.

The relay types described are always connected across the line, and the relay control impulses from the pilot device are sent along the line.

SERIES.—The series type relays (fig. 7-19) are operated by circuit current flowing through the coil or coils. This feature makes it possible to use the relay as a field decelerating relay, a field failure relay, or for any application where the relay operation is in response to changes in circuit current flow.

A one-coil series relay is shown in figure 7-19, A, and a two-coil relay is shown in figure 7-19, B. A two-coil relay is used for field accelerating service and serves as a full-field relay and afield accelerating relay.

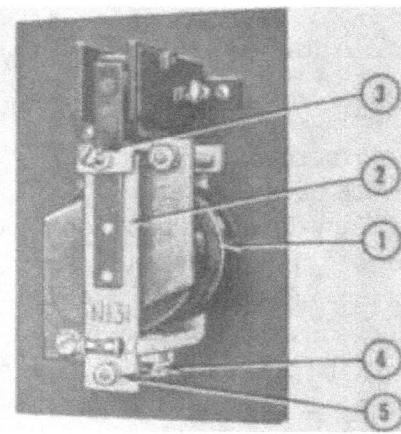

A-ONE COIL B-TWO COIL
1. Operating coil (on* or two 3. Differential adjusting screw coils) 4. Tension spring
2. Armature lever 5. Range adjusting nut
Figure 7-19.-Series type relays.

The electrical connections of a 2-coil relay, FA1 and FA2, are shown in the motor-starter diagram of figure 7-20. When connected in this manner, the relay is used to control motor speeds.

If the motor is used on a constant voltage supply, there is normally no means of speed variations by increasing the armature voltage. The field may be weakened by connecting resistance in series with the field circuit in order to get some control. Most standard d-c motors will permit a speed increase of 10 to 25 percent above normal by reducing the field voltage, but for any control beyond this range, a special motor must be used.

By connecting resistance in series with the field, the armature counter-voltage can be decreased, which will cause an increase in the armature current until the motor has accelerated to the new speed. If the amount of field resistance inserted is small, the increase in armature current will not cause difficulty. If, however, the inserted resistance is large, a means of limiting the armature

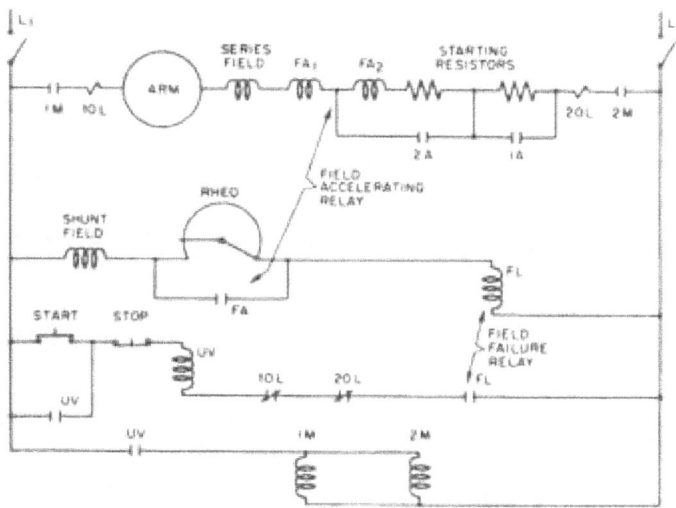

Figure 7-20.-D-c motor storting circuit with two-coil relay for excess-current protection.
current should be provided. The two-coil relay provides this protection.

The relay coils, FA1 and FA2 (fig. 7-20) are connected in series with the armature circuit. The relay contacts, normally open, are connected across the field rheostat. The relay is set to open at full-load current and close at approximately 25 percent over full-load current.

In operation, the relay armature alternately opens and closes the contacts across the field rheostat. This action cuts the field rheostat in and out of the circuit as the motor accelerates from its original speed to the increased speed.

The second relay coil, FA2 (fig. 7-20) has current flowing through it during the motor starting period. During this time, the relay is held closed, and the field rheostat is short circuited while the motor is being brought up to speed. This action makes it possible to have full field excitation and maximum starting torque.

The one-coil relay, FL (fig. 7-20), is used as a field failure protective device. In this application, the coil is connected in series with the shunt field, and the relay is factory adjusted to close on 65 percent and open on 30 percent of full-field current. The relay contacts are connected in series with the start-stop circuit. Thus, if the field fails for any reason, the relay coil releases the armature, which opens the relay contacts, and stops the motor.

There are two adjustments on the one-coil relay. The differential adjustment sets the difference between the opening and closing current values. The second adjustment sets the operating values. Usually, the operating adjustment is the only one required.

To adjust the differential, push the armature lever, @ (fig. 7-19, B) by hand until the contacts close. (Be sure the power is off.) Loosen the locknut holding the stationary contact and adjust the contact position to provide 1/4 of an inch spacing between lever @ and the face of the magnet core. Lock the contact in this position. When the contact is set in this manner,there is approximately 1/32 of an inch between the pin in the magnet core and the lever.

Release lever (2) and allow the contacts to open. Then

screw ③ until the gap

turn the differential adjusting

between the contacts is about 1/8 of an inch. Lock the screw in this position. This completes the differential adjustment, and the operating values should be within 20 to 25 percent of each other.

The range adjustment for operating values is made after the differential adjustment. To increase the range, increase the spring (J) tension by turning nut (5) clockwise (fig. 7-19, B). To decrease the range, turn the nut counterclockwise.

REVERSE CURRENT.—Two or more d-c generators may be connected in parallel to supply sufficient power to a circuit. Each d-c generator is driven by its own prime mover. If one prime mover fails, its generator will slow down and draw power from the line. The generator will then operate as a motor and instead of furnishing power to the line, it will draw power from the line. This can result in damage to the prime mover and overloading of the generator. To guard against this possibility, reverse-current relays are used. The reverse-current relay connections are such that when the reverse power reaches a definite percentage of the rated power output, it will trip the generator circuit breaker, disconnecting the generator from the line.

Normally, the reverse-power settings for d-c relays are about 10 percent of rated generator capacity for d-c generators. The reverse-current relay (one for each generator) is located on the generator switchboard. The mechanical construction of a d-c relay designed to limit reverse-current flow is illustrated in figure 7-21. Note that the construction is similar to that of a bipolar motor with stationary pole pieces and a rotating armature.

The potential coil is wound on the armature, and a current coil is wound on the stationary pole pieces. When used as a protective device, the current coil is in series with the load, and the potential coil is connected across the line, as shown in figure 7-22. If the line voltage exceeds the value for which the potential coil is designed, a dropping resistor is connected at point X in the circuit.

When the line is energized, current flowing through the series coil produces a magnetic field across the air gap. Voltage applied to the armature winding produces a current in the armature coil, which interacts with the magnetic field, thereby developing a torque that tends to

unit cu«if»» con.

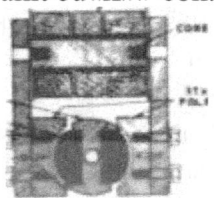

TIO«»lT

MH

o« HIUII

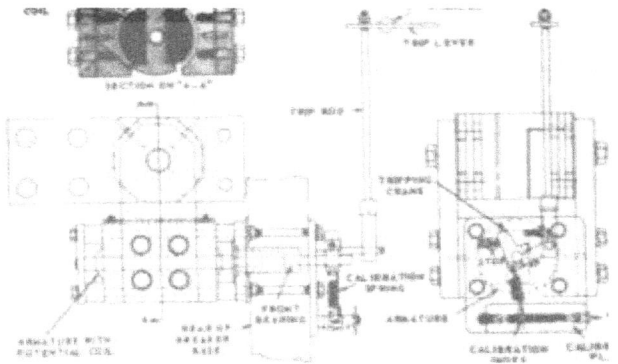

»«M»TUM «1TH POTIMTIM. cot.
CM.WUTWI
CU.lit.TIC C « t '» , AS°"

Figurt 7-21.-Mechanical construction of d-c reverse currant relay.

DC GENERATOR
AUX CONTACT
OPEN WHEN BREAKER
IS OPEN
RESISTOR (IF NEEDED) MECHANICAL TRIP

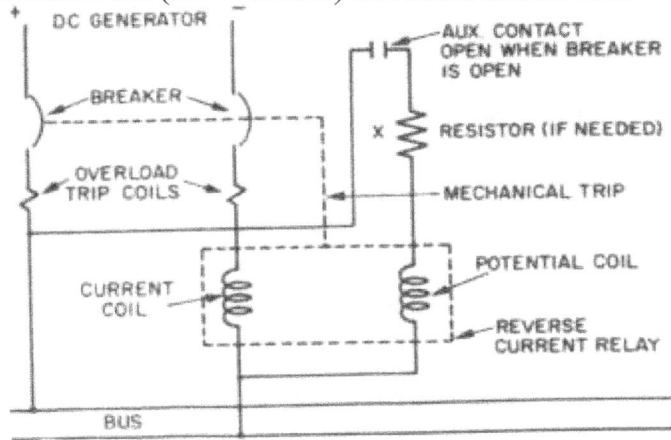

I POTENTIAL COIL

Figure 7-22.-D-C reverse-current relay connections.

rotate the armature in a given direction. The construction of the relay is such that the armature cannot turn through 360 degrees as in a motor. Instead, the torque produced by the two fields plus the force from the calibrated spring tends to hold the tripping crank on the armature shaft against a fixed stop. This pressure is maintained as long as current flows through the line in the right direction.

However, if one generator fails, the voltage output of that generator will drop. When the voltage drops below the terminal voltage of the bus to which it is connected, the generator terminal current (through the relay series coil) will reverse. However, the polarity of the voltage applied to the potential coil remains the same. As soon as the reversed current exceeds the calibration setting of the relay, the armature rotates, and through a mechanical linkage, trips the circuit breaker that opens the bus. This action disconnects the generator from the line.

OVERLOAD.—Relays designed to prevent damage to equipment due to overloads are overload relays. The relay disconnects the motor from the line when excessive loads occur.

A MAGNETIC TYPE overload relay is illustrated in figure 7-23. A pictorial view and a diagram identifying the various parts are shown in figure 7-23, A and B respectively.

In an installation, the operating (series) coil (6) is connected in series with the protected circuit (fig. 7^23, B). Normal current through the coil will have no effect on relay operation. If an overload occurs, increased current will flow through the coil and cause an increase in the magnetic flux around the coil. When the flux becomes great enough, the iron plunger @ will be lifted into the center of the coil, opening contacts ® and @ . This action opens the control circuit to the main contactor in series with the motor terminals, thereby disconnecting the motor from the line.

To keep the relay from operating when the motor is drawing a heavy, but normal, starting current, an oil dash-pot mechanism (T) and (2) is built in. This gives a time delay action that is inversely proportional to the amount of overload.

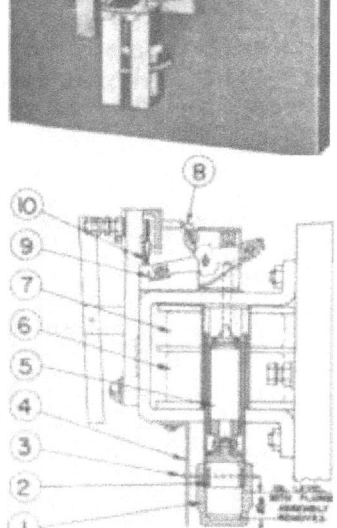

B

1. Dashpot 5. Plunger 8.
2. Pitton 6. Series ceil 9.
3. Indicating plat* 7. Shunt coil 10.
4. Calibration plat* (when used)
Figure 7-23.—Magnetic overload relay.
Latch
Movable contact Stationary contacts

Overload relays may use either single or double coils. In addition, the single-coil overload relay may be obtained with or without a manual "latching" control. Relays with manual latching are used on three-wire controls and reset automatically after an overload has occurred. Double-coil overload relays are used for two-wire control. They have a series coil (6) carrying the load current and a shunt holding coil © mounted above the series coil. These two coils are connected so that their respective fields aid each other. Then, when an overload occurs the plunger moves up into the shunt-connected coil field and is held in the tripped position until the shunt coil is de-energized by the pressing of a reset button or some other form of contact (switch) device.

Before placing the overload relay in service, raise the indicating plate (3) to allow the dashpot (T) to be unscrewed from the relay. Lift out the plunger (b) and make certain all of the internal parts are clean. Place about 9/16 of an inch of dashpot oil (furnished with the relay) in the dashpot. Replace the plunger and indicating

plate, and then screw the dashpot on the relay to the desired setting.

The relay is calibrated at the factory for the individual application, and the current values for which it is calibrated are stamped on the calibration plate ®. The marked values are minimum, maximum, and midpoint currents.

The operatinepoints may be set by first raising the indicating plate (3j which allows the dashpot to be turned. Then, to lower the tripping current, raise the dashpot by turning it. This action raises the plunger further into the magnetic circuit of the relay so that a lower current will trip the relay.

The current at which the relay trips may be increased by turning the dashpot in a reverse direction. This action reduces the magnetic pull on the plunger and requires more current to trip the relay. After the desired settings have been obtained, lower the indicating plate over the hexagonal portion of the dashpot to again indicate the tripping current and lock the dashpot in position.

THERMAL TYPE overload relays are designed to open a circuit when excessive current causes the heater coils to reach the temperature at which the ratchet mechanism releases. The heater coils are connected in series with the supply line to the load, and normal circuit current will have no effect on them.

The essential operating parts of the thermal overload relay (fig. 7-24) are the two heater coils (5) two solder tube assemblies (5) control contacts 1 compression springs @ and the ratchet mechanism. Under normal conditions the splitter arm (7) (so called because it splits the contacts) completes a circuit with the contacts. The spring is then under compression and the operating arm, (3) tends to rotate the splitter arm out of the circuit. This action is prevented by the ratchet assembly, which is held by the solder film between the outer and inner part of the solder tubes.

When current flows through the heater coils and produces enough heat to meit the solder film, the inner part of the solder tube assembly rotates and releases the ratchet mechanism to open the control circuits. When this happens, the circuit to the coil handling the power contacts (not shown in the figure) opens and disconnects

1. Contact structure 3. Operating arm 5. Solder tub*

2. Screw 4. Heater coil 6. Screw

7. Splitter arm Figure 7-24.-Thermal overload relay.

the load. As soon as the load is disconnected, the heaters cool, and the solder film hardens. When the hardening is complete, the relay is ready to be reset with the reset button.

An adjustable type thermal relay with magnetic reset is illustrated in figure 7-25. This relay may be adjusted to trip at a value between 90 and 110 percent of the rated coil current. To change the operating point, loosen the binding screws that hold the relay heater coil (5) (fig. 7-24) so that the coil position may be changed. Moving the coil away from the relay will increase the amount of current needed to trip the relay. Moving the coil closer to the relay will decrease the current needed to trip the relay.

The terminal plates and the underside of the slotted brackets of the heater coil assembly are serrated so that the coil is securely held in position when the binding screws are tightened. Some thermal overload relays have reset magnet assemblies attached. It may become necessary to replace the reset magnet coil. To do this, remove the heater coils from the relay. Then remove the four screws that hold the overload relay to the mounting plate. When removing the relay from the mounting plate, use care not to lose the phenolic pin and

an '

RESET MAGNET ASSEMBLY

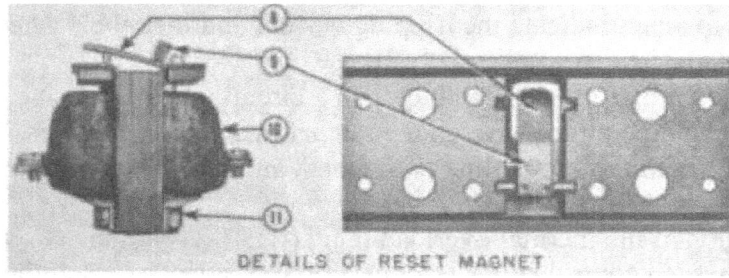

DETAILS OF RESET MAGNET

8. Uvm 10. CoH

9. Spring 11. Screw

Figure 7-25.-Adjustable thermal relay with magnetic reset.

bearing block located between the thermal blocks on the underside of the relay.

Next, remove the four large countersunk screws that hold the mounting plate and the reset magnet assembly to the square posts. Remove the four screws in the mounting plate, which support the reset magnet. Take care not to lose the lever and spring, items (§) and (5). Remove the two screws (fi) and pull out the plunger guides. Remove the old coil @ and install the new coil. Then insert the plunger guides and replace the screws (O) . Reassemble the magnet, spring, and lever to the mounting plate. Mount the plate on the posts and then mount the overload relay on the mounting plate. Replace the heater coils as the last operation.

A-C Relays

All complete electrical systems, whether a-c or d-c use relays for control and protection purposes. The basic difference in relays designed for use on a-c and those designed for d-c is in the armature and magnet core construction.

The armature and magnet cores of an a-c relay are made up of laminations; whereas, those of a d-c relay are of solid material. The use of laminations in an a-c relay reduces the heating due to eddy currents. In addition, a copper strap or ring is used near the end of the pole piece of an a-c relay to reduce "chatter" during operation. Because the alternating current is going through a peak, dropping to zero, and going through a peak in the opposite direction and then dropping to zero again during each complete cycle, the coil tends to release the armature each time the current drops to zero and attracts the armature each time it reaches a peak. The "shorted turn" acts as the secondary of a transformer, the primary of which is the relay operating coil. The current in the shorted turn is out of phase with the current of the operating coil because the copper ring has low inductive reactance. Thus, when the operating coil flux is zero, the flux produced by the shorted coil is different from zero, and the tendency of the relay to "chatter" is reduced.

SHUNT.—An a-c shunt relay is illustrated in figure 7-26. The basic function of the relay is to make or break an electrical control circuit when the relay coil is energized. To do this, voltage is applied to the operating coll ® (connected across the line), which attracts the movable armature (3). When the armature Is pulled down, it closes the main contacts (4).

The "pull-in" and "drop-out" current values may be adjusted. In figure 7-27 the various adjustment controls of the a-c shunt-type relay are identified. The spring and the setscrew, E, control the pick-up and drop-out values. Before the relay is adjusted, screw F should be set to clear the armature when the armature is in the closed position. Point B should be six threads from the extreme adjustment in the direction of point C.

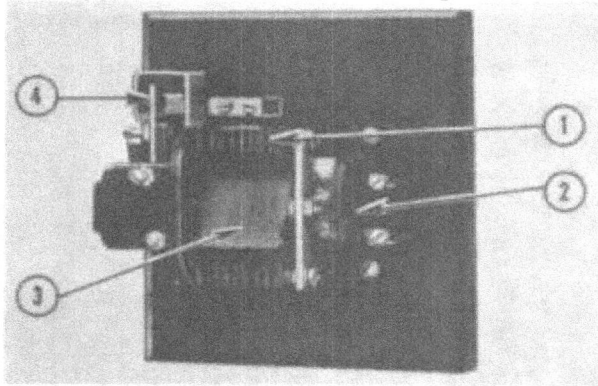

1. Magnet from* 3. Armatvrt
2. Operating ceil 4. Main contacts
Figure 7-26.-A-C shunt relay.

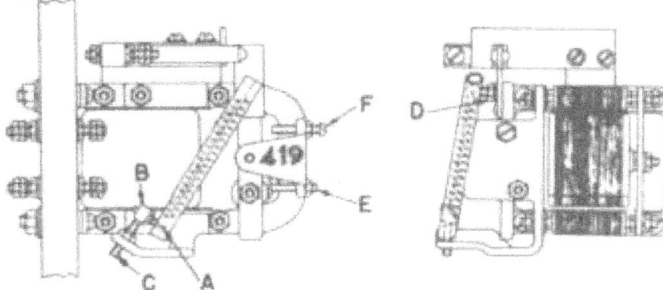

Figure 7-27.-Adjustment controls of a-c shunt relay.

The "pull-in" value can be raised by moving point A toward point B to increase the spring tension, or by turning screw E counterclockwise to increase the armature air gap. To raise the "drop-out" value, move point A toward point B; to lower the "drop-out" value, move point A away from point B.

OVERLOAD. —The a-c overload relay is designed to perform the same basic function as a d-c overload relay. That is, it prevents damage to motors and wiring when an overload occurs. Figure 7-28 shows two types of a-c

-NONLATCHING   B-LATCHING

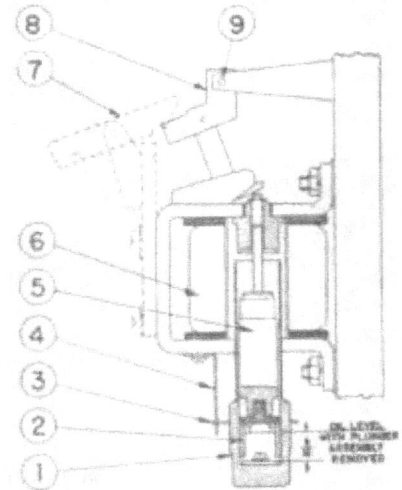

C ~ DIAGRAM

1. Da*hpot 4. Calibration plat* 7. Latch (whan u»ed)
2. Pitton 5. Plungar 8. Insulating splitter
3. Indicating plat* 6. Series coil 9. Contact*

Figure 7-28.-A-c overload relays. 378

overload relays. Part A is the nonlatching type, and part B is the latching type. In part C the various components are identified.

The operating (series) coil (6) is connected in series with the protected circuit (fig. 7^28, C). Therefore, the load current flows through the coil. If the circuit current rises above normal because of overload conditions, it will cause an increase in the magnetic lines of flux about the coil. The increased flux lifts the iron plunger @ into the center of the coil and opens the contacts (9). This, in turn, causes the main contactor (not shown) to open, and disconnects the motor or other device from the line. An oil dashpot mechanism (l) and @ is used to prevent the operation of the relay on motor starting current surges.

If the relay does not have manual latching, a three-wire control is provided to give automatic reset after an overload occurs. The manual-latch relay is generally used with two-wire control. The latch @ holds the contacts in the open position after an overload has occurred, and the circuits have been deenergized. It is necessary for the operator to manually reset the overload relay at the controller.

The operating adjustments and precautions are the same as for the d-c overload relay described previously.

REVERSE POWER.-Reverse-power protection is required when two or more a-c generators are connected to the same line. Unless such protection is provided, failure of a prime mover can result in the generator, running as a motor. An a-c reverse-power relay differs in construction from d-c reverse current relays. The a-c type consists of two induction disc-type elements. The upper element is the timer, and the lower one is the directional element.

Figure 7-29 shows the coil and induction disc arrangement in the induction type relay timer element. The disc is four inches in diameter and is mounted on a vertical shaft. The shaft is mounted on bearings for minimum friction.

An arm is clamped to an insulated shaft, which is geared to the disc shaft. The moving contact, a small silver hemisphere, is fastened on the end of the arm. The electrical connection to the contact is made through the arm and a spiral spring. One end of the spring is

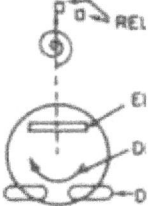

RELAY CONTACTS
ELECTROMAGNET DISC ROTATION RAGMAGNET
UPPER POLE FLUX
mm
DISC
DRAGMAGNET
SECONDARY CURRENT
TO SOURCE

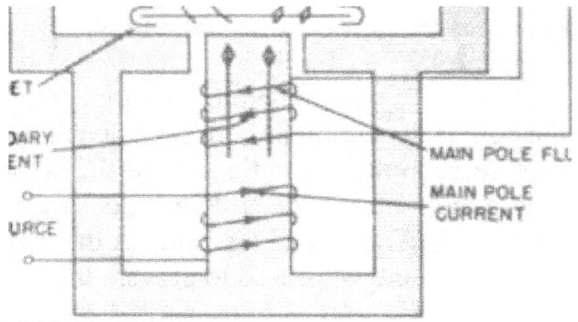

MAIN POLE FLUX
FRONT VIEW

Figur* 7-29. — Coil and disc arrangement, a-c reverse power relay.

fastened to the arm and the other end to a slotted spring adjuster disc fastened to a molded block mounted on the element frame. The stationary contact is attached to the free end of a leaf spring. The spring Is fastened to the molded block, and a set screw makes It possible to adjust the stationary contact position.

The main relay contacts (not shown In the figure) will safely handle 30 amperes at 250 volts d-c and will carry the current long enough to trip a breaker.

The induction disc is rotated by an electromagnet in the rear of the assembly. Movement of the disc Is damped by a permanent magnet In front of the assembly.

«OA

The operating torque of the timer element is obtained from the electromagnets (fig. 7-29). The main-pole coil is energized by the line voltage. This coil then acts as the primary of a transformer and induces a voltage in the secondary coil. Current then flows through the upper pole coils and thus produces a torque on the disc because of the reaction between the fluxes of the upper and lower poles.

The timer element cannot be energized unless the power flow is in the direction that will cause tripping. This interlocking action is accomplished by connecting the timer potential coil in series with the contacts of the directional element. Thus, the direction of power flow controls the timer relay.

The directional element is similar to the timing element, except that different quantities are used to produce rotation of the disc. There is also a different contact assembly. The two upper poles of the electromagnet are energized by a current that is proportional to the line current, and the lower pole is energized by a polarizing voltage. The fluxes produced by these two quantities cause rotation of the disc in a direction depending upon the phase angle between the current and voltage. If the line power reverses, the current through the relay current coils will reverse with respect to the polarizing voltage and provide a directional torque.

The contact assembly and permanent magnet construction are the same as that used for the timer element. The timer element is rated at 115 volts, 60 cycles. The minimum timer element trip voltage is 65 volts, and its continuous rating is 127 volts.

The directional element has a power characteristic such that when the current and voltage are in phase, maximum torque is developed. The potential coil is rated at 70 volts, 60 cycles.

The current coil rating is 5 amperes, and the minimum pickup current is 0.1 ampere through the coil. This current is in phase with 65 volts (minimum) across the potential coil. These are minimum trip values, and the timing characteristic of the timing relay may be erratic with low values.

For maximum protection and correct operation, the relay should be connected so that maximum torque occurs

for unity power factor on the system. Because the directional element has power characteristics, this connection may be accomplished by using line to neutral voltage for the directional element potential coil (polarizing voltage) and the corresponding line current in the series coils. If a neutral is not available, a dummy neutral may be obtained by connecting two reactors, as shown in figure 7-30. When connected in this manner, the directional element voltage coil forms one leg of a wye connection, and the reactors form the other two legs of the wye. The voltage-operated timer element should be connected across the outside legs of the transformer secondaries.

PHASE-FAILURE PROTECTIVE.-This type of relay is used to detect short circuits on alternating current propulsion systems for ships. Ordinary instantaneous trip relays cannot be used because, under certain conditions, for example when the motor is plugged (direction of rotation reversed on full voltage) during a crash stop, the momentary current maybe as great as the short circuit current.

The relay in use operates when there is a current unbalance. It is connected in the control circuit so that it will shut down the system instantly in case of an external or internal fault. However, operation of the relay is not limited to short-circuit detection, and the relay may be used as a phase-failure relay. A phase-failure relay is shown in figure 7-31. Part A is the arrangement of the parts in the complete assembly, and part B is a closeup of the contact assembly. The entire unit is enclosed in a cover to prevent dirt and dust from interfering with its operation.

The moving contact is the only moving element in the complete relay. There are two stationary contacts that make it possible to have the relay open or close a circuit when it operates.

Two coils are built into the relay. Each coil has two windings that are actuated by direct current from the two Rectox units. Four reactors are used to get sensitivity over a wide frequency range. Because variations in reactance are introduced during manufacture, two resistors are provided to balance the systems during the initial adjustment.

AC GEN

CONTROL BUS
o:
CD
OOO OOO
TRIP COIL
ABC
cv
-CD ii o
W CV CR W
TIMER ELEMENT
CR
©—M/VW—€>
^ '

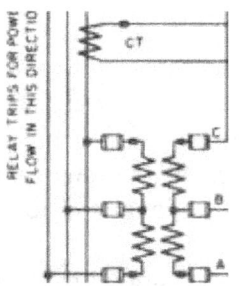

REAR VIEW
REVERSE POWER ELEMENT
REACTORS-
HI
OIRECTIONAL ELEMENT

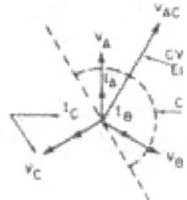

CV TIMING LEMENT
* CLOSING
ZONE
ZERO TORQUE LINE

Figure 7-30.-Schematic wiring diagram of an a*c reverse-power relay.

Figure 7-32 is the schematic wiring diagram for the phase-failure relay. The coils are represented by circles with numbers inside the circles that refer to numbered leads in the 3-phase bus. The coil (1-3) is connected to lines 1 and 3; the coil (1-2) is connected to

STATIONARY

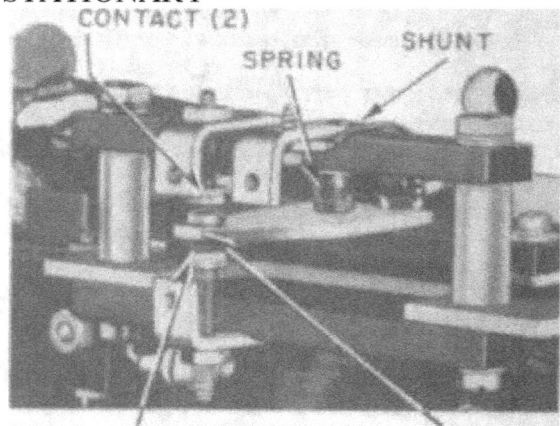

STATIONARY MOVING CONTACT CONTACT
Figure 7-31.-Phase-failure relay.
RCCTOX RCCTOX

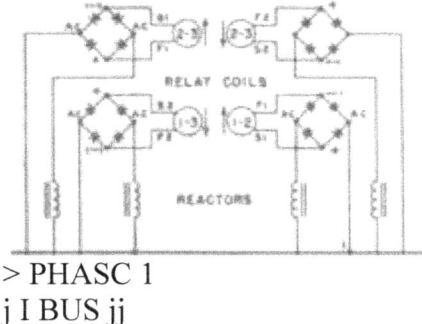

Figure 7-32.-Schematic wiring of phase-failure relay.

lines 1 and 2; and the two coils (2-3) are connected to lines 2 and 3. However, the coils are not directly connected to the bus lines. Instead, connection is made through the rectox units, which are connected to the line in series with a reactor.

When all three phase voltages are balanced, the flux produced by coil (1-2) is exactly equal and opposite to that produced by coil (2-3). The flux produced by coil (1-3) is exactly equal and opposite to that produced by the other (2-3) coil. Therefore, the resultant flux is zero, and no magnetic pull is exerted on the armature of the relay.

If a short circuit is placed across lines 1 and 2, no flux is produced by coil (1-2). This means that the flux produced by one of the (2-3) coils is no longer balanced, and there is a resultant flux, which exerts pull on the relay armature. The armature moves until the moving contact hits stationary contact 2 (fig. 7-31, B). This action opens the circuit between the moving contact and stationary contact 1. As soon as the short circuit is removed from lines 1 and 2, the resultant flux is zero, which allows the spring to return the armature to its original position. Similarly, if shorts occur on lines 2 and 3 or lines 1 and 3, the resultant flux is no longer zero, and the relay will operate.

A heavy copper tube is placed on each magnet core to damp out the triple-frequency ripple caused by the 120-degree displacement of the 3-phase voltages.

Figure 7-33 illustrates a wiring diagram of the relay, and figure 7-34 represents a vector diagram showing what happens to the line voltage when two lines are short

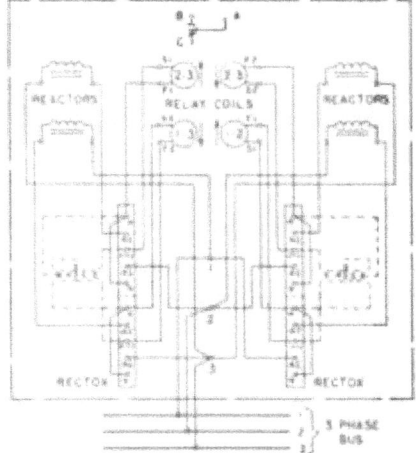

Figure 7-33.-Wiring connections for phase failure protection.

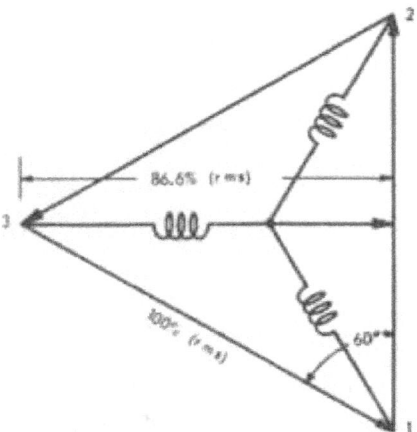

Figure 7-34. - Vector diagram showing perfect short circuit on 3-phase system.

circuited. Suppose a short occurs across lines 1 and 2. The voltage on coil 1-2 will then be zero and the average voltage on coils 1-3 and 2-3 will be 100% sin 60°/l.H = 78 percent of the normal rms value with no short. This voltage unbalance is enough to trip the relay.

The reactor in series with each coil keeps the coil current essentially constant over a frequency range of 12 to 90 cycles so that the relay sensitivity is as great at 12 cycles as at 90 cycles.

Mention was made of the use of resistors to achieve magnetic balance. If the reactor reactances are essentially equal, the resistors are not needed. However, if manufacturing tolerances are such as to introduce unbalance, the resistors are necessary. To determine if the resistors are needed, connect a 3-phase, 115-volt, 60 cycle voltage to terminals 1, 2, and 3, as shown in figure 7-33. Short coils 1-3 and 2-3 (on a common core), and then remove the armature spring for that relay. Put the armature against the core. No pull should be exerted on it. If there is pull, a resistance is needed. Repeat this test for coils 2-3 and 1-3 with the coils 2-3 and 1-2 shorted.

Never open the d-c circuit to the rectox rectifier while the voltage is being applied to the a-c side. This precaution is necessary because the voltage across the rectox is only a small portion of the total voltage drop due to the reactor being in the circuit. If the d-c side is opened, full voltage is applied across the rectox, which may cause it to break down.

Very little maintenance is required for this relay. No lubrication is needed. However, the relay should be kept clean so that dirt and dust will not interfere with its operation.

Because the relay operates only on a short circuit, which is a rare occurrence, it is advisable to check its operation every one or two months. To do this, short circuit the coil connections one at a time. Do this with a tool that has an insulated handle to avoid electrical shock. When the coil is shorted, the armature will drop because the coil is no longer energized. When the short is removed, the armature will return to its normal position.

CIRCUIT BREAKERS

Circuit breakers have two fundamental purposes: first, they are used to perform normal switching operations, and second, they are used to Isolate a defective circuit while repairs are being made.

Air circuit breakers are used on switch boards, switch gear groups, and distribution panels. The type groups that will be found on naval vessels are ACB, AQB. NQB, and ALB.

The maintenance of circuit breakers and safety precautions regarding them will be found in chapter 10 of this course.

Type ACB

This type of air circuit breaker has an open metallic frame construction and may be used for most protective functions. It is called an air circuit breaker because the main contacts open air.

When operated electrically, the operation is usually in conjunction with a pilot device such as a pressure switch. Electrically operated circuit breakers employ an electromagnet, used as a solenoid, to trip a release mechanism that causes the breaker contacts to open. The energy to open the breaker is derived from a coiled spring. The electromagnet is controlled by the contacts In the pilot device.

Figure 7-35 shows the external appearance of a type ACB circuit breaker.

Heavy duty type ACB circuit breakers are used to connect ship's service and emergency generators to the power distribution system. They are also used on all feeder circuits from the main switchboard. When used in these applications the pilot device may be a reverse-current relay or any of the control devices already discussed. The reverse-current relay Is usually mounted on the panel close to the circuit breaker. Other automatic controls may be located at remote points to give maximum protection to the circuit.

Circuit breakers designed for high currents have a double-contact arrangement. The complete contact assembly consists of the main bridging contacts and the

Figure 7-35.-Type ACB circuit breaker.

arcing contacts. All current carrying contacts are high-conductivity, arc-resisting silver or silver alloy inserts.

Each contact assembly has a means of holding the arcing to a minimum and extinguishing the arc as soon as possible. The arc control section is called an arc chute or arc runner. The contacts are so arranged that when the circuit is closed, the arcing contacts close first. Proper pressure is maintained by springs to ensure the arc contacts close first. The main contacts then close.

When the circuit opens, the main contacts open first. The current is then flowing through the arc contacts, which prevents burning of the main contacts. When the arc contacts open, they pass under the front of the arc runner. This causes a magnetic field to be set up, which

blows the arc up Into the arc quencher and quickly opens the circuit.

Type ACB circuit breakers are divided into manually and electrically operated types. The only difference is that the manually operated types must be operated by hand. However, all ACB circuit breakers may be operated electrically if the magnet and armature mechanism is installed. The heavy-duty types are electrically operated because it is then unnecessary for personnel to approach them in order to open or close the circuit.

No circuit breaker, regardless of type, should be worked on without opening the circuit. It must be kept in mind that certain terminals may have voltage applied to them even though the breaker is open.

Type AQB

Type AQB circuit breakers are mounted in supporting and enclosing housings of insulating material and have direct-acting automatic tripping devices. They are used to protect single-load circuits and all feeder circuits coming from a load center distribution panel.

The current capacity of the AQB circuit breakers is less than that of the type ACB. Where the requirements are low enough, the type AQB may be used on generator switchboards. Because they are electrically operated, they can be used in connection with any pilot device for circuit control. Figure 7-36 illustrates this type of circuit breaker.

Type NOB

Type NQB circuit breakers are nonautomatic and mounted in supporting and enclosing housings of insulating material. They are similar to the type AQB except that the type NQB has no automatic tripping devices and must be manually operated. They are used for circuit isolation and manual bus transfer applications. Because the type NQB has no tripping device, it is not suitable for use with pilot devices.

Figure 7-36.-Type AQB circuit breaker.

Type ALB

Type ALB circuit breakers are designated low-voltage, automatic circuit breakers. The continuous duty rating ranges from 10 to 50 amperes at 125 volts a-c or d-c. The breaker is provided with a molded enclosure, draw-out type connectors, and nonremovable and nonadjustable thermal trip elements.

The circuit breaker has three poles. The thermal overloads are in series with the two outer poles.

This circuit breaker is a quick-make, quick-break type. If the operating handle is in the tripped position, indicating a short circuit or overload, the operating handle must be turned to the off position, which automatically resets the overload unit.

aoi

QUIZ

1. Why must any trouble, whether due to overload or fault, be cleared before new protective devices (for example, fuses) are installed?

2. What type of action is obtained in fuses that employ a heater element either in parallel or in series with the fuse element to transfer heat to the fuse link from the heater in order to blow a fuse?

3. What is the function of materials, such as shredded asbestos and boric acid, used as filling in the tubes of cartridge fuses?

4. What is the function of the built-in neon bulb, which is normally shunted by the fuse element in some types of cartridge fuses ?

Refer to figure 7-3, A, for questions 5 through 10.

5. Which components protect the main feeder supply from heavy surges such as short circuits or overloads on the feeder cable?

6. Which components protect branch 1 ?

7. With SI closed, the full-phase voltage will appear across open fuse A-Al provided what condition exists across the associated branch circuit?

8. If voltage exists between terminals A and B but no voltage exists between terminals Al and B, where is the open circuit?

9. A resistance of 1 megohm or more between terminal X and ground indicates what condition of the lead from terminal X to the end of the circuit?

10. With the fuses removed from branch 1 and with switch SI closed, what will be the approximate ohm-meter reading across the switch terminals if there are ten 40-watt, 120-volt lamps connected across the circuit ?

Refer to figure 7-5 for questions 11 and 12.

11. If fuse Fl is open and normal voltage is developed across branch 3, what is the relative magnitude of the voltage across branch 1 and the voltage across branch 2 provided these branches have equal wattage ?

12. A blown phase fuse in one leg of the three-phase supply mains has what effect on the associated lighting load ?

13. What is the only condition that can prevent voltage across an open fuse, Fl (fig. 7-5) ?

14. Why should fuses always be removed and replaced with an approved fuse puller?

15. The various types of switches encountered in naval equipment fall into what three general types?

16. If the motor field circuit is opened (fig. 7-8), what automatic action will occur in the faceplate starter to protect the motor?

17. What type of switch is suitable for use with multi-speed, reversing, and nonreversing controllers in Navy applications ?

18. How is the motion of the float (fig. 7-11) transmitted to the switch operating arm?

19. How does the pilot device (fig. 7-12) stop the motor when SI is in the automatic position?

20. What general type of thermal unit (fig. 7-14) is used to control the temperature of (a) liquids and (b) gases ?

21. What type (series or shunt) d-c relay has (a) a small number of turns of large wire and (b) a large number of turns of small wire ?

22. How are the main contacts (fig. 7-15) of the d-c contactor connected (with respect to the voltage supply) to the controlled circuit (series or shunt)?

2 3. What safety feature is obtained in the double-pole, shunt-type d-c contactor (fig. 7-16) as a result of the use of two movable armatures of equal mass connected together by levers so that they move simultaneously in opposite directions?

24. How does the two-coil field accelerating relay (fig. 7-20) prevent excessive motor current during the starting period?

2 5. What type of relay is used to prevent the motoring of one or more d-c generators when they are operating in parallel ?

26. What are the normal reverse-power settings for d-c generators (percent of rated capacity)?

27. In which coil (series or shunt) of the relay (fig. 7-22) will the current reverse if the voltage generated in the machine falls below the bus voltage ?

28. What feature in the overload relay (fig. 7-2 3) prevents the relay from operating when the motor with which it is being used is drawing a heavy but normal starting current?

29. What will be the effect on the current required to trip the relay (fig. 7-24) if the heater coil is moved away from the relay?

30. What is the purpose of the copper strap or ring mounted near the end of the pole piece of an a-c relay?

31. Before the time element (fig. 7-30) can be energized the power flow must be in a direction to cause what action in the directional element of the reverse power relay?

32. What type of relay is used to protect circuits and motors in a-c propulsion systems that, under certain conditions, may demand momentary currents as great as short-circuit currents?

33. What will be the effect of a short circuit across lines 1 and 2 (fig. 7-32) on the operation of the phase-failure relay?

34. What will be the magnitude of the average voltage on coils 1-3 and 2-3 of the phase-failure relay in the event that a short circuit occurs between lines 1 and 2 (fig. 7-34)?

35. Why is it important never to open the d-c side of the rectox rectifiers in the phase-failure relay (fig. 7-32) when voltage is applied from the a-c side ?

36. Electrically operated circuit breakers derive the energy to open the breaker contacts from what source ?

37. What is the essential difference between type AQB circuit breakers and type NQB breakers?

CHAPTER

CONTROL DEVICES

INTRODUCTION

In large electrical systems, such as those on naval vessels, it is necessary to have a variety of controls for operation of the equipment. These controls range from simple START buttons to heavy-duty circuit breakers that are designed to control the operation of large motors.

The previous chapter explained how certain devices were used to operate other controls. When this is done, the basic control is called a pilot device. The pilot device then controls the operation of another device, usually a circuit breaker, which controls the motor. Because many pilot devices are a form of protective device, they were described in chapter 7, Protective Devices. In this chapter some additional pilot devices, such as timer controls, will be introduced.

In studying this chapter, you should remember that many controls will be actuated by pilot devices, and it will be helpful to review chapter 7 from time to time to refresh your memory concerning specific basic controls.

In addition to basic controls in the electrical circuits, some electrically operated mechanical controls are used. These controls and their operation are described in this chapter.

Rheostats

It is often necessary to reduce the voltage that is being applied to a motor, a generator field, a lighting circuit, or some other electrical device.

One method of reducing voltage is to introduce resistance in series with the load. The amount of resistance in the circuit may be a fixed value, or it may be varied mechanically. Where mechanical variation of resistance is provided and the unit is connected in series with a load, it is called a rheostat.

MANUAL.—Rheostats are classified as continuously variable or step types. In the continuous type, a moving arm makes contact with a resistance wire. One end of the wire is connected to one circuit lead, and the moving arm is connected to the other circuit lead. Thus, movement of the arm over the resistance wire varies the effective circuit resistance from zero up to the full resistance of the resistance element.

In the step type, fixed resistors are connected in series, and the junctions between adjacent resistors are connected to contacts arranged either in a straight line or a circular form. When the contacts are arranged in a straight line, the moving contact is driven by a worm gear. When the contacts are arranged in circular form, the moving contact is pivoted in the center of the circle.

Because considerable current may be present in some applications, the moving contact is arranged to make contact with one fixed contact before breaking the connection to another fixed contact. This prevents arcing and momentary interruptions of current.

When a rheostat is used to lower voltage, the difference between the source voltage and the load voltage will appear across the rheostat terminals. This voltage is sometimes called a voltage drop and is equal to the resistance in OHMS times the current in amperes flowing through the resistance. The voltage-dropping action is illustrated in figure 8-1. The rheostat (fig. 8-1, A) is set for zero resistance; all resistance is cut out by the moving arm. Under this condition, there is no current flowing through the resistance and no voltage drop across the rheostat. Therefore, the full source voltage is applied to the load.

The rheostat (fig. 8-1, B) is set at 10 ohms. This resistance is in series with the 10-ohm load; therefore, the total circuit resistance is 20 ohms. The circuit current found by Ohm's law 1 - E/R, is 6 amperes. This current flows through the rheostat, and theIR drop across

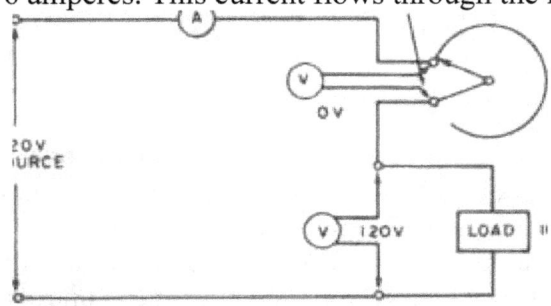

120V SOURCE

A -0-OHMS RESISTANCE

RHEOSTAT

10 OHMS

6a

10 OHMS

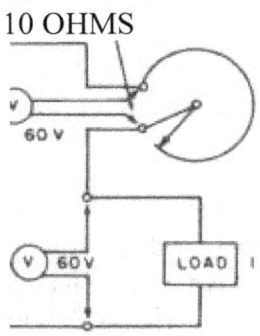

RHEOSTAT

120V SOURCE

10 OHMS

B-IO-OHMS RESISTANCE

Figure 8-1.— Voltage-dropping action.

its terminals is 6 amperes times 10 ohms or 60 volts. This voltage drop must be subtracted from the source voltage so that the load will be supplied with 60 volts.

This type of voltage reduction might be used to dim lights. The rheostat could be either the continuously variable type or the step type. It would control the lamp voltage as long as it was connected in series with the load.

MOTOR-DRIVEN.—As the name implies, a motor-driven rheostat is a variable resistor with a motor-driven arm. This rheostat is usually mounted on a switchboard and is used to help maintain stable excitation voltage to an a-c generator field winding. The motor that drives the rheostat arm may be controlled by a manually operated switch, or it may be controlled by a voltage regulator.

A voltage regulator is an automatic control, which is covered in the training courses, Basic Electricity and Electrician's Mate 1 and C.

The schematic for a motor-driven rheostat used to control generator field voltage is illustrated in figure 8-2.

I?0 V DC

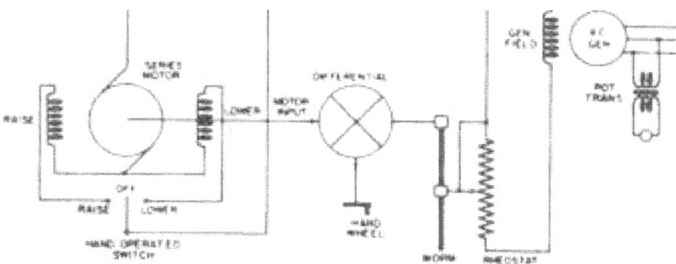

Figure 8-2.-Connections for a motor-driven rheostat used to control generator field voltage.

The manually operated switch has three positions labeled, RAISE, OFF, and LOWER, which refer to generator voltage changes. The rheostat motor has two oppositely wound field windings. The switch completes the circuit to the armature and one field winding in series. The change in direction of operation is accomplished by changing fields through the switch connections.

The rheostat arm is mounted on a worm drive, which is mechanically connected to the motor and a hand wheel through a differential gear. The hand wheel may be used in case of power failure to the motor.

Any change in generator loading will cause the generator output voltage to change. These changes in the voltage being supplied to other devices cause faulty performance of the equipment. Because the motor-driven rheostat controls the field excitation voltage to the generator, the output voltage can be adjusted by changing the field exciting voltage. This is done by adjusting the amount of resistance in use in the motor-driven rheostat. To see how this is accomplished, assume that the generator voltage has dropped because of an increase in load. To Increase the generator output voltage, you must

Increase the field excitation voltage by turning the manual control switch to the RAISE position. This action completes the circuit to the series motor through the RAISE field.

The motor then starts and drives the differential gear, which, in turn, drives the worm gear that moves the rheostat arm down the row of contacts so that less resistance is in the circuit. Less resistance in the circuit means more current in the generator field, which increases the field strength. This increase in field strength causes an increase in generator voltage so that the load voltage is brought to the correct value. When the voltage has reached the desired value, as indicated by the voltmeter on the switchboard, the switch control should be returned to the OFF position.

The procedure for high generator output voltage is the reverse of that just described; therefore, you must turn the control to the LOWER position.

Because a number of moving parts are involved, proper maintenance will help prevent generator failure. The worm screw should be lubricated with a good grade of cup grease. The motor is an important part of the control, and it should be checked frequently for faulty commutation. Contacts should be checked for loose connections. Contact buttons should always be clean and bright. Dirty, dark, or discolored buttons usually indicate loose connections or poor contact. The buttons maybe polished with fine sandpaper (00 grade).

Heat develops in equipment that handles heavy current. Because the brushes expand under heat, check them frequently to make sure that they are free to move in the brush holders.

D-C Brakes

Many motor-operated devices require brakes to stop the action or to hold the load when power to the motor is shut off. In addition, brakes are required to slow down operation in certain applications.

Brake systems used with d-c motors are divided into two groups: dynamic and magnetic. The dynamic brakes use electrical principles, and the magnetic brakes use a combination of mechanical and electrical principles.

DYNAMIC—Dynamic braking maybe compared to the use of an auto engine to slow a car going down hill. As you no doubt know, pressure is generated by the compression, which acts as a brake. This comparison is possible because a d-c motor, if its field remains excited while it is being driven by a load, will act as a generator and return power to the line. The generator action holds the load.

However, in an actual system that uses dynamic braking with a series d-c motor, the motor is disconnected from the line, and its armature and field are connected in series with a resistor to form a loop. The field connections to the armature are reversed so that the armature counter-voltage will maintain the field with its original polarity.

Figure 8-3 shows the connections used for dynamic braking of a series wound d-c motor. The field switching is done by switches SI, S2, and S3. The three switches are part of a 3PDT circuit breaker assembly and are magnetically operated from a controller described later in this chapter.

With the switch arms in position 1, the motor operates from the line. When the switch arms are in position 2,

LI

L2

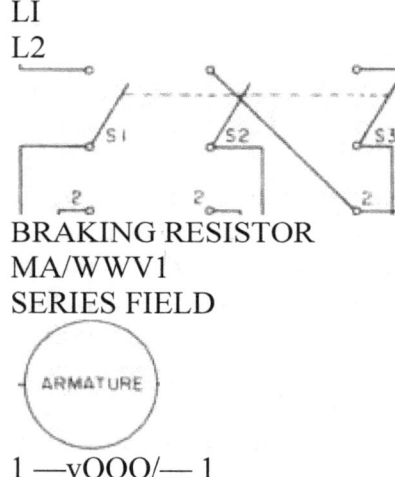

BRAKING RESISTOR
MA/WWV1
SERIES FIELD

1 —vQQQ/— 1

Figure 8-3.-Connections for dynamic braking of a series-wound d-c motor.
Ar\r\

the resistor is connected in series with the field, and, at the same time, the field coil connection to the armature is reversed. Thus, as long as the armature turns, it will generate a counter-voltage, which will force current through the resistor and the series field. Although the direction through the armature is reversed (because of the counter-voltage), the direction through the series field coil is the same as it was when operating as a motor. When operating in this manner, the machine is essentially a generator that is being driven by the momentum of the armature and mechanical load. This energy is quickly consumed in forcing current through the resistor, and the armature stops turning.

The time required to stop the motor may be varied by using different resistor values. The lower the resistance, the faster the braking action. If two or more resistors are connected by switches, the braking action can be varied by switching in different load resistors. Usually, the same braking resistors that are used to stop the motor are also used to reduce the line voltage during acceleration. This feature has been omitted on the drawing for simplification.

When dynamic braking is used with a d-c shunt-type motor, resistance is connected across the armature (fig. 8-4).

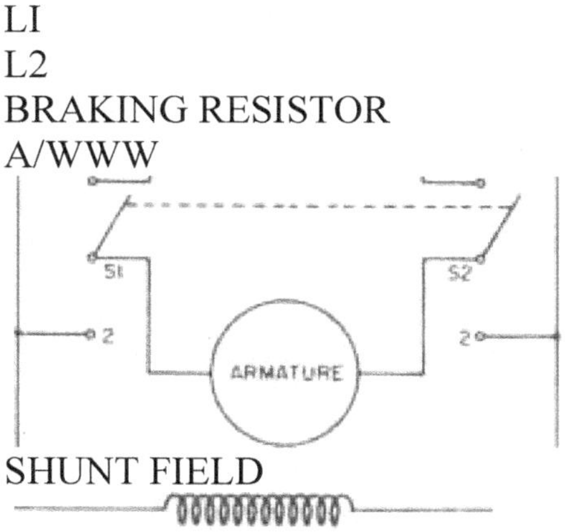

LI

L2

BRAKING RESISTOR

A/WWW

Figure 8-4.—Connections for dynamic braking of a shunt-wound d-c motor.

The switches, SI and S2, are parts of a DPDT circuit breaker assembly. When the switch arms are connected to position 2, the armature is across the line, and motor operation is obtained. When the switch arms are in position 1, the armature is disconnected from the line and connected to the resistor. The shunt field remains connected to the line. As the armature turns, it generates a counter-voltage, which forces current through the resistor. The remainder of the action is the same as described for the circuit in figure 8-3.

Although dynamic braking provides an effective means of slowing motors, it is not effective when the field excitation fails or when an attempt is made to hold heavy loads; without rotation the counter-voltage is zero, and no braking reaction can exist between the armature and the field.

MAGNETIC—Magnetic brakes are used for complete braking protection. In the event of field excitation failure, they will hold heavy loads. The essential parts of a d-c magnetic brake system are the brake shoes or bands, the brake wheel or drum, the electromagnet, spring mechanism, and mounting frame. The spring applies the brakes, and the electromagnet releases them.

Disc brakes are arranged for mounting directly to the motor end bell. The brake lining is riveted to a steel disc, which is supported by a hub keyed to the motor shaft. The disc rotates with the motor shaft.

The band-type brake has the friction material fastened to a band of steel, which encircles the wheel or drum and may cover as much as 90 percent of the wheel surface. Less braking pressure is required and there is less wear on the brake lining when the breaking surface is large.

D-c brakes may be operated by a solenoid or by a direct operating magnet. The operating coil of either type may be wound for either shunt or series connection to the line. Shunt coils are wound with a large number of turns of small wire, and series coils are wound with a small number of turns of large wire.

Normally, magnetic brakes with shunt coils do not release as fast as those with series coils. This is due to the increased inductance, L, of a shunt coil. The collapsing field of the magnet induces a voltage in the coil, which holds the relay armature for a short time after the magnet is deenergized. Coils are wound to operate at a voltage that is lower than the line voltage to speed up the operation of a shunt-type magnet. With lower voltage coils, fewer turns are required and this lowers the ratio of inductance to resistance. A resistance is used in series with the coil to drop the line voltage and prevent damage to the brake coil. Usually, the coil is wound to operate on one-half the normal line voltage, but to get extremely fast release, it may be designed to operate on voltages as low as one-tenth the normal line voltage.

Series brake coils have advantages over shunt coils. They are faster in operation because they have less inductance, are wound with heavy wire (not likely to burn out), and, because the voltage per turn is low, the insulation is not likely to break down. In addition, these coils are in series with the motor armature, and if the armature circuit opens, the brake will set. However, if the load varies over a considerable range, the armature current may not always be large enough to keep the brake released. In this case, a shunt coil is preferred. Series brakes are not in general use on naval vessels.

As previously mentioned, magnetic brakes are applied when the coil is not energized. A spring or weight holds the band, disc, or shoes against the wheel or drum. When the coil is energized, the armature or solenoid plunger overcomes the spring tension and releases the brake.

A magnetic brake assembly with the various parts numbered is illustrated in figure 8-5. When the coil, 187, is energized, armatures 188 and 189 are pulled horizontally into the coil. The armatures are mechanically linked to the levers, 178. The levers pivot on the pins, 179. When the magnetic pull overcomes the pressure of the coil springs, 192, the pressure of the brake shoes on the drum releases and allows it to turn. The drum is mechanically coupled to the motor shaft or the shaft of the device driven by the motor. The coil is connected to the voltage supply lines. The method of connecting the coil (series or parallel) is determined by the coil design.

D-C CONTACTORS

Basically, a contactor is a device that closes or opens a circuit. The d-c relays described in chapter 7 are forms of d-c contactors. However, they are primarily

Figure 8-5.-Magnetic brake assembly.

used with pilot devices to operate the larger contactors or circuit breakers.

The complete contactor is composed of an operating magnet, which is activated by either switches or relays, fixed contacts, and moving contacts. It may be used to handle the load of an entire bus, or a single circuit or device. However, when heavy currents are to be interrupted, larger contacts must be used. The contacts must snap open or closed to reduce contact arcing and burning.

At\A

In addition to these precautions, other arc-quenching means are used.

Magnetic Blowout

When a circuit carrying appreciable current is interrupted, the collapse of the flux linking the circuit may induce a voltage, which will cause an arc. If the spacing between the opened contacts is small, the arc will continue once it is started. The arc, if it continues long enough, will either melt the contacts or weld them together. To correct this condition, make the open contact spacing great enough to extinguish the arc. This may be impracticable because of space limitations and the fact that long stroke operating magnets require heavy current for operation.

Magnetic "blowouts" were developed to overcome these objections. These devices provide a magnetic field, which blows out the arc in much the same manner as you would blow out a match (fig. 8-6).

FORCE ON CONDUCTOR

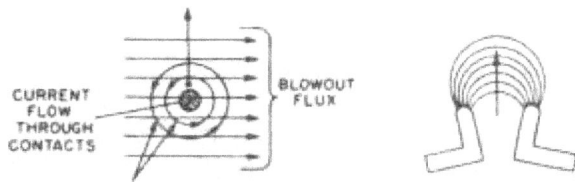

FLUX OF THE CURRENT THRU THE CONTACTS

A - DIRECTION OF FLUXES B " ACTION OF ARC Figure 8-6.-A magnetic field is set up for arc blowout.

The conductor is represented by the shaded circle (fig. 8-6, A), and conventional current in the conductor is assumed to be flowing toward you as you look at the drawing. The direction of the magnetic flux around the wire is then counterclockwise, as shown by the circled arrows, and the blowout flux is established from left to right, as shown by the horizontal straight arrows.

The flux around the conductor and the blowout flux oppose each other above the conductor and combine into a strong field below the conductor. This action produces a force on the conductor in an upward direction, as shown

by the vertical vector (fig. 8-6, B). If the conductor (fig. 8-6, A) is formed by contacts and the contacts are opened, the air between them becomes a conductor while the contacts are arcing. The resultant magnetic field created by the arc and the blowout field forces the conducting air upward. This lengthens the conductive path (fig. 8-6, B) and extinguishes the arc. The effective arc gap is several times as large as the gap measured straight across from one contact to the other.

The magnetic blowout operation is illustrated in figure 8-7. It is Important that the fluxes remain in the proper relationship. Otherwise, if the direction of the current is changed, the blowout flux will be reversed and the arc will actually be pulled into the space between the contacts.

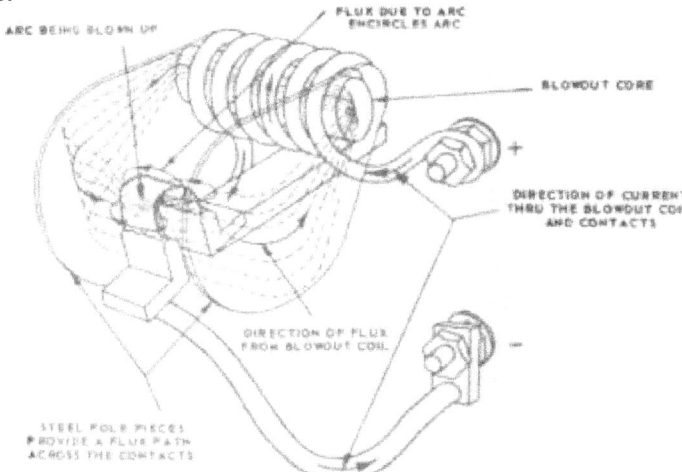

Figu.» 8-7.-Action of magnetic blowout coil.

When the direction of conventional current and flux are as illustrated in figure 8-7, the blowout force is upward. The blowout effect varies with the magnitude of the current and with the blowout flux. The blowout coil should be chosen to match the current so that the correct amount of flux will be obtained. The blowout flux across the arc gap is concentrated by a magnetic path provided by the

steel core in the blowout coil and the steel pole pieces extending from the core to either side of the gap.

Arcing Contacts

Because arcing at contacts wears away the usable surface, a second set of contacts (arcing contacts) is shunted across the main contacts.

A shunt-type contactor that will handle 600 amperes at 230 volts is pictured in figure 8-8. The blowout shield is removed in the pictorial view (fig. 8-8, A). The diagram (fig. 8-8, B) shows the main sections of the contactor. The arcing contacts (?) are made of rolled copper, which has a heavy protective coating of cadmium. These contacts are self-cleaning because of a sliding or wiping action after the initial contact is made. The wiping action keeps the surface bright and clean and thus maintains a low contact resistance.

The contactor is operated by connecting the coil (l) directly across a source of direct voltage. When the coil is energized, the movable armature @ is pulled toward the stationary magnet core @. This action causes

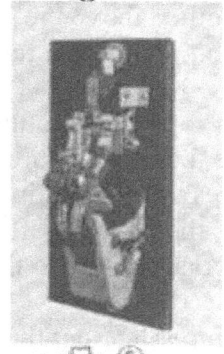

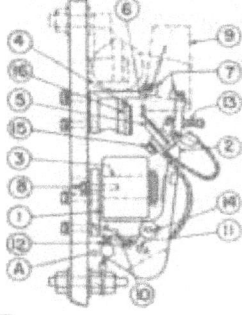

B
1. Coil
2. Armature
3. Magnet cor*
4. Stationary contact
5. Main brush contact
6. Brush arcing tip
7. Arcing contacts
8. Cor* holding screw
9. Blowout assembly
10. Tail spring
11. Spring adjustment nut
12. Armature stop pin
13. Finger spring
14. Hinge pin
15. Bolt

16. Plat*

A. Location of auxiliary contacts

Figure 8-8.-Shunt-type contactor.

the contacts that carry current (4), ®, ($) , and (T) to close with a sliding action.

The main contacts, (4) and ©, are called brush contacts. They are made oithin leaves of copper, which are backed by several layers of phosphor bronze spring metal. A silver brush arcing tip (6) is attached to the copper leaves and makes contact slightly before the leaf contact closes. The stationary contact consists of a brass plate @, which has a silver-plated surface. The plating lowers the surface resistance, and therefore the contact surfaces should never be filed or oiled. If excessive current has caused high spots on the contact, the high places may be smoothed down by careful use of a fine ignition-type file.

A detailed side view of the contacts is given in figure 8-9. Operation and contact spacing may be checked by manually closing the contactor (be sure the power is off). The lowest leaf of brush contact © in figure 8-8, B should just barely touch plate (?) in figure 8-8, B. If the lower leaf hits the plate too soon, the entire brush assembly should be bent upward slightly.

The contact dimension, A (fig. 8-9), should be measured with the contactor in the OPEN position. The

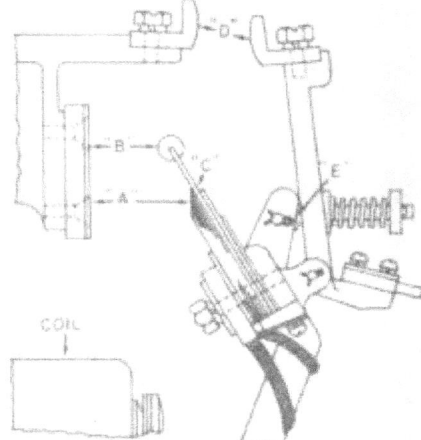

Figure 8-9.-Detailed view of contacts.

distance should be 1-1/2 Inches and is adjusted by moving stop @ in figure 8-8, B up or down.

The arcing tip dimension, B, should be 1-1/4 inches and is adjusted by bending the arcing tip at point C. The arcing contact dimension, D, should be 1-1/4 inches and is set by the addition of spacer washers or shims at point E.

Tail spring @ in figure 8-8, B should be set by means of the adjustment nut @ so that with the contactor inclined 45° backward, the armature @ will return to its open position when closed and then quickly released. This adjustment gives the tail spring the minimum tension required. If the contactor cannot be tilted 45° backward, the tail spring should be adjusted so that the contactor will pick up at 70 percent of the line voltage.

Main Contacts

The main contacts,® and©, in figure 8-8, A, carry the current drawn by the load during the operating cycle. The contact dimensions vary in accordance with the current handled by them. However, in most cases the same principles are used and the only difference is in physical size.

The moving main brush contact @ is hinged to the armature. A rod fastened to the armature extends through the slot in the moving contact arm. A finger spring is held under compression between the washer at the end of the rod and the moving contact arm. With the armature open, the spring holds the moving contact arm against the armature, which drops back and opens the contacts. When the armature is attracted, the contacts touch before the armature is pulled against the pole piece. The continued pull of the magnet compresses the spring so that it adds to the pressure, holding the moving contact against the fixed contact.

Electric Interlock

The voltages present in most contactor enclosures are dangerous. An electrically operated interlock (fig. 8-10) is used to safeguard personnel and prevent opening of the enclosure while the contactor is energized.

1. Coll 5. ControlUr door hook
2. Armature 6. Core holding tcrew
3. Magn«t cor* 7. Adjustment screw
4. Latch

Figure 8-10. —Electric interlock.

A pictorial view and a diagram of the electric interlock are shown in figure 8-10, A and B respectively.

The components of this electric lock are identified in part B. The coil of the lock is connected in parallel with the contactor coil. Therefore, it is energized when the contactor is energized. As soon as the voltage is applied to the lock coil (T) ,the armature (2) is pulled toward the magnet core (3), and the latch (J) engages the controller door hook © in the enclosure door. The door of the enclosure cannot be opened as long as the coil is energized.

When the contactor is deenergized, the lock magnet releases the armature, which disengages the latch from the door hook. The door can be opened.

Mechanical Interlock

When contactors are designed for two-speed operation, it is possible to actuate the FAST control at the same time the SLOW control is in operation unless precautions are taken. Simultaneous operation of two controls is preventedby the use of a mechanical interlock (fig. 8-11). The interlock may take the form of a lever pivoted at its center. The SLOW button is located at one end of the lever along with a STOP button. A second stop button and the fast button are located at the opposite end of the lever.

SLOW

SHOULDERS

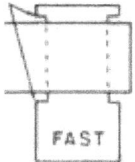

SHOULDERS V_>^PlVOT TOP
PRESSING THIS BUTTON RAISES FAST BUTTON p|V0T
I—^

FRONT

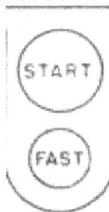

Figure 8-11.-Mechanical interlock.

If the slow button is pressed, that end of the lever is pushed down, and the opposite end of the lever rises and presses against the bottom of the fast button. If the fast button is pressed, that end of the lever will be pushed down, and the opposite end will rise. In rising, it raises the slow button, which opens the slow circuit. When the fast button is all the way down, the fast circuit is energized, and the slow button is held up by lever pressure.

Two stop buttons are provided because of the mechanical arrangement of the lever. To stop the machine, pr ess the stop button on the side of the control that is in use. Thus, a mechanical interlock can be used to prevent improper operation of an electrical control.

Coils

The coils used in contactors operate as electromagnets. The basic theory of electromagnet operation is given in Basic Electricity, NavPers 10086. Study or review this material so that you will be familiar with the operating principles.

DASHPOT.—One type of d-c overload relay is called a dashpot relay because of its construction (fig. 8-12).

Part A is a photograph of a dashpot relay and part B is a diagram showing the various parts.

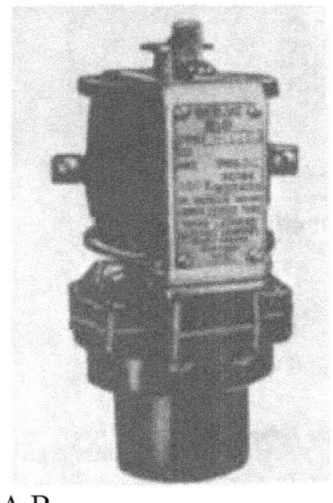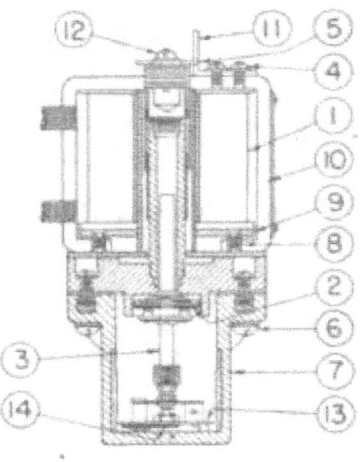

A B
1. Coil 6. Scr.w, 11. Flog
2. Normally closed contact* 7. Deskpot assembly 12. Screw
3. Plungor assembly 8. Screws 13. Dash
4. Screws 9. Coil clamp 14. Bi-motallic
5. Rating adjuster head 10. Frame strip of metal
Figure 8-12.-Dashpot overload relay.

The contacts ® are normally closed, and any circuit in which the relay is connected will be closed. The relay coil (T) is normally connected in series with the line to be protected so that changes in the line current will influence the magnetic flux of the coil.

Dashpot assembly (7) is filled with oil, and the dash (Q> is a disc that has a slightly smaller diameter than the inside diameter of the pot. The dash is connected to a plunger assembly @, which is a part of the solenoid assembly. The contacts @ are mounted on this plunger.

When the current in relay coil (T) exceeds the normal rated current, the magnetic force on plunger (3) pulls it slowly upward. The upward motion is slowed down because of the opposition offered the metal disc by the oil in the dashpot.

The overload current keeps exerting a pull on the plunger until the rising plunger opens the normally closed contacts, which open the control circuit. When the circuit is opened, current no longer flows through the coil and there is no pull on the plunger. The relay then resets itself as the plunger and contacts return to their normal positions.

The use of oil in the dashpot (1) acts as a time delay medium because it slows the movement of the disc; (2) because the contacts are immersed in the oil, serves as an arc-quenching or cooling medium; and (3) acts as a shock absorber because it slows the motion of dash @ and contacts @ under shock impacts.

The relay is set, at the factory, to trip when the line current is slightly above the current rating of the motor it protects. This, together with the time delay feature, prevents tripping of the overload under normal starting current surges.

Shunt D-C Contactors

Contactors for motor control are designed to energize or deenergize a circuit repeatedly without failure. Figure 8-13 is a d-c shunt contactor. Part A is the external view, and part B illustrates the mechanical construction.

The contactor is operated by energizing coil (l), which is composed of a large number of turns of wire and connected directly across a source of voltage. Armature (2) is pulled against core (§), which closes the main contacts, (4) and (5). The contacts are shaped to give a sliding motion when they close. Shunt-type d-c contactors are generally used as main-line or accelerating contactors. However, they are not restricted to this usage and can be used for many other power-control applications.

Series Lockout D-C Contactors

A delay in circuit closing is desirable in some motor-control applications. The contactor used for this purpose is called a lockout type because the delay action is obtained by using a lockout coil. This coil gets its name from the

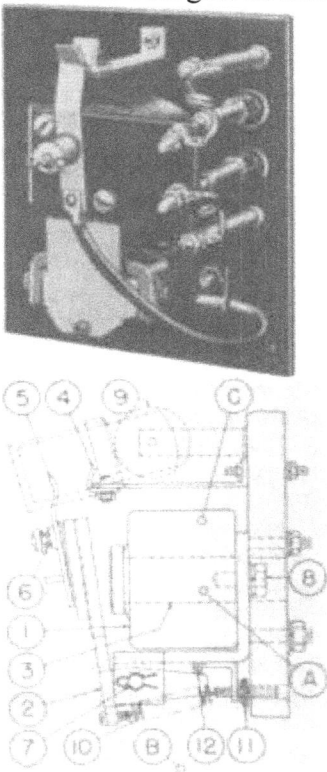

A B

1. Coll 7. Hinge pin A. Location of normally
2. Ararat*!* 8. Cera holding screw op on auxiliary contact*
3. Magna* cara 9. Blowout assembly B. Location of normally
4. Post contact 10. Tall spring closed auxiliary contacts
5. Finger contact 11. Spring adjuitmtnt nut C. Location of lamp switch
6. Finger spring 12. Armature step pin

Figure 8-13.-D-C shunt contactor.

fact that it locks the main contacts open on initial surges of heavy currents and then allows the contactor to close after the current has fallen to a predetermined value. A lockout-type contactor is used as an accelerating contactor on current-limit acceleration controllers.

SINGLE-COIL LOCKOUT.—As may be seen from figure 8-14 the upper part of the contactor is very similar to other magnetic contactors. However, the lower part incorporates the lockout mechanism, which consists of an armature (2) (an integral part of the closing armature @); a lockout coil (T); and a means of adjusting the air gap so as to obtain the desired operation with respect to motor current. The contactor has two magnetic circuits with the magnetic paths shown by the dashed lines in figure 8-14, B. The upper part of the control consists of the closing coil (4), the closing armature (§), and the contacts.

A

B

1. Lockout coil
2. Lockout armoturo
3. Closing armoturo
4. Closing coil
Figure 8-14.-D
5. Lockout coil coro
6. Adjusting scrow disc
7. Main contact spring
lockout contactor.

In operation, both coils are energized at the same time. The closing coil (4) may be connected across the line (shunt or in series) with the motor which is being controlled. The lockout coil (T) is connected either across part of the motor circuit or in series with the motor circuit.

The high starting current surge causes a strong magnetic field in the lockout. This strong field overcomes the field about the closing coil (?) and holds the contactor in its open position. As the motor accelerates, the heavy starting current decreases, and, at a predetermined current value, releases the lockout armature (2) and allows the contacts to close. The adjustment of the magnetic air gap, B, determines the current at which the lockout coil releases the armature.

This current is set by the screw disc (6). Turning the disc clockwise decreases the magnetic air gap, B, which

means that fewer ampere turns will be needed to lock open the contactor. Thus, a longer interval is required to allow the current to reduce to the lower release point. The motor will then accelerate more slowly.

DOUBLE-COIL LOCKOUT.-Although the contactor just described has two coils (lockout and holding), it is known as a single-coil lockout contactor. The type that will now be described is a double-coil lockout contactor (fig. 8-15).

m

s

■

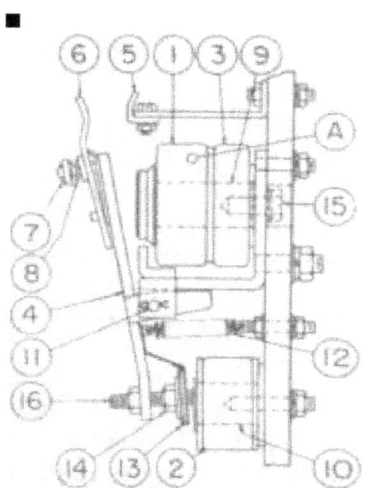

B

1. Series up-coil

2. Series down-coil

3. Shunt coil

4. Armature

5. Post contact

6. Contact finger

7. Cotter pin

8. Fingor spring

9. Upper magnetic core

10. Lower magnetic core

11. Hinge pin

12. Tail spring

13. Adjustment disc

14. Lock nut

15. Core holding screw

16. Disc stud

A. Location of auxiliary contacts

Figure 8-15.-Double-coil lockout contactor.

As shown in the figure, the double-coil unit has a series up-coil and a series down-coil in addition to a shunt coil. The series up-coil (T) and the series down-coil (2) are connected in series with each other and the motor armature. The series down-coil serves to hold the contactor armature (4) in the open position during the

motor starting high-current surge. When the current drops to the predetermined value, the combined magnetic pull of the up-coil and the shunt coil is greater than the magnetic pull of the down-coil, and the armature (?) closes the main contacts, @ and ©. The shunt coil (3) is used to hold the contactor closed when the load on the motor is light. Because it is designed for connection across the line, it has many turns of fine wire. The contacts of this relay have the same shape and construction as those used on contactors previously described.

D-C Controllers

Because of the inherent characteristics of a-c and d-c motors, d-c motors require different controls than do a-c motors.

For safety reasons, all controllers are enclosed in metal cabinets, and the cabinets usually contain equipment other than the controllers. The equipment and its relation to the complete control system will be described.

Ordinarily, direct current motors are not started on full voltage. The exceptions to this statement are small horsepower motors or motors of special design.

D-c controllers insert a resistance in series with the motor armature during the starting cycle to cut down the heavy initial current that would flow in the absence of armature counter-voltage. As the motor accelerates, the resistance is taken out of the circuit, and, at normal speed, the motor operates directly from the line.

D-c motors have an armature and usually two or more field windings. One of the windings is called the series field because it is connected in series with the armature. The other is called the shunt field because it is connected in shunt (parallel) with the series-connected field and armature. The connections for a d-c motor and controls are illustrated in figure 8-16.

When the START button is pressed, the path for current is from line terminal L2 through the STOP button, the START button, and the line contactor coil LC to line terminal LI. Current flowing through the contactor coil causes the armature to pull in and close the line contacts, LCI,

LC2, LC3, and LC4.

When contacts LCI and LC2 close, motor-starting current flows through the series field, SE, the armature, A,

o m
LCI
Hl-
SM
-Ms
SE
SR R

LC3
OL LC2
4—V-lr-f^> U
AC2 HI-
AC
SR
ACl HI-
STOP ~ Q 1 Q -
START-EMCRG.
_L
—o o
OL
LC
■Ar
LC*

Figure 8* 16.-Schematic of d-c control connections to motor.

the series relay coll, SR, the starting resistor, R, and the overload relay coil, OL. At the same time, the shunt field winding, SH, is connected across the line and establishes normal shunt field strength. Contacts LC3 close and prepare the circuit for the accelerating contactor coil, AC. Contacts LC4 close the holding circuit for the line contactor coil, LC.

The armature current flowing through the series relay coil causes the armature to pull in' t opening the normally closed contacts, SR. As the motor speed picks up, the armature current drawn from the line decreases. At approximately 110 percent of normal running current, the series relay current is not enough to hold its armature in; therefore, it drops out and closes its contacts, SR. These contacts are in series with the accelerating relay coil, AC, and cause it to pick up its armature, closing contacts ACl and AC2.

Auxiliary contacts ACl on the accelerating relay keep the circuit to the relay coil closed while the main contacts, AC2, short out the starting resistor and the series relay coil. The motor is then connected directly across the line, and the connection will be maintained until the STOP button is pressed.

418

Reversing Controllers

Certain applications make it necessary to reverse the direction in which a motor turns. In a d-c motor this is accomplished by reversing the connections of the armature with respect to the field. The reversal of connections can be done in the motor controller by adding two electrically and mechanically interlocked contactors.

A suitable motor reversing connection is given in figure 8-17. Note that there are two START buttons—one marked, START EM ERG FORWARD and the other marked, START EMERG REVERSE. These buttons serve as master switches, and the desired motor rotation is obtained by pressing the proper switch.

It should be noted that the emergency run feature (holding the start button in on overload to shunt the OL contact) is not included on auxiliary equipment aboard ships that have been designed and built since 1948 with the exception of manual starter installations. However deck equipment, the anchor windlass, davits, and other essential loads will have the emergency run feature but in a different form.

In some cases the emergency run feature will be obtained by holding in the reset button on the overload trip during the emergency operation. In all of the newer ships the "Em Run" switch will be separate from the start button.

There is danger of damaging the motor, the cable, or the controller when the motor is operated by the "Em Run" switch. Therefore you must know that an emergency does exist before you use the emergency run feature.

Assuming that the FORWARD button has been pressed, the line voltage will be applied through the button to the forward contactor coil, F, which pulls in its armature and closes the normally open contacts in the motor armature circuit, F1,F2, the forward contactor holding circuit, F3, the line contactor operating circuit, F4, and opens its normally closed contact, F5, in the reverse contactor circuit. The normally closed contact, F5,is an electrical interlock. With the forward contactor operated, the reverse contactor cannot be energized.

After the line contactor is energized, acceleration is accomplished in the manner described for the previous starter.

LC

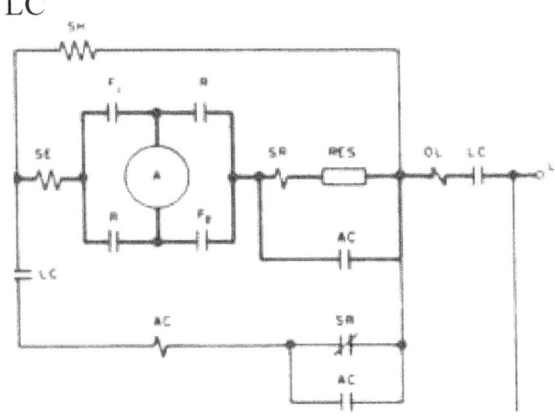

STOP — p I Q -
START-EMERG FQPWARO
01
START-EMERC -L_ -REVERSE
O-
F R
-N
R >»
-\ Yr
LC

Figure 8-17.-Motor reversing connections.

Speed Regulation

It is often necessary to vary the speed of a d-c motor. One way to accomplish this is to vary the strength of the shunt field. Rheostats are used for this purpose. By properly connecting the rheostat it can be made to add or remove resistance in series with the shunt field of the motor.

Figure 8-18 illustrates how a rheostat is added to the basic controller circuit to obtain varying speed.

420
SM
I—wv
RMEO.
HI-
FA
SE S \ FA, FA t SR
~ LC
AC

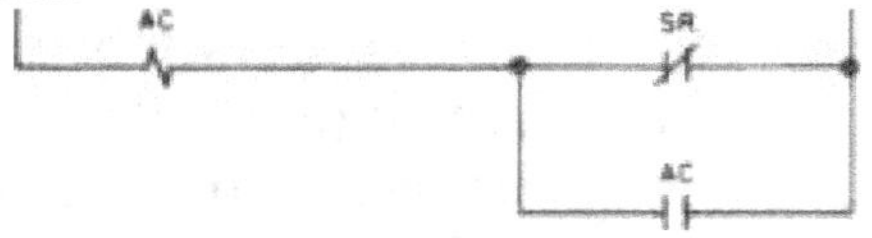

OL LC
START-EMERG.
STOP
O I Q <
-o o •
LC
LC OL

Figure 8-18.-Field rheostat connected to control motor speed.

If resistance is added in series with the field, the field will be weakened and the motor will speed up. If the amount of resistance in series with the field is decreased, the field strength will increase and the motor will slow down.

In the circuit (fig. 8-18) the rheostat is operated manually. However, it is possible to use a motor-operated rheostat and control it, as described in the section on motor-operated rheostats. The field accelerating relay, FA, was described in Chapter 7, Protective Devices, of this training course.

Corrective Maintenance

To ensure proper mechanical operation and to decrease the possibility of failures at critical times, keep the motor control equipment clean.

The presence of foreign material will cause binding of latches, plungers, armatures, and lever mechanisms, which, in turn, will lead to coil burnouts, contact burnouts, short circuits between phases, and other electrical failures. The foreign material may be dust, dirt, gum, filings,grease, moisture,and rust or corrosion. A regular cleaning schedule should be followed to ensure dependable operation.

A troubleshooting chart for contacts is given in table 6. Contact burnouts or overheating of contacts may be due to foreign matter on the contact surfaces. Generally, foreign matter can be removed by using a cloth moistened with an approved cleaner. Where oxides or other corrosions are present, they can be removed by smoothing over the contact surfaces, using a fine file or 00 grade sandpaper.

Slow making and breaking of the load circuit may be caused by binding of the armature, plunger or lever mechanism. To eliminate this condition, remove rust, gum, and filings from the shafts, bearings, or guides. Shafts and bearings should be lubricated with two drops of light oil (S.A.E. 10) semiannually.

Insulation breakdowns are often caused by metal filings or excessive moisture. Short circuits across insulating barriers, grounded circuits, and high voltage breakdowns can often be avoided by regular cleaning.

Paint should never be used to refinish shafts, bearings, guides, sliding parts, magnetic pole faces, ground surfaces, contact members, or phenolic or ceramic parts.

A-C Brakes

Brake systems used on a-c equipment perform the same work as those used on d-c systems. However, there are differences in brake construction.

SOLENOID.—The a-c solenoid brake is designed the same as the d-c brake (fig. 8-5), except that the brake frame and solenoid are of laminated construction to reduce eddy currents. Because the magnetic flux passes through zero twice each cycle, the magnet pull is not constant. To overcome this, shading coils are used to provide pull during the change of direction of the main

Toble 6.-Troubleshooting chart for contacts

♦ WARNING: Short circuits or grounds in the associated equipment necessitates a thorough inspection of the controller, as the contacts may be damaged by these conditions.

/lOQ

flux. However, even with shading coils it is difficult to design a solenoid that is completely free from vibration and quiet in operation.

The principal disadvantage of an a-c solenoid is that it draws a heavy current when voltage is first applied. This effect is caused by the large air gap that exists before the solenoid is energized. With a large air gap, the flux density and inductive reactance, X_i , are low. Because the current is limited principally by the ohms, of the coil, the inrush current is high as the air gap closes. When the solenoid is operated, the flux density and X_l ohms increase, thereby lowering the current to the operating value. In general, a-c solenoids are used only for small brake applications. It is possible, however, to use polyphase solenoids to brake larger motors.

TORQUE MOTOR.—The torque motor brake uses a specially wound polyphase squirrel-cage motor in place of the brake release solenoids. The motor may be stalled without injury to the winding and without drawing heavy currents. The mechanical arrangement of a torque-motor brake assembly and the ball jack assembly are given in figure 8-19.

The mechanical connection between the torque-motor shaft and the brake operating lever is through a device called a "ball jack" (fig. 8-19, B), which converts the rotary motion of the torque-motor shaft to a straight-line motion.

When power is applied to the torque motor, the shaft turns in a clockwise direction. The thrust element (T) in the jack pushes against the operating lever @ in figure 8-19, A, and the upward movement of the jack screw (3) (fig. 8-19, B) moves the operating lever to release the brake. As soon as the brake is fully released, the torque motor stalls across the line. This holds pressure against the spring (?) in figure 8-19, A, and keeps the brake released.

When the voltage supply to the torque motor is opened, the torque spring forces the brake shoes against the brake wheel. This stops and holds the motor shaft.

The torque motor brake can be released manually by raising lever©. However, if the lever is not held manually in the UP position, the brake will be applied.

A BRAKE ASSEMBLY

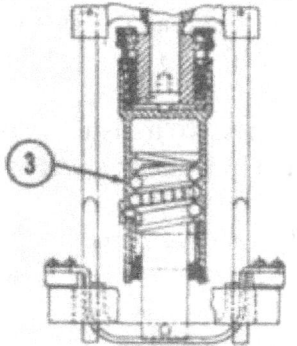

B BALL-JACK ASSEMBLY

Figure 8-19.-Torque-motor brake and ball jack assembly.

A-C CONTACTORS

Basically, construction, operation, and application of an a-c contactor is the same as the d-c contactor described earlier in this chapter. However, because of the characteristics of a-c, there are differences in
construction. Some of these have been detailed in preceding sections of this training course.

A-C Magnets

The cores used for a-c magnets are of laminated construction so that eddy currents may be reduced.

Shading Coil

Because the flux in a single-phase magnet is in phase with the coil current, the magnetic pull at the air gap varies with the frequency of the supply voltage. Therefore, twice each cycle, the pull will go to zero, and the armature will chatter against the core. The noise is objectionable, and the chattering causes excessive wear on the magnet core face.

This noise and wear is overcome by a shading coil imbedded in one of the magnet faces (fig. 8-20).

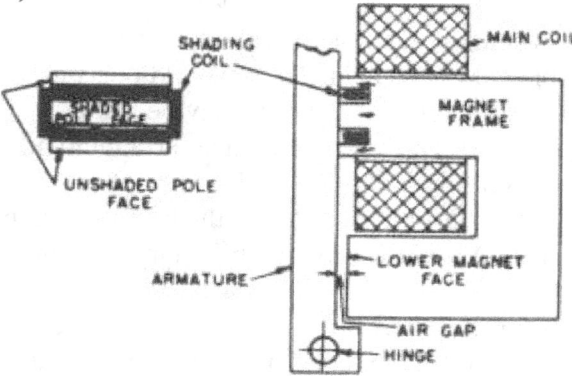

Figure 8-20.—Core construction for a-c
contactor operation.

The shading coil acts as the secondary of a transformer, as explained in chapter 7. The effect of the shading coil is to cause the main coil flux to be out of phase with the shading coil flux so that when the main coil flux goes through zero, the shading coil flux will be high enough to maintain pull on the armature and keep it closed.

Most magnets have a second air gap, shown above the binge in figure 8-20. This gap must not be closed when the armature is pulled in because its purpose is to prevent residual magnetism from holding the armature closed after the coil has been deenergized.

Magnetic Blowout

Because the voltage drops to zero twice during each a-c cycle, an arc formed as the contacts open will be reduced when the voltage drops. However, there is usually an inductive effect, caused by the collapsing field, which generates an induced voltage that attempts to maintain, the arc. The arc is extinguished as quickly as possible by magnetic blowout coils. The principle of operation is the same as for d-c blowouts, but the core is laminated. The purpose of the blowout magnetic field is to lengthen the arc to the point where it can no longer jump from one contact to the other.

Coils

The design of an a-c magnet is more complicated than that for a d-c magnet. Long stroke magnets are likely to have a pull, which, at first, increases as the gap decreased, and then stays fairly constant until the gap becomes nearly closed. The pull then increases sharply as the gap decreases to zero. Figure 8-21 illustrates how the pull increases as the gap becomes smaller. Note that as the pull becomes greater, the coil current and wattage decrease.

The current decreases because the reluctance of the magnetic circuit decreases as the air gap becomes smaller; hence, the flux density and reactance increase, and as stated before, the principal opposition to the coil current is the inductive reactance of the coil. With a rising inductive reactance and a constant potential source, the current will decrease.

Shunt A-C Contactors

Shunt type a-c contactors have laminated frame construction, and the coils are wound with many turns of fine wire. The coils are designed for connection directly across the line and are rated for either continuous or

Ann

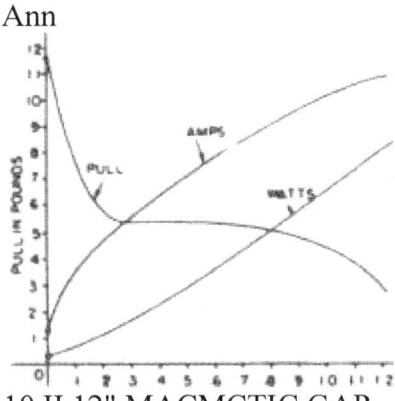

10 II 12" MACMCTIC GAP

Figurt 8-21.-Curvts showing relationship botwoon mogootic pull, coil currtnt, and wattago.

intermittent operation. Usually, shunt-type a-c contactors are used to control motor circuits. The initial control impulses may come from a pilot device or from a manual control, which serves as a master switch for the contactor.

STARTING.—Magnetically operated starters for a-c motors may be connected to give low-voltage protection, as shown in figure 8-22.

The circuit used for low-voltage protection with manual restart employs momentary contact pushbuttons (fig. 8- 22, A). The motor is started by pushing the START button. This action completes the circuit from LS through the control fuse, STOP button, START button, the overload relay contacts, OL, and the contactor coil,M, to LI. When the coil is energized, it closes line contacts Ml, M2, and MS, which connect the full-line voltage to the motor. The line contactor auxiliary contact, MA, also closes and completes a holding circuit for energizing the coil circuit after the START pushbutton has been released.

The motor will continue to run until the contactor coil is deenergized by the STOP pushbutton, failure of the line voltage, or tripping of the overload relay, OL.

Low-voltage protection (L.V.P.) with manual restart is provided by the use of momentary contact pushbuttons. If coil M is deenergized, the contactor will not reclose and start the motor when voltage is restored, or when the overload relays are reset until the START pushbutton is

TO 3 PM A C SUPPLY S

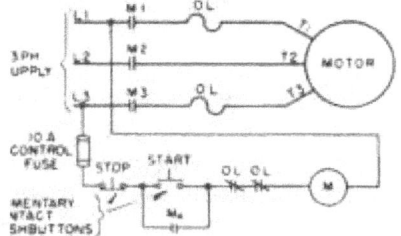

MOMENTARY CONTACT PUSHBUTTONS
A LOW-VOLTAGE PROTECTION

TO JPM A C. SUPPLY

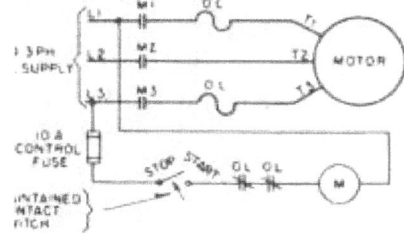

MAINTA'NE 0 CONTACT SWITCH
B LOW-VOLTAGE RELEASE

Figure 8-22.—Low voltage protection and low-voltoge release circuits.

again depressed. The overload relays, OL, are reset by a reset button on the outside of the motor starter cabinet.

A low-voltage release circuit with automatic restart employs a maintained contact, STOP-START, switch (fig. 8-22, B). The motor is started by turning the switch to the START position. This action completes a path from L3 through the fuse and maintained contact of the STOP-START switch to the overload relay contacts,OL,and the contactor coil, M, to LI. When the contactor coil, M, is energized, it closes line contacts M1,M2, and M3, which applies full-line voltage to the motor. The motor will continue to run until the coil, M, is deenergized by loss of voltage, tripping of the overload relay, or by turning of the STOP-START switch to the STOP position.

Low voltage release (L.V.R.) with automatic restart is provided by the use of maintained contacts on the STOP-START master switch. If the operating coil, M, becomes deenergized through loss of voltage, the

contractor will close and restart the motor as soon as voltage is restored or when the reset button for the overload relay is pressed.

REVERSING.—Certain equipment, such as hoists, requires a driving motor that can be reversed. It has been explained how d-c motor armature rotation may be changed by reversing the connections to either the field or armature winding. The reversal Of an a-c induction motor is accomplished by interchanging any two of the three leads to the motor. The connections for an a-c reversing-type controller are illustrated in figure 8-23. The STOP, REVERSE, and FORWARD pushbutton controls are all of the momentary-contact type. Note the connections to the REVERSE and FORWARD switch contacts. The circles with F and R inside represent contactor coils.

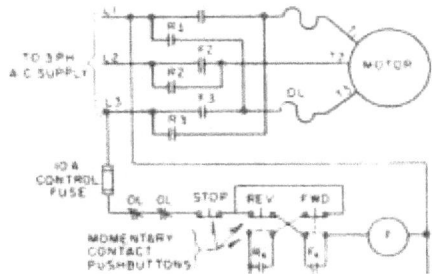

i &

Figure 8-23.-A-c reversing controller.

If the FORWARD pushbutton is pressed (solid to dotted position), coil Fwill be energized and will close its holding contacts, FA. These contacts will remain closed as long as coil F is energized. When the coil is energized, it closes line contacts Fl, F2,and F3, which applies full-line voltage to the motor. The motor then runs in a forward direction until the voltage supply is interrupted.

If either the STOP button or the REVERSE button are pressed, the circuit to the Fcontactor coil will be broken and the coil will release and open line contacts Fl, F2, F3, and holding contact FA.

If the REVERSE pushbutton is pressed (solid to dotted position) coil R will be energized and will close holding

contacts RA and line contacts Rl, R2, and R3. Note that R contacts reverse the connections of lines 1 and 3 to motor terminals Tl and T3. This causes the motor rotor to rotate in the reverse direction. TheF andR contactors are mechanically interlocked to prevent both being closed at the same time.

Momentary contact pushbuttons provide low-voltage protection with manual restart in the circuit shown in figure 8-23. If either the F or R operating coil is de-energized, the contactor will not reclose and sfcxt the motor when voltage is restored unless the FORWARD or REVERSE pushbutton is pressed. The circuit arrangement of the pushbuttons constitutes an electrical interlock that prevents energizing both coils at the same time.

SPEED REGULATION.-When it is desired to operate at different speeds while using an a-c motor, you may use a circuit such as that in figure 8-24.

An a-c induction motor that has been designed for two-speed operation frequently has two separate windings, one for each speed. In figure 8-24 the motor slow winding is connected to terminals Tl, T2, and T3. The motor fast winding is connected to terminals Til, T12, and T13. Overload protection is provided by the S-OL coils and

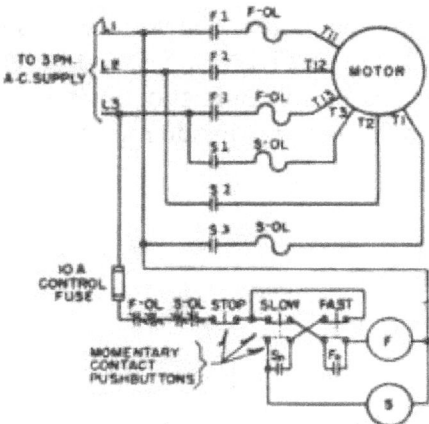

Figure 8-24.-Speed control circuit for an a-c motor.

contacts for the slow winding and the F-OL contacts and coils for the fast winding.

The control pushbuttons are of the momentary contact type. Pressing the FAST pushbutton (solid to dotted position) closes the fast contactor by energizing coil F. This coil remains energized, after the pushbutton is released, through holding contacts FA. The coil, F, closes main line contacts F1,F2,and F3, which apply full-line voltage to the motor fast winding. The motor will then run at fast speed until coil F is deenergized.

Pressing the SLOW pushbutton closes the slow contactor by energizing coil S. The coil remains energized, after the button is released, through holding contacts SA. The coil, S, closes main-line contacts SI, S2, and S3, which apply full-line voltage to the motor slow winding. The motor will then run at slow speed until coil S is de-energized. The F and S contactors are mechanically interlocked to prevent both being closed at the same time.

A single-phase autotransformer has a tapped winding on a laminated core. Normally, only one coil is used on each core, but it is possible to have two autotransformer coils on the same core.

Figure 8-25 shows the connections for a single-phase autotransformer being used to step down voltage. Part

Autotransformers

C

o-

9.22 AMPERES

Ei -440 VOLTS
b-7 AMPERES

4.78 AMPERES

b

E₂ — 140 VOLTS

a

LOAD
= S.S2 AMPERES

Figure 8-25.-Single-phase autotransformer.

of the winding, that between a and b, is common to both primary and secondary and carries a current that is equal to the difference between the load current and the supply current.

Any voltage applied to the terminals, a and c, will be uniformly distributed across the winding in proportion to the number of turns. Therefore, any voltage that is less than the source voltage can be obtained by tapping the proper point on the winding between terminals a and c.

One commercially available autotransformer is the Variac and is designed so that a knob-controlled slider makes contact with wires of the winding in order to vary the load voltage.

The positive directions for current flow through the line, transformer winding,and load are shown by the arrows in figure 8-25. Note that the line current is 2.22 amperes and that this current also flows through the part of the winding between b and c. In the part of the winding that is between a and b, the load current of 7 amperes is opposed by the line current of 2.22 amperes. Therefore, the current through this section is equal to the difference between the load current and the line current. If you subtract 2.22 amperes from 7 amperes you will find the secondary current is 4.78 amperes.

A-C Controllers

Two common uses for autotransformers are to start 3-phase induction and synchronous motors and to furnish variable voltage for test panels. Figure 8-26 shows an autotransformer motor starter, which incorporates starting and running magnetic contactors, an autotransformer, a thermal overload relay, and a mercury timer to control the duration of the starting cycle.

The autotransformer is a dual unit with separate windings on the outside legs. The windings are connected in open delta to supply reduced voltage to a three-phase motor during the starting period. Each winding is tapped to provide starting voltages at the motor of 50 percent, 65 percent, or 80 percent of the line voltage. It is possible to change the connections to the transformer taps to get higher or lower starting voltage. It is necessary to select the tap for the desired starting voltage when making the original connections.

Figure 8-26. - Autotransformer motor starter.

A complete schematic diagram of a motor starter using autotransformers as well as other control devices Is given In figure 8-27. Because a number of separate controls are shown in schematic form, the action will be traced from the time the START button Is pressed. This control is a springloaded button, which completes the circuit between two contacts.

The START button, when pressed, completes the circuit from LI and the button contacts through the normally closed contacts CR1 and MR4, and through the starting coil MS to L2. When the starting coil, MS, is energized, it attracts an armature, which operates seven contacts. Four of these contacts (MSI, MS2, MS3, and MS4) connect the autotransformer in open delta across the line.

The fifth contact, MS5, completes the holding circuit to the starting coll. The holding circuit is from LI through the STOP button, to normally closed contacts 10L and 20L, through contact MS5, to normally closed contacts CR1 and MR4, and then through coil MS to L2.

MS4
wywywww^
JOB 3 |2 Tl
65B SOB

4 CRi M O 0 O-f
R4 MS
T R
Hl-
T.C.
CR
rn
MS7

Figure 8-27.-Schematic of motor starter using autotransforrners and other control devices.

The sixth contact, MS6, completes the circuit to the timer relay, TR. This circuit path is from LI through the STOP button, through the normally closed contacts 10L and 20L, through contact MS6,and through the timer relay coil TR to L2.

The seventh contact, MS7, is normally closed, and when relay MS operates, it opens the circuit to the coil of motor running contactor, MR, so that MS7 acts as an electrical interlock. Thus, contactMS7 prevents coil MR from being energized when MS is operated. This action prevents a portion of the autotransformer winding from being short circuited by contacts MR1 and MSI or MR3 andMS3. After a time interval for which the timing relay is set, the mercury timer, TR, closes the circuit to

4^

the control relay, CR, governing the running contactor, MR.

The control relay contacts, CR1 open the holding coil circuit to the main starting coil, MS, permitting the starting contactor to open. Contacts CR2 close and energize the holding circuit for relay coil CR. Contacts CR3 close in preparation for energizing the running contactor coil, MR.

During the transition from reduced voltage to full voltage relay MS opens and temporarily disconnects the transformer and motor from the circuit. During this time the motor is free-running with no connection to the line. However, this interval is of short duration because the opening of MS completes the circuit to the main running coil, MR, through the seventh or interlock contact, MS7. The MR coil then closes contacts MR1, MR2, and MR3 to connect the motor directly across the line. Electrical interlock MR4 opens to prevent the starting coil, MS, from being energized at the same time that MR is energized.

The starter previously described is called an OPEN TRANSITION type because the motor is disconnected from the line during the transition from low (starting) voltage to line (running) voltage.

When the starter does not completely disconnect the motor from the line at any time during the starting period, the starter is called a CLOSED TRANSITION type.

Trouble Analysis

The ability to make a fast analysis of conditions when faults develop is not acquired overnight. You must study individual circuit controls and then determine which part of section could, if defective, cause the trouble.

In simple circuits where the control is a simple make and break contact arrangement, motor stoppage could be due to a lack of voltage (generator failure, or an open or shorted supply line), a defective contactor (not closing or broken), or a defective motor.

Simple voltage tests at the controller input and output terminals will show if voltage is available from the line and if the contacts are in good condition. Of course, the contacts must be closed when the output terminal test is

made. Voltage present at the controller output indicates that the power source is working and that the contactor is closing the circuit. These two possible trouble spots are therefore eliminated. The remaining possible trouble spots are the leads between the controller and the motor. A measurement for voltage at the motor terminals will show the condition of the leads. If voltage is present at the motor terminals and the motor does not run, the motor must be defective. As you gain experience in troubleshooting, you will develop your own testing techniques.

No matter how complex a complete circuit may appear, each individual section can be broken down into three parts: the source of voltage; the load (motor, relay coil, or other device); and the wires between the source and the load. Make it a habit to trace circuits when you encounter them. Start at one of the voltage supply terminals and then trace through the circuit or diagram to the other terminal. Once you are able to trace a circuit through on a diagram, you will find it easier to trace actual wiring. Eventually, you will find yourself forming a mental picture of a schematic diagram as you trace wiring. However, it takes a great deal of practice so don't become discouraged if you cannot trace a circuit the first time you try. You should find Chapter 9, Blueprint Reading and Sketching, NavPers 10077-A, helpful.

There are times when a motor will fail to start and no signs of trouble will be readily apparent. In such cases, operate the overload reset button and again press the start button. Listen for the sound of the main contactor closing. If there are no results, press the stop button and check to determine if power is available at the input terminals of the controller. If power is available, the trouble is in the section from the controller input to the contact operating means (relay coil or leads). If no power is available, check the supply line and the associated fuses.

If the motor stopped because of tripping of the overload relay, and the relay tripped again after being reset, some possible causes of the overload are:

1. Jammed machinery (frozen bearings or foreign material in gears).
2. Overheated or stiff bearings in motor or motor driven device.
3. Blown fuse in the supply circuit, causing single phasing of a three-phase supply.

If no apparent cause for overload tripping can be found, the motor current should be carefully checked with an ammeter and the measured current compared with current value marked on the motor nameplate, the controller or overload relay coil or heater rating, and the current values specified on drawings for the equipment. Many overload relays have an adjustment, which allows the trip settings to be adjusted and thus take care of minor variations. Any changes should be made in small steps in order for the relay to give maximum protection to the motor.

Corrective Maintenance

Many equipment failures can be prevented by correcting possible trouble-making conditions as soon as they are discovered. For example, if you notice worn insulation on a wire, protect it by taping the worn spot as soon as possible. If moisture has accumulated in an outlet or switch box, dry it out before trouble develops. By correcting conditions that are potential trouble spots, you can save lost time and make your job easier.

MAINTENANCE

An ideal electrical installation is one in which the equipment always operates correctly when the controls are operated and one in which there are no breakdowns or stoppages while the equipment is in use.

Ideal operation of this sort is our aim. However, it cannot be approached unless proper maintenance procedures are followed. Trouble-free operation requires a knowledge of the various circuits, a knowledge of the equipment, an understanding of the purpose of each item, and a willingness to do jobs before they become absolutely necessary.

The first step in an adequate maintenance program should be a thorough inspection of the equipment after it is installed and before it is operated electrically. An inspection of this type should include cables, connections, fuses, and equipment mountings. All oil and grease fittings should be checked to make sure the equipment has been lubricated.

Most motor control equipment failures are due to mechanical faults, not to electrical troubles. Dirty equipment, improperly placed parts, andparts that bind or stick are all basic mechanical faults that can cause coil burnouts, welded contacts, and circuit overloads.

PERIODIC TEST AND INSPECTIONS

Controllers for motors and the associated equipment should be inspected monthly. During this inspection you should:

1. Disconnect the device or apparatus from its source of power.
2. Open the cover of the enclosure.

3. Make a visual check for:

a. Dirt, oil, grease or moisture on component parts on or in the enclosure.

b. Burned or worn contacts, charred insulation, or varnish extrusions.

c. Torn gaskets, bent or missing cover bolts, corrosion, or rust.

4. Check manually for:

a. Bind in plungers, armatures, fingers, levers, and other moving parts.

b. Loose connections, parts, and cotter pins.

5. Check for grounds on both the main circuit and the pilot circuit.

6. If permissible, connect the equipment to a source of power and operate it in a normal manner.

a. Check for proper operation.

b. Check for excessive noise.

7. Shut off the machine and close the cover. Then tighten the cover bolts or screw pins so that the cover is held securely in place.

In addition to regular monthly test inspections, a more complete inspection should be made semiannually. The following tests should be made:

1. Check all contacts for excessive wear, proper dimensions, and proper spring pressure. If necessary, remove contact hoods or arc shields in order to reach the contacts.

2. Clean all parts thoroughly. Disassemble or partially disassemble the device, when necessary, for proper cleaning.

3. Lubricate all shaft bearings with two drops of light weight (S.A.E. 10) oil. DO NOT LUBRICATE CONTACTS, PLUNGERS, AND POLE FACES OR USE EXCESSIVE QUANTITIES OF OIL ON THE SHAFT BEARINGS.

4. Check the mechanical and electrical adjustments of each device. Make any necessary readjustments.

Of course, the regular monthly and semiannual inspections do not eliminate the need for careful observation during the time the equipment is operating and the need to correct conditions that could cause trouble as soon as they are noticed.

SAFETY PRECAUTIONS

One of the first things a prospective EM should learn is that electricity is potentially a killer I A conductor carrying several hundred volts is no different in appearance than a conductor disconnected from a source of voltage. The lethal voltage that may be present on the wire has no odor and makes no sound. Yet, under certain conditions, it can snuff out your life.

However, properly handled, electricity is not dangerous. Almost all deaths caused by electric shock could have been prevented if a few simple rules had been followed. Therefore, you should memorize these rules and then ALWAYS FOLLOW THEM.

The first rule is to always assume, until tests prove differently, that all conductors are hot. Do not assume that low-voltage circuits (115 volts or less) are not dangerous. A number of shipboard fatalities have been reported due to contact with 115-volt circuits. Shipboard conditions are particularly conducive to shock because of the metal structure of the ship. Also, the body resistance may below due to perspiration or damp clothing.

Do not lose your sense of caution because you are in a hurry to get the job done. Often taking time to make certain will save time. Except in cases of emergency, never work on an energized circuit. Safety precautions for the special cases are outlined later in this section. All circuits must be considered as energized until a

personal check has been made to see that the switch has been opened and tagged and that the circuit has been tested with a voltmeter or voltage tester.

Navy specifications for portable tools require a separate ground wire, which grounds the frame of the tool. Normally, this is taken care of by using a three-wire cable with a special plug and receptacle. Two of the wires in the cable connect the tool to the line. The third wire, distinctively marked, is connected to the tool frame at one end and to the ground contact of the plug on the other end.

Where grounded receptacles have not been installed or where the tool does not have the internal ground wire connection, the tool should be grounded by obtaining an additional wire and connecting it between the tool frame and the ship's metal structure.

Portable cables should be carefully selected and maintained. They should be of the proper length and cross-sectional area. Spliced portable cables are extremely dangerous and should not be used unless the emergency justifies the risk involved.

Always use a fuse puller when changing fuses. This precaution is necessary because one end of a fuse is hot, and, in some cases, both ends are hot even though the fuse is blown. When a fuse is blown, it should be replaced with a good fuse having the same rated voltage and current capacity.

Extreme care should always be taken to ensure safety for the operator and other personnel when operating switches and circuit breakers. Before closing any circuit braker or switch, be sure that:

1. The circuit is ready with all equipment connected to it in condition to be energized.

2. Men working on the circuit are clear and are notified that it is to be energized.

3. All circuit protective devices (fuses, circuit breakers, and overload protectors) are in good working condition.

4. Only one hand is used to operate a circuit breaker or switch unless operating arrangements specifically provide for the use of both hands. Keep the other hand clear because metal parts of the circuit breakers or switches may be at circuit potential.

5. For manually operated knife switches the opening or closing motion is made positive and rapid.

Live-Circuit Precautions

Military necessity may require repair or maintenance work on energized circuits. When this condition exists, extreme precautionary measures MUST BE TAKEN. Such work should be accomplished only by adequately supervised personnel fully aware of the dangers involved. Every precaution should be taken to adequately insulate the person performing the work from ground. In addition the following precautions should be taken:

1. The person doing the work should remove wrist-watches, rings, watch or key chains, metal articles, or loose clothing, such as neckties or unbuttoned sleeves which might accidentally catch and throw his body into contact with live parts. All clothing and shoes should be as dry as possible.

2. Ample illumination should be provided.

3. The worker must be insulated from ground by covering nearby metal that he might contact with dry wood, rubber mats, dry canvas, or even several thicknesses of dry paper. Be sure that any material used for insulation has no holes in it and no conducting material imbedded in it.

4. Working metal tools must be covered with insulating rubber tape (not friction tape) as far as practical.

5. Insulating barriers must be provided between the work and any live metal parts

immediately adjacent to the work to be done.

6. The work should be accomplished, using one hand only, if possible.

7. If the work permits, rubber gloves should be worn on both hands. If this is not possible, a rubber glove should be worn on the hand not used in handling tools.

8. Men must be stationed at circuit breakers or switches, and telephones must be manned, if necessary, so that the circuit or switchboard can be deenergized immediately in case of emergency.

9. A man qualified in first aid for electric shock should be immediately available while work is being done.

Charged-Circuit Precautions

In certain cases, electrical machinery may retain a charge even though it is secured. This charge can be great enough to cause severe shock. Therefore, before touching the terminals of an apparently dead machine, discharge it to ground, using an insulated portable wire or a tool with an insulated handle.

Capacitors used to suppress radio interference and other purposes store electrical energy. Before touching the terminals of a capacitor, which is connected to a de-energized circuit or has been disconnected entirely, short circuit the terminals.

Cleaning Precautions

Never use toxic materials for cleaning in closed spaces. Carbon tetrachloride is an excellent cleaner, but extremely poisonous and dangerous to use in closed spaces. It is no longer authorized for cleaning purposes and methyl chloroform should be used in its place.

Gasoline, benzine, ether, and other similar inflammable fluids should never be used for cleaning electrical equipment even though it has been deenergized. Fumes from these fluids form a highly explosive mixture when mixed with air, and a slight spark can set them off.

Because many cleaning fluids are detrimental to good commutation, they should never be poured or sprayed on commutator surfaces. Instead, any cleaner should be applied with a clean cloth, and the surface should be dried with another cloth.

QUIZ

1. A rheostat having a resistance of 10 ohms at a particular setting is in series with a 30-ohm load across a 120-volt source, (a) What current flows through the rheostat? (b) What is the voltage drop across the rheostat? (c) What is the voltage across the load?

2. If the a-c generator output voltage (fig. 8-2) is above normal, in what direction (up or down) should the motor-driven rheostat arm be moved to return the voltaee to normal?

3. When the switch (fig. 8-3) is operated from position 1 to position 2, what is the relation of the series field connection with respect to the armature?

4. What is the source of the voltage used to establish the field and the current through the braking resistor (fig. 8-3)?

5. In magnetic brakes used for complete braking protection (a) what applies the brakes and (b) what releases them?

6. What is the effect on speed of release if brakes are designed to operate on one-tenth the normal line voltage compared with normal line voltage?

7. Besides employing contacts that snap open, how is the arc extinguished in the larger contactors or circuit breakers?

8. How are the contacts in the contactor (fig. 8-8) made self-cleaning ?

9. Identify the main current-carrying contacts in the contactor shown in figure 8-8.

10. If the contactor (fig. 8-8) cannot be tilted backward, the tail spring should be adjusted

so that the contactor will pick up at what percent of the line voltage?

11. What device is used to secure contactor enclosures and thereby safeguard personnel from dangerous voltages by preventing the opening of the enclosure while the contactor is energized?

12. In the electric door lock (fig. 8-10) how is the operating coil of the door lock connected with respect to the operating coil of the contactor in order to prevent the door of the enclosure from being opened as long as the contactor coil is energized?

13. What device is used to prevent the simultaneous operation of the FAST and SLOW control (fig. 8-11)?

14. What provides the time delay feature in the overload relay shown in figure 8-12?

15. Shunt-type contactors (fig. 8-13) have what two general applications?

16. What prevents the contactor (fig. 8-14) from closing its main contacts on the initial surge of motor -starting current?

17. What prevents the contactor (fig. 8-15) from opening when the load on the motor is light?

18. What causes the simultaneous closure of contacts LCI. LC2, LC3, and LC4 (fig. 8-16)?

19. At approximately 110 percent of normal running current (fig. 8-16), what causes contacts AC1 and AC2 to close ?

20. What contact comprises an electrical interlock (fig. 8-17) that prevents the reverse contactor from being energized when the forward contactor is operated?

21. How is the speed of the motor (fig. 8-18) increased after the accelerating period?

22. What corrective maintenance will eliminate slow making and breaking of the load circuit contacts caused by binding of the armature, plunger, or lever mechanism of the contactor?

2 3. Why does an a-c solenoid draw a heavy current when voltage is first applied?

24. When the voltage supply to the torque motor (fig. 8-19) is opened, what component forces the brake shoes against the brake wheel?

25. What is the purpose of the second air gap shown above the hinge in figure 8-20?

26. As the pull of a long stroke magnet becomes greater (with the air gap approaching zero, fig. 8-21), how do the magnitudes of the coil current and wattage vary?

27. Low-voltage release with automatic restart (fig. 8-22) is provided by the use of what type of contacts on the STOP-START master switch (maintained or momentary) ?

28. Closing the R contacts instead of the F contacts (fig. 8-23) reverses the connections of which two lines with respect to motor terminals Tl and T3?

29. How are the F and S contactors (fig 8-24) prevented from closing at the same time?

30. What are two common uses for autotransformers in the Navy?

31. Which four contacts connect the autotransformer in open deltaacross the line (fig. 8-27) when the START button is pressed?

32. Why is the starter shown in figure 8-27 called an "open transition" starter?

33. What are the three parts into which a circuit (no matter how complex) can be subdivided for ease in circuit tracing?

34. How often should motor controllers and the associated equipment be inspected?

35. What is the first safety rule to follow when working on an electrical circuit?

36. What is the rule about grounding portable tools?

37. What is the rule about wearing apparel such as wrist-watches, rings, watch or key chains, metal particles, or loose outer clothing for example neckties and unbuttoned sleeves,

when working on live circuits?

38. Why should the terminals of an apparently dead machine be grounded before proceeding to work on it?

CHAPTER

ELECTRICAL SYSTEMS IN SMALL CRAFT

Small craft perform an important function in the daily routines of all naval vessels. Underway, they serve as duty life boats and must be ready at all times to operate instantly and efficiently. In port, they must be prepared to perform a variety of tasks, such as transporting stores and liberty parties. On some vessels, such as troop transports, small craft are vital to the accomplishment of the ship's mission and the condition of these boats affects the condition of readiness of the ship.

The capability of small craft to carry out these services depends, to a great extent, on the assigned Electrician's Mates, who are responsible for the proper operation and upkeep of the electrical systems installed in these boats. The electric plant consists essentially of an engine starting system and lighting system (fig. 9-1). The voltage of the system depends on the battery voltage required for the starting motor. The power for the auxiliary loads is distributed from the engine control and distribution panel.

ENGINE STARTING SYSTEM

The engine starting system essentially consists of a storage battery, battery-charging generator, starting motor, and suitable controls for regulating and protecting the electrical system. The engine starting system will vary, depending on whether the boat is equipped with a

CONTROL AMD DISTRIBUTION PANEL

INSTRUMENT PANEL

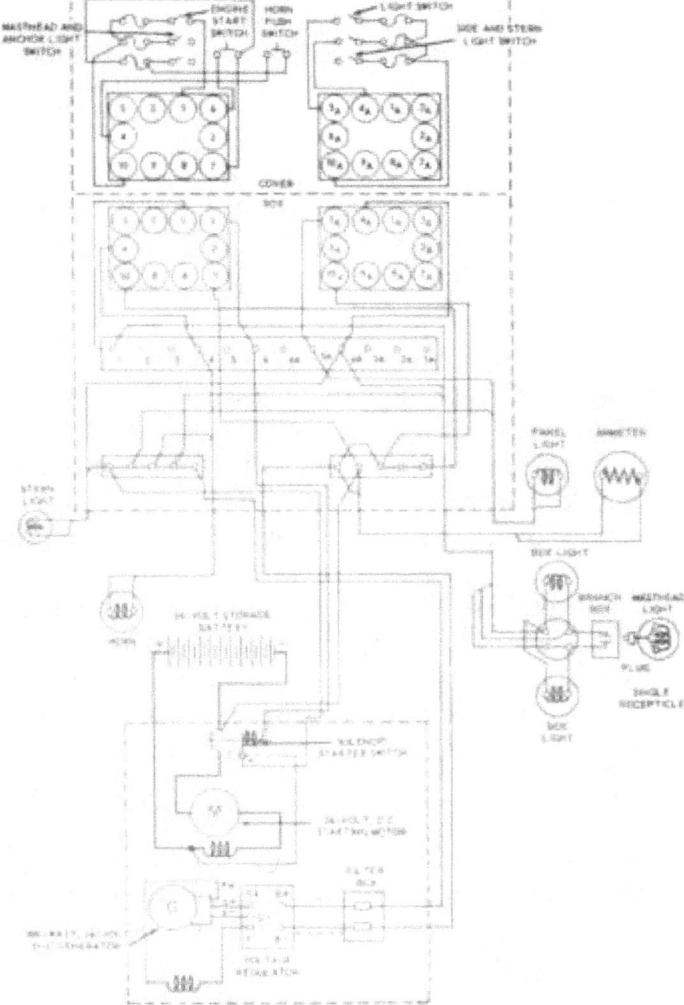

OlEJEL PROPULSION ENGINE

Figure 9-1.-Typical wiring diagram of a 40-foot utility boat.

gasoline engine or a diesel engine. For example, a gasoline engine-driven boat includes an electrical ignition system; whereas, a diesel engine-driven boat often includes an electric air heater.

Storage Battery

The lead-acid storage battery provides a source of current for starting the engine, functions as a voltage stabilizer in the electrical system, and supplies current for a limited time when the electrical demands of the boat exceed the output of the generator. The battery is located close to the engine to avoid long cable runs and thus minimize the voltage drop in the system.

The storage battery requires very little attention, but periodic inspections are essential to obtain the maximum efficiency and life of the unit. The operation and maintenance of engine-starting batteries are discussed in chapter 6 of this training course, and additional information is contained in the manufacturers' instruction books furnished with the equipment.

Battery-Charging Generators

In order to maintain the battery in a fully charged condition, it is necessary that the discharge current be balanced by a charging current supplied from an external source, such as a battery-charging generator. If the discharge current exceeds the charging current for an appreciable period, the battery will gradually lose its charge and then will not be capable of supplying the necessary current to the electrical system.

Battery-charging generators sometimes are flange mounted on the rear of the engine and driven from the timing gear train, but usually they are cradle mounted on the side of the engine and driven by a V-belt from the crankshaft pulley. These generators are rated according to the particular application and are designed for clockwise or counterclockwise rotation. They are supplied for use with either 6-, 12-, or 24-volt systems.

Battery-charging generators employed with the electrical systems in small craft are usually d-c generators. However, the recent use of additional electrical apparatus

that requires considerable power at low engine speeds has appreciably increased the output requirements of these generators. Therefore, the a-c generator (alternator) is being installed in some small craft because it can be designed to produce sufficient power over a speed range that varies from idle to top engine speed without materially increasing its physical size.

The types of battery-charging d-c generators are the third-brush generator and the shunt generator (fig. 9-2). Electrical systems in small craft were formerly grounded

REGULATOR CONTACTS
RCMATMI
CONTACTS

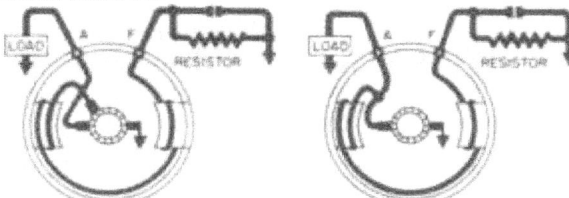

A -EXTERNALLY GROUNDED THIRD- B'EXTERNALLY GROUNDED SHUNT BRUSH GENERATOR (STANDARD) GENERATOR (STANDARD)

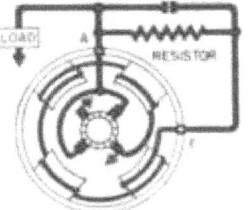

. REGULATOR CONTACT-

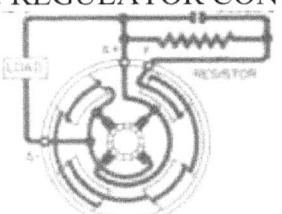

REGULATOR CONTACTS
•INTERNALLY GROUN0E0 SHUNT GENERATOR (HEAVY DUTY)
D-UNGROUNOED SHUNT GENERATOR (HEAVY DUTY)

Figure 9-2.-Battery-charging d-c generators.

utilizing the externally grounded and internally grounded types of generators illustrated in figure 9-2, A, B, andC. Many of these systems are now ungrounded, utilizing the type of generator illustrated in figure 9-2, D, in which the generator grounding feature is eliminated.

THIRD-BRUSH GENERA TOR.-The third-brush generator (fig. 9-2, A) utilizes a third brush and armature reaction to control the amount of current through the field, and thus the maximum output of the generator. One end of the generator field is connected to a third brush, which is placed behind a main brush so that maximum armature voltage is not impressed on the generator field. The output of the generator depends on the position of the third brush. On many third-brush generators the third brush is fixed and cannot be adjusted. This type generator is used with a step-voltage control or voltage regulator to provide the necessary control of the generator output for various operating conditions. On most third-brush generators having an adjustable third brush, the output can be increased by moving the third brush in the direction of armature rotation and decreased by moving the third brush in the opposite direction.

The third-brush generator is not used extensively because it is slow to build up to its maximum output at the lower speeds, and its output tends to taper off due to armature reaction at the higher speeds.

SHUNT GENERATOR.-The shunt generator (fig. 9-2, B, C, and D) is widely used with electrical systems in small craft because of its improved performance at low speeds and because its output does not taper off at high speeds. Shunt generators and many third-brush generators require some form of regulation to control the output current and prevent the generator from exceeding its maximum rated voltage when the speed increases.

Regulators operate on the principle of inserting a control resistance in series with the generator field circuit to reduce the voltage and output of the generator as required by the operating conditions. Generators are classified as EXTERNALLY GROUNDED, INTERNALLY GROUNDED, and UNGROUNDED according to the system used to connect the control resistance in the field circuit.

The EXTERNALLY GROUNDED GENERATOR (fig. 9-2, A and B) usually has one end of the field circuit connected to the ungrounded generator brush and the other end brought out through terminal F to ground through the regulator. Thus, the field circuit is grounded outside the generator.

The INTERNALLY GROUNDED GENERATOR (fig. 9-2, C) usually has one end of the field circuit connected to a grounded generator brush and the other end brought out through terminal F and connected to terminal A through the regulator. Thus, the field circuit is grounded INSIDE the generator through the grounded brush connection.

The UNGROUND GENERATOR (fig. 9-2, D) very often has one end of the field circuit connected internally to terminal A— and the other end connected externally through terminal F and the voltage regulator to terminal A+. The essential difference between the internally grounded generator and the ungrounded generator is that in the grounded system the circuits are completed through ground (generator frame and engine); whereas, in the ungrounded system the circuits are completed through a second ungrounded wire.

Generators with the field circuits externally grounded are generally used with STANDARD regulators common to most automotive applications; whereas, internally grounded and ungrounded generators are used with HEAVY DUTY regulators. The type of generator must be determined before any generator troubles can be analyzed because the procedures for testing these generators are different for the two types.

The battery-charging a-c generator is a 3-phase alternator that employs a rectifier to convert the a-c output to direct current (fig. 9-3). A pictorial diagram is illustrated in figure 9-3, A, and the corresponding schematic wiring diagram in figure 9-3, B. The alternator is the revolving field type.

The alternator rotor contains the d-c field winding. When the ignition switch is turned on, the d-c field is energized. When the motor is started, the alternator d-c field is rotated and the lines of force will cut the stator windings, thereby generating 3-phase voltages across the alternator terminals.

These voltages are applied to the 3-phase, full-wave rectifier, and the output from the rectifier is used to charge the battery and operate the electrical accessories. The electrical equipment is designed to operate at a specific voltage irrespective of the speed of the motor and the alternator. The regulator controls the alternator

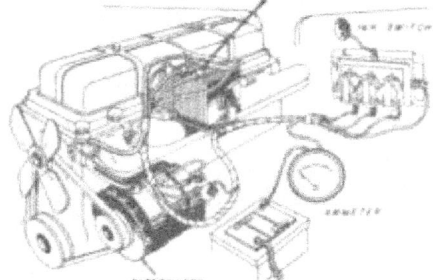

A PICTORIAL DIAGRAM

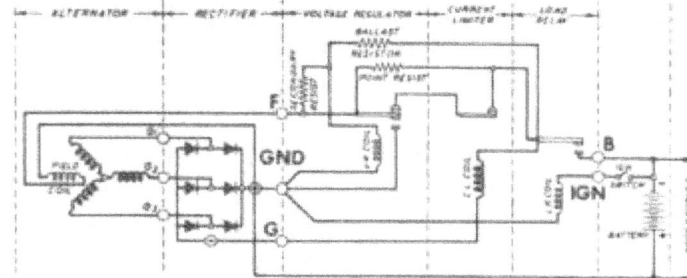

I SCHEMATIC WIRING DIAGRAM

Figure 9-3.-Battery-charging a-c generator.

output. It consists of the (1) load relay, (2) voltage regulator, and (3) current limiter.

The LOAD RELAY acts as a switch to connect the battery to the alternator system when the motor is running, and to open the circuit when the motor is not in operation. The load relay is operated by the ignition switch. When the switch is turned on it connects the load-relay coil to the battery and the contacts close.

The VOLTAGE REGULATOR controls the alternator field strength and thereby regulates the output voltage. The point resistor and contacts are connected in series with the alternator rotor d-c field coil. The voltage regulator (VR) coil and its ballast resistor are connected

across the d-c output terminals of the rectifier. Therefore, any variation in output voltage will change the magnetic strength of the VR coil.

When the alternator speed is lowand/or the electrical load is heavy, the VR coil will cause the VR armature to vibrate on the upper contact. This action inserts periodically a small amount of resistance in the field circuit to control the output voltage. This condition is called top contact operation.

When the alternator speed is high and/or the load is light, the VR armature vibrates on the bottom contact and periodically inserts resistance in series with the bottom, and at the same time shorts out the entire d-c field coil of the alternator. This action intermittently weakens the field magnetism sufficiently to hold the output voltage to the proper value for this load and speed condition.

The CURRENT LIMITER protects the alternator and rectifier by limiting the current output of the alternator. The current limiter (CL) coil is connected in series with the negative terminal of the rectifier and the load circuit. When the current exceeds the rating of the alternator, the CL coil will cause the CL armature to vibrate and insert enough resistance intermittently in the d-c field to limit the output current to the proper value.

On 6-volt regulators of this type, a temperature compensator is used to adjust the voltage regulator action to different temperatures. The temperature compensator consists of a bimetallic strip, located at the spring end of the VR armature. When the temperature changes the bimetallic strip bends, thereby causing a change in the armature spring pressure.

MAINTENANCE.—Battery-charging generators in small craft should be inspected periodically to obtain optimum performance. The frequency of these inspections is determined by the type of generator and the application.

As previously mentioned, the OUTPUT of the ADJUSTABLE THIRD-BRUSH GENERATOR is determined by the position of the third brush. To check the output of this generator, connect an ammeter and a variable resistance in series with the charging circuit at the A terminal and voltmeter between terminal A and the generator frame

(fig. 9-2, A). If the generator is used with a cutout relay or regulator, connect the ammeter and variable resistance to the BAT terminal and the voltmeter across the BAT terminal and the generator frame. Ground the F terminal temporarily to prevent any regulator action that would reduce the generator output.

With the generator at operating temperature, increase the generator speed to the specified value to obtain the specified voltage, and note the generator output current. If the generator is checked with a fully charged battery in the circuit, the variable resistance may not be required. However, if the battery is not fully charged, the resistance will be necessary to increase the voltage to the specified value.

The third-brush generator will produce excessive output if the position of the third brush is advanced too far in the direction of armature rotation. This condition can be corrected by checking the generator output and readjusting the third brush, as previously described.

If a manually or thermostatically controlled field resistance is used with the third-brush generator, be certain to short-circuit the field resistance before checking or adjusting the generator output.

The output of the SHUNT GENERATOR is determined by the current setting of the current regulator, which is discussed later in this chapter.

If a battery-charging shunt generator does not produce rated output or if it produces excessive output and the trouble has been isolated in the generator itself, disconnect the leads from the A and F terminals and remove the generator from the engine. Then inspect the commutator and field coils to locate any visual trouble indications. If no indications of trouble are visible, conduct further tests for grounds, opens, and shorts in the generator.

If the shunt generator is the externally grounded type (fig. 9-2, B), remove ground from the armature by inserting a piece of cardboard between the brush and the commutator. Test the generator for an internal ground with a megger or similar resistance-measuring device between terminal A and the generator frame. If a ground is indicated, insulate the other brush from the commutator and repeat the test. If a ground is still indicated, it will

be in the field circuit. If no ground is indicated, the ground will be in the armature. To locate the ground in the armature, conduct a bar-to-bar test, as described in chapter 5 of this training course.

If a ground is not indicated in the generator, test the field for an open circuit with an ohmmeter between the A and F terminals (brushes in contact with armature).

If the field is not open circuited, test for a shorted field by passing normal current through the field circuit and measuring the voltage drop across each field coil. The shorted coil will be indicated by having the lowest reading on the voltmeter.

If the trouble is not in the field, test the armature for open and short circuits by conducting a bar-to-bar test in accordance with the information contained in chapter 5.

Accidental internal grounding of the field circuit of an externally grounded generator will prevent normal regulation so that the generator may produce excessive output. The test for the accidental ground in the field circuit is to first insulate the grounded brush from the commutator and then to connect a megger between terminal F and the generator frame.

If the shunt generator is the INTERNALLY grounded type (fig. 9-2, C), insulate the grounded brushes from the commutator. Test the armature circuit for a ground with megger between terminal A and the generator frame. U a ground is indicated, insulate the ungrounded brushes from the commutator and repeat the test.

If a ground is now indicated it will be between the ungrounded brushes and terminal A. If no ground is indicated, the ground will be in the armature. To locate the armature ground, conduct a bar-to-bar test. To test the field circuit for a ground, disconnect the field lead from the grounded brush and connect an ohmmeter between terminal F and the generator frame. Be certain that the lead disconnected from the grounded brush is not inadvertently grounded against the generator frame.

When a ground is not indicated in the field circuit, test the field for an open circuit with an ohmmeter between the F terminal and the field lead that was disconnected from the grounded brush.

If the field is not open circuited, test for a short-circuited field by passing normal current through the
field circuit and measuring the voltage drop across each field coil.

When the trouble has not been located, test the armature for open and short circuits by conducting a bar-to-bar test.

Excessive output produced by an internally grounded shunt generator may result from a short-circuited field coil. The test for this condition is to establish normal current through the field circuit and to measure the voltage across each coil, as previously described.

If the shunt generator is the UNGROUNDED type (fig. 9-2, D), determine which brush is connected to the field winding and insulate it and the brush diametrically opposite it from the commutator. Test the armature circuit for grounds with a megger between terminal A+ and the generator frame. If a ground is indicated, it will be in the armature or the circuit between terminal A+ and the commutator. To locate a ground in the armature, conduct a bar-to-bar test.

When the trouble is not located in the armature circuit, test the field circuit for grounds with the megger between terminal A— and the generator frame. If a ground is indicated, it will be in the field winding or in the circuit between terminal A — and the field winding.

If aground is not indicated in the field circuit, test the field for an open circuit with an ohmmeter between terminal F and terminal A—.

If the field is not open circuited, test for a short-circuited field by passing normal current through the field circuit and measuring the voltage drop across each coil.

When the trouble has not been located, test the armature for open and short circuits by conducting a bar-to-bar test.

If a battery-charging a-c generator does not produce rated output or if it produces excessive output and the trouble has been isolated in the generator proper, disconnect the leads from the generator terminals and remove the generator from the engine. Then inspect the rotor and stator windings for any visual trouble indications. When no indications of trouble are visible, conduct further tests for grounds, opens, and shorts in the generator. The procedures to be followed in conducting

these tests are contained in chapter 5 of this training course. Also, more detailed information concerning these tests is listed in the manufacturers' instruction books furnished with the specific equipment.

An UNSTEADY or LOW OUTPUT can result in any of the generators and is usually caused by a loose drive belt, or a condition of the brush rigging that prevents good contact between the brushes and commutator. Also, a rough, out-of-round, or burned commutator; high mica; or dirt in the commutator slots can cause low or unsteady output. To correct these conditions, turn down the commutator in a lathe and undercut the mica in accordance with the instructions contained in chapter 5.

A NOISY GENERATOR can be caused by a loose mounting, drive pulley or gear. Also worn or dirty bearings, improperly seated brushes, or a bent brush holder may be the cause.

Before installing a new or repaired battery-charging, d-c generator in a boat, the generator must be POLARIZED so that it will have the correct polarity with respect to the electrical system. Failure to polarize the generator can result in burned relay contacts, a discharged battery, and even serious damage to the generator. The correct procedure to follow in polarizing a generator depends on whether the generator field circuit is externally or internally grounded, or ungrounded (fig. 9-4).

If the generator has the field circuit externally grounded and uses a standard regulator (fig. 9-4, A), ascertain the polarity of the lead from the engine battery to the regulator BAT terminal. The arrows indicate the direction of electron flow (opposite to conventional current flow). In this example the polarity is negative. Insulate the generator brushes from the commutator and connect one lead from a spare battery to terminal A of the generator. The polarity of this lead must be the same as the polarity of the regulator BAT terminal (negative in this example). Momentarily touch the other lead from the spare battery to terminal F of the generator. This action allows a momentary surge of current to flow from the spare battery through the generator field to correctly polarize the generator.

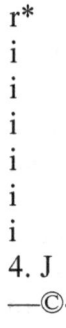

r*
i
i
i
i
i
i
4. J
—©-

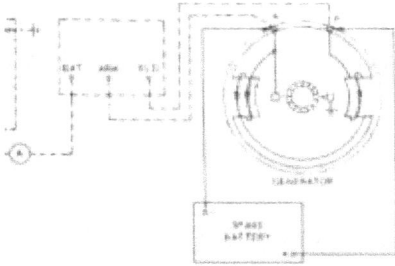

A HTEKNM.lt GOOUNOfD STANDATOI GtXEHATOR

I 1

I

4~

-©---

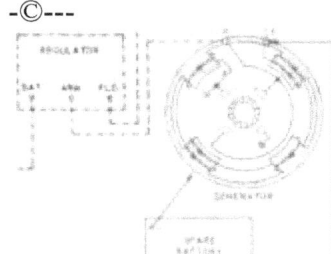

I MTERNALLT GROuNDCO (HEAVY-DUTY) GENERATOR

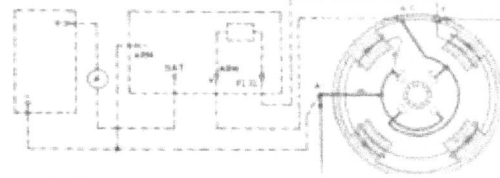

y>A»i

Q UNCAXXM>CD(neAVT.CXJTTi GCNCDATOD

Figure 9-4.-Polarizing generator fields.

If an internally grounded generator is provided with a heavy-duty regulator (fig. 9-4, B), determine the polarity of the regulator BAT terminal. In this example the polarity is positive. Insulate the generator brushes from the commutator and connect one lead from a spare battery to terminal F of the generator. Be certain that the polarity of the spare battery lead corresponds to the polarity of the regulator BAT terminal (positive in this case). Clean a spot on the generator frame to ensure a good connection and momentarily touch the other lead from the spare battery to the generator at this spot. This action allows a momentary surge of current to flow from the spare battery through the generator (frame, grounded brush, and field windings) to correctly polarize the generator.

If the ungrounded generator is provided with a heavy-duty regulator (fig. 9-4, C), determine the polarity of the regulator BAT terminal (positive in this case). Insulate the generator brushes and connect one lead from a spare battery to terminal F of the generator. The polarity of this lead must be the same as that of the regulator battery lead (positive). Momentarily, touch the other lead from the spare battery to terminal A — of the generator. This action allows a momentary surge of current to flow from the spare battery through the generator field to correctly polarize the generator.

Be certain that when the generator is installed on the engine that the polarity of the leads remains the same as for the polarizing procedure. Do not attempt to polarize a battery-charging generator after it is installed in the boat because of the fire hazard. There is always the possibility of explosive vapors in a small boat, which can be ignited by the arc when the generator field is flashed.

Regulators

Regulators are usually employed with battery-charging generators to provide complete control of the generator voltage and output at all times. As previously mentioned, regulators are classified as standard and heavy-duty regulators for use with externally grounded, internally grounded, and ungrounded generators.

A standard or heavy-duty regulator consists of two or more units mounted on a common base and located in the proximity of the associated generator. The units, or components, in a regulator usually include a cutout relay, voltage relay, and current relay, depending on the specific application.

STEP VOLTAGE CONTROL (STANDARD).-The step -voltage control is designed for use with the third-brush generator with externally grounded field circuit. It consists of a cutout relay and a step-voltage control unit (fig. 9-5).

CUTOUT RELAY

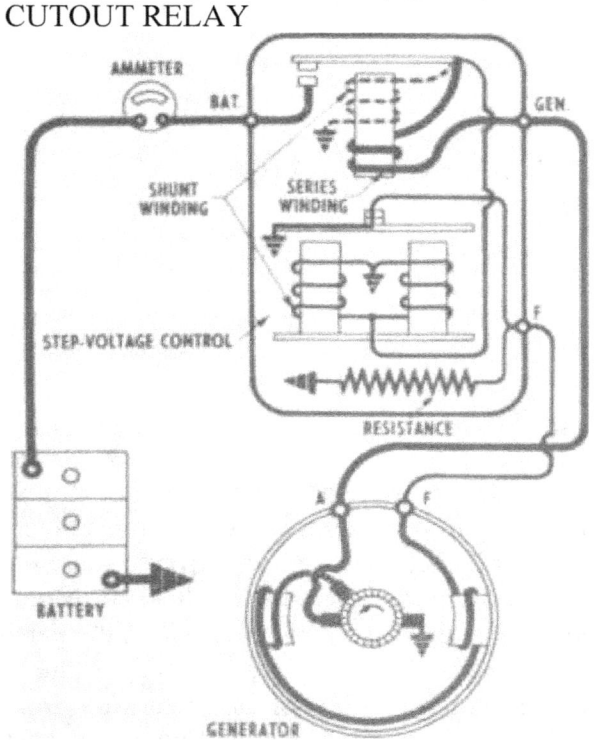

Figure 9-5.-Step-voltage control wiring diagram.

The CUTOUT RELAY is used to close the circuit between the generator armature and the battery when the generator is operating at sufficient speed to charge the battery. It opens the circuit when the generator slows or stops to prevent the battery from discharging through the generator and motoring it when the battery voltage exceeds the generator voltage. The action of the cutout relay is described in Basic Electricity, NavPers 10086.

The STEP VOLTAGE CONTROL is used to decrease the generator output by inserting a resistance in the generator field circuit when the generator voltage reaches a predetermined maximum value.

This control usually has two shunt coils connected across the generator output and assembled on separate cores. The cores include a magnetic circuit that is completed through the armature of the control element. A spring holds the armature contacts normally closed, thereby externally grounding the field circuit. This action places the field effectively in shunt with the generator armature via the third brush, and bypasses the field circuit resistor (connected across the control element armature control). With the control element contacts closed, the resistance of the field circuit is reduced and the generator voltage and output is increased. When the control element contacts open, the resistor is inserted in the field circuit, thereby reducing the field current, generator voltage, and output.

When the electrical load is heavy and the battery is in a low state of charge, the generator and circuit voltage will be comparatively low. The step-voltage control contacts will remain closed and the generator output will increase to its maximum, as determined by the generator voltage and speed, and position of the third brush. On the other hand, if the electrical load is decreased and the battery approaches a fully charged condition, the circuit voltages will increase and cause the control element contacts to open. This action inserts the resistance in the generator field circuit to reduce the generator output. The contacts remain open with the resistance in the field circuit, and the generator output remains at a low value as long as the battery is fully charged and the electrical load is small. The effect is similar to that of two generators; one, a full-output generator when the

battery and line voltage are low, and the other, a low-output generator when the battery and line voltage are high.

VOLTAGE REGULATOR (STANDARD).-The voltage regulator (fig. 9-6) is a two-unit regulator designed for use with the third-brush generator with externally grounded field circuit. It consists of a cutout relay and a voltage relay. The construction and operation of the cutout relay is similar to that previously described with the step-voltage control.

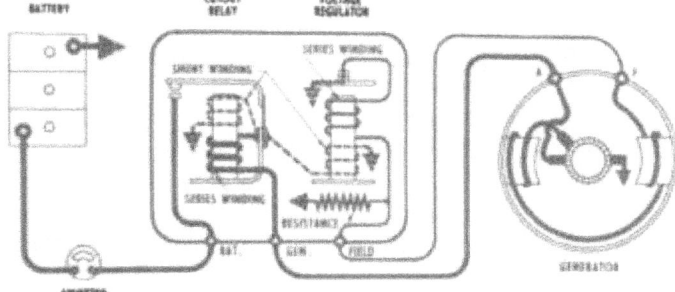

Figure 9-6.-Two-unit voltage regulator (standard) wiring diagram.

The voltage relay is of the vibrating type. It has a shunt coil and a series coil assembled on one core. The shunt coil is connected across the generator output, and the series coil is connected in series with the generator field circuit when the relay contacts are closed. The core with coils is attached to the frame, and the hinged armature is centered above the core. The armature has a contact located below a stationary contact that is connected to the generator field circuit. When the voltage regulator is not operating, the armature is held away from the core by the tension of a coil spring so that the contacts are closed, and the field circuit resistor is bypassed to ground through the series coil of the regulator unit.

When the voltage of the generator rises to a certain value, the pull of the magnetic field produced by the shunt

and series coils (acting additively) overcomes the spring tension and opens the regulator armature contacts. This action inserts resistance in the generator field and also opens the series coil circuit. Reduced generator voltage on the shunt coil and the open-circuited series coil combine to reduce the magnetic pull on the relay armature to allow the spring to again close the regulator points. This action bypasses the field current around the generator field resistor and energizes the voltage regulator series winding, thereby increasing the generator voltage and inc r easing the pull on the regulator armature produced by the shunt and series coils acting together. Thus, the tension of the spring is overcome and the points are again opened. The cycle of action takes place repeatedly at varying rates of 50 to 200 times per second, depending on the excitation requirements of the generator, and the generator supplies varying amounts of current, depending on the state of charge of the battery and the demands of the electrical load.

CURRENT AND VOLTAGE REGULATOR (STANDARD).—The current and voltage regulator (fig. 9-7) is a three-unit regulator designed for use with a shunt generator with externally grounded field circuit. It consists of a cutout relay, voltage regulator, and current regulator. The construction and operation of the cutout relay and voltage regulator are similar to those described with the two-unit regulator.

cutout OW* KXIMi
urrtn hut tKuuiot ttauNM

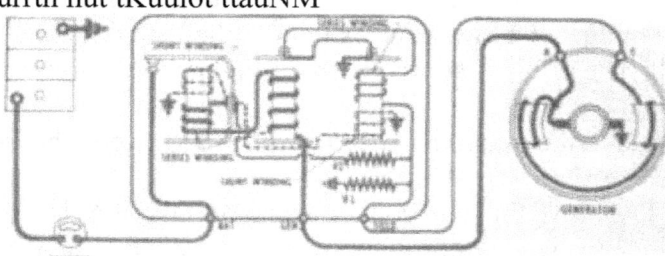

Figure 9-7.-Three-unit current and voltage regulator (standard)

wiring diagram.

The current regulator and the third-brush effect are both current-limiting devices that limit the generator output and prevent it from exceeding its maximum safe value.

This regulator has a series coil assembled on a core. The generator output flows through this coil at all times. The core with the coil is attached to the frame, and the hinged armature is centered above the core. The armature has a contact located below a stationary contact that is connected to the generator field circuit. When the current regulator is not operating, the armature is held away from the core by the tension of a coil spring so that the contacts are closed, and the generator field circuit is completed to ground through the current regulator contacts in series with the voltage regulator contacts.

When the generator is operating and the electrical demands are heavy, the voltage may not increase sufficiently to operate the voltage regulator. Hence, the generator output will continue to increase until the generator reaches its maximum rated value at which the current relay is set. The output current flows through the series coil and produces a magnetic field that overcomes the spring tension and pulls the armature toward the core to open the contacts. This action inserts resistor Rl in the generator field circuit so that the generator output is reduced. The reduced output weakens the magnetic field produced by the series coil, and the spring tension pulls the armature away from the core to close the contacts. This action short-circuits Rl, thereby strengthening the field and permitting the generator output to again increase. This cycle continues at a rate of from 50 to 200 times a second to limit the generator output to its maximum rated value.

When the electrical load on the generator is reduced, the voltage increases so that the voltage regulator operates and reduces the generator output. This action prevents the current regulator from operating. Either the voltage regulator or the current regulator operates, but not simultaneously.

A common resistor (Rl) is used in the current and voltage regulator circuits (fig. 9-7), which is inserted in the field circuit when either the current or voltage regulator operates. A second resistor (R2) is connected

between the F terminal of the regulator and the frame of the cutout relay so that it is connected effectively in parallel with the generator field windings. The sudden reduction in field current, which occurs when either the current or voltage regulator contacts open, is accompanied by a surge of induced voltage in the field windings as the strength of the magnetic field changes. These surges are partially dissipated by the two resistors (Rl and R2) to prevent excessive arcing of the contacts.

CURRENT AND VOLTAGE REGULATOR (HEAVY DUTY).—The current and voltage regulator (fig. 9-8) is a three-unit regulator designed for use with the ungrounded shunt generator. It also consists of a cutout relay, voltage regulator, and current regulator.

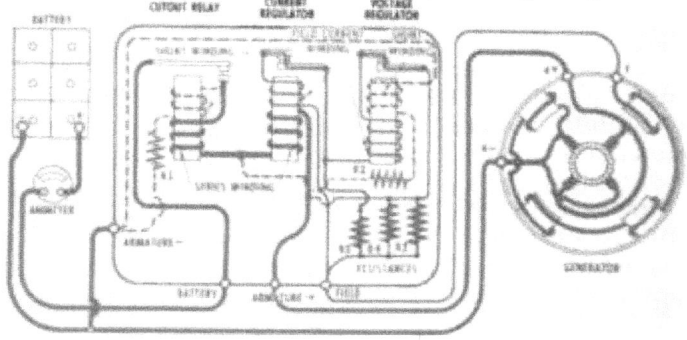

Figure 9-8.-Three-unit current and voltage regulator (heavy-duty) wiring diagram.

The internal wiring is similar for all regulators of this type, except that the internal connections vary according to whether the regulator is designed for use with a grounded (one wire) system or an ungrounded (two wire) system (fig. 9-8). The principal difference between the two systems is that in the grounded system the circuits are completed through ground or the engine; whereas, in the ungrounded system the circuits are completed through a second wire, and all the electrical circuits are insulated rom ground or the engine.

The units in the heavy-duty regulator are similarly constructed and perform the same functions as the corresponding units in the standard regulator previously described.

The cutout relay (fig. 9-8) has a shunt coil and a series coil assembled on one core. The shunt coil and resistor Rl comprise a series circuit that is connected across the generator so that generator voltage is impressed on this circuit at all times. The series coil is connected in series with the charging circuit so that the generator output passes through it. When the generator is not operating, the armature is held away from the core by the tension of a coil spring so that the contacts are open and the circuit is incomplete between the generator and battery.

When the generator is operating and the voltage reaches the value for which the cutout relay is adjusted, the magnetic field produced by the shunt coil overcomes the spring tension and pulls the armature toward the core to close the contacts of the cutout relay and complete the circuit between the generator and the battery. If the generator slows down, or stops, and the voltage decreases to a value less than the battery voltage, a reverse current will flow through the series coil and the mmf produced by the series coil opposes the mmf produced by the shunt coil. When the reverse current increases to the value for which the cutout relay is adjusted, the spring tension overcomes the magnetic field and pulls the armature away from the core to open the contacts and open the circuit between the generator and the battery.

The voltage regulator (fig. 9-8) has a shunt coil and a field current coil assembled on one core. The shunt coil and resistor R2 comprise a series circuit that is connected across the generator, thereby placing generator voltage across this circuit at all times. The field-current coil is connected in series with the generator field circuit when the current regulator contacts are closed so that field current passes through it. The stationary contact is assembled into a flat spring that rises slightly above the fiber mounting bracket when the contacts are closed. This arrangement provides a wiping action between the contacts as they close and open to assure better contact.

When the generator is not operating, the armature is held away from the core by the tension of a coil spring so that the contacts are closed and the resistance is shorted out of the generator field circuit, permitting the generator voltage to increase. The external field circuit connection to the generator armature by way of the regulator (when both the current-regulator and voltage -regulator contacts are closed) is from terminal F of the generator to the field terminal of the regulator, through the field-current coil and contacts of the current regulator, through the field-current coil and contacts of the voltage regulator, to the armature terminal of the regulator, and back to terminal A+ of the generator.

When the generator is operating and the voltage reaches the value for which the voltage regulator is adjusted, the magnetic field produced by the shunt and field-current coils overcomes the spring tension and pulls the armature toward the core to open the contacts of the voltage regulator. This action inserts parallel resistors R3 and R4 in the field circuit to reduce the generator field current and voltage. The external field circuit connection to the generator armature by way of the regulator (when the voltage regulator contacts are open) is from terminal F of the generator to the field terminal of the regulator, through parallel resistors R3 and R4, to the armature terminal of the regulator, and back to terminal A+ of the generator.

The current regulator (fig. 9-8) has a series coil and a field-current coil assembled on one core. The series coil is connected in series with the charging circuit when the cutout relay contacts are closed so that full generator output passes through it. The field-current coil is connected in series with the field circuit when the current-regulator contacts are closed (assuming that the voltage-regulator contacts are also closed) so that field current flows through it.

When the generator is not operating, the armature is held away from the core by the tension of a coil spring so that the contacts are closed, and resistors R3 and R4 are shorted out of the field circuit to permit the field current and generator output to increase. The external field circuit connection to the generator armature byway of the regulator (when both current-regulator and

voltage-regulator contacts are closed) is from terminal F of the generator to the field terminal of the regulator, through the field-current coil and contacts of the current regulator, through the field-current coil and contacts of the voltage regulator, to the armature terminal of the regulator, and back to terminal A+ of the generator.

When the generator is operating and the output reaches the value for which the current regulator is adjusted, the magnetic field produced by the series and field-current coils overcomes the spring tension and pulls the armature toward the cord to open the contacts of the current regulator. This action inserts the effective parallel combination of resistors R3, R4, and R5 in series with the field circuit to reduce the generator output. The external field circuit connection to the generator armature by way of the regulator (when the current regulator contacts are open and the voltage regulator contacts are closed) is from terminal F of the generator to the field terminal of the regulator, through resistor R5, and the field current coil and contacts of the voltage regulator, in parallel with resistors R3 and R4, to the armature terminal of the regulator, and back to terminal A+ of the generator.

MAINTENANCE.—Proper maintenance of generator controls and regulators is necessary to ensure continued satisfactory performance. This maintenance consists of periodic tests and adjustments to the various components that comprise the control or regulator. The correct air-gap measurements and voltage values for these tests and adjustments are specified in the manufacturers' instructions furnished with the specific equipment. Do not attempt these tests and adjustments without first obtaining the manufacturers' instruction books.

Battery Ignition System

Ignition of the fuel-air mixture in the cylinders of a gasoline engine is initiated by an electric spark. The spark is produced by the battery ignition system that functions to step up the relatively low voltage of the battery and to deliver the high voltage to the spark plugs at the proper time. The high voltage is capable of forcing current through the high resistance set up by the pressure in the combustion chamber and across the electrodes

of the spark plugs. The hot spark created across the electrodes ignites the fuel-air mixture.

A battery ignition system consists of (1) the battery, (2) the ignition coil, (3) the ignition distributor, and (4) spark plugs (fig. 9-9). The battery and generator supply the voltage and current for the ignition system. The battery is required for starting, but after the engine starts, the generator carries the ignition load.

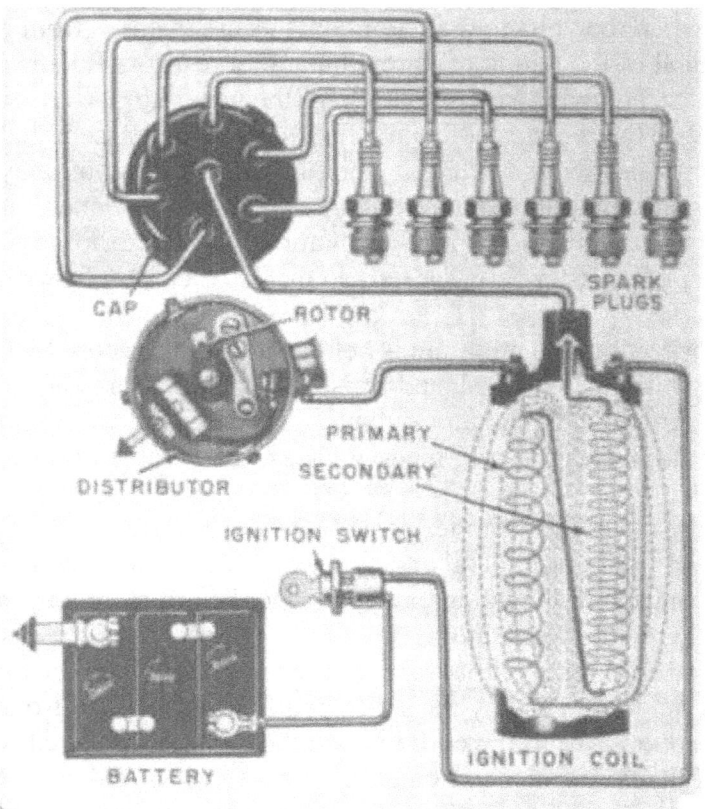

Figure 9-9.-Battery ignition system.

IGNITION COIL. -The ignition coil (fig. 9-10) is a pulse transformer that steps up the low battery, or generator,

I

u

r.

SEALING NIPPLE
HIGH TENSION TERMINAL
COIL CAP
PRIMARY TERMINAL
SPRING WASHER
SEALING GASKETS
SECONDARY WINDING
PRIMARY WINDING
COIL CASE
LAMINATION
PORCELAIN INSULATOR

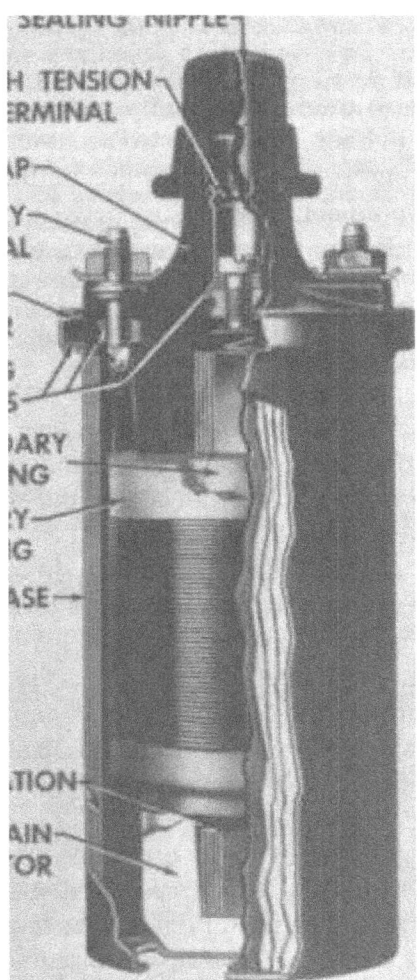

Figure 9-10.-Ignition coil.

An 1

voltage to the high voltage necessary to jump the gaps at the spark plugs in the engine cylinders. It consists of a primary winding having a few hundred turns of relatively heavy wire, a secondary winding having many thousand turns of very fine wire (up to 100 times as many turns of wire as the primary), and a laminated soft-iron core. The secondary winding is usually wound around the soft-iron core, and the primary winding surrounds the secondary winding. This subassembly is enclosed in a laminated soft-iron shell that serves to concentrate the magnetic field. The entire subassembly is placed in a steel case, and the remaining space is usually filled with oil to improve insulation and reduce the effects of moisture. The case is hermetically sealed with a molded insulating cap that carries two primary terminals and the high-tension terminal.

The primary circuit of the ignition system (fig. 9-9) is from the battery, through the ignition switch to the primary terminal of the ignition coil, through the primary winding, and out the other primary terminal to the distributor. When the ignition switch is in the ON position and the contacts in the distributor are closed, current flows from the battery, through the primary winding of the ignition coil, through the contacts of the distributor, and back to the battery through ground. The current produces a magnetic field around the windings of the ignition coil. When current begins to flow through the winding, self-induction occurs and prevents the current and consequently the magnetic field from reaching their maximum values instantly. A small fraction of a second (calledbuild-up time) is required for this action to occur. During this time energy is being stored in the magnetic field of the coil. When the cam lobe strikes the breaker-lever rubbing block, the breaker contact points open, thereby interrupting the primary circuit.

When the breaker contacts in the distributor begin to open, the primary current tends to continue to flow because of the self-induction of the winding. If it were not for the ignition capacitor connected across the breaker contacts, current would continue to flow between the separating points. The current would form an arc that vould burn the points badly and would also drain away

most of the energy stored in the coil. Thus insufficient energy would be left in the coil to produce the necessary high voltage surge in the secondary.

The ignition capacitor, however, provides a path around the points during the instant they begin to separate. Thus, the capacitor acts as a storage reservoir for the energy otherwise dissipated as an arc across the points, and also as a check on the current, quickly bringing it to a stop in the primary circuit. As a result of this action the magnetic field produced by the current quickly collapses. The rapid collapse of the magnetic field in cutting the windings of the ignition coil induces a high voltage in both the primary and secondary windings. The voltage in the primary winding may reach 250 volts (further charging the capacitor); whereas, the secondary winding (which may have 100 times as many turns as the primary) may reach 25,000 volts.

The voltage normally increases to a value sufficient to produce a spark across the spark-plug gap connected to the secondary of the ignition coil through the distributor rotor, cap insert, and high tension lead. This voltage is usually from 4,000 to 18,000 volts, depending on such variables as engine speed, engine compression, mixture ratios, width of spark-plug gap, and others.

As a spark appears at the spark-plug gap, the energy in the ignition coil drains from the coil through the secondary circuit to sustain the spark for a small fraction of a second, or for several degrees of crankshaft revolution. During this interval the capacitor discharges back through the primary circuit, producing an oscillation of the current in the primary circuit during the brief interval required for the primary circuit to return to a state of equilibrium. This sequence is repeated as each lobe of the breaker cam moves under and past the rubbing block on the breaker lever to cause the contacts to close and open.

Normally, ignition coils do not require any service except to keep all terminals and connections clean and tight. If the performance of the ignition system is not satisfactory and the trouble has been isolated in the ignition coil, it should be removed and tested for open circuits and grounds.

ATI

Test the primary for an open circuit with anohmmeter between the two primary terminals. Test the secondary for an open circuit with the ohmmeter between the high-tension terminal and either one of the primary terminals.

Test the primary for a ground with a megger between a primary terminal and the container (ground) of the ignition coil. Test the secondary for a ground with the megger between the high-tension terminal and the container.

The test for grounds does not usually apply to the secondary windings of ignition coils used with two-wire (insulated) systems because one end of the secondary winding is grounded to the metal container. However, this ground test does apply to the primary windings of ignition coils used with two-wire systems.

IGNITION DISTRIBUTOR.-The ignition distributor (fig. 9-11) closes and opens the primary ignition circuit and distributes the high-voltage surge to the proper spark plug at the correct time in the engine cycle. The distributor includes a drive shaft with breaker cam and spark advance mechanism, a breaker plate with contacts, and a rotor. This assembly fits into a cast-iron housing and is usually supported on a bronze bearing. A molded phenol-resin cap, equipped with terminals for the high-tension secondary is clamped on the top of the housing (fig. 9-11, A).

The shaft with the breaker cam is usually driven by the engine cam shaft at one half the engine speed through spiral gears. The distributor contacts are held closed by spring pressure and opened by the breaker cam, which usually has the same number of lobes as there are cylinders in the engine.

The primary circuit through the ignition distributor includes the distributor contacts and capacitor (fig. 9-9). As the breaker cam rotates, the cam lobes move around under the contact arm, causing the distributor contacts to open and close (fig. 9-11, B). Thus, the distributor contacts open and close once for each cylinder with every rotation of the breaker cam, and one high-voltage surge is produced by the ignition coil for each cylinder with every two revolutions of the crankshaft.

As mentioned before, the capacitor connected across \e distributor contacts provides a quick collapse of the

CAP
ROTOR
DUST SEAL
CENTRIFUCAL
ADVANCE MECHANISM
HOUSING

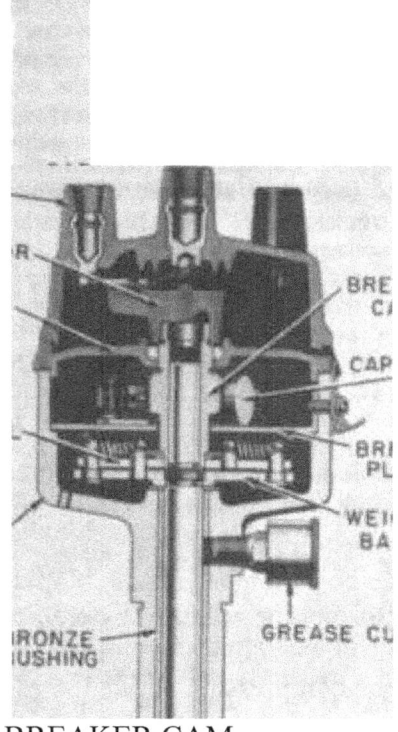

BREAKER CAM
CAPACITOR

BREAKER PLATE
BRONZE BUSHING
WEIGHT BASE
GREASE CUP
CAPACITOR
COUPLING —L A SIDE VIEW (SECTIONAL)
^CAP SPRING
BREAKER CAM

TERMINAL
ECCENTRIC SCREW

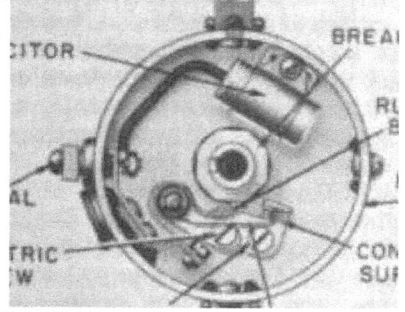

RUBBING BLOCK
BREAKER PLATE
CONTACT SUPPORT
LOCK ^ w r BREAKER LEVER SCREW V CAp SPR | NG
B TOP VIEW (CAP REMOVED)

Figure 9-11.-Ignition distributor.

magnetic field in the ignition coil in order to produce the high-voltage surge; it also protects the distributor contacts from arcing.

The secondary circuit through the ignition distributor includes the distributor rotor, distributor cap with high-tension terminals, and the spark plugs (fig. 9-9). The high voltage surge in the ignition coil secondary is conducted through the high-tension lead by way of the center terminal of the distributor cap to the rotor. The rotor is mounted on the breaker cam and rotates with it. During each revolution, a metal spring and segment on the rotor (fig. 9-11, A) connect the center terminal of the distributor cap with each outside terminal in turn. This action directs the high-voltage surges from the ignition coil to the various spark plugs in the engine according to the firing order.

SPARK ADVANCE MECHANISMS.-To obtain efficient performance of an internal-combustion gasoline engine for varying conditions of operation, the spark must be timed so that it will occur at the proper time in the compression stroke. For example, when the engine is operating at part throttle, the spark must appear early (before the piston has reached top dead center) to allow sufficient time to ignite the smaller amount of fuel-air mixture (by weight), which is less highly compressed and consequently slower burning. Similarly when the engine is operating at higher speeds, the spark must appear earlier in the compression stroke in order to give the fuel-air mixture ample time to ignite, burn, and give up its energy to the piston as it starts down on the power stroke. The timing of the spark is usually accomplished by either the centrifugal advance mechanism or the vacuum advance mechanism or both.

The CENTRIFUGAL ADVANCE MECHANISM is located in the distributor housing below the breaker plate that carries the distributor contacts (fig. 9-12). It consists of an advance cam (which is integral with the breaker cam), two weights, and two weight springs assembled on a weight base that is integral with the distributor shaft (fig. 9-12, A). Each weight is mounted on a stud of the weight base and linked by a weight spring to the advance cam. At low speeds the tension of the weight springs holds the weights in the NO ADVANCE position so that

there is no spark advance, and the spark appears in the cylinder just before the piston reaches top dead center on the compression stroke.

As the speed increases, the centrifugal force developed by the rotating shaft tends to move the weights outward against the tension of the springs. The faster the distributor shaft rotates, the greater the centrifugal force, and the greater the movement of the advance weights. The movement of the weights pushes the advance cam and thus the breaker cam ahead (to an advanced position in the direction of rotation) of the distributor shaft (fig. 9-12, B). This action causes the lobes of the breaker cam to open and close the contacts earlier in the compression stroke so that the spark is advanced. The broken line denotes the idle NO ADVANCE position; whereas, the solid line denotes the high-speed FULL ADVANCE position.

The centrifugal advance for any particular engine is determined initially by operating the engine at full throttle on a dynamometer and varying the spark advance at each engine speed until the advance is determined that will provide maximum power at that speed. The centrifugal

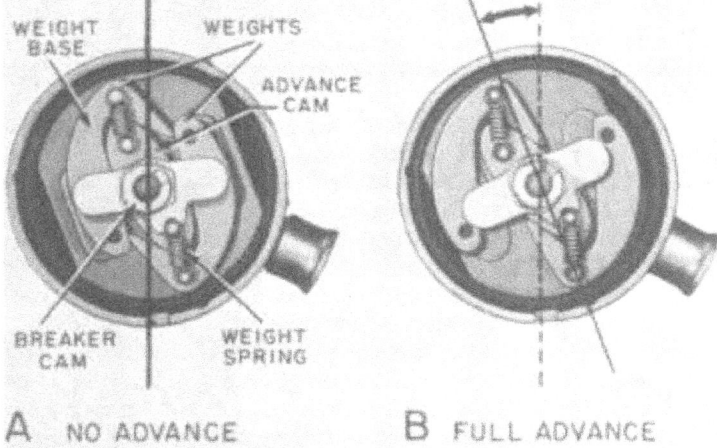

Figure 9-12.— Centrifugal advance mechanism.
Ann

advance weights, advance cam contours, and weight springs are designed to supply this advance through the speed range of the engine that varies the spark advance at each engine speed until the advance is determined that will provide maximum power at that speed.

The VACUUM ADVANCE MECHANISM utilizes the vacuum in the intake manifold to provide the additional spark advance required under part-throttle operation (fig. 9-13).

The vacuum advance mechanism in which the complete distributor is rotated consists of an airtight, spring-loaded diaphragm that is linked to the ignition distributor (fig. 9-13, A). The spring-loaded side of the diaphragm is airtight and is connected by a tube to the carburetor. This opening is on the atmospheric side of the throttle valve when the throttle is in the idle position so that there is no vacuum advance. However, when the throttle is opened it swings past the opening of the vacuum passage; the intake-manifold vacuum can then act on the diaphragm to compress the spring and rotate the distributor housing in its mounting. This action moves the breaker plate and contacts along with the distributor housing in the direction of rotation of the breaker cam so that the contacts close and open earlier in the compression stroke to provide the desired vacuum spark advance. The amount of throttle opening determines, in part, the amount of intake manifold vacuum, and thus the amount of spark advance obtained.

The vacuum advance mechanism in which only the breaker plate is rotated (the distributor housing is stationary) has the advance mechanism mounted on the side of the distributor, and the diaphragm is linked to the breaker plate (fig. 9-13, B). The breaker plate is supported on bearings so that it can rotate independently of the distributor housing.

At any particular engine speed there is a definite amount of spark advance, resulting from the operation of the centrifugal advance mechanism due to speed, and an additional amount of advance resulting from the vacuum advance mechanism due to vacuum conditions in the intake manifold.

SPARK PLUGS.-The spark plug (fig. 9-14) in a spark ignition system provides the gap across which the

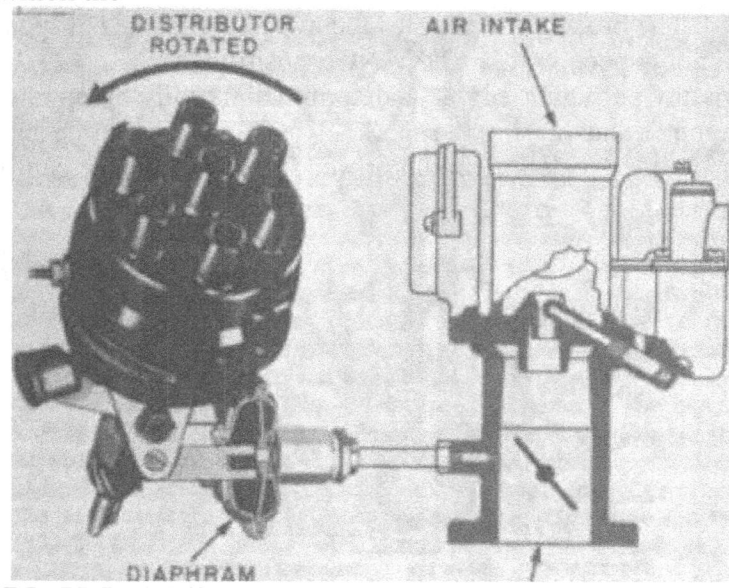

DIAPHRAM (SPRING LOADED)
TO MANIFOLD
ROTATED DISTRIBUTOR

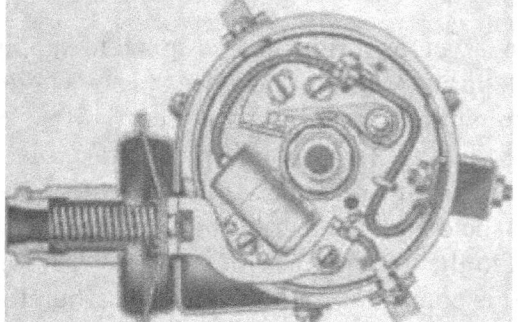

B ROTATED BREAKER PLATE
Figure 9-13.-Vacuum advance mechanism.

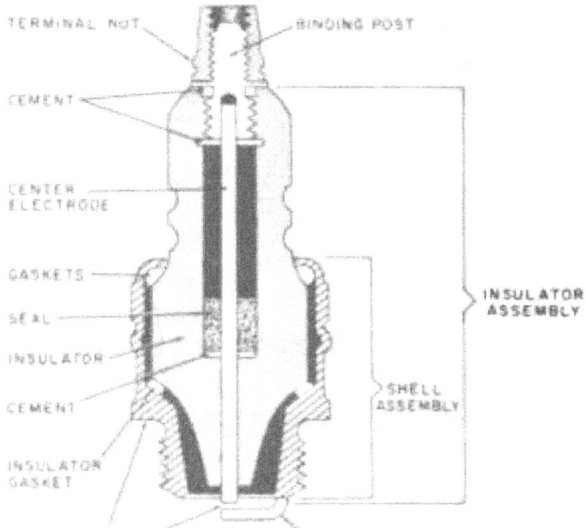

GASKET SEAT SPARK GAP GROUND ELECTROOE

Figure 9-14.-Spark plug.

high-tension voltage jumps to create a spark that ignites the compressed fuel-air mixture. It consists of a center (insulated) electrode that is connected to the secondary of the ignition coil through the distributor, and a side (grounded) electrode. The center electrode extends through a porcelain insulator that is supported by a circular metal shell. The side electrode protrudes from the edge of the metal shell and is positioned so that a gap exists between it and the center electrode. The base of the shell is threaded to allow it to be screwed into a tapped hole in the cylinder head.

The size of the spark-plug gap depends on the engine compression ratio, the characteristics of the combustion chamber, and the ignition system. At one time a gap of 0.025 inch was practically standard on all engines. However, manufacturers now specify gaps of from 0.03 to 0.04 inch to permit more readily igniting the increased mixtures (by weight) used in the higher horsepower engines.

Spark plugs are designed so that the temperature of the firing end is sufficiently high to burn off carbon and other combustion deposits, but not high enough to cause preignition and deterioration of the insulator and electrodes.

The temperature of the insulator depends on the characteristics of the spark plug and on the burning fuel in the combustion chamber. The temperature of the burning fuel in the combustion chamber varies with the engine design, compression ratio, fuel-air ratio, and cooling system. The heat absorbed by the tip of the spark plug from the burning fuel travels up the insulator to the metal shell, to the cylinder head, and to the water jacket. As the temperature in the combustion chamber increases, the heat absorbed by the insulator increases. The tip of the spark plug will have a lower temperature if the length of the path that the heat must travel to reach the cooling system is short as compared to the heat transfer when the path is long. Hence, plugs with short paths are called COLD PLUGS; plugs with long paths are called HOT PLUGS.

Engine manufacturers select plugs that will provide good performance for average operating conditions. However, if the engine is operated for long periods under full-load or overload conditions, the standard-equipment spark plug will operate at too high a temperature and preignition will result. Hence, it will be necessary to install a colder plug to carry off the heat more rapidly. On the other hand, if the engine is operated for long periods at part throttle, the standard-equipment plug may tend to foul due to the accumulation of carbon at reduced temperature, thereby resulting in poor engine performance. Hence, it will be necessary to install a hotter plug to concentrate the heat and burn off the accumulated products of combustion.

MAINTENANCE.—A regular inspection and maintenance procedure is necessary to attain maximum service with minimum trouble. For the ignition distributor, the procedure includes periodic lubrication (where required) and inspection of the breaker contacts, centrifugal and vacuum advance mechanisms, connections, distributor cap, and rotor.

Test the vacuum advance mechanism to be certain it operates freely. On the type that rotates the complete distributor, turn manually the distributor in its mounting and then release it. The vacuum advance spring should return it to its original position without sticking. On the type that rotates the breaker plate only, manually turn the plate. The breaker plate should return to its original position when released.

Test the centrifugal advance mechanism for freedom of movement by turning the breaker cam in the direction of shaft rotation and then releasing it. The advance springs should return the cam to its original position without sticking.

Wipe out the inside of the distributor cap with a clean, lint-free cloth and inspect both the cap and the rotor for chips, cracks, and carbonized paths that would allow leakage of the high-tension voltage to ground. These defects require replacement of the affected part.

After installing new points, test the breaker contact opening with a feeler gage. Do not use the feeler gage to check the opening of old contacts because the contour of these contacts will be irregular, and the gage will measure only the distance between the high spots instead of between the high and low spots, which is the actual opening. To adjust the contact opening, loosen the lock screw and turn the eccentric screw (fig. 9-15). Tighten the lock screw after the adjustment is completed.

The cam (contact) angle is the number of degrees that the breaker cam rotates from the time the contacts close until they again open (fig. 9-15). The cam angle increases as the contact opening is decreased, and conversely, it decreases as the contact opening is increased.

Test the breaker contact pressure with a brush tension gage (hooked to the breaker lever and the spring secured to its terminal in the distributor) and exert a pull at an angle of 90° with the contact surface. The reading should be taken just as the contacts separate. Adjust the pressure by bending the breaker lever spring. If the pressure is excessive, it can be decreased by pinching the spring carefully. If the pressure is not sufficient, remove the breaker lever from the distributor and bend the spring away from the lever. It is important to remember that excessive pressure causes rapid wear of the rubbing

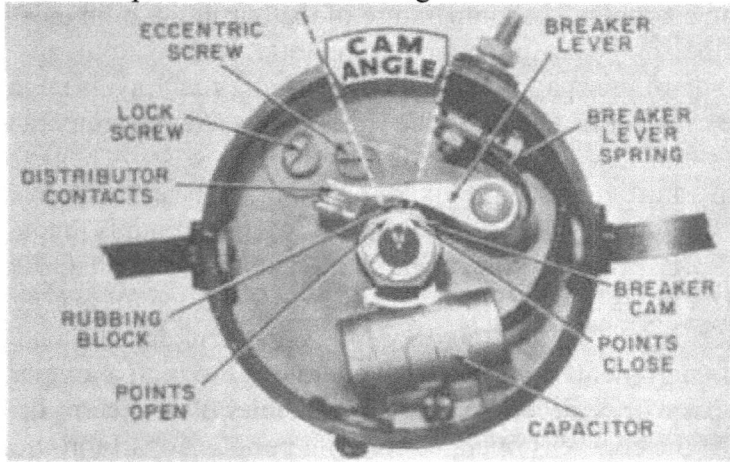

Figure 9-15.-Distributor adjustments.

block, cam, and contacts. Conversely, insufficient pressure permits high-speed bounce of the contacts that causes arcing and burning of the contacts with consequent missing of the engine.

If necessary, clean the contacts with a clean, fine-cut contact file. Remove only the scale or dirt and do not remove all roughness or dress down the contact surfaces. Never use emery cloth or sandpaper because imbedded particles will cause rapid burning of the contacts.

If the contacts burn or pit, they will soon become unsatisfactory for service. They must be replaced and the ignition system and engine must be checked to determine, and eliminate the cause of the trouble. Contact burning will result from high voltage, a defective capacitor, improper contact adjustment, and oil or foreign matter.

High voltage can result from a maladjusted or inoperative voltage regulator, or an excessively advanced third brush on a third-brush generator without a regulator.

Oil or crankcase vapors that work up into the distributor and deposit a film on the contact surfaces will cause them to burn rapidly. This condition can be caused by clogged engine breather pipes, resulting in a crankcase

pressure that forces the oil or vapors up into the distributor.

If the contact opening is too small (cam angle too large), the contacts will be closed over a greater portion of the total operating time. The current through the contacts will be too high, causing them to burn rapidly and arcing will occur between the contacts, resulting in a low secondary voltage and engine miss.

The capacitor circuit connections should be tested to detect excessive resistance because a high series resistance in the capacitor circuit will prevent normal capacitor action and result in rapid burning of the contacts. This high resistance can be caused by a loose (capacitor) mounting or lead connection. Other capacitor tests should include allowable voltage, insulation resistance, and capacity checks.

Contact pitting will result from an unbalanced condition in the ignition system, causing a transfer of tungsten from one contact to the other to form a tip on one and a pit in the other.

If the tungsten transfers from the negative to the positive contact, one or more of the following corrections may be made: increase the capacity of the capacitor, shorten the capacitor lead, separate the low- and high-tension leads of the distributor-to-coil leads, or move these leads closer to ground (engine block or frame).

If the tungsten transfers from the positive to the negative contact, reduce the capacity of the capacitor, move the distributor-to-coil leads closer together, move these leads away from ground, or lengthen the capacitor lead.

The spark plugs should be removed and inspected for cracked porcelain and wear. If they can be retained in service, file the end of the center electrode to a flat surface, clean both electrodes and then readjust the spark gap in accordance with the manufacturers' specifications. When readjusting the gap, bend the side electrode only. Do not bend the center electrode because it may crack the insulator.

More detailed information concerning the maintenance and repair of the ignition system is contained in the manufacturers' instruction books furnished with the equipment.

Air Heaters

Ignition in a diesel engine is accomplished by a combination of fuel injection and compression. Diesel engines normally require a longer cranking period than gasoline engines and at low ambient temperatures, they are more difficult to start because the heat of compression may not be sufficient to ignite the fuel-air mixture. Therefore, at low temperatures it is necessary to preheat the engine by means of an electric air heater, or to furnish an auxiliary low-ignition temperature fuel during the starting period by means of a pressure primer system. The types of air heaters and primers used for starting diesel engines include the (1) grid resistor, (2) glow plug, (3) flame primer, and (4) ether capsule primer.

GRID RESISTOR.—The grid resistor usually consists of a 1200-watt resistance grid mounted on a frame and supported by insulating blocks in the engine air-intake manifold. The grid is preheated by current from the starting battery before the engine is cranked and is operated during the cranking period until the engine is running smoothly.

GLOW PLUG.-The glow plug consists of a coil of resistance wire enclosed in a stainless-steel tube and is installed in the combustion chamber of each cylinder. The glow plugs are preheated by current from the starting battery and require about the same amount of power and preheat period as the grid resistor.

The glow-plug and grid-resistor type heaters are not used extensively because of the heavy load imposed on the starting battery at the time that the cranking-motor load is heavy.

FLAME PRIMER.-The flame primer (fig. 9-16) is the most widely used type of air heater for preheating diesel engines when starting at low temperatures. It is essentially a small, pressure oil burner with electric ignition. The fuel oil is sprayed into the engine air-intake manifold with a manually operated pump, and ignited by means of a spark plug, ignition coil, and vibrator. The device consists of two assemblies. One unit contains the burner, ignition coil, and vibrator. The other unit comprises the pressure pump and ignition switch. The principal

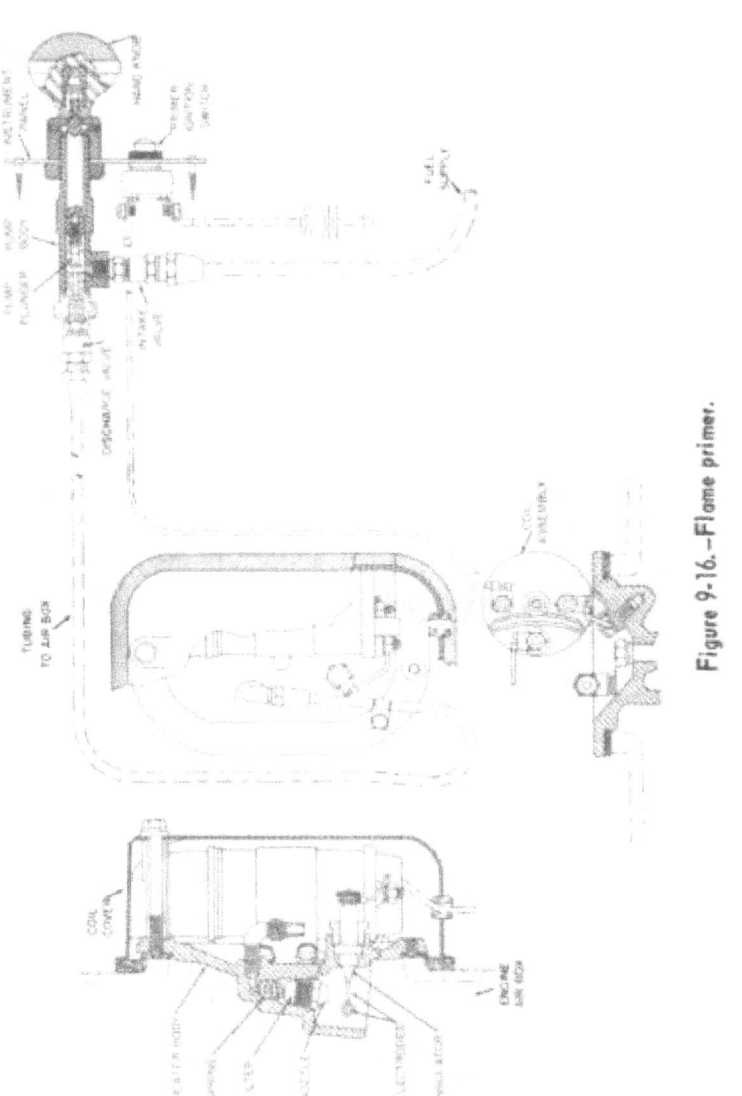

Figure 9-16.—Flame primer.

advantage of the flame primer is that it imposes a negligible load on the starting battery.

The HEATER UNIT consists of a nozzle, filter, ignition electrodes, and ignition coil with vibrator. One side of the heater body contains the filter, nozzle, and ignition electrodes; and the other side contains the ignition coil, terminals, and connection for the fuel supply. The unit is designed to replace one of the hand-hole cover plates nearest the center of the engine air box. The entire heater unit is provided with a protective metal cover. The air necessary for combustion is obtained from the charging blower, and the products of combustion including the flame-heated air are discharged into the engine cylinders (with practically no heat loss), resulting in an immediate response of the engine.

The PRESSURE PUMP and IGNITION SWITCH (fig. 9-16) are mounted on the instrument panel near the engine starter switch so that both the ignition switch and engine starter switch can be depressed simultaneously with one hand, thereby leaving the other hand free to operate the pump. The pump supplies fuel under pressure to the heater unit where the charge is filtered before reaching the nozzle. The suction side of the pump is either connected directly to the main fuel tank, or to the engine supply line between the main tank and the engine transfer pump. When the pump plunger is not in use, it is held in the IN position by a spring mechanism. The pump plunger is designed so that a pressure of about 10 pounds on the knob will deliver (from the nozzle) a finely atomized fuel. This fuel is readily ignited by the spark at the electrodes in the heater unit. The rate of travel of the plunger on the pumping stroke is determined by the flow of oil from the discharge nozzle, and normally requires 3 or 4 seconds per stroke.

The IGNITION SWITCH is connected in the line between t'he starting battery and one terminal of the ignition coil in the heater unit. The other terminal (primary) of the ignition coil is grounded to the engine.

If the engine fails to fire after two or three strokes of the primer pump while cranking the engine with the throttle wide open, stop cranking and examine the flame primer for possible causes of failure. If the engine is operating satisfactorily and the cranking speed is 80rpm

or more, test the flame primer for (1) ignition failure and (2) poor oil spray. If possible, remove the burner element from the engine air box and reconnect it outside the engine so that the burner operation can be readily observed.

To test the flame primer for IGNITION FAILURE, depress the ignition switch to the ON position and observe the action of the ignition-coil interrupter inside the coil assembly. It should vibrate rapidly and a continuous hot spark should occur between the ignition electrodes. If the interrupter does not vibrate, examine the interrupter contacts for dirt or carbon. Also examine the wiring for loose or broken connections. Clean the contacts with fine sandpaper and reset the air gap to one-eighth inch with the vibrator armature held against the coil.

If the spark jumps across the porcelain of the electrode, check the gap and reset it to about one-eighth inch (if necessary) by loosening the set screw and moving the grounded electrode. If the air gap is correct, remove the threaded gland on the porcelain electrode and withdraw the electrode assembly. Wash the porcelain with a cleaning solvent and scrape off any accumulation of carbon. Reassemble the porcelain electrode after it has been cleaned properly.

ETHER CAPSULE PRIMER.—The ether capsule primer (fig. 9-17) consists of a (1) discharger cell, (2) discharger nozzle, and (3) pressure primer bulb that contains a liquid ether mixture. The discharger cell and the discharger nozzle are connected together by a suitable length of 3/16" tubing. The discharger cell is a metal enclosure containing a piercing pin and provided with a removable cap for inserting the pressure primer (capsule) bulb. The cap is equipped with a discharger lever. When this lever is operated, it forces the capsule bulb against the piercing pin.

The discharger cell is installed at the control station in a vertical position so that the neck of the capsule bulb is always down toward the piercing pin. The discharger nozzle is installed through a 1/4" pipe connection at the forward end of the intake manifold.

When the ether capsule primer is used for cold-weather starting, press the engine starter switch. As soon as the starting motor brings the engine up to

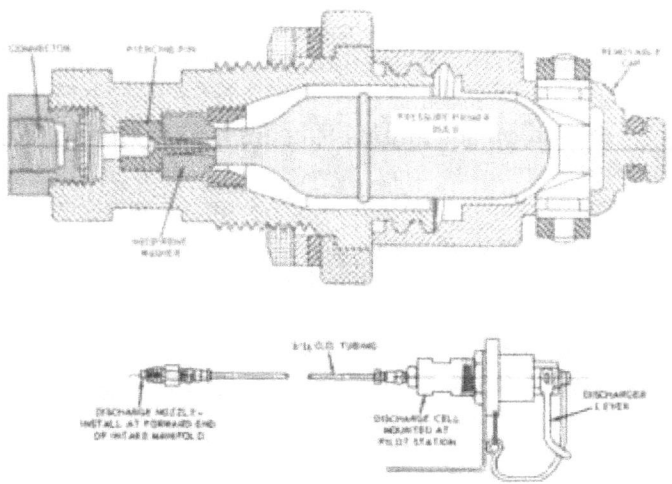

Figure 9-17.-Ether capsule primer.

cranking speed, operate the discharger lever to discharge the capsule bulb. Continue cranking while the ether mixture is being sucked rapidly through the connecting tube to the intake manifold. The capsule bulb requires about 15 seconds to discharge, and the dlesel engine should start during this interval.

Starting Motors

The starting, or cranking, motor is a low-voltage d-c series motor used to start internal combustion engines by rotating the crankshafts. It is flange mounted on the engine flywheel housing and is supplied with current from the battery. All starting motors are very similar in design and consist essentially of a frame, armature, brushes, field windings, and drive mechanism. The armature shaft is supported on bronze bearings equipped with wick oilers. The number of field poles and brushes vary according to the cranking requirements, and the operating voltage corresponds to that of the generator.

The starting motor has low resistance; it is designed to operate under heavy load with relatively high horsepower for short periods of time. The high horsepower is accompanied by a high current that creates considerable heat, and if operated for any considerable length of time will result in failure of the motor due to overheating. Hence, the starting motor must be operated for not more than 30 seconds at a time, and at about two-minute intervals to allow the heat to dissipate. A 4-pole, 8-brush, 12-volt diesel starting motor will draw 570 amperes maximum at 2.3 volts with the armature locked and develop a torque of 20 pound feet.

The starting motor is equipped with a drive mechanism that transmits the power from the motor to the engine. The function of the drive mechanism is to (1) engage the drive pinion with the flywheel for cranking the engine, (2) provide a gear reduction between the drive pinion and flywheel, and (3) disengage the drive pinion and flywheel after the engine is started.

When the starting motor is operated, the drive mechanism causes the drive pinion to mesh with the teeth of the flywheel ring gear, thereby cranking the engine.

The gear reduction is necessary because the starting motor must rotate at a relatively high speed with respect to the engine cranking speed to produce sufficient output power to crank the engine. Thus, a gear reduction ratio of 15 to 1 will permit the starting motor to rotate at 1500 rpm while cranking the engine at 100 rpm.

As soon as the engine is started, the drive mechanism causes the drive pinion to disengage from the flywheel. The engine speed increases immediately and may soon attain speeds up to 1000 rpm. If the drive pinion is allowed to remain in mesh with the flywheel, the engine would drive the starting motor at speeds up to 15,000 rpm, resulting in serious damage to the motor.

The types of drive mechanisms provided on starting motors are the (1) Bendix drive, (2) overrunning clutch drive, and (3) Dyer drive.

BENDIX DRIVE.-The Bendix drive provides an automatic means of engaging the drive pinion of the starting motor with the engine flywheel ring gear for cranking the engine and for disengaging the drive pinion from the flywheel after the engine starts. This drive mechanism (fig. 9-18) consists of a drive pinion mounted on a threaded sleeve, or hollow shaft, which has spiral threads that match the internal threads in the drive pinion. The sleeve

fits loosely on the armature shaft of the starting motor. One end of the sleeve is bolted to the drive spring. The other end of the drive spring is keyed and bolted to the armature shaft through the drive head.

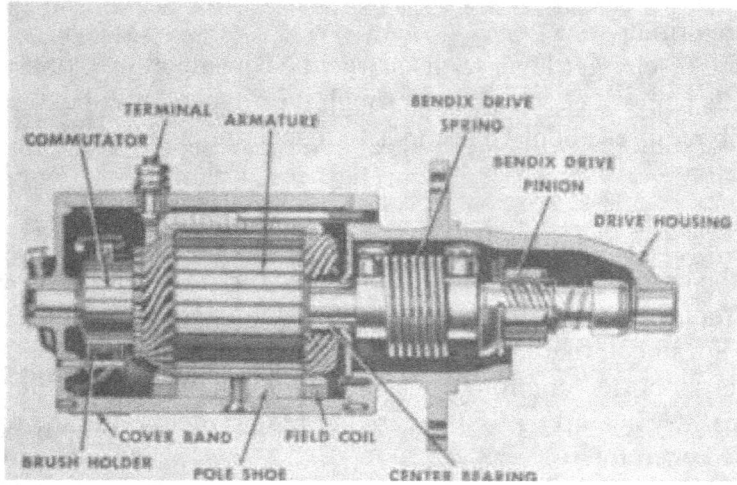

Figure 9-18.-Bendix drive.

When the starting motor is not in operation, the pinion is disengaged from the flywheel ring gear. As soon as the starting' switch is closed, the armature begins to rotate and its speed increases very rapidly. The threaded sleeve picks up speed with the armature as it is driven through the drive spring. The drive pinion does not pick up speed instantly because it fits loosely on the sleeve. Hence, the sleeve turns within the drive pinion, forcing the pinion along the shaft and into engagement with the flywheel ring gear. As the drive pinion reaches the stop on the end of the sleeve, it then rotates with the sleeve and the armature so that the engine is cranked.

The drive spring compresses slightly. to absorb the shock when the gears are meshed to prevent the shock from being transmitted back through the starting motor. Bendix heavy-duty drives are provided with a friction clutch interposed between the drive spring and drive pinion. The clutch comprises a series of spring-loaded

clutch plates that slip momentarily during the engagement to absorb the shock.

After the engine starts, the flywheel drives the pinion at a higher speed than the speed at which the armature and the threaded sleeve are revolving. This action causes the pinion to be turned relative to the threaded sleeve and in such a direction that the pinion is backed out of engagement with the flywheel ring gear.

Some Bendix drives are provided with a small antidrift spring between the drive pinion and the pinion stop. This device prevents the pinion from drifting into mesh when the engine is running. Other Bendix drives use a small antidrift pin and spring inside the pinion. This device provides sufficient friction to keep the pinion from drifting into mesh.

OVERRUNNING CLUTCH DRIVE.-The overrunning clutch drive provides positive engaging and disengaging of the starting-motor drive pinion and the flywheel ring gear. This drive mechanism (fig. 9-19) utilizes a shift lever that slides the clutch and drive pinion assembly along the armature shaft so that it can be engaged and disengaged with the flywheel ring gear. The clutch transmits cranking torque from the starting motor to the engine flywheel, but permits the pinion to overrun the armature after the engine starts. Thus, power can be transmitted through the overrunning clutch in only one direction. This action protects the starting motor from excessive speed during the brief interval that the drive pinion remains engaged with the flywheel ring gear after the engine has started.

The overrunning clutch consists of a shell-and-sleeve assembly (fig. 9-19) that is splined internally to match the splines on the armature shaft. Thus, both the shell-and-sleeve assembly and the armature shaft must turn together. A pinion-and-collar assembly fits loosely into the shell. The collar is in contact with four steel rollers that are assembled into notches cut in the inner face of the shell. The notches taper inward slightly so that less space is in the end away from the rollers. The rollers are spring loaded by means of springs and plungers.

When the shift lever is operated, the clutch assembly is moved along the armature shaft until the pinion engages with the flywheel ring gear. If the teeth should butt instead

CONTACTS SOLENOID
PLUNGER AND LINKAGE
RETURN SPRING
SHIFT LEVER
RETAINER RING
OVERRUNNING CLUTCH
BUSHING

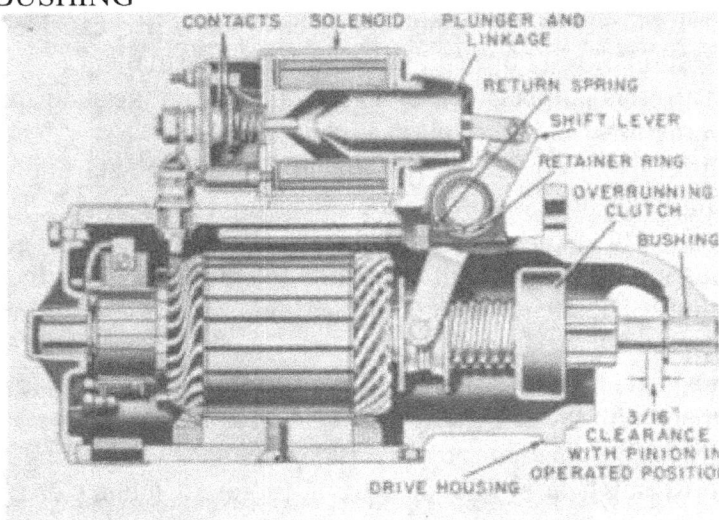

DRIVE HOUSING
r
3/16" CLEARANCE WITH PINION IN OPERATED POSITION
PINION AND COLLAR ASSEMBLY CLUTCH SPRING ■ COLLAR
; I
m
■

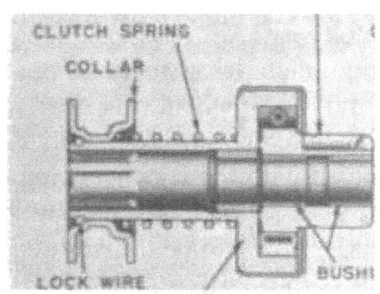

OUTER RACE CONNECTED TO STARTER MOTOR SHAFT
SPRING PLUNGER ROLLER
LOCK WIRE
BUSHING

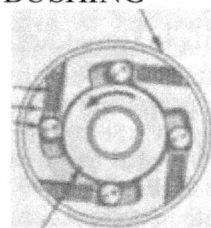

SHELL AND SLEEVE ASSEMBLY
ROTOR CONNECTED TO PINION

Figure 9-19.-Overrunning clutch drive with solenoid-operated switch.

of engage, the clutch spring compresses so that the pinion is spring loaded against the teeth of the ring gear. Thus, engagement of the teeth takes place immediately when the armature begins to rotate.

The starting-motor switch is closed when the movement of the shift lever is completed, causing the armature to rotate. This action rotates the shell-and-sleeve assembly, causing the rollers to move forward (in the direction of rotation) and jam tightly in the smaller sections of the notches. The rollers jam between the pinion collar and the shell so that the pinion is forced to rotate with the armature, and thereby crank the engine.

After the engine starts, it spins the pinion faster than the armature of the starting motor is rotating. This action causes the pinion to rotate with respect to the shell so that it overruns the shell and starting-motor armature. The rollers are now rotated backward into the larger sections of the notches where they are free, and thus they permit the pinion to spin independently of, or overrun, the shell and sleeve assembly. This feature protects the armature for the brief interval after the engine starts until the operator releases the starting-motor switch. When the starting-motor switch is opened, the shift lever releases, causing the drive spring to pull the overrunning clutch drive pinion out of engagement with the engine flywheel ring gear.

DYER DRIVE.-The Dyer drive provides for positive engagement of the starting-motor drive pinion with the engine flywheel ring gear before the starting-motor switch is closed. It is used for heavy duty applications where it is important to avoid clashing of gears because of the relatively high horsepower developed in cranking these engines. As soon as the engine starts,the flywheel spins the drive pinion more rapidly than the armature and shaft assembly are turning. This action backs the pinion out of mesh with the flywheel ring gear so that the armature of the starting motor is not subjected to excessive speeds.

This drive mechanism (fig. 9-20) consists of a shift lever, a shift sleeve, a pinion guide, a pinion spring, a pinion, a pinion stop, and thrust washers. The pinion guide fits snugly on the spiral splines of the armature shaft, and the pinion (which has internal splines that match the armature splines) fits loosely on the splines of the armature shaft. When the drive assembly is at rest, the drive pinion is retained in the disengaged position by the pinion guide, which drops into milled notches in the armature splines. The pinion can be released from this position only by movement of the pinion guide through operation of the shift lever.

When the shift lever is operated, the movement causes the shift sleeve, pinion guide, pinion spring, and pinion to be moved along the armature shaft so that the pinion

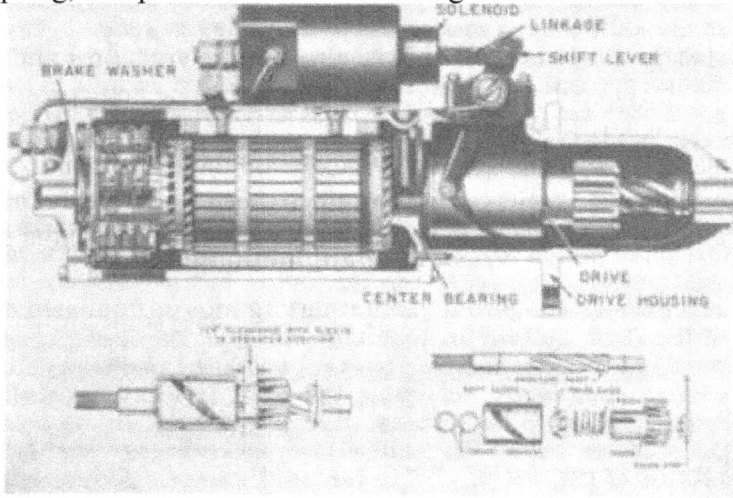

Figure 9-20.—Dyer drive.

engages with the flywheel ring gear If the teeth are aligned properly. Further movement of the shift lever closes the starting-motor switch, causing the armature to rotate, and thereby crank the engine. If the teeth are not aligned and engagement cannot take place at once, the pinion is rotated against the ring gear teeth until the teeth align and engagement is accomplished.

The pinion rotates because the pinion fits loosely on the armature-shaft splines, and the pinion guide fits tightly on the spiral splines of the armature shaft. The continued forward movement of the pinion guide causes it to rotate as it follows the spiral splines on the armature shaft. This rotation is transmitted by means of the two lugs on the pinion guide to the pinion. The pinion rotates without any forward movement until alignment of the teeth takes place, then the pinion is thrust forward into engagement with the flywheel ring gear.

The pinion stop limits the forward movement of the pinion. As the shift lever completes its travel, it closes the starting-motor switch, which is linked mechanically with the shift lever, and the armature of the starting motor

begins to rotate. As soon as the armature of the starting motor begins to rotate, the friction of the armature shaft in the shift sleeve causes the shift sleeve to rotate. The stud on the end of the shift lever is in the slot in the shift sleeve so that rotation of the shift sleeve causes it to move back out of the way from the drive pinion. Cranking takes place as the shaft splines cause the drive pinion to rotate with the armature shaft.

As soon as the engine begins to operate, it spins the drive pinion faster than the armature is rotating, causing the drive pinion to be spun back out of engagement with the flywheel ring gear. As the pinion moves back out of engagement, the pinion guide drops into the milled section of the shaft splines to lock the pinion in the disengaged position. It is impossible to start another cranking cycle without completely releasing the shift lever. The shift lever must drop back to the disengaged position so that the stud can rotate the shift sleeve and reengage the flat section of the spiral slot in the shift sleeve. Movement of the shift sleeve will then cause the cranking cycle to again take place.

The drive pinion will not engage with the flywheel ring gear while the engine is operating because when the pinion teeth touch the teeth of the moving flywheel ring gear, the shift sleeve will be rotated, and the pinion will follow the splines on the armature shaft back to the locked position.

CONTROLS.—The controls used with starting motors in small craft are the (1) magnetic switch and (2) solenoid switch.

The MAGNETIC SWITCH (fig. 9-21) is used to close the motor switch on some starting motors equipped with Bendix drive. It is usually mounted on the motor frame and consists of a coil provided with a spring-loaded plunger. A heavy contact disk is attached to one end of the plunger. The coil is connected in series with a starter switch located on the instrument panel. When the starter switch is operated, the coil is energized from the battery and pulls the plunger so that the contact disk is forced across two contacts to complete the circuit between the battery and the starting motor.

The SOLENOID SWITCH (fig. 9-22) is used on some starting motors equipped with overrunning clutch drives

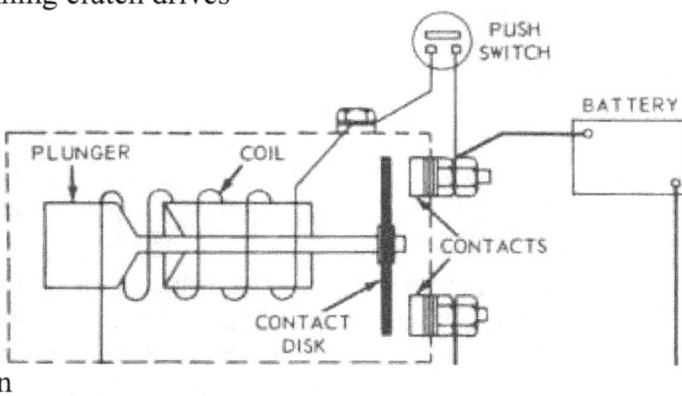

n
STARTING
MOTOR TERMINALS
Figure 9-21.—Magnetic switch.

and Dyer drives to close the circuit to the starting motor and also to engage the pinion with the flywheel ring gear. It is mounted on the motor frame and consists of a pull-in coil and a hold-in coil provided with a spring-loaded plunger. A heavy contact disk is attached to one end of the plunger, and the other end is connected by linkage to the shift lever. Both coils are connected in series with a starter switch located on the instrument panel. When the starter switch is operated, both coils are energized (from the battery) and the plunger is pulled so that the pinion engages with the flywheel ring gear. The pull-in coil draws a comparatively heavy current necessary to complete the plunger movement. The hold-in coil aids the pull-in coil. Continuation of the plunger movement closes the switch contacts, thereby permitting the starting motor to crank the engine. As soon as the solenoid switch is closed (and the pinion shifted),the pull-in coil is shorted by the switch contacts in the starting-motor circuit so that only the hold-in coil is energized to retain the plunger in the operated position.

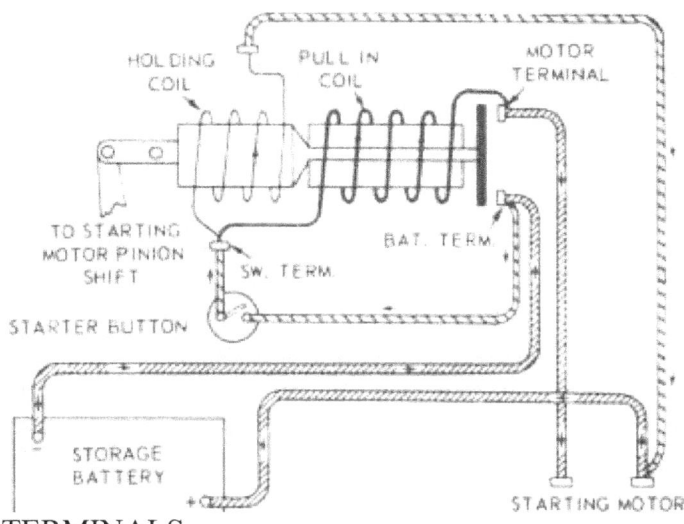

TERMINALS

Figure 9-22.-Solenoid switch.

When the starter switch is released, the tension of the return spring in the drive assembly actuates the plunger to open the circuit to the starting motor.

MAINTENANCE.—A regular inspection and maintenance procedure should be followed to obtain the maximum service with minimum trouble from starting motors in small craft. Periodic inspections of the brushes, commutator, drive assembly, and lubrication (where required) are essential. Also disassembly and thorough overhauling of the starting motor at periodic intervals are necessary to ensure against failures caused by accumulations of grease and moisture and normal wear of the parts. The external circuit between the starting motor and the battery, including the magnetic switch or solenoid switch, must be properly maintained because defective wiring, such as loose or corroded connections, will prevent normal performance of the starting system.

The following intervals of inspection are based on average operating conditions but can be adapted for •pecialized operations.

EACH WEEK, visually check the connections and cables between the starting motor and the battery. Also, check the mounting bolts to be certain they are secure.

EACH MONTH, lubricate the starting motor by adding a few drops of light engine oil to the hinge-cap oiler. On motors equipped with grease cups, turn the grease cups down one turn. Be certain that the cups are filled with medium cup grease. Remove the cover band and inspect the commutator and brushes. If the commutator is dirty, clean it with a strip of No. 00 sandpaper held against it with a piece of soft wood while operating the starting motor with the ignition switch in the OFF position. Blow out the dust with compressed air or a bellows. If the commutator is rough, out-of-round, or has high mica, turn it down in a lathe in accordance with the procedures described in chapter 5 of this training course.

EACH YEAR, remove and disassemble the starting motor so that all parts can be cleaned and the worn parts replaced. Check the brush holders to be certain they are free on their pivots and have the correct spring tension. Clean the bearings and repack them with the proper grease. On motors equipped with oil wicks, saturate the wicks with oil before reassembly. On motors equipped with oilless type bushings, supply a few drops of light engine oil on the bushings at any time the starting motor is disassembled for repair or service. Avoid excessive lubrication.

The Bendix drive should be washed in kerosene and then lubricated by applying a small amount of light engine oil. Do not over oil this mechanism because it might cause the pinion to stick. Replace the pinion if the pinion teeth are burred. Inspect the drift-pin spring because, if the spring is weak, the pinion might tend to drift into engagement with the flywheel ring gear while the engine is running.

The overrunning clutch drive is packed with a special high melting-point grease when the clutch is initially assembled and requires no further lubrication. Do not clean the clutch by grease-dissolving or high-temperature methods because this would cause the clutch to lose its lubricant. Replace the clutch if the pinion does not turn freely in the overrunning direction, if the pinion tends to slip in the cranking direction, or if the pinion is

excessively loose. When the overrunning clutch is operated by means of a solenoid switch, check the clearance (fig. 9-19) between the pinion and the bearing housing when the pinion is in the operated position. This clearance should be three-sixteenths inch and is adjusted by turning the stud in the solenoid plunger in or out as required.

The Dyer drive should be lubricated with a small amount of light engine oil after reassembly. Do not use heavy oil or grease as it might retard or prevent normal drive performance. The adjustment of the drive mechanism must be checked to be certain that the travel of the pinion against the pinion spring is correct when the pinion is in the cranking position. When the shift lever is in the cranking position, it should be possible to push the pinion one-eighth to three-sixteenths inch back against the pinion spring pressure (fig. 9-20). This free travel adjustment can be checked on the solenoid-operated starting motor by disconnecting the lead between the solenoid and starting motor and connecting a battery of the specified voltage to the two terminals on the solenoid.

Manually operate the shift lever until the switch is closed. The current from the battery will maintain the switch in the operating position so that the pinion travel can be checked. The travel is adjusted by turning the stud in the solenoid plunger in or out as required. More detailed information concerning the adjustments, tests, and maintenance of starting motors is contained in the manufacturers' instruction books furnished with the equipment.

LIGHTING SYSTEM

The lighting system in small craft includes the (1) compartment lighting, (2) hand lanterns, (3) portable multipurpose signaling light, and (4) navigational lights.

General Lighting

The general lighting includes the compartment lighting and hand lanterns.

The COMPARTMENT LIGHTING for general illumination consists of units similar to that shown in figure 9-23. This fixture is nonadj us table and consists of a steel enclosure and a conical bowl-shaped globe. The lamp-holder is equipped with a bayonet candelabra double -contact base.

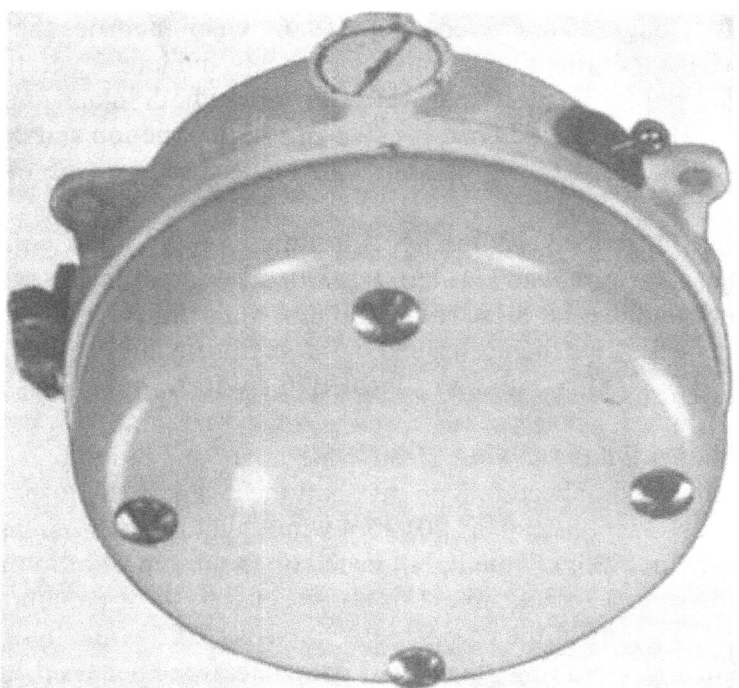

Figure 9-23.-Compartment lighting fixture.

The HAND LANTERN for general use is a manually operated type K-10A lantern. This lantern is identical to those described in chapter 3 of this training course except they are not connected to «relays.

Signaling Light

The portable multipurpose signaling light is similar to that shown in figure 9-24. It consists of a metal housing provided with a black phenolic, louvered signaling tube and a metal pistol-grip handle with trigger switch for blinker communications. A spotlight that has a 5-inch diameter parabolic-type reflector with a clear glass lens and a rheostat, for varying the light intensity, are mounted inside the metal housing. The light uses a Navy type TS-74 candelabra, single-contact lamp with bayonet-type pre-focusingbase. A 25-foot cable with plug is connected to the Figure 9-24.-M u 11 i p u r p o s e light through the bottom of the signaling light. handle for the power supply and remote signaling key. The power supply consists of a transformer and provision for a type BA-207/U battery mounted in a carrying kit. The remote signaling key is equipped with a cable and plug.

Also included with the multipurpose light are a buzzer for an audible signal, a short signal tube, a long signal tube, an open-type sight, and a spare parts box with two lamps and a remote signaling key.

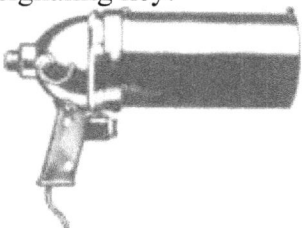

Navigational Lights

The navigational lights installed in motorboats of any length, but less than 40 tons, must comply with Public Law 552, 84th Congress, approved 4 June 1956, which modifies the Motorboat Act of 25 April 1940. This modification permits small boats and landing craft to display the navigational lights required for international waters for use in both inland and international waters. However, the 19-foot rescue boat, because of its construction and because it only operates in inland waters, is provided with navigational lights in accordance with the Motorboat Act.

RUNNING LIGHTS.—The prescribed running lights for motorboats comprise the (1) side lights (port and starboard), (2) bow light, and (3) stern light. In all weather from sunset to sunrise every motorboat must carry and exhibit these running lights when underway. During this time no other lights that may be mistaken for the prescribed lights shall be exhibited.

The SIDE LIGHTS are 10-point (112-1/2°) lights located in the forward part of the boat. They can be a

combination light or individual lights showing green to starboard and red to port, so constructed as to throw the light from right ahead to 2 points abaft the beam on the respective sides. The side lights must be at least 3 feet below the 20-point white light on the foremast.

The BOW LIGHT is a 20-point (225°) white light located on the foremast or in the forward part of the boat as near the stem as practicable. It is constructed to show an unbroken light over an arc of the horizon of 20 points of the compass, so fixed as to throw the light 10 points on each side of the boat—that is, from right ahead to 2 points abaft the beam on either side. The 20-point white light must be at least 9 feet above the gunwale.

The STERN LIGHT is a 12-point (135°) white light located in the after part of the boat. It is constructed to show an unbroken light over an arc of the horizon of 12 points of the compass, so fixed as to throw the light 6 points on each side of the boat—that is, from right astern to 2 points abaft the beam on either side.

ANCHOR LIGHT.-The anchor light isa32-point (360°) white light shown in the fore part of the boat when at anchor. It is so constructed as to throw an unbroken, uniform light all around the horizon.

The running and anchor lights are controlled from switches located on the control and distribution panel.

SPOTLIGHT.-Spotlights installed in small boats for navigational purposes consists of a spraytight, chromium-plated copper housing equipped with a 7-inch diameter sealed-beam lamp. These spotlights are the standard Navy type, K-31A and K-24A. The type K-31A light is provided for deck mounting and is controlled locally at the light; whereas, the type K-24A light is provided for mounting on top of a canopy and is controlled from beneath the mounting surface.

QUIZ

1. What type of equipment is used to provide a source of current for starting the engine in small craft?

2. What type of equipment is used to maintain the source in a fully charged condition?

3. What type of generator is being installed to handle the appreciably increased output requirements of small craft electrical systems?

4. What is the effect of the limitation of the output capabilities of the third-brush generator at low and high engine speeds on the extensive use of this generato r ?

5. Do internally grounded and ungrounded generators employ standard, or heavy-duty regulators?

Refer to figure 9-3 for questions 6 through 9.

6. What is the effect on the d-c field when turning on the ignition switch ?

7. What is the relation of the battery and alternator system when the load relay is energized?

8. (a) When the alternator speed is low and/or the electrical load is heavy, the voltage regulator armature vibrates on which contact (upper or lower)? (b) This action periodically inserts what component in the d-c field circuit?

9. (a) When the alternator speed is high/or the load is light the voltage regulator armature vibrates on which contact (upper or lower)? (b) This action does what to the d-c field coil periodically?

10. To check the output of the generator (fig. 9-2, A), how are the ammeter and variable resistance connected ?

11. In testing for a shorted field coil by passing normal current through the field circuit and measuring the voltage drop across each field coil, what is the relative magnitude of the voltage across the shorted coil ?

12. What two sources of trouble should be looked for when an unsteady or low output occurs in any one of the several types of generators?

13. In the polarizing test (fig. 9-4, A), what is the relation between the polarity of the lead from a spare battery to terminal A of the generator and the polarity of the regulator BAT terminal?

14. Why should a generator never be polarized after it is installed in a small craft?

Refer to figure 9-7 for questions 15 through 17.

15. When the generator is ope rating with a heavy load and the voltage does not increase sufficiently to

operate the voltage regulator, what current regulator action will occur to limit the generator output?

16. When the voltage regulator operates, what action occurs in the current regulator?

17. (a) How is R2 connected with respect to the generator field?

(b) What is the function of R2 ?

Refer to figure 9-8 for questions 18 through ZO.

18. What is the effecton the cutout relay when a reverse current flows through the series coil?

19. When the generator is operating and the voltage reaches the value for which the voltage regulator is adjusted, what is the effect on the regulator contacts and the generator voltage?

20. When the generator is operating and the output reaches the value for which the current regulator is adjusted, what is the effect on the regulator contacts and the generator output?

21. What ignites the fuel-air mixture in the cylinders of a gasoline engine ?

22. What are the four principal parts of a battery ignition system ?

2 3. What may be the peak magnitude of the voltage induced in the secondary winding of the induction coil when the breaker contacts in the distributor open ?

24. One high-voltage surge is produced by the ignition coil for each cylinder with how many revolutions of the crankshaft? 2 5. The timing of the spark is usually accomplished automatically by either or both of what two general types of mechanism?

26. (a) With an increase in engine speed the breaker cam on the distribution is positioned in what direction with respect to the direction of rotation of the distributor shaft?

(b) How does this action affect the timing of the spark in the engine cycle ?

27. What is the relative length of the path that the heat must travel from the tip of the spark plug to reach the cooling system when the engine employs (a) cold plugs and (b) hot plugs ?

2 8. What is the name of the angle represented by the number of degrees that the breaker cam rotates from the time the contacts close until they againopen ?

2 9. Name four types of airheaters and primers used for starting diesel engines.

30. What type of motor is used for starting internal combustion engines?

31. A 4-pole, 8-brush, 12-volt diesel starting motor will draw what current at what voltage with the armature locked, developing a torque of 20 pound feet?

32. What is the gear ratio between the drive pinion and flywheel that will permit the starting motor to develop sufficient power to crank the engine?

33. What are three types of drive mechanisms provided on starting motors?

34. In the overrunning clutch (fig. 9-19) if the teeth of the pinion should butt instead of engage the teeth of the flywheel ring gear, what component will force engagement of the teeth when the armature begins to rotate ?

35. In heavy-duty applications requiring high horsepowe r, why does the Dyer drive (fig. 9-20) provide positive engagement of the starting motor drive pinion with the engine flywheel ring gear before the starting motor switch is closed?

36. On motors equipped with oil wicks, what action should be taken before reassembly?

37. What four classes of electric lighting are included in small craft applications ?

CHAPTER

ELECTRIC POWER DISTRIBUTION

Electric power furnishes the numerous services that are indispensable to the operation of a modern naval vessel. It trains and elevates the guns, turns the rudder; runs important auxiliaries; furnishes light; and operates the interior-communications, fire-control, radio, radar, and sonar systems. Because of the extensive and highly diversified use of electricity aboard ship, the electric system comprises a great many different items of equipment for the generation, distribution, and utilization of electric power.

The type of ship determines whether the electric power system is an a-c or a d-c system. The general practice is to install a-c systems in new construction battleships, cruisers, aircraft carriers, destroyers, escort vessels, submarine chasers, and numerous auxiliary vessels. D-c systems are usually installed in new construction submarines, small surface vessels, and large surface vessels with considerable deck machinery that warrants the use of direct current. D-c systems are also used in numerous vessels converted for Navy use, and in battleships, cruisers, destroyers, and other vessels built before the general adoption of a-c ship's service systems.

This chapter describes electric power distribution exclusive of electric propulsion systems. The latter systems are described in the training course, EM 1 and Chief.

A-C ELECTRIC POWER SYSTEMS

The a-c electric power system installed in surface vessels consists of the (1) electric plant and (2) electric power distribution system. The entire system is designed to provide maximum continuity of service with minimum interruption.

Electric Plant

The electric plant includes the ship's service electric plant and the emergency electric plant.

SHIP'S SERVICE ELECTRIC PLANT.-The ship's service electric plant comprises the ship's service generator sets, which are driven by steam turbines, diesel engines, or gas turbines. These generator sets supply power to the ship's service power distribution system, which is the normal source of power to the ship's electrical equipment and machinery.

The number, type, and capacity of the ship's service generators are determined by the load imposed on the generating plant by all of the connected auxiliaries under the various operating conditions.

EMERGENCY ELECTRIC PLANT.—The emergency electric plant comprises the emergency generator sets, which are driven by diesel engines or gas turbine engines. If power is not available from the ship's service generators, the emergency generators supply a limited amount of power to the vital auxiliaries through the emergency power distribution system.

The number and rating of the emergency generators must be sufficient to carry the loads that are designated to receive emergency power, but not to supply all of these loads simultaneously. The emergency generator capacity is determined by a particular combination of these loads, the choice of which depends on the type of ship.

The ship's service generators are required to operate in parallel; whereas, the emergency generators are not operated in parallel with each other or with the ship's service generators. The electric plant is ungrounded except as required for ground detectors, instrument grounds, equipment frame grounds, personnel protection, or other special grounding requirements.

Power Distribution System

The power distribution system is the connecting link between the generators that supply electric power and the electrical equipment that utilizes this power to furnish the various services necessary to operate the ship. The power distribution system comprises the (1) ship's service power distribution system, (2) emergency power distribution system, and (3) casualty power distribution system.

The majority of a-c power distribution systems in naval vessels are 450-volt, 3-phase, 60-cycle, 3-wire systems. The lighting distribution systems are 120-volt, 3-phase, 60-cycle, 3-wire systems supplied from the power circuits through transformer banks.

SHIP'S SERVICE POWER DISTRIBUTION SYSTEM.-The ship's service power distribution system is the electrical system that normally supplies electric power to the ship's equipment and machinery. The switchboards and associated generators are located in separate engineering spaces to minimize the possibility that a single hit will damage more than one switchboard.

The ship's service generator and distribution switchboards are interconnected by BUS TIES so that any switchboard can be connected to feed power from its generators to one or more of the other switchboards. The bus ties also connect two or more switchboards so that the generator plants can be operated in parallel (or the switchboards can be isolated for split plant operation).

Power distribution is direct from the ship's service generator and distribution switchboards to large and important loads, such as the steering gear and turrets, and to loads near the switchboard. In large installations (fig. 10-1), power distribution to other loads is from the generator and distribution switchboards or switchgear groups to load centers, to distribution panels, and to the loads, or directly from the load centers to some loads.

On certain new construction, such as large aircraft carriers, a system of zone control of the ship's service and emergency distribution is provided, as described in chapter 2 of this training course (fig. 2-23). Essentially, the system establishes a number of vertical zones, each of which contains one or more load center switchboards

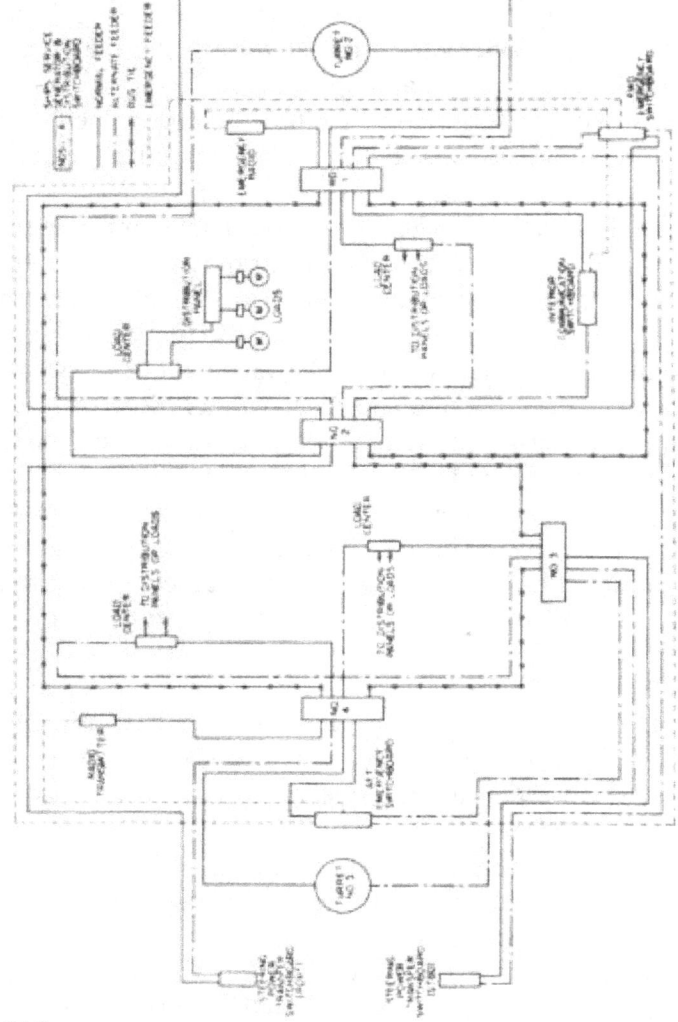

510

supplied through bus feeders from the ship's service switchgear group. A load center switchboard supplies power to the electrical loads within the electrical zone in which it is located. Thus, zone control is provided for all power within the electrical zone. The emergency switchboards may supply more than one zone, the number of zones depends on the number of emergency generators installed.

In small installations (fig. 10-2), the distribution panels are fed directly from the generator and distribution switchboards. The distribution panels and load centers (if any) are located centrally with respect to the loads that they feed. This arrangement simplifies the installation and requires less weight, space, and equipment than if each load were connected to a switchboard.

The GENERAL POWER CIRCUITS supply power to motors and appliances that are not essential during battle and are therefore nonvital. The BATTLE POWER CIRCUITS comprise all other circuits and are energized under battle conditions. The feeders in the general power circuits and those in the battle power circuits are designated by the letters, F and FB, respectively.

At least two independent sources of power areprovided for selected vital loads. The distribution of this dual supply is accomplished in several ways: by a NORMAL and an ALTERNATE ship's service feeder; NORMAL ship's service feeder and an EMERGENCY feeder; or NORMAL and ALTERNATE ship's service feeder and an EMERGENCY feeder (figs. 10-1 and 10-2).

The normal and alternate feeders to a common load run from different ship's service switchboards and are located below the waterline on opposite sides of the ship to minimize the possibility that both will be damaged by a single hit.

BUS TRANSFER EQUIPMENT is installed at load centers, distribution panels, or loads that are fed by both normal and alternate and/or emergency feeders. This equipment is used to select either the normal or alternate source of the ship's service power, or to obtain power from the emergency distribution system if an emergency feeder is also provided.

Automatic bus transfer equipment is used for loads that require two power supplies, except for cold-ship

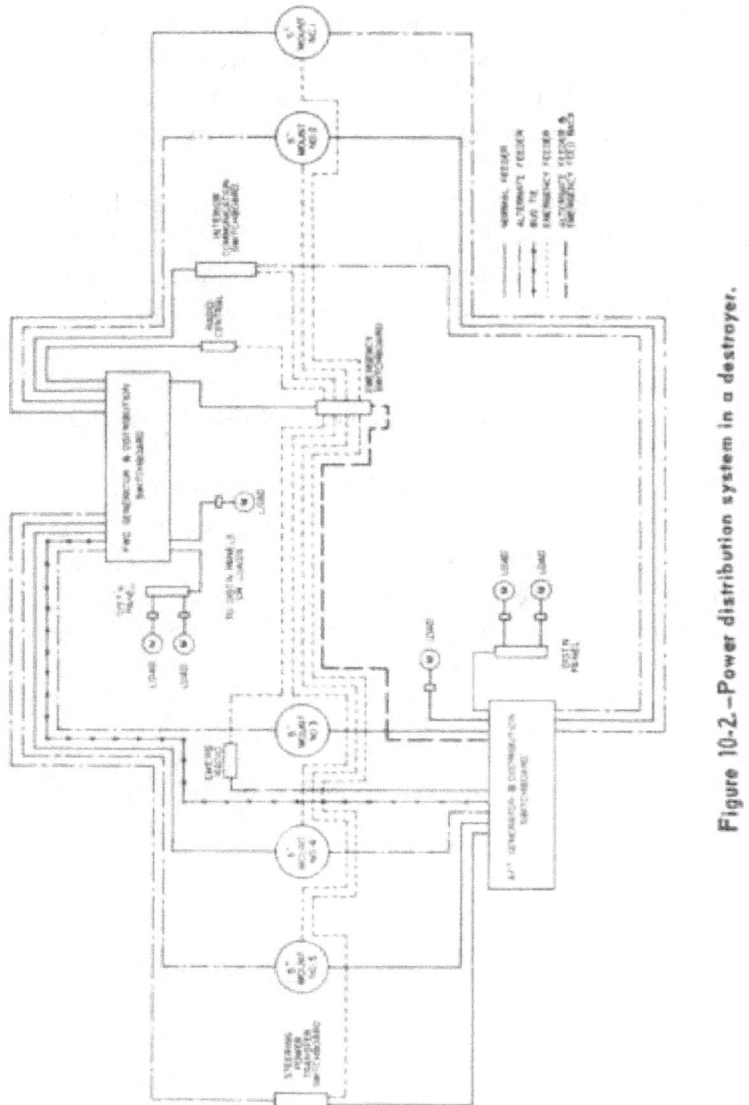

Figure 10-2.—Power distribution system in a destroyer.

starting auxiliaries and fire pumps, which have manual bus transfer equipment. On the steering power switchboard, which is provided with a normal, alternate, and emergency power supply, manual bus transfer equipment is used to select between the normal and alternate supplies, and automatic bus transfer equipment is used to select between the ship's service and emergency supplies.

The lighting circuits are supplied from the secondaries of 450/120-volt transformer banks connected to the ship's service power system. In large ships the transformer banks are installed in the vicinity of the lighting distribution panels located at some distance from the generator and distribution switchboards. In small ships the transformer banks are located near the generator and distribution switchboards and energize the switchboard buses that supply the lighting circuits.

The lighting distribution system feeders, mains, and submains are 3-phase circuits, and the branches are single-phase circuits. The single-phase circuits are connected so that under operating conditions the single-phase loads on the 3-phase circuits are as nearly balanced as possible.

The PHASE SEQUENCE in naval vessels is ABC; that is, the maximum positive voltages on the three phases are reached in the order: A, B, and then C. Phase sequence determines the direction of rotation of 3-phase motors. Reversal of the phase sequence reverses the direction of rotation of electric motors. The phase sequence of the power supply throughout a ship is always ABC, irrespective of whether power is supplied from any of the switchboards or from the shore power connection. This condition ensures that 3-phase, a-c motors will always run in the correct direction.

Phase identification is denoted by the letters, A, B, and C in a 3-phase system. Switchboard and distribution-panel bus bars and terminate on the back of switchboards are marked to identify the phase with the appropriate letters, A, B, or C. The standard arrangement of phases in power and lighting switchboards, distribution panels, feeder distribution boxes, feeder junction boxes, and feeder connection boxes is in the order A, B, and C from top to bottom, front to back, or right to left when facing the front of the switchboard, panel, or box and left to

right when facing the rear of the switchboard, panel, or box. The color coding on cables for a-c and d-c electric power systems is described in chapter 2 of this training course.

A SHORE POWER CONNECTION is provided at, or near, a suitable weather-deck location to which portable cables from the shore or from a vessel alongside can be connected to supply power for the ship's distribution system when the ship's service generators are not in operation. This connection also can be used to supply power from the ship's service generators to a vessel alongside.

EMERGENCY POWER DISTRIBUTION SYSTEM.-The emergency power distribution system is provided to supply an immediate and automatic source of electric power to a limited number of selected vital loads in the event of failure of the ship's service power distribution system. The system, which is separate and distinct from the ship's service power distribution system, includes one or more emergency distribution switchboards. Each emergency switchboard is supplied by its associated emergency generator. The emergency feeders run from the emergency switchboards (figs. 10-1 and 10-2) and terminate in manual or automatic bus transfer equipment at the distribution panels or loads for which emergency power is provided. The emergency power distribution system is a 450-volt, 3-phase, 60-cycle system with transformer banks at the emergency distribution switchboards to provide 120-volt, 3-phase power for the emergency lighting system.

The emergency generators and switchboards are located in separate spaces from those containing the ship's service generators and distribution switchboards. As previously stated, the normal and alternate ship's service feeders are located below the waterline on opposite sides of the ship. The emergency feeders are located near the centerline and higher in the ship (above the waterline). This arrangement provides for horizontal separation between the normal and alternate ship's service feeders and vertical separation between these feeders and the emergency feeders, thereby minimizing the possibility of damaging all three types of feeders simultaneously.

The emergency switchboard is connected by feeders to at least one and usually to two different ship's service

switchboards. One of these switchboards is the preferred i source of ship's service power for the emergency switchboard and the other is the alternate source (fig. 10-3). The emergency switchboard and distribution system are normally energized from the preferred source of ship's service power. If this source of power should fail, bus transfer equipment automatically transfers the emergency switchboard to the alternate source of the ship's service power. If both the preferred and alternate sources of ship's service power fails, the diesel-driven emergency generator starts automatically, and the emergency switchboard is automatically transferred to the emergency generator.

When the voltage is restored on either the preferred or alternate source of the ship's service power, the emergency switchboard is automatically retransf erred to the source that is available, or to the preferred source if voltage is restored on both the preferred and alternate sources. The emergency generator must be manually shut down. Hence, the emergency switchboard and distribution system are always energized either by a ship's service generator or by the emergency generator. Therefore, the emergency distribution system can always supply power to a vital load if both the normal and alternate sources of the ship's service power to this load fails. The emergency generator is not started if the emergency switchboard can receive power from a ship's service generator.

A FEEDBACK TIE from the emergency switchboard to the ship's service switchboard (fig. 10-2) is provided in most ships. The feedback tie permits a selected portion of the ship's service switchboard load to be supplied from the emergency generator. This feature facilitates starting up the machinery after major steam alterations and repairs, and provides power to operate necessary auxiliaries and lighting during repair periods when shore power and ship's service power are not available.

CASUALTY POWER DISTRIBUTION SYSTEM.-The casualty power distribution system is provided for making temporary connections to supply electric power to certain vital auxiliaries if the permanently installed ship's service and emergency distribution systems are damaged. The system is not intended to supply circuits to all the

LX" ' I

r i

:—c

:l_c

~t j . ,rt-

in;

ship's 5l«vCE GrNC«4T0R a msTH'Hi/'ioN sw'CMBOioo

IMIHGCNCt SwTCmBO»«0

K CGI 1 K'MShiP S Srfl.'tCI-r I vl.. a %c> •.«iTCM80»hO

Mo'lt PfTwf (N S«*S SCRV'Ct

'.a tf^o -AWT'.

Figure 10-3.-Emergency and ship's service distribution system interconnections.

electrical equipment in the ship but is confined to the facilities necessary to keep the ship afloat and to get it away from a danger area. The system also supplies a limited amount of armament, such as antiaircraft guns and their directors that may be necessary to protect the ship when in a damaged condition. The casualty power system for rigging temporary circuits is separate and distinct from the electrical damage control equipment, which consists of tools and appliances for cutting cables and making splices for temporary repairs to the permanently installed ship's service and emergency distribution systems.

The casualty power system includes portable cables, bulkhead terminals, risers, switchboard terminals, and portable switches. Portable cables in suitable lengths are stowed throughout the ship in convenient locations. The bulkhead terminals are installed in watertight bulkheads so that horizontal runs of cables can be connected on the opposite sides to transmit power through the bulkheads without the loss of watertight integrity. The risers are permanently installed vertical cables for transmitting power through decks without impairing the watertight integrity of the ship. A riser consists of a cable that extends from one deck to another with a riser terminal connected to each end for attaching portable cables.

Suitable terminals are provided at switchboards and some distribution panels for connecting portable cables at these points to obtain power from or supply power to the bus bars. Casualty power circuit breakers are installed at switchboard so that the terminals can be de-energized when connecting the cables. The portable switches are stowed in repair party lockers and are used when necessary for connecting and disconnecting the circuits. The locations of the portable cables, bulkhead terminals, and risers are selected so that connections can be made to many vital electrical auxiliaries from any of the ship's service or emergency generators. Casualty power cables should be rigged only when required for use, or for practice in rigging the casualty power system. When rigging the casualty power cables, the connections should be made from the load to the supply to avoid handling energized cables.

The riser terminals, bulkhead terminals, and portable cables are marked to identify the A, B, and C phases both visually and by touch when illumination is insufficient for visual identification.

Optimum continuity of service is ensured in vessels provided with ship's service, emergency, and casualty power distribution systems. If one generating plant should fail, a remote switchboard can be connected by the bus tie to supply power to the switchboard that normally receives power from the generator or generators that have failed.

If a circuit or switchboard fails, the vital loads can be transferred to an alternate feeder and source of ship's service power by means of a transfer switch near the load.

If both the normal and alternate sources of the ship's power fail because of a generator, switchboard, or feeder casualty, the vital auxiliaries can be shifted to an emergency feeder that receives power from the emergency switchboard.

If the ship's service and emergency circuits fail, temporary circuits can be rigged with the casualty power distribution system and used to supply power to vital auxiliaries if any of the ship's service or emergency generators can be operated.

D-C POWER in ships with a-c power systems is furnished either by oversize exciters for the ship's service generators or by separate motor-generator sets. The principal d-c loads are carbon-arc searchlights, degaussing installations, battery-charging stations and the interior-communications and fire-control systems. The use of the 24-inch carbon-arc searchlight has been discontinued aboard destroyers with a consequent reduction in the d-c power requirements. Metallic rectifiers are used as d-c power sources in the latest ship's provided with a-c power systems.

MULTIPURPOSE POWER CIRCUITS are provided to supply 450-volt, 3-phase power for portable hoists; portable tools that require 450-volt power, portable welding units for repair, maintenance, and damage repair purposes, including underwater welding and cutting; and portable submersible bilge pumps. The multipurpose power outlets are of the grounded type and are used with grounded plugs and cables having a ground wire that grounds the metallic case and exposed metal parts of the tool or equipment when the plug is inserted in the receptacle. The ground wire provides a conducting path of low resistance between the metal housing of the tool and the ship's structure. In the event of a casualty to the insulation of the tool, the ground wire will shunt the operator, thereby protecting him from shock.

The outlets are located so that two portable pumps can be operated in any compartment by using 75 feet of cable for each pump. The outlets are fed from battle power distribution panels whenever practicable. A minimum number of outlets are fed from any one panel in order to provide as great a diversity of supply as possible. An adapter is provided with the 75-foot extension cables for making connections to the casualty power system if power is lost from the outlets.

D-C ELECTRIC POWER SYSTEMS

The d-c electric power systems installed in surface vessels are the (1) low-voltage system, (2) 120-volt, 2-wire system, and (3) 120/240-volt, 3-wire system.

Low-Voltage System

The low-voltage (12-volt or 24-volt) system is installed in motor torpedo boats, small landing craft, and small boats. Power is supplied by generators that are driven by the propulsion engines or by small auxiliary engines. Storage batteries are used to supply power to the system when the generators are not operating. These systems are described in chapter 8 of this training course.

120-Volt, 2-Wire System

The 120-volt, 2-wire system is usually installed in ships in which the total electrical load is small or in ships in which the 120-volt load is the major part of the connected load. The power loads are usually supplied by feeders running directly from the switchboard. The lighting in the forward and after part of the ship is usually supplied by feeders running to lighting distribution boxes located near the centers of the forward and after loads.

120/240-Volt,3-Wire System

The 120/240-volts, 3-wire system was formerly installed in all large surface vessels. This system is still in use in older ships and in converted merchant ships with large deck machinery loads that warrant the use of d-c power. The 120/240 volt, 3-wire system is lighter in weight, smaller in size, and more efficient in operation than the 120-volt, 2-wire system.

In the 3-wire system the power is generated by a 120/ 240-volt, 3-wire, d-c generator or a converted 240-volt, 2-wire, d-c generator provided with a transformer balancer (center tapped autotransformer) connected through slip rings to tapping points on the generator armature, which establishes a neutral. The present Navy practice

is not to ground the neutral (or either leg) of a 3-wire system. The positive,neutral, and negative polarities of bus bars and terminals are indicated + (black), ± (white), and - (red) respectively, as previously described in chapter 2 of this training course.

Power distribution from the switchboards to the power loads is a 240-volt, 2-wire system. The 120-volt distribution from the switchboards to the distribution panels is a 120/240-volt, 3-wire system and from the distribution to the branch circuits is a 120-volt, 2-wire system. The 120-volt, 2-wire branch circuits for lighting and appliances are balanced on the two sides of the neutral.

EMERGENCY POWER DISTRIBUTION SYSTEM.-The emergency power distribution system is generally similar to the corresponding a-c system previously described. The principal difference is that in a few of the older ships, emergency power is provided by storage batteries instead of by d-c generators driven by diesel engines.

CASUALTY POWER DISTRIBUTION SYSTEM.-The casualty power distribution system is also generally similar to the corresponding a-c system except that portable switches are not provided for d-c casualty power systems, and all fittings are suitably modified for use with d-c power.

SWITCHBOARDS

A generator and distribution switchboard or switch-gear group is provided for each generator or group of generators to control the operation of the associated generators and to control, through appropriate switching equipment, the distribution of electric power. Many of the components that comprise a distribution system are mounted on the switchboards or switchgear groups and include measuring instruments (voltmeters, ammeters, etc.); switching equipment (circuitbreakers,bus transfer equipment, switches, current and voltage-sensitive relays); voltage regulation equipment; protective equipment (circuit breakers and fuses); and conductors (bus bars and cables).

A switchboard may consist of a single section or of several sections that are physically separated and are connected by cables to form a switchgear group. This

arrangement provides sufficient separation between sections to minimize damage from shock, to localize damage from fire, and to permit easy removal of damaged sections for repairs or replacement.

Switchboards are of the live-front and the dead-front types of construction and consist of a welded steel framework for supporting the various panels and equipment. The panels associated with the supporting structure differ for the two types of switchboards.

The LIVE-FRONT SWITCHBOARD (fig. 10-4) utilizes panels composed of an insulating material on which are mounted the knife switches, circuit breakers, fuses, meters, indicating devices, and other components. This type of construction was formerly used for all vessels but is now generally limited to low-voltage, d-c systems of 120-volts or less.

Figure 10-4.-Live-front switchboard.

The DEAD-FRONT SWITCHBOARD (fig. 10-5) utilizes sheet-steel panels or enclosures from which only the meters and operating handles of the switches and circuit breakers protrude to the front of the switchboard. This type of construction is used for all a-c distribution systems and for the d-c distribution systems in some large vessels.

On modern switchboards the equipment is grouped to form a number of units, each complete with a separate front panel and all the required appurtenances, such as the a-c generator control unit, a-c bus tie unit, power distribution unit, and lighting distribution unit. A number

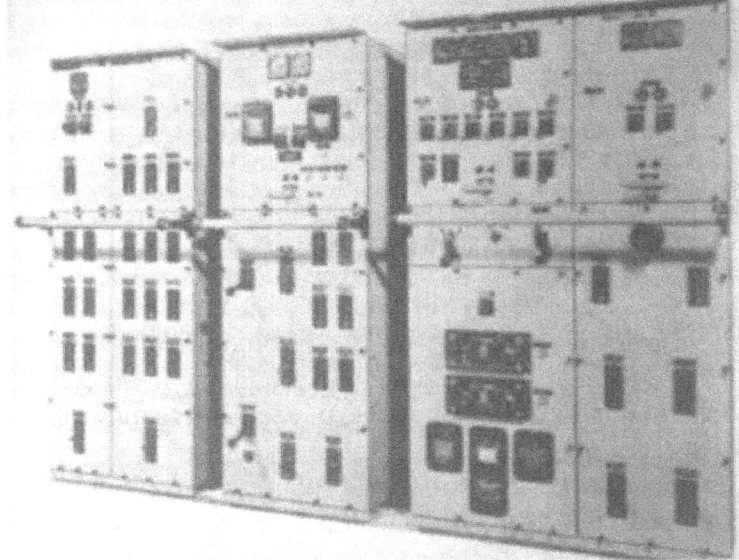

Figure 10-5.-Dead-front switchboard.

of units mounted on a common base comprise a section or several sections that are physically separated and are connected by cables to form a switchgear group.

A separate CONTROL BENCHBOARD (fig. 10-6) is provided in the switchgear groups for battleships, cruisers, and aircraft carriers. This benchboard mounts all the generator control equipment, all measuring instruments, and all the control for electrically operated equipment. This arrangement provides for a centralized control of the generators and major switching operations. The control benchboard in ships equipped with four ship's service switchgear groups are provided with a mimic bus that has indicating lights to show which generator circuit breaker and which bus tie circuit breakers are closed throughout the ship. In ships not provided with control benchboards the metering and control equipments are mounted on the front panels of the units in the switchboards or switchgear groups.

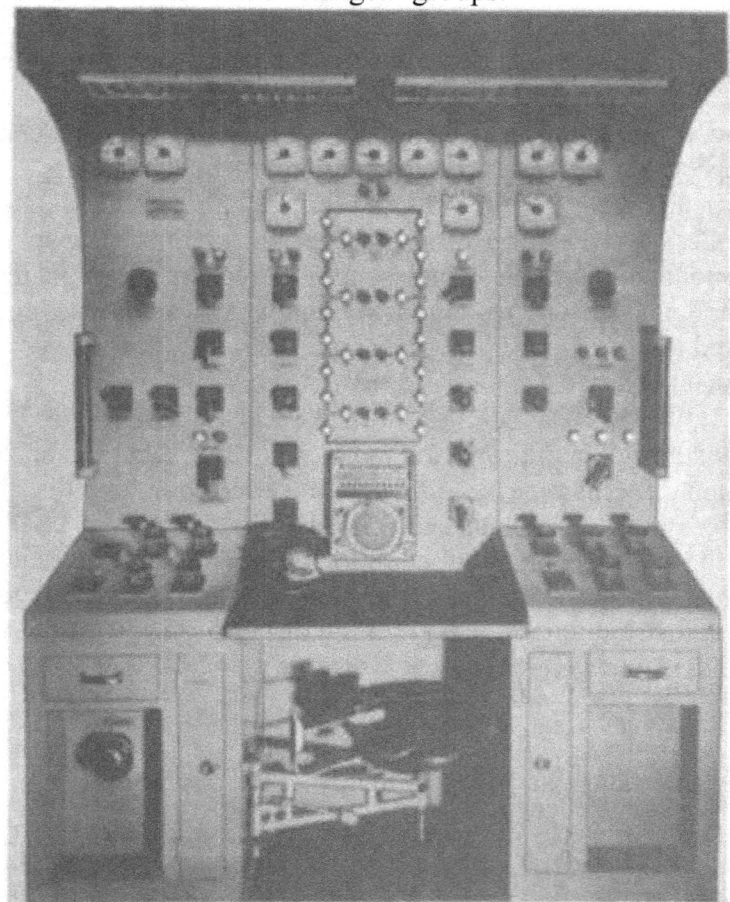

Figure 10-6.-Control benchboard.

A-C Switchboards

The electric plant in a 692-class destroyer consists of two turbine-driven ship's service generating units (groups 1SG and 2SG) located one in each engine room, and two dies el-driven emergency generating units (groups 1EG and 2EG) located one in each emergency generator room.

The ship's service a-c generators are 400-kw, 450-volt, 3-phase, 60-cycle units. Each ship's service generating unit is equipped with a direct-connected oversize exciter to supply excitation for its associated a-c generator and the d-c power system. The exciters are 50-kw, 120-volt, stabilized-shunt, d-c generators.

The emergency a-c generators are 100-kw, 450-volt, 3-phase, 60-cycle units. Each emergency generating unit is equipped with a direct-connected exciter to supply excitation for its associated a-c generator. The exciters are 120-volt, shunt-wound, d-c generators.

The switchboards for controlling the generating units and for the distribution of electric power include the ship's service and the emergency switchboards.

SHIP'S SERVICE SWITCHBOARDS.-The ship's service switchboards, which consist of switchgear groups IS and 2S (fig. 10-7), are located in the forward and after engine rooms respectively.

The forward ship's service switchgear group (fig. 10-7,

A) is designated as the control switchboard. It is provided with instruments and controls for the after generator to allow for dividing the load. All paralleling of the generators is accomplished at the ship's service switchboard associated with the incoming generator.

The after ship's service switchgear group (fig. 10-7,

B) is very similar to the forward ship's service switch-gear group and consists of the same number of corresponding designated panels.

Generator switchboards are equipped with meters to indicate the generator voltage,current, watts, frequency, and power factor. Synchroscopes and synchronizing lamps are provided for paralleling generators. Indicator lamps are provided for visual indication of the operating conditions of various circuits.

The frequency is controlled by generator speed. The speed is automatically controlled by the speed governor of the prime mover. The speed governors for large machines can be set to the required speed by a governor motor that is controlled from the switchboard.

To prevent the generator from operating as a motor when running in parallel with other generators, the generator circuit breaker is equipped with a reverse power relay (described in chapter 7 of this training course) that

trips the breaker and takes the generator off the line when power is fed from the line to the generator instead of from the generator to the line.

Protection against overspeed is provided in the governing mechanism of the prime mover.

A voltage regulator is mounted on each switchboard and operates automatically to vary the field excitation in order to maintain the generator voltage constant throughout normal changes in load. The voltage regulator operates a field-rheostat motor that varies the resistance in the generator field. A standby regulator and transfer switch are also located on each switchboard to provide automatic voltage regulation if the regulator in operation fails. If both regulators fail, the field-rheostat motor can be operated by a manual control switch, and if the field-rheostat motor fails, the rheostat can be operated directly by a manual control on the switchboard. The voltage of the d-c generator (exciter) does not require automatic regulation and is therefore controlled by a manually operated field rheostat. A description of the motor-driven field rheostat is given in chapter 8 of this training course.

EMERGENCY SWITCHBOARDS.-The emergency switchboards, which consist of switchgear groups IE and 2E (fig. 10-8), are located in the forward and after emergency generator rooms respectively. The emergency switchboards provide electric power to certain vital auxiliaries and a limited amount of lighting in the event of failure of the ship's service power. The diesel generating unit starts automatically within 10 seconds after failure of the ship's service power. The emergency power system is arranged for automatic transfer of the circuit for the steering power panel to the emergency generator. The forward and after emergency switchgear groups are normally energized from the forward and after ship's service switchgear groups respectively.

The forward emergency switchgear group (fig. 10-8, A) controls the forward emergency diesel generating unit; the 15-kva emergency lighting transformer bank; the forward ship's service and emergency bus transfer equipment; the local emergency power bus feeders, the remote emergency power bus feeders; and the emergency lighting feeders.

5 a • ∎
M M * I
0 0 o «^
• • m .
- - o J
" I T x x w • £
u u ! 1! - O ;
k * i I A 1 £
o O ./∎> <✓>'c •
0 O . • f e 0
u o u ^ ^ u ?
« a a < « « Z
1 i i i i i i
« a i -1 - - - !
2 O
z
z S
I I
i x
0 2
1 3
o <
* * 0
o <
I f.
i
"G O
E 2 z
w D 3
| o o
<::
O 0
— - n n n n
a t
O J
> <
x 2
^ o
* *
x -* u » * O
t f f !
* s < i

" t to

«, O O u » Z Z «.

2 a ~ o o o o -

If ??

v >- v

23

i o

z 2 o «

32

< ∎

w 6

< >

^ o 5x2

ill

< 2 «

< ?.

° V

S u z

o « y

5 2

i

32*

3 O ∎*

o t

o

5 >

o o > o

5 O

Z • s

5 i

« <-» O

»- £ !£ • •* >

o

z

. °.

O ^ O

5 - « i ∎ > ∎

*5

« 3 *

i * ▶ •

£ C y O

o o >>

<< Q Q « O

o _

I s - I < > s <

2 i

z ∎
S -
> r£ 0:9
o < 2
H
I!
a o
u "5
i
22
g 22
z z
o o
o z
Z •« 1
- 2 23
x t ^ 2 z z o ^ 2 « ^ o o <
< < < 3 < o s
- = = 2 >
z z z < < <
S z < <

a
o U _ I
z
Ml
o i u r
x
3 a
< £
z o
S z
o o <
<
o «

h
o »>
- O 1
o *
1 1 :
o
7
7
o
o
J Z <M
z
- o
* o
0 "
1 i
5 3
i
0
t:
• >> ~ > 3 1 o o
« S V * S C * * ^ i. 2 = 5 I 2 2
< <
o a
z z
^* y
z z
22
r i
- - = < Z
O O u " o
» « « or U
° ° «
< <
o o
z z
o o
z « ~
5 ^ * _ =
o o < <
6 o
a a f
o
z
o o
< <

o o o o 3
nr,nnt^^*i
o 6 > >
00 CO
3 uj tu m
on:
5 " S *
u O O J
a * » *
^ u u u
C Q Q O
z » o 5 *f
o
I 5
8
I §
2 2« s
y >
o w o o
a o ? O
O — I* m
5 S J. JS 1
• f ; 5 cp
O O j « -
u u w u
« o o <
i
^ 1 K 1
c o •
? •
■ J • o • I * E ■»
41 4| Mi 7
- _ Z - ~ &
< « < _ > o J; " o x
22
O a o Z
lis!
uuu;
< 4 4 2
22
m m m
°°s
< < 5
a «
z z o — E 7
o o<

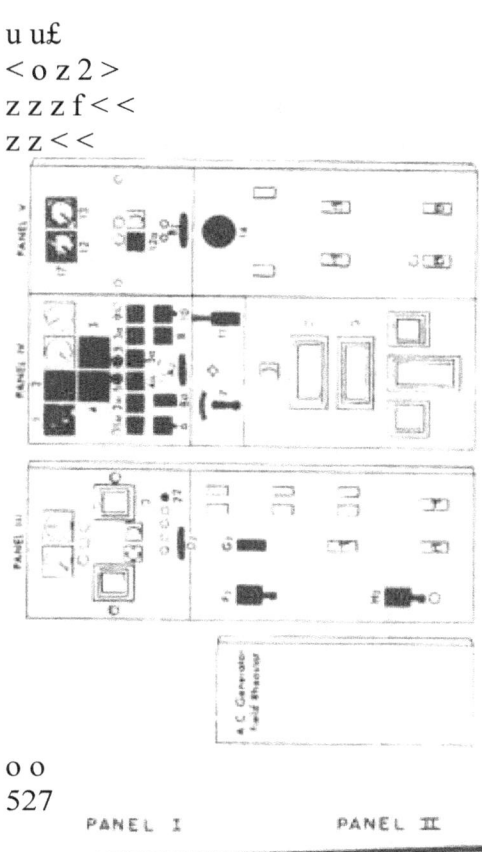

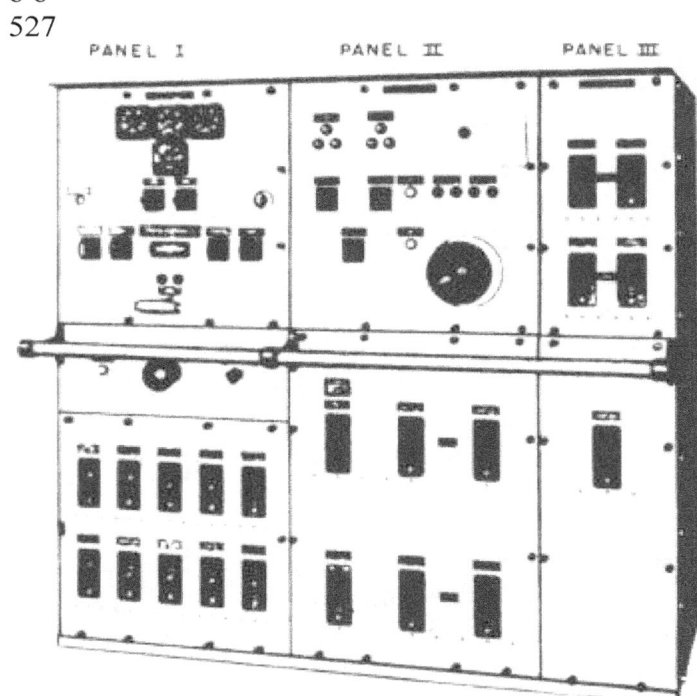

Figure 10-8A.—Forward emergency switchboard.

The after emergency switchgear group (fig. 10-8, B) controls the after emergency diesel generating unit; the 15-kva emergency lighting transformer bank; the after ship's service and emergency bus transfer equipment; the special emergency power bus feeder; the emergency power bus contactor; the local emergency power bus; the remote emergency power bus feeders;

and the emergency lighting feeders.

The generator panels of both the forward and after emergency switchboards are provided with an a-c voltmeter and a voltmeter transfer switch to permit reading the (three) phase-to-phase voltages in the a-c generators, or the voltage of the local emergency power bus on the forward emergency switchboard and the special emergency power bus on the after emergency switchboard.

PANEL I PANEL 31

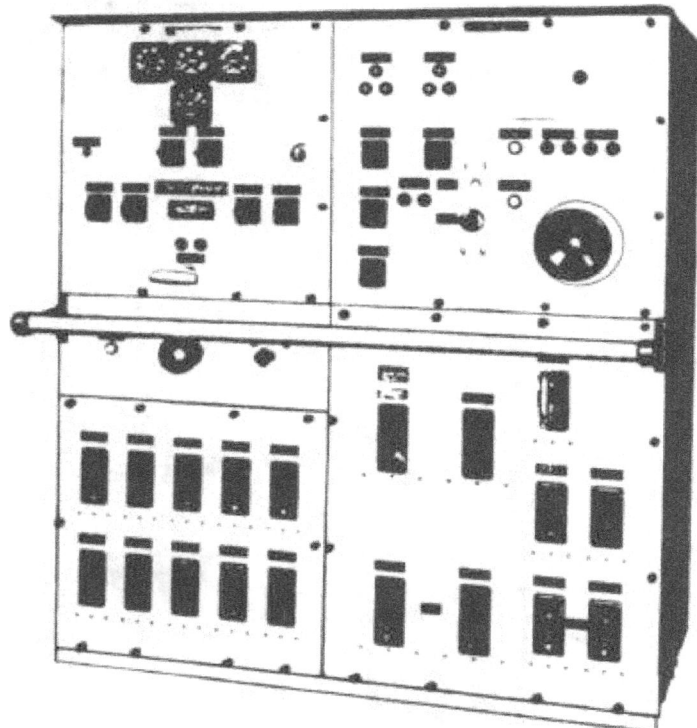

Figure 10-8B.-After emergency switchboard.

Both emergency switchboards are also provided with an a-c ammeter and an ammeter transfer switch to permit reading the current in the three phases of the emergency generator, an a-c wattmeter for reading the power of the emergency generator, and a frequency meter for reading the frequency of the local power bus on the forward emergency switchboard and the special emergency power bus on the after emergency switchboard.

INTERCONNECTIONS.-The connections between the ship's service and the emergency generating units and their associated switchboards and the interconnections between the switchboards are illustrated in the schematic

line diagram in figure 10-9. The a-c buses on the forward and after ship's service switchboards can be connected together, and the d-c buses on these switchboards can also be connected together. This arrangement enables one generating unit to supply power to both ship's service switchboards when the other unit is out of service,and also provides for parallel operation of the two ship's service generating units (1SG and 2SG). However, when operating SPLIT PLANT the generators are operated separately, each unit supplying power for its own section of the ship.

Each emergency generator (1EG and 2EG)is connected to the bottom studs of a generator circuit breaker. The top studs of the generator circuit breaker are connected to one side of the ship's service and emergency generator automatic bus transfer equipment (ABT). The other side of this bus transfer equipment is connected to the feeder from the forward ship's service switchboard (IS) for the forward emergency switchboard (IE) and to the feeder from the after ship's service switchboard (2S) for the after emergency switchboard (2E). The midpoint of the bus transfer equipment is connected to the 450-volt local bus on the forward emergency switchboard and to the 450-volt special bus on the after emergency switchboard. The special bus on the after emergency switchboard supplies power to the steering power transfer switchboard (not shown) through a type AQB circuit breaker and is connected to the 450-volt local bus through a bus contactor.

A 450-volt remote power bus is provided on each emergency switchboard (for example, IE) and is supplied from the local bus of the second emergency switchboard (2E) to allow for the transfer of emergency power from one emergency generator to the other emergency switchboard to supply certain auxiliaries. This tie feeder is controlled by a type AQB circuit breaker connected to the local bus of one emergency switchboard (2E) and by a type NQB circuit breaker connected to the remote bus of the other emergency switchboard (IE). Certain feeders, such as gun mounts, I. C. switchboards, and emergency lighting transformers, are connected by means of mechanically interlocked type AQB circuit breakers to either the local or remote emergency power buses. The

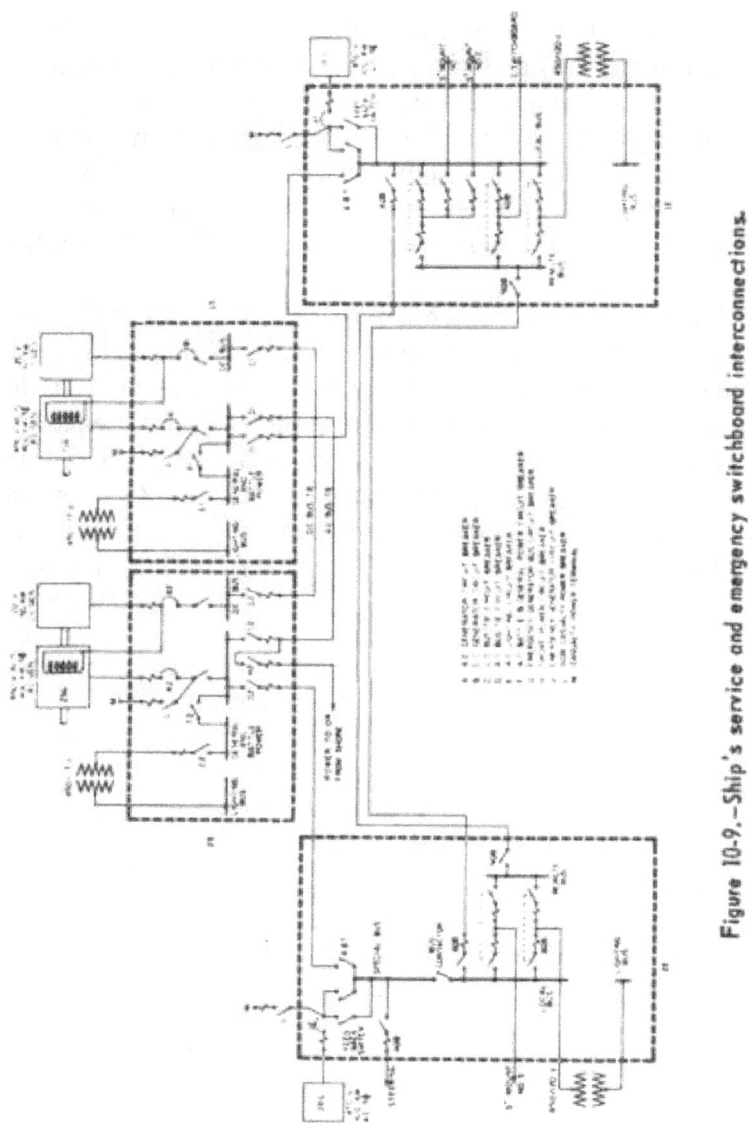

Figure 10-9.—Ship's service and emergency switchboard interconnections.

mechanical interlocks prevent both circuit breakers from being closed at the same time so that the emergency generators cannot be paralleled.

The type NQB feedback circuit breaker must NOT be closed if the white indicator light (not shown) is lighted to indicate potential on the feeder from the ship's service switchboard connected to the particular emergency switchboard. Otherwise, the ship's service generator could supply power to the emergency generator. Also, when an emergency generator (for example, 1EG) is used to supply limited power to the associated ship's service switchboard (IS) through the feedback circuit, the type ACB ship's service generator circuit breaker (A1) and the type ACB bus tie circuit breaker (D1) to the remote ship's service switchboard (2S) must NOT be closed. If the a-c generator circuit breaker is closed, power could be fed to the 1SG ship's service generator, which is not operating. If the a-c bus tie circuit breaker is closed, the emergency generators can be paralleled or power can be fed to circuits connected to the after ship's service switchboard.

A casualty power terminal and a type AQB casualty power circuit breaker are located on panel in of the forward emergency switchboard and on panel II of the after emergency switchboard. The terminals are used for connecting casualty power cables to supply power to the emergency switchboards. The casualty power circuit is connected to the local emergency power bus on the forward emergency switchboard and to the special emergency power bus on the after emergency switchboard.

A shore power circuit breaker, H2, is located on the after ship's service switchboard. A connection box is provided on deck so that the ship's service switchboards can be energized from the shore or from another ship. This connection can be used also to supply power from either ship's service switch to the shore or to other ships alongside.

BUS TRANSFER EQUIPMENT.-Power to the emergency switchboards can be supplied from the associated ship's service switchboard or from either emergency generator. An automatic bus transfer controller (ABT, fig. 10-9) is mounted on each emergency switchboard and is normally set to select power from one of the ship's

service switchboards for the emergency distribution section. If a loss of power occurs from the associated ship's service switchboard supply, the automatic bus transfer controller will start the diesel emergency generator, remove the ship's service supply from the local emergency power bus, and connect the emergency diesel generator to the local emergency power bus. The vital loads are shifted either automatically or manually at the loads, to the emergency supply. When the ship's service voltage is restored, the emergency generator is automatically disconnected, and the emergency switchboard is again energized from the ship's service switchboard.

The standard bus transfer controller (fig. 10-10) consists of (l)a motor-operated contactor unit and (2) a separate control panel designed for mounting in dead-front switchgear.

The CONTACTOR UNIT (fig. 10-10, A) consists essentially of two 3-pole, cam-operated contactors for main line connection; six auxiliary cam-operated contacts for control and indicating light circuits; two pilot motors for automatic operation; and a hand wheel for manual operation. The two sets of main contactors are operated from a cam shaft that is driven by the two pilot motors. When the cam shaft moves from one extreme position to the other, the cams open one contactor and close the other contactor by means of springs. Both contactors cannot be closed at the same time but both can be open. One pilot motor drives the shaft in one direction and the other motor drives it in the opposite direction. The auxiliary contacts are used to control the contactor-position indicator lights, and to deenergize either motor when the shaft reaches the limit of its travel. An automatic-manual transfer switch permits manual operation of the bus transfer controller. The control panel (fig. 10-10, B) consists of an insulating base on which are mounted two voltage-sensitive relays with associated rectifiers and transformers.

The setup for automatic operation of the bus transfer equipment located on the after emericy switchboard in a 692-classdestroyer is illustrated infigure 10-11. The voltage-sensitive relays, VN and VE, have two operating coils connected in series and four (three a and one b) contacts. When the relay is energized, the a contacts are

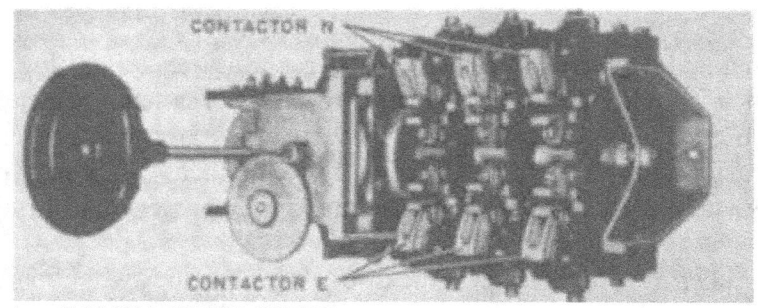

A - CONTACTOR UNIT

VOLTAGE-SENSITIVE RELAYS

B ~ CONTROL PANEL Figure 10-10.— Automatic bus transfer controller.

closed and the b contacts are open. Conversely, when the relay is deenergized, the a contacts are open and the b contacts are closed. Relay VN (fig. 10-11) is shown in the energized position and relay VE in the deenergized position. The relay is d-c operated by means of a 450/160-volt transformer and copper oxide rectifier. Relay VN is operated by transformer Tl and rectifier CR1 and relay VE is operated by transformer T2 and rectifier CR2. The VN relay, connected in the after ship's service supply, is adjusted to pick up at 395 volts and drop out at 290 volts. The VE relay, connected in the emergency supply, is adjusted to pick up at 420 volts.

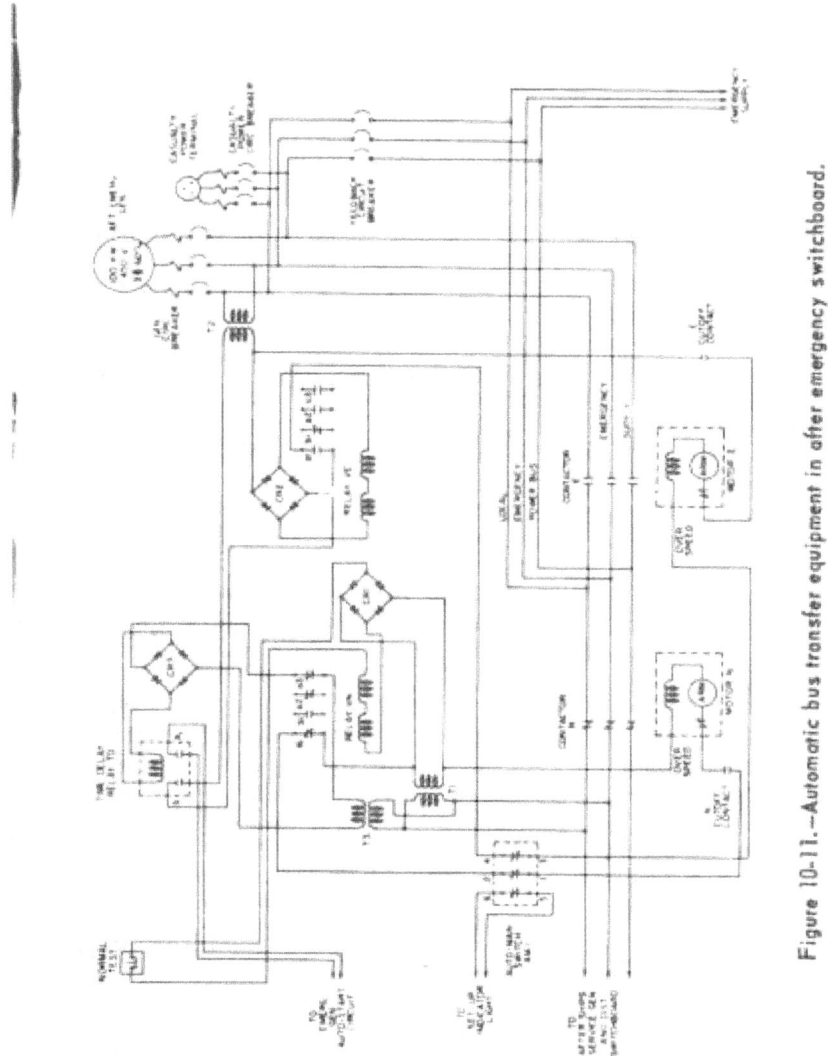

Figure 10-11.—Automatic bus transfer equipment in after emergency switchboard.

The automatic-manual transfer switch (AMI) has three contacts. When the transfer switch is in the AUTOMATIC position the contacts are closed, and when the switch is in the MANUAL position the contacts are open and prevent operation of the pilot motors. This switch is shown in the AUTOMATIC position. One pair of contacts (1-2) is in the coil circuit of motor N, one pair (3-4) in the circuit of motor E, and the other pair (5-6) is in the circuit of the transformer supplying the setup indicator light (not shown). This light indicates when the setup is properly made for automatic operation of the emergency diesel generator and the bus transfer equipment.

A time delay relay, TD, having a single coil and two pairs of contacts, b and bl, is connected in the circuit of the bus transfer equipment to delay the starting of the emergency diesel generator, and to delay the energizing of the rectifier for relay VE. The time delay on dropout is approximately one second (not adjustable). The relay is d-c operated by means of a 450/160-volt transformer (T3) and copper oxide rectifier (CR3).

Under normal conditions, with either or both ship's service turbine generators in operation, the two emergency diesel generators will not be running, and the emergency switchboards will be supplied with power from the forward and after ship's service switchboards through the ship's service and emergency bus transfer equipments located on the emergency switchboards. Hence, when the supply from the after generator switchboard is 290 volts or above, relay VN will be energized, contactor N will be closed, and the 450-volt local emergency power bus will be fed from the after ship's service switchboard. With contactor N normally closed, the pilot motor for contactor N will be deenergized as the cutoff contact for contactor N will be open. If the supply voltage should fall to 290 volts or below, relay VN will drop out and open its a3 contact, which opens the circuit to rectifier CR3 that supplies relay TD.

After a brief interval, relay TD closes its b contact to complete the circuit to rectifier CR2 for relay VE. At the same time contact bl closes to complete the starting circuit to the emergency diesel generator (not shown). The b contact of relay TD is also in the circuit to the pilot motor for contactor E. However, this motor is not

energized until contact al of relay VE closes. When the emergency generator voltage increases to 420 volts or above, contact al of relay VE closes to complete the circuit to the pilot motor for contactor E. Pilot motor E will rotate the cam to open contactor N and close contac -tor E. At the same time, auxiliary cutoff contact, N, closes. When contactor E closes, its cutoff contact opens the circuit to the pilot motor for contactor E. The 450-volt local emergency power bus on the after emergency switchboard is now supplied from the emergency generator.

If the voltage of the preferred supply should rise to 395 volts or above, relay VN will pick up and close itsal contact to complete the circuit to the pilot motor for contactor N. Motor N will rotate the cam to open contactor E and close contactor N. When contactor N closes, its cutoff contact opens the circuit to the pilot motor for contactor N. Because contact a3 is now closed, CR3 and relay TD will be energized. Contact b of relay TD will open and deenergize the circuit to rectifier CR2, thereby deenergizing relay VE and further opening the circuit to pilot motor E. However, the emergency diesel generator will continue to run until it is shut down manually.

SELECTIVE TRIPPING.—The overcurrent devices incorporated in the circuit breakers protect the circuits against extreme conditions of overcurrent. The overcurrent tripping devices of all circuit breakers (described in chapter 7 of this training course) are set to obtain selective tripping on all faults, within practical limits, to protect the distribution system and generators against large overcurrents. The purpose of selective tripping is to isolate the faulty section of the system and at the same time to maintain power on as much of the system as possible. Selective tripping of circuit breakers is secured by coordination of the time-current characteristics of the protective devices so that the breaker closest to the fault will open first and the breaker farthest from the fault and closest to the generator will open last.

A portion of a distribution system with circuit breakers employing selective tripping is illustrated in figure 10-12, A. The tripping-time characteristics are indicated in figure 10-12, B. The so-called instantaneous tripping time is the minimum time required for a breaker to open

and clear a circuit when the operation of the breaker is not intentionally delayed. Thus, each circuit breaker will trip in less than 0.1 second (almost instantaneously) when the current exceeds the instantaneous trip current setting of the breaker. The individual breakers in a group (generator breakers, bus tie breakers, feeder breakers, etc.) may differ slightly with each other because it is necessary to allow some tolerance in the adjustment, but all breakers in the group will have characteristics that fall within the band. The band is the relation between current and tripping time for a group of circuit breakers that have the same nominal setting.

Generator circuit breakers are usually set to trip instantaneously at a current that is GREATER than the short circuit current of the generator, and therefore will not trip instantaneously on the generator short circuit current. However, it will trip on the generator short circuit current at some definite interval of time within the tolerance of the breaker (points Pi to P).

Bus tie circuit breakers are usually set to trip almost instantaneously at a current that is equal to the short-time (instantaneous) rating of the breaker unless the total generator capacity that can be connected to the buses is insufficient to produce a current greater than the breaker short-time rating. If this condition exists, no instantaneous trip is used on the bus tie breakers.

The construction of circuit breakers for selective tripping for currents less than the instantaneous trip current setting is such as to cause an intentional delay in the operation of the breaker. The time delay is greater for small currents than for large currents and is therefore known as an inverse time delay. The current that would trip the AQB load circuit breaker instantaneously and clear the circuit will not trip the ACB feeder circuit breaker unless the current flows for a greater length of time. The same sequence of operation occurs for the other groups of circuit breakers adjusted for selective tripping in the system. The difference between the tripping times of the breakers is sufficient to permit each breaker to trip and clear the circuit before the next breaker starts to operate.

Assume that a fault or defect develops in the cable insulation at point A (fig. 10-12, A) and allows an

1SG I GENERATOR
2 SO GENERATOR
IS Ox SWITCHBOARD J'
TYPE ACB 6 \ FEEDER) BREAKERo '
TYPE ACB GENERATOR BREAKERS
■TYPE ACB BUS TIE BREAKERS

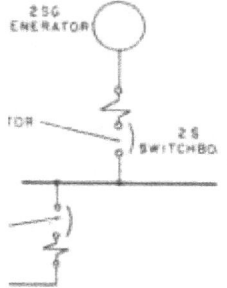

SWITCHBOARD
DISTRIBUTION PANEL
A-CIRCUIT

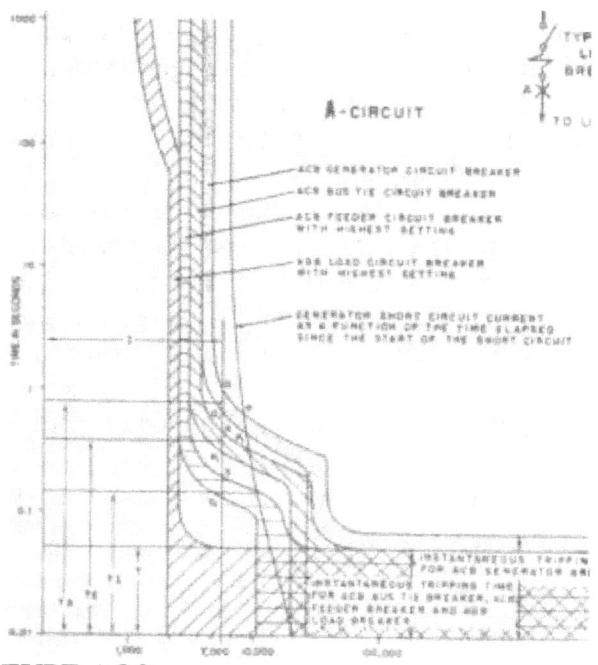

A-CIRCUIT

TYPE AO8
LOAD BREAKER
TO LOAD
ACS Of HCRATOR CIRCUIT BatAKCIt ACB BUS Tit CIRCUIT BRCAKC*
ACB FtfOC" CIRCUIT BRCAKCR WITH HISHCST BCTTIN8
AOB LOAD CIRCUIT BRCAKCR WITH HIBHCST BCTTINB
cincrator short circuit current
AS A FUNCTION Of TMt TI«»C ClARSIO SINCC TH€ START Of THE SNORT CIRCUIT
INSTANTANEOUS TRiP»NS T IMC 'OR ACB ((NCAATOR BRCAKCR
T.000 10.000 100,000 CURRENT IN AMPERES
B-TRIPPINO TIME CHARACTERISTIC
Figure 10-12.-Selective tripping of circuit breakers.

overcurrent, I (fig. 10-12, B) to flow through the AQBload circuit breaker and the ACB feeder circuit breaker. The AQB load breaker will open the circuit and interrupt the current in an interval of time, T, that is less than the time, Tl, required to open the ACB feeder circuit breaker. Thus, the ACB feeder breaker will remain closed when the AQB breaker clears the circuit. However, if the fault current should exceed the Interrupting capacity of the AQB load breaker (for example, an excess of 10,000amperes), this breaker would be unable to interrupt the fault current without damage to the breaker. To prevent damage to the AQB load breaker, the ACB feeder breaker (on switchboard IS) serves as a BACK UP breaker for the AQB load breaker and will open almost instantaneously in time, T.

A fault at point B with overcurrent I would trip the ACB feeder breaker in time Tl but not the ACB generator or bus tie breakers, which require longer time intervals, T3 and T2 respectfully, in which to trip. A fault at point C with overcurrent I would trip both ACB bus tiebreakers at time T2. A fault at D with overcurrent I on switchboard IS would trip the associated ACB generator breaker in time T3 and one or both of the ACB bus tie breakers in time T2. In each case, the faulty section of the system is isolated, but power is maintained on as much of the system as possible with respect to the location of the fault.

The attainment of selective tripping requires careful coordination of the time-current characteristics for the different groups of circuit breakers. For example, if the system illustrated in figure 10-12 is operating split plant (bus ties open) and if the time-current characteristics of the ACB feeder breaker and the ACB generator breaker were interchanged, a fault at B with overcurrent I would trip generator 1SG off the line in time Tl but would leave the feeder connected to the switchboard (the time required to trip the ACB feeder breaker would be T3). This action would disconnect power to all equipment supplied toy switchboard IS and also would not isolate the faulty sec -tion. Therefore, no unauthorized changes should be made to circuit breaker trip settings because these changes may completely disrupt the scheme of protection based on selective tripping. The adjustments for selective

tripping of the circuit breakers are made and sealed at the factory.

It is not feasible to provide system protection by selective tripping of circuit breakers in all types of naval vessels or for all circuits. For example, d-c distribution systems in older vessels and all lighting circuits use fuses to a great extent. Time delay can be incorporated only to the extent that is permitted by the characteristics of the fuses. The use of progressively larger fuse sizes from the load to the generator provide some degree of selectivity for overload or limited fault protection.

GROUND DETECTOR LAMPS.-A set of three ground detector lamps (fig. 10-13), connected (through transformers) to the main bus of each ship's service switch-gear group and to the emergency bus, enables the Electrician's Mate to check for grounds on any phase of the 3-phase bus.

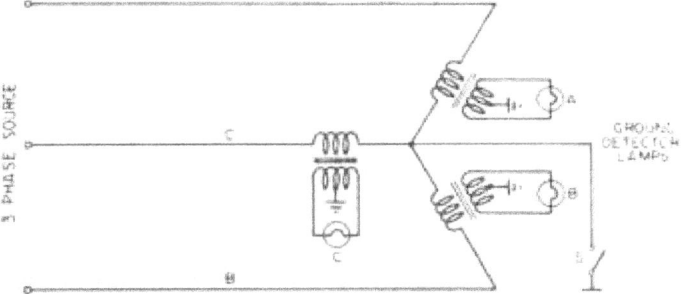

3 PHASE BUS

Figure 10-13.—A-c ground detector lamp circuit.

To check for a ground, turn switch S on and observe the brilliancy of the three lights. If the lights are equally bright, no grounds exist, and all lights receive the same voltage. If lamp A is dark and lamps B and C are bright, phase A is grounded. In this case, the primary of the transformer in phase A is shunted by ground and receives no voltage. Similarly, if lamp B is dark and lamps A and C are bright, a ground will exist on phase B. If lamp C is dark and lamps A and B are bright, a ground will exist on phase C.

D-C Switchboards

Although a-c power systems have replaced d-c ship's service systems in combat ships, d-c power is still widely used in auxiliary vessels having considerable deck machinery. The d-c power systems in auxiliary vessels are not ordinarily required to maintain the high degree of reliability required for combat ships. However, the degree of reliability is carried out within the limits of practicability and depends on the type and size of the vessel.

As previously explained, an outstanding feature of reliability in combat ships is the installation of two or more independently operated ship's service switchboards and associated generators located in different parts of the ship. On the other hand, auxiliary vessels are usually provided with one switchboard and associated generators located in the main engineroom or adjacent auxiliary machinery space.

The electric plant in an AP-type auxiliary vessel consists of three turbine-driven snip's service generating units located in the main engineroom. The ship's service d-c generators are300-kw, 120/240-volt, 3-wire units equipped with center-tapped autotransformers connected across slip rings.

The main generator and distribution switchboard is the dead-front type and consists of three combined generator and power-feeder panels, one feeder panel, one backup ACB panel, and two combined lighting and power feeder panels (fig. 10-14). The switchboard provides for single and parallel operation of the main generators and for control of the power and lighting feeders emanating from the back-up power and lighting buses.

The generators are controlled from panels I, n, and III and furnish 120/240-volt power to the ACB back-up circuit breakers Bl, B2, B3, and B4 on panels I and VI. Panels I and VI furnish 120/240-volt, 120-volt, and 240-volt power to panels IV, V, and VII and the lower portions of panels II and HI. All circuits are protected with manually operated circuit breakers. These circuit breakers are equipped with overcurrent devices that operate to open the respective breaker on short circuit or sustained overload. In addition to the overcurrent devices, each generator breaker is equipped with a direct-acting reverse

current device, an undervoltage device, and anoverspeed device to trip the breaker on current reversal, undervoltage, or overspeed, respectively.

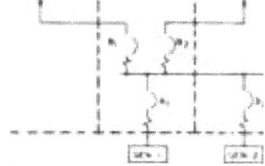

4-

o— r

V i _[

1. Generator field rheostat
2. Generator voltmeter
2a. Voltmeter •witch
3. Trip and hold-in do vie*
4. Ground dotector lamps
5. Negative gen. ammeter
6. Positive gen. ammeter
7. Trip button
8. Mechanical position indicator

LEGEND

A] Circuit breaker for gen. no. 1
A2 Circuit breaker for gen. no. 2
A3 Circuit breaker for gen. no. 3
Bj Backup breaker for feeder breakers on panel VII
B2 Backup breaker for feeder breakers on panel II
B3 Backup breaker for feeder breakers on panels III and VI
B4 Backup breaker for feeder breakers on panels IV and V
Panel I Gen. 1 control and 120/240-volt power

Panel II Gen. 2 control and
120/240-volt 240-volt power
Panel III Gen. 3 control and
120/240-volt 240-volt power
Panel IV 120/240-volt, 120-volt, and 240-volt power
Panel V 120/240-volt lighting and 120/240-volt and 240-volt power
Panel VI 240-volt backup panel
Panel VII 120/240-volt lighting and power

Figure 10-14.—D-c generator and distribution switchboard.

The backup circuit breakers should be closed at all times. If the feeder breaker and backup breaker open on a short circuit, the feeder breaker should be checked to determine that it is in satisfactory operating condition before closing the backup breaker.

The switchboard meters installed for observing the performance of d-c generators usually include only voltmeters and ammeters. The generator speed is controlled by the governor of the prime mover. Remote speed control is not required, and no governor motor control switch is used on the switchboard as with a-c generators. Manual voltage control is obtained by a rheostat in the shunt field. This rheostat is operated at the switchboard. D-c generators are usually not provided with voltage regulators. The inherent characteristics of the generators provide all the voltage regulation that is necessary in most applications. Disconnect links, located on the back of the switchboard are provided for each generator circuit to isolate the generator circuit breaker and associated devices for maintenance and repairs.

GROUND DETECTOR LAMPS.-The neutral of the main power system is grounded in this AP type of auxiliary vessel. The ground connection (fig. 10-15) is made through a ground contactor (normally closed) that is connected in series with a parallel combination comprising a current-limiting resistor and a ground circuit breaker (normally closed). The ground circuit breaker has an instantaneous trip device. If a ground occurs on either the positive or negative side of the system, the instantaneous trip device will open the ground circuit breaker and insert the resistor in the neutral ground circuit. The resistance is of sufficient magnitude to limit the ground current to one percent of the rated full-load current of the generator.

The purpose of the ground contactor is to disconnect the neutral ground when ground tests are being made (by means of the ground detector voltmeter or the ground lamps) to determine if any grounds exist on the system.

A ground lamp test pushbutton, PB, is provided for making ground tests with the ground detector lamps, A and B. When the pushbutton is held closed, the operating coil, G, of the ground contactor is energized and opens contacts G to remove the neutral ground. At the same

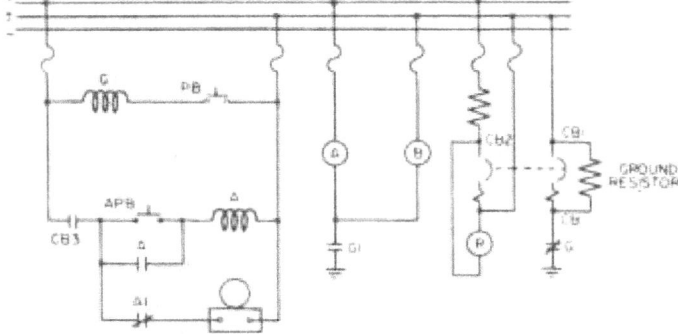

ALARM

Figure 10-15.-D-c ground detector lamp circuit.

time, contacts Gl close and place ground on one side of the ground lamp circuit.

An alarm system is provided to indicate operation of the ground circuit breaker, CB. When the ground circuit breaker is tripped by a ground fault, it opens contacts CBI and CB2 and closes contacts CB3. When contacts CBI open, the mechanical interlock opens contacts CB2, which removes a shunt from the ground alarm lamp, R, allowing it to light. When contacts CB3 close, the circuit to the alarm bell is energized.

The alarm system includes an alarm reset relay, A, and pushbutton APB. When the reset relay pushbutton is operated, contacts A close to maintain the operating coil A energized, and contacts Al open to deenergize the alarm bell. The alarm lamp, R, remains lighted at all times when the ground circuit breaker, CB, is open. The alarm bell does not sound when the ground circuit breaker is opened manually. For manual operation, the red alarm lamp, R, is the only indication that the ground circuit breaker is open.

Under normal operating conditions with no ground except through the grounding system (contacts G closed), the main bus is energized; the ground circuit breaker contacts, CBI andCB2are closed and contacts CB3 are open; the ground lamp test pushbutton is in the normal (open) position; and the ground lamps, A and B, are lighted dimly.

When the positive bus becomes grounded and the ground lamp test switch is operated, ground lamp A will be dark, or nearly dark, depending on the extent of the ground. Ground lamp B will be bright.

When the neutral bus becomes grounded and the ground lamp test switch is operated, ground lamp A will be dim, and ground lamp B will be dark, or nearly dark, depending on the extent of the ground.

When the negative bus becomes grounded and the ground lamp test switch is operated, both ground lamps A and B will be lighted brightly.

PREPARING GENERATORS FOR OPERATION

Both a-c and d-c ship's service generators should be inspected before starting an idle generator and also before applying load to a running generator to minimize the possibility of any machinery derangement that may be caused by negligent operating personnel.

Idle Generators

Before starting a d-c generator or an a-c generator at any time (except automatically started emergency power generators):

1. See that the generator circuit breaker and field switch (if furnished) are open.

2. See that the field rheostat handle is set to cut in all the field resistance.

3. Make sure that the voltage regulator control switch (if furnished) is turned to the MANUAL position.

4. Make sure that the manual voltage control is set at the position that gives the lowest generator voltage.

5. Examine both the prime mover and electrical ends for evidence of obstruction to moving parts.

6. See that the oil reservoir for oil ring lubricated bearings contains the proper amount of oil as indicated by the oil sight gage.

Running Generators

Before applying load to a d-c generator or an a-c generator at any time:

1. See that the unit is started and brought up to speed in accordance with the instructions for the type of prime mover used. Be sure that the overspeed governor and trip and all other safety devices are tested and are functioning as required by these instructions.

2. If vibration is apparent, search for possible misalignment, sprung shafting, loose foundation bolts, or something chafing the rotating elements.

3. Check lubrication of bearings; check oil and bearing temperatures.

4. See that commutator and collector rings run true, that all brushes ride freely in the brush holders, and that there is no chattering of brushes.

5. See that there is ample clearance between all rotating and stationary parts.

6. For totally enclosed machines with coolers, see that water is flowing through the cooler, as prescribed in the manufacturer's instruction books and that there is no leakage of water.

7. If the machine is equipped with heaters, see that the heaters are turned off.

8. Replace covers and all cover-holding bolts if any have been removed.

OPERATION OF GENERATORS

Standard operating procedures are established in accordance with good engineering practice to familiarize the assigned Electrician's Mates in the proper operation of a-c and d-c ship's service generators.

Nonparallel Operation

D-C GENERATOR.—To connect a single d-c generator and/or exciter to the bus for nonparallel operation:

1. Check before starting, start, and bring the generator up to speed in accordance with the procedures for starting an idle generator.

2. Close the shunt field switch and gradually cut out the field resistance until normal voltage is obtained.

3. Close the generator circuit breaker.

4. Close the line switch (if provided).

A-C GENERATOR.—To connect a single a-c generator to the bus for nonparallel operation:

1. Check before starting, start, and bring the generator up to speed in accordance with the procedures for starting an idle generator.

2. Adjust the generator voltage to the rated value by the manual voltage control. Observe the generator voltmeter while manually adjusting the voltage and do not allow it to build up above its rated value. Be certain that the voltmeter indicates the voltage of the generator being regulated. Exercise care when regulating the generator because the voltage may build up high enough to damage the windings.

3. Check the frequency on the frequency meter and raise or lower the speed of the prime mover, as necessary, by means of the governor motor control switch until the normal frequency of 60 cycles is indicated. Observe the frequency meter while adjusting the frequency and be certain that the meter indicates the frequency of the generator being adjusted.

4. Place the voltage regulator in control by turning the control switch to the NORMAL or AUTOMATIC position. Check the generator voltage and, if necessary, adjust it to the desired value by means of the voltage adjusting rheostat. Normally, the voltage regulator should be placed in, and then out of, service only under steady-load or no-load conditions.

5. Turn the voltmeter switch to read the voltage on the bus. Check the bus voltage for all three phases if the connections to the voltmeter switch will permit.

6. If the bus is deenergized, turn the handle of the synchronizing switch to the ON position. The synchroscope is not needed for synchronizing because the bus is dead, but in many installations an interlock necessitates that the synchronizing switch must be in the ON position

before the generator circuit breaker can be closed.

7. Close the generator circuit breaker to energize the bus and then apply load to the generator.

8. Turn the synchronizing switch to the OFF position.

SECURING.—To secure an a-c generator connected alone to the bus, hence, not operating in parallel with any other machine:

1. Reduce the load on the generator as much as practicable by opening feeder circuit breakers on the power and lighting circuits.

2. Trip the generator circuit breaker by pushing its trip button.

3. Turn the voltage regulator control switch to the MANUAL position.

4. Turn the manual voltage control as far as it will go in the DECREASE VOLTAGE direction.

To secure a single d-c generator and/or exciter connected along to the bus, hence, not operating in parallel with any other machine:

1. Reduce the loadby opening feeder circuit breakers.

2. Trip the generator circuit breaker.

3. Secure the prime mover in accordance with the prime mover instructions.

4. Cut in all the shunt field resistance.

5. Turn on heaters (if any) within the generator enclosure.

6. Secure water through air coolers, if any.

Parallel Operation

D-C GENERATORS.-If the bus is already energized and it is desired to connect another generator and/or exciter to the line:

1. Check before starting, start, and bring the generator up to speed in accordance with the procedures for starting an idle generator.

2. Close the shunt field switch (if installed).

3. Gradually cut out field resistance until the generator voltage is from 1 to 4 volts higher than the bus voltage.

4. Close the generator circuit breaker.

5. Close the line switch (if open).

6. Adjust the field rheostat to cause the generator to take its proper share of the load. Turning the field rheostat in the INCREASE VOLTAGE direction will cause a generator to take more load.

A-C GENERATORS.—Do not close the generator circuit breaker unless the incoming generator and the bus have approximately equal voltages and are in synchronism. To synchronize generators for parallel operation, proceed as follows:

1. Bring the generator up to approximately normal speed and voltage, and place the voltage regulator in control, as previously described for connecting a single generator to a dead bus.

2. Turn the voltmeter switch on the incoming generator to indicate the generator voltage on the voltmeter.

3. Turn the voltmeter switch on the adjacent generator panel to indicate the bus voltage.

4. Turn the voltage adjusting rheostat of the voltage regulator until the incoming generator voltage is equal to the bus voltage.

5. Compare the frequency of the incoming generator with that of the bus and adjust to correspond with the bus frequency.

6. Turn the synchroscope switch to the ON position. The synchroscope will rotate in one direction or the other. Adjust the speed of the generator by means of the governor motor control switch until the synchroscope rotates very slowly in the clockwise direction.

7. Be certain that the voltages of the bus and the incoming generator are still equal, and close the generator circuit breaker just before the synchroscope pointer passes very slowly through the zero position (pointing vertically upward).

8. When synchronizing lamps are used instead of the synchroscope, close the circuit breaker just before the midpoint of the dark period of the lamps is reached. The midpoint of the dark period corresponds to the vertical position of the synchroscope pointer.

9. Turn the synchronizing switch to the OFF position.

When a-c generators are operated in parallel, the kilowatt and reactive kva load should be divided between them in proportion to the generator ratings. The desired division of the kilowatt load is obtained by adjusting the governor, which controls the generator speed. To balance the reactive kva load, the generator line currents should be equal for equally rated generators and divided in propor -tion to generator ratings for unequally rated generators. Where power factor meters are provided, the power factors for all generators in parallel should be equal. Equality of power factor or correct division of generator line currents is obtained by adjusting the voltage-adjusting

rheostats of the voltage regulators. These adjustments are made as follows:

1. Turn the governor control switches until the wattmeters have equal readings (if the generators have the same ratings) or the kilowatt load is divided in proportion to the generator ratings (if the generator ratings differ from each other).

a. If the frequency is above normal, turn the governor control switches for the heavily loaded generators in the DECREASE SPEED direction.

b. If the frequency is below normal, turn the governor control switches for the lightly loaded generators in the INCREASE SPEED direction.

c. If the frequency is normal, make the adjustment in small steps, turning the governor control switches for the lightly loaded generators in the INCREASED SPEED direction and for the heavily loaded generators in the DECREASE SPEED direction.

2. Turn the voltage adjusting rheostats of the voltage regulators until generator line currents are divided in proportion to generator ratings or power factor readings are equal.

a. If the voltage is above normal, turn the voltage adjusting rheostats for the generator with the highest line current, (lowest power factor) in the DECREASE VOLT -AGE direction.

b. If the voltage is below normal, turn the voltage adjusting rheostats for the generator with the lowest line current (highest power factor) in the INCREASE VOLTAGE direction.

c. If the voltage is normal, make the adjustments in small steps, turning the voltage adjusting rheostats for the generators with the highest line current (low power factor) in the DECREASE VOLTAGE direction, and for the generators with the lowest line current (highest power factor) in the INCREASE VOLTAGE direction.

The reason for the procedure in step (b) above may be explained by considering two equally rated generators operating in parallel (fig. 10-16). Generator Gl (fig. 10-16, A) is in parallel with generator G2. The power factors are not equal; Gl has a power factor of 0.99, and G2 has a power factor of 0.67. Although both generators are assumed to have the same rating, the load current, Ij (fig.

10-16, B) of Gl Is smaller than the load current, I 2 , of G2. Both generators supply inductive loads. The voltage, E2\ is larger than E 2 by an amount equal to E r . Voltage E r causes an undesirable circulating component of current I s to flow between the two generators. Current I s lags E r almost 90° because of the large ratio of reactance to resistance in the local circuit between the two generators. Note that I s lags E 2 ' by 90° and leads E t by 90°.

0 0
A PARALLEL CONNECTION
It

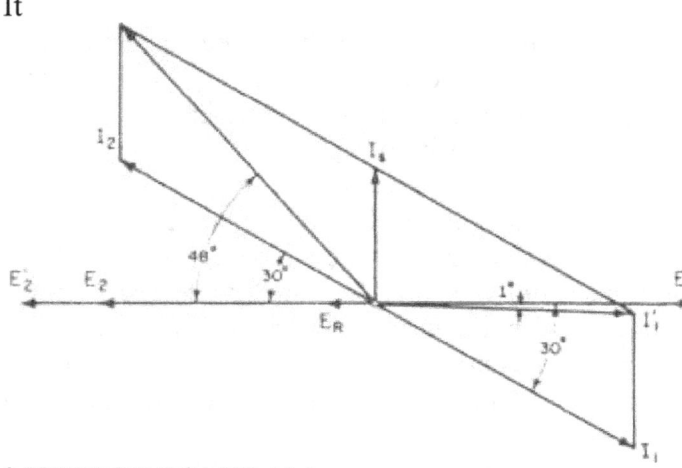

8 VECTOR DIAGRAM
Figure 10-16. — Equalizing the power factor between generators when the voltage is above normal.

The circulating current, I s , comprises a part of the total current of each generator. Thus, in Gl the total current, l x \ is made up of components I s and I I . Current

i! is the current of Gl when the generator voltages are equalized, and there is no unbalanced voltage, E r . Current I2' is made up of components I s and I 2 . Current I 2 is the current of G2 when E r is zero.

Equalization of the generated voltage may be obtained by lowering the field strength of G2 until E 2 is equal to Ei. For this condition I s is reduced to zero. The current of Gl is increased from If to Ii, and the current of G2 is decreased from I 2 ' to I 2 .

The power factor of G2 is increased from cos 48° = 0.67 to cos 30° = 0.866, and the power factor of Gl is decreased from cos 1° = 0.99 to cos 30° = 0.866. Thus, if the voltage is above normal and G2 has the lower power factor, turn the voltage adjusting rheostat of G2 in the DECREASE direction. This action weakens the field of G2 and lowers its generated voltage until the circulating current is reduced to zero and the power factors are equalized.

Note that the circulating current that exists when the power factors are unequal cannot be read on either generator ammeter. However, there will always be a circulating current when the power factors are unequal, and the ammeter readings are not the same (for equal kva ratings of both generators). This current is undesirable because it causes heating and limits the output of the generator having the higher generated voltage.

SECURING.—To secure an a-c generator that is operating in parallel with another generator or other generators;

1. Turn the governor motor control switch of the generator being secured in the DECREASE SPEED direction and the governor motor control switch (or switches) of the other generator (or generators) in the INCREASE SPEED direction until all the load is shifted from the generator being secured. This condition can be checked by the indications on the wattmeters. If the total connected load is greater than can be carried by the generators that are to continue in

operation, it will be necessary to decrease the load by opening feeder circuit breakers.

    2. Trip the circuit breaker of the generator being secured.

    3. Turn the voltage regulator control switch to the MANUAL position.

    4. Turn the manual voltage control as far as it will go in the DECREASE VOLTAGE direction.

    To secure a d-c generator that is operating in parallel with another generator, or other generators:

    1. Gradually cut in the shunt field resistance to reduce the load to a minimum.

    2. Trip the generator circuit breaker.

    3. Secure the prime mover in accordance with the instructions for the type of prime mover used.

    4. Cut in all the shunt field resistance.

    5. Turn on the heaters (if any) within the generator enclosure.

    6. Shut off the flow of water through the air coolers (if any).

Emergency Generator Operation

    Emergency diesel generators can be stopped only from within the space where the generator is located. Also, it is usually necessary to set up manually the lube oil alarm after the diesel starts. In some instances, emergency diesel generators have started because of momentary loss of ship's service voltage when emergency power was not required for the security of the ship, and when the emergency generator spaces were not manned. Hence, the emergency generators have run for some time with the consequent result that they have been damaged while operating unattended. Therefore, it is considered desirable to secure the automatic starting of the diesel generators when the immediate availability of emergency power is not required for the security of the ship and when a watch cannot be provided in the emergency diesel generator space.

    An example of the conditions under which the automatic starting of emergency diesel generators should be secured is when shifting from ship to shore power or vice versa. Ship's service and emergency generators must not be operated in parallel with each other or with shore power. Therefore, at the instant of transferring a ship's service switchboard from ship's service power to shore power or vice versa, the switchboard will be dead. As previously stated, emergency switchboards are always energized from the ship's service switchboards, and when loss of voltage occurs from the ship's supply to any emergency switchboard, the emergency generator associated with the switchboard will start automatically. Therefore, before making a transfer that will cause momentary loss of ship's service voltage on the emergency switchboard, the automatic diesel starting circuit should be made inoperative for the transfer period. After the transfer has been made, the switchboard should be set up for automatic starting if desired.

CONNECTING EMERGENCY GENERATOR TO MAIN SWITCHBOARD.—The emergency diesel generator provides emergency power and lighting automatically on starting, and, through the use of feedback switches and manual operation of the ABT switch, can be used to supply selected portions of the ship's load by feeding through the main ship's service switchboards. This additional load must be limited to vital andsemivital services within the capacity of the emergency diesel generators.

To place the emergency diesel generator on the main switchboard, take the following action at this switchboard (fig. 10-9):

1. Open the a-c generator circuit breaker, A.

2. Open the d-c generator circuit breaker, B.

3. Open the a-c bus tie circuit breaker, D.

4. Open the d-c bus tie circuit breaker, C.

5. Open the battle and general power circuit breaker, F.

6. Open the emergency generator bus AQB circuit breaker, G.

7. Strip (remove load from) the general power bus except for the vital engine room or fireroom auxiliaries and lighting. At general quarters, the gun mounts should be stripped to prevent all gun mounts from being thrown on one emergency generator and overloading it.

8. Notify the connecting emergency switchboard over the 2JV sound-powered telephone circuit to CLOSE the feedback switch to the main switchboard to which it is tied.

9. When the main switchboard receives power (indicating light on), CLOSE the emergency switchboard feeder and feedback circuit breaker, G, to supply power to the restricted and battle power bus.

10. Close the battle and general power bus circuit breaker, F.

11. Cut in only vital circuits and exercise care not to exceed the capacity of the emergency diesel generator. Check the emergency diesel load before cutting in each circuit.

12. If necessary, limit the enginerooms and firerooms to exhaust ventilation and secure the supply vents.

When the diesel engine starts, the following action is necessary at the emergency switchboard:

1. Turn the diesel starting and lockout switch on the emergency switchboard to the LOCKOUT position to prevent restarting.

2. Turn the automatic manual control switch for the automatic bus transfer on the emergency switchboard to the MANUAL position.

3. Close the feedback switch on the emergency switchboard. If the emergency switchboard is supplying power, trip emergency generator circuit breaker before manually operating the bus transfer equipment to prevent arcing of the contacts.

4. Manually turn the automatic bus transfer on the emergency switchboard to the normal N contacts.

5. Notify the main switchboard over the 2JV sound-powered telephone circuit that power is available.

DISCONNECTING EMERGENCY GENERATOR FROM MAIN SWITCHBOARD.-
To place the ship's service generator back on the main switchboard and set the plant in normal operation, take the following action at this switchboard:

1. Bring up the d-c and a-c generator voltages and set the voltage regulator and frequency the same as for parallel operation.

2. Trip the emergency switchboard feeder and feedback circuit breaker, G.

3. Close the a-c and d-c generator circuit breakers, A and B.

4. Check with the opposite main switchboard. If power is needed, it can be provided by closing the a-c and d-c bus tie circuit breakers.

5. To set up the electrical plant for split plant opera-f ion, be certain that the a-c and d-c bus tie circuit breakers
are OPEN on the forward switchboard. Close thea-c and d-c bus tie circuit breakers on the after switchboard.

6. Direct the emergency switchboard to standby to receive power.

7. When the word is received that the feedback switch is open and the emergency switchboard is ready for power, close the emergency feeder and feedback circuit breaker, G.

8. Direct the emergency switchboard to set up the emergency diesel and generator for automatic operation.

At the emergency switchboard proceed as follows:

1. When the word is received at the emergency switchboard to standby to receive power, manually rotate the automatic bus transfer back to the emergency E contacts.

2. Open the feedback switch.

3. Turn the automatic manual control switch for the automatic bus transfer to the AUTOMATIC position.

4. Report to the main switchboard that the feedback switch is open, ready for power.

5. After power and the word are received to set up the emergency diesel generator for automatic operation, slowly decrease the speed (by means of the hand throttle) of the diesel engine and stop it.

6. Return the hand throttle to the RUN position.

7. Turn the diesel starting and lockout switch back to the NORMAL position, and any additional switches required to set up the diesel for automatic operation.

8. Report to the main switchboard that the emergency diesel generator is set up for automatic operation.

OPERATION OF ELECTRIC PLANTS

The ship's electric power and lighting systems are designed to provide a high degree of flexibility to ensure continuity of service to vital power and lighting loads under normal and casualty conditions. The distribution systems in most naval vessels are arranged so that the electric plants can be operated in parallel (cross plant) or separately (split plant).

Cross Plant Operation

The setup for cross plant operation is to CLOSE all the bus ties between the ship's service switchboards with
the generators running in parallel so that any switchboard or several switchboards can supply electric power to any other switchboard. However, when the plant is operating CROSS PLANT, a casualty to one switchboard or load center may cause a short circuit that could trip all the generators off the line and result in temporary loss of all ship's service power.

Split Plant Operation

The setup for split plant ope ration is to OPEN the bus ties between the ship's service switchboards so that each switchboard with its generators and loads forms a system that is independent of the others. When the plant is operating SPLIT PLANT, a casualty to one switchboard will result in loss of power for the loads fed from this switchboard but will not affect the loads fed from the other switchboards. Hence, split plant ope ration should be used under battle or other conditions where maximum assurance against loss of all ship's service power is desired.

If auxiliaries are provided with normal and alternate power supplies, the feeder circuit breakers are CLOSED for both the normal and alternate supplies. Thus, if there is a casualty to one generator plant, power will be immediately available at the manual transfer switch for this vital equipment by means of the alternate power supply from the other generator plant. Battle loads that are not provided with alternate feeders can be supplied over the bus tie feeder if a generator casualty occurs after isolating the damaged equipment.

The circuit breakers are closed on the bus tie feeders between the ship's service switchboards and the emergency switchboards to permit the utilization of the emergency system as an additional means of distributing power from the ship's service generators. If a loss of ship's service power occurs on these bus tie feeders, automatic starting of the emergency generators provides emergency power for the vital loads.

Hence, during war cruising the normal setup is to operate each generating plant separately, each one feeding its associated switchboard or switchboards. If there is more than one forward and one after switchboard, the bus ties between the associated switchboards are CLOSED,

but the bus ties between switchboards associated with more than one generator must be OPEN. This setup provides independent generating plants, each receiving steam from an associated group of boilers, and one or more emergency diesel generators standing by to take over the emergency load. The diesel emergency generators are set for automatic starting.

Operation Under Casualty Conditions

If a switchboard is fed by two or more generators and if some of the generators are lost, split plant operation can be continued by using the remaining generators to supply power for some of the loads fed from the switchboard, and by shifting other loads normally fed from the switchboard over the alternate feeders that connect to other switchboards.

If all the generating capacity for a switchboard is lost, the bus tie circuit breakers can be closed to energize the switchboard from one of the other switchboards. The principal use of the bus tie connection is obtained when all generating capacity for one switchboard is lost when operating under split plant conditions. The bus tie is always in use when one switchboard is supplying power for the entire ship.

If only a part of the generating capacity for a switchboard is lost, it is good practice to continue split plant operation instead of using the bus tie to parallel the remaining generators with another switchboard. When operating split plant and part of the generators for a switchboard are lost, the remaining generators can be used to supply power to a part of the loads that normally receive power from the switchboard. Other loads can be shifted over to an alternate feeder to decrease the load on the switchboard and those loads on its generators that are still in service.

If a 692-class destroyer (fig. 10-9) is operating split plant (a-c and d-c bus tie circuit breakers OPEN on the forward switchboard and a-c and d-c bus tie circuit breakers CLOSED on the after switchboard) and if one of the ship's service generators loses its load:

1. Open the a-c generator circuit breaker of the affected generator.

2. Open the d-c generator circuit breaker of the affected generator.

3. If the casualty is in the forward generator, close the a-c and the d-c bus tie circuit breakers.

4. If the casualty is in the after generator, notify the forward switchboard over the 2JV sound-powered telephone circuit to close the a-c and d-c bus tie circuit breakers.

5. Turn the voltage regulator control switch to the MANUAL or OFF position.

6. Decrease the a-c generator voltage by means of the rheostat motor control switch. If the d-c generator voltage decreases to less than 95 volts, discontinue use of the rheostat motor control switch and reduce the voltage by means of the MANUAL control.

7. Decrease the d-c generator voltage by means of the d-c generator field rheostat.

8. Open the field switch.

9. If the generator is out of commission for an indefinite period, OPEN the generator disconnect switches located in the rear of the switchboard. If the generator is out of commission temporarily, the generator disconnect switch can remain CLOSED in order to speed up placing the generator back in commission.

RIGGING CASUALTY POWER CABLES.-When rigging casualty power cables:

1. Ensure that power is not available at the damaged panel or switchboard. The engineer officer will designate the switchboard to be used as the source of supply.

2. Ensure that all supplies are tagged, "opened." The Electrician's Mate making the connections must be provided with a voltage tester, rubber gloves, and rubber boots.

3. Ensure that no short circuits exist in the panel or equipment. If the supply cables are damaged and no switch is available, disconnect the leads.

4. Lay out the portable casualty power cables ready for making the connections.

5. Connect all horizontal cables starting at the riser or bulkhead terminal at the casualty (load) and work toward the riser or bulkhead terminal entering the space that is to provide power (supply).

6. Test, then connect the damaged equipment to the riser or bulkhead terminal leaving the space.

7. Never use a riser terminal for a connection block unless the other end of the riser is to be used to supply some piece of equipment. A portable switch should be connected in the line near the casualty to deenergize the circuit in the event of an emergency, or for reversing the leads in case of reverse-phase rotation.

8. Notify damage control central immediately when all the cables have been connected (including the horizontal connections) to the panel or equipment to be supplied to the riser leading to the space that is designated as the power supply. The damage control assistant requests the bridge to pass the word, "stand clear of casualty power cables rigged on main deck from frame 82 to frame 168 starboard side." When the word has been passed, the damage control assistant notifies main engine control, "connect and energize casualty power to riser or bulkhead terminal 2-82-1 in the forward engine room."

9. Ensure that the casualty power circuit breaker is OPEN and that the casualty power terminal is deener-gized. Test and connect the casualty power cable to the designated riser or bulkhead terminal. Test and connect the other end of the casualty power cable to the switchboard casualty power terminal.

10. Close the casualty power circuit breaker, test, and report to main engine control, "casualty power riser or bulkhead terminal 2-82-1 is energized."

11. Ensure that the rotation is in the correct direction. If not, reverse the direction of rotation by deenergizing the circuit at the portable switch or at the switchboard and by reversing any two of the three leads.

UNRIGGING CASUALTY POWER CABLES.-The damage control assistant notifies main engine control, "de-energize and disconnect casualty power from the riser or bulkhead terminal 2-82-1." The switchboard electrician opens the casualty power circuit breaker, tests, and disconnects the casualty power cable from the switchboard terminal. When unrigging casualty power cables:

1. Test and disconnect the cable from the riser or bulkhead terminal leaving the space.

2. Report to the main engine control, "casualty power deenergized and disconnected from riser or bulkhead

terminal 2-82-1." The main engine control notifies damage control central. When the word is received at damage control central that the casualty power has been de-energized and disconnected, the word is passed to the repair party to unrig the casualty power cables.

3. In unrigging casualty power cables, test each riser or bulkhead terminal before removing the cable.

4. Start first by disconnecting the cables from the switchboard (supply) and the casualty (load) last.

5. Keep the leads separated between the fingers and when all three connections are broken, remove the cable. When the casualty has been restored and the cables unrigged, the word is relayed to damage control central, and the damage control assistant will inform the engineer officer to energize the appropriate supplies.

6. Notify the bridge when the casualty has been restored and tested.

OPERATING PROCEDURES

Electrician's Mates should study the electrical system installed in their ship so that they know the physical locations of the generators, switchboards, distribution panels, and cables, and thoroughly understand the functions and relations of the various components of the system. They must observe the procedures of system operation to obtain optimum performance and maximum reliability of the electric installation. The operating procedures include the (1) general rules and (2) safety precautions.

General Rules

In the operation of any shipboard electrical installation the assigned Electrician's Mates should watch the switchboard instruments because they show how the system is operating. The instruments reveal overloads and the improper division of the kilowatt load or reactive current between generators operating in parallel and other abnormal operating conditions.

The frequency (on a-c systems) and the voltage must be maintained at the correct values to obtain satisfactory operation of all equipment supplied with electric power.

Low voltage results in a marked decrease in illumination; whereas, high voltage materially shortens the life of electric lamps. The operation of vital electronic, interior communications, and fire-control equipment is also seriously affected. Therefore, it is necessary to carefully adjust the voltage regulators and the prime mover governors to obtain satisfactory performance of this equipment.

Operating personnel must realize that no automatically operated devices are installed to protect the distribution system or the generators from damage caused by an overcurrentor power overload that is only moderately in excess of the rated capacity. The switchboard electrician must be vigilant in observing the ammeter and wattmeter readings to detect the presence of a moderate over-current or power overload, which, of long continued, would cause excessive heating of the generators.

If a switchboard controls two or more generators and less than the full number is being used to supply power, the load on the switchboard may increase to a point that will overload the generators in use. When the switchboard instruments reveal this condition, another generator should be placed in service.

Emergency switchboards are connected by feeders to loads which at some time may need emergency power. The capacity of the emergency generators is not sufficient to provide power for the simultaneous operation of all loads that can be connected to the emergency switchboard. Hence, if the ship's service power is lost, an indiscriminate use of emergency power can easily overload an emergency generator and possibly stall its diesel engine. The only way for the switchboard electrician to reduce the load on the emergency generator is to trip the circuit breakers on some of the feeders. To utilize the emergency power most effectively and to ensure that it will be available where needed the most, it is important for the engineering force to establish an operating procedure for the emergency switchboard so that the switchboard electrician will know which loads should have preference, and also know the order of preference for the additional loads that can be carried if one of the preferred loads is lost due to derangement of equipment.

Because of the limited capacity of the emergency generators, the feedback tie installed in some ships is not to be generally utilized to supply power to the ship's service distribution system. The use of the feedback tie should be limited to special circumstances such as when the ship is alongside a dock and it is necessary to secure the ship's service generators, or when it is necessary to feed power through the ship's service switchboards to certain auxiliaries to start the ship's service generators. When the feedback tie is used, increased vigilance of the switchboard electrician is necessary to prevent overloading the emergency generators.

Always operate switchboards and distribution system equipment as if no automatic protective devices are installed. These devices are not designed or intended to protect the system from damage caused by careless operating practices, but to afford protection against damage caused by equipment failure.

Exercise care when reclosing circuit breakers after they have tripped automatically. If a circuit breaker trips immediately on the first reclosure, investigate the cause before again reclosing the breaker. However, the circuit breaker may be closed a second time without investigation if the immediate restoration of power to the circuit is important and if the interrupting disturbance that tripped the breaker was not excessive. Remember that repeated closing and tripping may result in damage to the circuit breaker and thereby increase the repair or replacement work required to place the breaker back in operation.

Use the hold-in device on circuit breakers only when it is absolutely necessary. The hold-in device enables the operator to hold a trip-free circuit breaker closed when the current is in excess of the tripping value. The circuit breaker will open automatically as soon as the hold-in device is released if the current is above the tripping current. In an emergency, it may be vitally important to obtain power even at the risk of burning out equipment by using the hold-in device. However, when a circuit breaker is held closed, the circuit is deprived of protection against damage by excessive current, and the longer the circuit breaker is held closed, the greater is the possibility of permanent damage to circuits or

equipment. Do not hold a circuit breaker closed unless an emergency exists to justify the risk.

Do not parallel ship's service a-c generators until they have been synchronized. Do not close the bus tie circuit breakers to parallel the buses on two energized switchboards until the buses have been synchronized. Do not close the bus tie circuit breaker to restore power to a switchboard that has lost power because of failure of its associated generator until the generator circuit breaker has been tripped manually, or it has been determined that the generator circuit breaker is already in the OPEN position. Do not parallel ship's service generators with shore power. Do not parallel an emergency generator or a casualty power generator with any other generator.

Always check the phase sequence before making any connection to a shore power supply and be certain to make the connections so that the phase sequence in the ship will be ABC. If the shore power connection is made so that it gives the wrong phase sequence in the ship, the a-c motors will run in the wrong direction.

Do not adjust a ventilation opening (for the personal comfort of watchstanders) toaposition that permits spray or solid water (entering the system through weather openings) from the ventilating system to be discharged on switchboards, distribution panels, busbars, or other electrical equipment.

Safety Precautions

Protective grab rods and guard rails around switchboards and other equipment should always be in position when the equipment is energized unless emergency repairs are necessary while the equipment is in service. Grab rods and guard rails should be carefully maintained to ensure that they are secure and will not become dislodged accidentally. The insulating mattings provided for covering the deck in the front and in the rear of switchboards should always be in place.

When performing maintenance work on a circuit, be careful to ensure that the circuit is dead and that it cannot be inadvertently energized by closing a remote circuit breaker. All circuit breakers or switches that would

energize the circuit should be opened, and the circuit should be tested with a voltmeter or voltage tester. These switches should be tagged, "WARNING: Do not change position of switch except by direction of NAME Rate/Rank." Warning tag NavShips 3950 (5-51) maybe used for this purpose.

If more than one repair party is engaged in repair work on an electrical circuit, a tag for each party should be placed on the supply switches. After the work has been completed, each party should remove its own tag but no other. As a further precaution, metal locking devices are available that can be attached to the switch handles to prevent accidental operation.

When checking to ascertain if circuits aredeenergized, check the metering and control circuits in addition to the power circuits because these circuits are often connected to the supply side of a circuit breaker. A check of the power circuits on the load side of a circuit breaker may indicate that these circuits are dead after the circuit breaker is opened, but the associated metering and control circuits may or may not be dead, and should be checked also to be certain they are dead.

MAINTENANCE

A routine maintenance program should be established to ensure maximum performance with minimum interruption of switchboards and control equipment. It is impracticable to set up a rigid schedule of tests and inspections that will be equally applicable to every ship because the service, age, and the condition of equipment differ for each ship. However, the engineering force should set up a schedule based on past experience and the following suggested schedule. A frequent recurrence of trouble indicates that the interval between tests and inspections should be decreased.

EVERY WATCH, use the ground detector voltmeter or ground detector lamps to test for grounds.

EACH WEEK, test the circuit breakers; test the bus transfer equipment; test the control circuits; and test the emergency switchboards.

AFTER FIRING, if practicable, inspect the switchboards and distribution panels.

EACH YEAR AND AFTER EACH OVERHAUL, inspect and clean the switchboards and distribution panels; and check overload relays.

EVERY FOUR YEARS, check the overload relays for tripping time.

Switchboards

Numerous derangements of electrical equipment have been caused by loose electrical connections or mechanical fastenings. Loose connections can be readily tightened, but a thorough inspection is necessary to detect them.

INSPECTION.—At least once a year and during each overhaul, each switchboard propulsion control cubicle, distribution panel, and motor controller should be de-energized for a complete inspection and cleaning of all bus work equipment. The inspection of deenergized equipment should not be limited to visual examination but should include grasping and shaking electrical connections and mechanical parts to be certain that all connections are tight and that mechanical parts are free to function. Be certain that no loose tools or other extraneous articles are left in or around switchboards and distribution panels.

Check the supports of bus work and be certain that the supports will prevent contact between bus bars of opposite polarity or contact between bus bars and grounded parts during periods of shock. Clean the bus work and the creepage surfaces of insulating materials, and be certain that creepage distances (across which leakage currents can flow) are ample. Check the condition of control wiring and replace if necessary.

Be certain that the ventilation of rheostats and resistors is not obstructed. Replace broken or burned out resistors. Temporary repairs can be made by bridging burned out sections when replacements are not available. Apply a light coat of petrolatum to the face plate contacts of rheostats to reduce friction and wear. Be certain that no petrolatum is left in the spaces between the contact buttons as this may cause burning and arcing. Check all electrical connections for tightness and wiring for frayed or broken leads. Service commutators and brushes of potentiometer-type rheostats in accordance with the instructions for d-c machines.

The pointer of each switchboard instrument should read zero when the instrument is disconnected from the circuit. The pointer may be brought to zero by external screwdriver adjustment. Caution: This should not be done unless proper authorization is given. Repairs to the switchboard instruments should be made only by the manufacturers, shore repair activities, or tenders.

Be certain that fuses are the right size; clips make firm contact with the fuses; lock-in devices (if provided) are properly fitted; and that all connections in the wiring to the fuses are tight.

In addition to the foregoing inspections, switchboards and distribution panels should be deenergized after firing, if practicable, and thoroughly inspected for tightness of electrical connections and mechanical fastenings.

CLEANING.—Bus bars and insulating materials can usually be cleaned sufficiently by wiping with a dry cloth. A vacuum cleaner, if available, can also be used to advantage. Be certain that the switchboard or distribution panel is completely dead and will remain so until the work is completed; avoid cleaning live parts because of the danger to personnel and equipment.

Use a dry cloth, not soap and water, on the front panels of live-front switchboards or on other panels of insulating material. The front panels of dead-front switchboards can be cleaned without deenergizing the switchboard. These panels can usually be cleaned by wiping with a dry cloth. However, a damp, soapy cloth can be used to remove grease and finger prints. Then wipe the surface with a cloth dampened in clear water to remove all soap and dry with a clean, dry cloth. Cleaning cloths must be wrung out thoroughly so that no water runs down the panel. Clean a small section at a time and then wipe dry.

Control Equipment

CIRCUIT BREAKERS.-Circuit breakers should be carefully inspected and cleaned at least once a year (more frequently if subjected to unusually severe service conditions). The oil should be changed in oil-film type over-current tripping devices every six months. A special inspection should be made of the contacts after a circuit breaker has opened a heavy short circuit. Before working

on a circuit breaker, deenergize all control circuits to which it is connected. Before work is performed on draw-out circuit breakers, they should be switched to the open position and removed. Before working on fixed-mounted circuit breakers, open the disconnecting switches ahead of the breakers. If disconnecting switches are not provided for isolating fixed-mounted circuit breakers, de-energize the supply bus to the circuit breaker, if practicable, before inspecting, adjusting, replacing parts, or doing any work on the circuit breaker.

Contacts in circuit breakers, contactors, relays, and other switching equipment should be clean and bright, free from severe pitting or burning, and properly aligned. Remove surface dirt, dust, or grease with a clean cloth moistened, if required, with a small amount of methyl chloroform. Be certain that ample ventilation is provided if methyl chloroform is used. Remove all traces of residue left by the methyl chloroform.

When cleaning and dressing contacts, maintain the original shape of the contact surface and remove as little material as possible. Inspect copper contact surfaces for black, copper-oxide film and clean with fine sandpaper (No. 00), if required. Severely burned or pitted copper contact surfaces should be dressed with a fine file or fine sandpaper.

Slight discoloration of solid silver or silver alloy contacts is normal and should not be removed. Remove heavy, black deposits and any high spots caused by burning with very fine sandpaper or with a fine file. Opening and closing laminated, silver-plated contacts will aid in cleaning. If necessary, use a cloth moistened with methyl chloroform to remove dirt and grease but do not use a file or sandpaper.

The function of arcing contacts is not necessarily impaired by surface roughness. Remove excessively rough spots with a fine file. Replace arcing contacts when they have been burned severely and cannot be properly adjusted. Make a contact impression and check the spring pressure in accordance with the manufacturers' instructions. If information on the correct contact pressure is not available, the contact pressure should be checked with that of similar contacts. When the force is less than the designed value, the contacts either require replacing

because they are worn down, or the contact springs should be replaced. Always replace contacts in sets; not singly, and replace contact screws at the same time. Do not use emery paper or emery cloth to clean contacts, and do not clean contacts when the equipment is energized.

Clean all surfaces of the circuit breaker mechanism, particularly the insulation surfaces, with a dry cloth or air hose. Before directing the air on the breaker, be certain that the water is blown out of the hose, that the air is dry, and that the pressure is not over 30 psi. Check the pins, bearings, latches, and all contact and mechanism springs for excessive wear or corrosion and evidence of overheating. Replace parts if necessary.

Slowly open and close circuit breakers manually a few times to be certain that trip shafts, toggle linkages, latches, and all other mechanical parts operate freely and without binding. Be certain that the arcing contacts make before and break after the main contacts. If poor alignment, sluggishness, or other abnormal conditions are noted, adjust in accordance with the manufacturers' instructions for the particular circuit breaker.

Oil-piston type overcurrent tripping devices (grade B timers) are sealed mechanisms and normally do not require any attention. When oil-film (dashpot) overcurrent tripping devices are used, the oil should be removed and the interior of the oil chambers should be cleaned with kerosene and refilled with new oil every six months. Be certain that the dashpot is free of dirt, which destroyes the time-delay effect, and that the tripping device is clean, operates freely, and has sufficient travel to trip the breaker. Do not change the air-gap setting of the moving armature because this would alter the calibration of the tripping device. Lubricate the bearing points and bearing surfaces (including latches) with a drop or two of light machine oil. Wipe off any excess oil.

Before returning a circuit breaker to service, inspect all mechanical and electrical connections, including mounting bolts and screws, draw-out disconnect devices, and control wiring. Tighten where necessary. Give the breaker a final cleaning with a cloth or compressed air. Operate manually to be certain that all moving parts function freely. Check the insulation resistance.

The sealing surfaces of circuit-breaker contactor and relay magnets should be kept clean and free from rust. Rust on the sealing surfaces decreases the contact force and may result in overheating of the contact tips. Loud humming or chattering will frequently warn of this condition. A light machine oil wiped sparingly on the sealing surfaces of the contactor magnet will aid in preventing rust.

Oil should always be used sparingly on circuit breakers contactors, motor controllers, relays, and other control equipment, and should not be used at all unless stated in the manufacturers' instructions or unless oil holes are provided. If working surfaces or bearings show signs of rust, disassemble the device and carefully clean the rusted surfaces. Light oil can be wiped on sparingly to prevent further rusting. Oil has a tendency to accumulate dust and grit, which may cause unsatisfactory operation of the device, particularly if the device is delicately balanced.

Arc chutes or boxes should be cleaned by scraping with a file if wiping with a cloth is not sufficient. Replace or provide new linings when they are broken or burned two deeply. Be certain that arc chutes are securely fastened and that there is sufficient clearance to ensure that no interference occurs when the switch or contactor is opened or closed.

Shunts and flexible connectors, which are flexed by the motion of moving parts, should be replaced when worn, broken, or frayed.

Operating tests that consist of operating the circuit breakers in the manner in which they are intended to function in service should be conducted regularly. For manually operated circuit breakers, simply open and close the breaker to check the mechanical operation. To check both the mechanical operation and the control wiring, electrically operated circuit breakers should be tested by means of the operating switch or control. Exercise care not to disrupt any electric power supply that is vital to the operation of the ship, or to endanger personnel by inadvertently starting motors and energizing equipment under repair.

BUS TRANSFER EQUIPMENT. -For manual bus trans -fer equipment, manually transfer a load from one power

source to another and check the mechanical operation and mechanical interlocks. For automatic bus transfer equipment, check the operation by means of the control push-switches. The test should include operation initiated by cutting off power (opening a feeder circuit breaker) to ascertain if an automatic transfer occurs. The precautions for circuit breaker operating tests should be observed when testing bus transfer equipment.

OVERLOAD RELAYS.—During periodic inspections of motor controllers, or at least once a year, the overload relays should be examined to ascertain that they are in good mechanical condition and that there are no loose or missing parts. The size of the overload heaters should be checked to determine that they are of the proper size indicated by the motor nameplate current and heater rating table. Proper allowance should be made for short-time and intermittent duty motors by using undersized coils. Any questionable relays should be checked for proper tripping at the next availability and repaired if necessary. Each relay should be checked for tripping time at 150, 300, and 600 percent rated current by a naval shipyard at intervals not exceeding four years.

CONTROL CIRCUITS.—Control circuits should be checked to ensure circuit continuity and proper relay and contactor operation. Because of the numerous types of control circuits installed in naval vessels, it is impracticable to set up any definite operating test procedures in this training course. In general, certain control circuits, such as those for the starting of motors or motor-generator sets, or voltmeter switching circuits, are best tested by using the circuits as they are intended to operate under service conditions.

Protective circuits, such as overcurrent, reverse power, or reverse current circuits, usually cannot be tested by actual operation because of the danger involved to the equipment. These circuits should be visually checked, and, when possible, relays should be operated manually to be certain that the rest of the protective circuit performs its intended functions. Exercise extreme care not to disrupt vital power service or to damage electrical equipment.

EMERGENCY SWITCHBOARDS.-Emergency switchboards should be tested regularly in accordance with the

instructions furnished with the switchboard in order to check the operation of the automatic bus transfer equipment and the automatic starting of the emergency generator. This test should be made in connection with the weekly operating test of emergency generators.

QUIZ.

1. What two principal plants are included in the electric plant aboard ship?

2. What types of prime movers are used to drive the ship's service generators?

3. What types of prime movers are used to drive the emergency generators?

4. What is the requirement regarding the capability for parallel operation of (a) ship's service generators and (b) emergency generators?

5. The majority of a-c power distribution systems in naval vessels have what voltage, frequency, phase, and number of wires in the main feeders?

6. Why are the switchboards and associated generators located in separate engineering spaces ?

7. In large installations (fig. 10-1) power distribution to loads other than large and important loads is from the generator and distribution switchboards or switchgear groups to what points before going to the load itself?

8. In small installations (fig. 10-2) the distribution panels are fed from what units?

9. The feeders in the general power circuits are designated by what letter?

10. The feeders in the battle power circuits are designated by what letters ?

11. What is the smallest number of independent sources of power provided for selected vital loads ?

12. What is installed at load centers, distribution panels, or loads that are fed by both normal and alternate and/or emergency feeders for selectionof the associated power supply?

13. How many phases are used in (a) a-c lighting distribution system feeders, mains, and submains and (b) branch circuits ?

14. What is the correct phase sequence in naval vessels ?

15. What facility is provided in most ships between the emergency switchboard and the ship's service switchboard that permits a selected portion of the ship's service switchboard load to be supplied from the emergency generator?

16. What distribution system is provided for making temporary connections to supply electric power to certain vital auxiliaries if the permanently installed ship's service and emergency distribution systems are damaged ?

17. What are the five principal components of the casualty power system?

18. When rigging casualty power cables, why should the connections be made from the load to the supply?

19. If both the normal and alternate sources of the ship's power fail because of a generator, switchboard, or feeder casualty, how can the vital auxiliaries be fed?

20. If the ship's service and emergency circuits fail, how can power be supplied to vital auxiliaries?

21. In ships with a-c power systems, d-c power is furnished in what three possible ways?

22. How is the operator of a portable tool protected in case of a casualty to the insulation?

23. What are three general types of d-c power systems used in naval vessels?

24. Positive, neutral, and negative polarities of bus bars and terminals of the 120/240-volt, 3-wire d-c system are indicated by what color coding?

25. What is present Navy practice with regard to grounding the neutral or either leg of the 120/240-volt, 3-wire d-c system?

26. (a) What does the electric plant consist of in a 692-class destroyer? (b) What is the rating of the a-c generators? (c) What is the rating of the oversize exciters directly connected to the ship's service generators ?

27. What device is employed that trips the generator breaker and takes the generator off the line when power is fed from the line to the generator instead of from the generator to the line?

28. What is the purpose of the mechanical interlocks on type AQB circuit breakers that are used to connect feeders supplying vital loads to either the local or remote emergency power buses?

29- If a loss of power occurs from the associated ship's service switchboard supply (fig. 10-9), what component will start the diesel emergency generator, remove the ship's service supply from the local emergency power bus, and connect the emergency diesel generator to the local emergency power bus?

30. The mechanical interlock on the automatic bus trans-fer switch imposes what restrictions on the opening and closing of the two sets of main contactors? Refer to figure 10-11 for questions 31 and 32.

3 1. If the (preferred) supply voltage falls to 290 volts or below, what will be the effect on (a) relay VN, (b) contact a3, (c) the circuit to rectifier CR3, (d) the b contact of relay TD, (e) the circuit to rectifier CR2, (f) relay VE, and (g) the bl contact of relay TD?

32. When the emergency generator has started and its voltage has increased to 420 volts or above, what will be the effect on (a) contact al of relay VE, (b) the circuit to the pilot motor for contactor E, (c) contactor N, (d) contactor E, (e) auxiliary cutoff contact N, (f) the cutoff contact of contactor E, and the circuit to the pilot motor for contactor E?

33. The overcurrent tripping devices of all circuit breakers are set to obtain what system of tripping on all faults within practical limits ?

34. What is the relation between the breaker location with respect to the fault and the operating time required for the breaker to open?

35. What is the relative magnitude of the current for which generator circuit breakers are set to trip open (expressed in terms of generator short circuit current) ?

36. What is the inverse time delay feature employed in selective tripping?

37. Assume that a fault or defect develops in the cable insulation at point A (fig. 10-12, A) and allows an overcurrent, I (fig. 10-12,B) to flow through the AQB load circuit breaker and the ACB feeder circuit breaker, (a) Which breaker will open first? (b) If the fault cur rent should exceed 10,000 amperes, what action will prevent damage to the AQB load breakers ?

38. Assume that a fault develops at point B (fig. 10-12, A) with overcurrent, I (fig. 10-12, B) flowing through the ACB feeder circuit breaker, the ACB generator circuit breaker and the ACB bus tie circuit breaker, (a) Which breaker opens first? (b) A fault at point C with overcurrent I would trip both ACB bus tie breakers in what time interval? (c) A fault at point D with overcurrent I on switchboard IS will trip the associated ACB generator circuit breaker in what time interval? (d) If the system is operating split plant (bus ties open) and time-cur rent characteristic s of the ACB feeder breaker and the ACB generator circuit breaker are interchanged, a fault at B with overcurrent I will trip generator 1SG off the line in what time interval? (e) Would this action isolate the fault ?

39. In the ground detector circuit of figure 10-13, what is the condition of the three-phase bus with regard to the existence of grounds if switch S is closed and (a) all three lamps are equally bright ? (b) lamp A is dark and lamps B and C are bright?

40. If a feeder breaker and backup breaker open on a short (fig. 10-14), what action should be taken regarding the feeder breaker before closing the backup breaker?

41. What components are provided on the back of the switchboard that isolate the generator circuit breaker and associated devices for maintenance and repairs ?

Refer to figure 10-15 for questions 42 through 45.

42. If a ground occurs on either the positive or the negative side of the system, what action will automatically occur to limit the ground current to one percent of the rated full-load current of the generator?

43. What indication is provided when the ground circuit breaker is tripped (a) by a ground fault and (b) manually?

44. Under normal operation with no ground except through the grounding system (the main bus energized, the ground circuit breaker closed, and the ground test switch open), what is the condition of the ground test lamps ?

45. If the positive bus becomes grounded and the ground lamp test switch is operated, what will be the condition of the ground test lamps?

46. Before starting a d-c generator or an a-c generator at any time, what conditions should

be determined with respect to (a) the generator circuit breaker and field switch, (b) the field rheostat, and (c) the oil reservoir for oil-ring lubricated bearings?

47. After a d-c generator or an a-c generator is brought up to speed and before applying load, what conditions should be determined with respect to the commutator, collector rings, and brushes?

48. After a d-c generator is brought up to speed and before connecting it to the bus for nonparallel operation, what action should be taken to obtain normal voltage ?

49. When adjusting the frequency of the a-c generator in preparation for nonparallel operation prior to loading, what precaution must be taken?

50. To secure an a-c generator connected alone to the bus, hence not operating in parallel with any other machine, how is the load reduced on the generator?

51. When operating d-c generators in parallel, what will be the effect of turning the field rheostat of one of the generators in the INCREASE VOLTAGE direction, on the loading of that generator?

52. When a-c generators are operating in parallel, the kilowatt load should be divided in what way with respect to their kilowatt rating?

53. What components on the prime movers are adjusted to obtain the desired load division between a-c generators operating in parallel?

54. What components are adjusted to obtain equality of power factor between generators operating in parallel ?

55. In what direction should the voltage adjusting rheostat of generator G2 (fig. 10-16) be turned to equalize the power factors?

56. Why is it considered desirable to secure the automatic starting of the diesel generators when the immediate availability of emergency power is not required for the security of the ship and when a watch cannot be provided in the emergency diesel generate- r space ?

57. What restriction is imposedonthe operation of ship's service and emergency generators with regard to paralleling them with each other and with shore power ?

58. The distribution systems in most naval vessels are arranged so that the electric plants can be operated in what two ways ?

59. Should the bus ties between the ship's service switchboards be opened or closed for (a) cross plant operation and (b) split plant operation?

60. Why should split plant operation be used under battle conditions ?

61. If all the generating capacity for a switchboard is lost while under cruising conditions (split plant), what action should betaken with regard to the bus tie circuit breakers in order to restore power to the switchboard ?

62. Why should switchboards and distribution system equipment be operated as if no automatic protective devices are installed?

63. How often should circuit breakers be inspected and cleaned ?

64. Why should the sealing surfaces of circuit-breaker contactor and relay magnets be kept clean and free from rust?

CHAPTER

ELECTRIC AUXILIARIES

Electrician's Mates 3 and 2 are required toman electrical equipment in the steering gear room, anchor windlass room, at the degaussing switchboard, and at electric elevator and hoist stations. They are also required to service electric galley equipment and appliances by taking resistance and current measurements, cleaning switches and relays, replacing defective heating units, thermostats, and control or unit wiring.

The purpose of this chapter is to explain the operating principles of these electric auxiliaries and to describe the proper methods and procedures for maintaining this equipment.

DEGAUSSING INSTALLATIONS

The detonation of a magnetic mine or torpedo depends on a localized disturbance or distortion of the earth's magnetic field. This distortion is caused by a ship passing over or near the mine or torpedo. The purpose of the degaussing installation is to reduce the magnitude of the distortion caused by the ship, and thus minimize the danger of firing the magnetic mine or torpedo.

A ship is a magnet because of the presence of magnetic material in its hull, machinery, and other equipment. Therefore it is surrounded by a magnetic field, which is large near the ship and small at a considerable distance from the ship. The earth's magnetic field is uniform at any particular point on the surface of the

earth. When a ship passes over a point on the earth's surface, the ship's magnetic field is superimposed on the earth's magnetic field, thus tending to distort the earth's field around the ship. If the ship is close to a magnetic mine or torpedo, the distortion caused by the ship's field will activate the firing mechanism to detonate the mine or torpedo.

Degaussing equipment is installed aboard ship to neutralize the disturbance of the earth's magnetic field caused by the ship and thus to reduce the possibility of detonating a magnetic mine or torpedo. A shipboard degaussing installation consists of one or more coils of electric cable in specific locations inside the ship's hull, a d-c power source to energize these coils, and a means of controlling the magnitude and polarity of the current through the coils. Compass-compensating equipment, consisting of compensating coils and control boxes, is also installed as a part of the degaussing system to compensate for the deviation effect of the degaussing coils on the ship's magnetic compasses.

Naval vessels are tested periodically at magnetic range stations to determine the configuration (signature) of the ship's magnetic field. The magnetic range station measures and records the ship's magnetic field. A ship is ranged by passing over measuring coils located at or near the bottom of the channel in which the ship travels. Sensitive measuring and recording equipment respond to signals induced in the coils when the ship passes over them. These measurements are used to determine the values of the degaussing coil currents, which give a minimum magnetic lield below the ship.

Earth's Magnetic Field

The earth's magnetic field is illustrated in figure 11-1. The magnetic field of the earth can be represented by considering the earth as a sphere with the imaginary lines of force leaving the south magnetic pole and entering the north magnetic pole (fig. 11-1, A). The direction of the field at any point is the direction of the line of force through the point. The field is strong where the lines of force are close together (near the magnetic poles) and weak where they are far apart (away from the magnetic

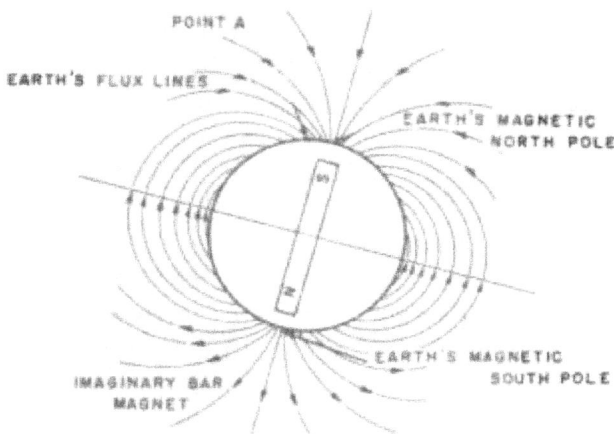

A EARTH AS A MAGNET

Figure 11-1.-Earth's magnetic field.

poles). The lines of force in the earth's magnetic field extend vertically outward at the south magnetic pole, vertically inward at the north magnetic pole, horizontally midway between the magnetic poles, and at varying angles at all other points.

The unit of magnetic flux density is the gauss, and the unit of magnetic field intensity is the oersted. The relation between the gauss and oersted is expressed by the equation

B = fx H

where B is the magnetic flux density in gausses, H the magnetic field intensity in oersteds, and the permeability of the surrounding media. For all practical purposes with respect to degaussing, the permeability p, is taken as 1, and the magnetic field strength (intensity) H, is measured in terms of the gauss (rather than oersteds), which is incorrect but does not lead to serious misunderstanding in degaussing terminology. One milli-gauss is one-thousandth of a gauss. At any point on the earth's surface the earth's magnetic field can be resolved into a horizontal component and a vertical component.

HORIZONTAL COMPONENT.-The horizontal component (also designated by the symbol, H) of the earth's magnetic field is always directed from the magnetic south to the magnetic north, is zero at the magnetic poles, and is maximum at the magnetic equator.

VERTICAL COMPONENT.-The vertical component (Z) of the earth's magnetic field is directed radially outward (upward) south of the magnetic equator, is directed radially inward (downward) north of the magnetic equator, is zero at the magnetic equator, and is maximum at the magnetic poles.

The angle at which the flux lines strike the earth's surface (horizontal) depends on the magnetic latitude. Hence, for a given total field strength, the magnitudes of the horizontal and vertical components depend on the angle that the field makes with the horizontal. For example, the horizontal and vertical components of the earth's magnetic field at point A (fig. 11-1, B) are H = R cos 6 and Z = R sin 6 respectively, where H is the horizontal component, Z the vertical component, R the total earth's field, and 9 the angle at which the flux lines strike the earth's surface.

Ship's Magnetic Field

The magnetic field of a ship is the resultant of the algebraic sum of the ship's permanent magnetization and the ship's induced magnetization. The ship's magnetic field may have any angle with respect to the horizontal axis of the ship, and any magnitude.

PERMANENT MAGNETIZATION.-The process of building a ship in the presence of the earth's magnetic field develops a certain amount of permanent magnetism in the ship. The magnitude of the permanent magnetization depends on the earth's magnetic field at the place where the ship was built, the material of which the ship is constructed, and the orientation of the ship at time of building with respect to the earth's field. This magnetization after construction is independent of the varying or immediate earth's magnetic field, the heading, and the roll and pitch of the ship. The permanent magnetism of a ship may vary over long periods of time.

The ship's permanent magnetization can be resolved into the (1) vertical permanent field component and (2) horizontal permanent field component. The horizontal permanent field component comprises the longitudinal permanent field component and athwartship permanent field component. The vertical, longitudinal, and athwartship permanent field components are constant, except for slow changes with time, and are not affected by changes in heading or magnetic latitude.

All ships that are to be fitted with a shipboard degaussing installation and some ships that do not require degaussing installations are depermed. Deperming is essentially a large scale version of demagnetizing a watch. The purpose is to reduce permanent magnetization and bring all ships of the same class into a standard condition so that the permanent magnetization, which remains after deperming, is approximately the same for all ships of the class.

INDUCED MAGNETIZATION.-After a ship is built, its very existence in the earth's magnetic field causes a certain amount of magnetism to be induced in it. The ship's induced magnetization depends on the strength of the earth's magnetic field and on the heading of the ship with respect to the inducing (earth's) field.

The ship's induced magnetization, similar to the ship's permanent magnetization, can be resolved into the (1) vertical induced field component and (2) horizontal induced field component. The horizontal induced field component also comprises the longitudinal induced field component and athwartship induced field component.

The magnitude of the vertical induced magnetization depends on the magnetic latitude. The vertical induced magnetization is maximum at the magnetic poles and zero at the magnetic equator. The vertical induced magnetization is directed down when the ship is north of the magnetic equator and up when the ship is south of the magnetic equator. Hence, the vertical induced magnetization changes with magnetic latitude, and to some extent, when the ship rolls or pitches. The vertical induced magnetization does not change with heading because a change of heading does not change the orientation of the ship with respect to the vertical component of the earth's magnetic field.

The LONGITUDINAL INDUCED MAGNETIZATION changes when either the magnetic latitude or the heading changes, and when the ship pitches. If a ship is headed magnetic north, the horizontal component of the earth's magnetic field induces a north pole in the bow and a south pole in the stern (fig. 11-2). In other words, the horizontal component of the earth's field induces a longitudinal or fore-and-aft component of magnetization. The stronger the horizontal component of the earth's magnetic field, the greater will be the longitudinal component of

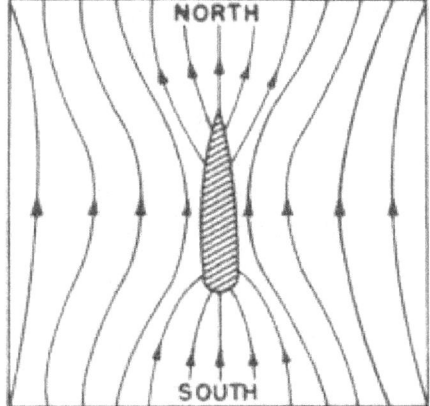

Figure 11-2.-Horizontal component of earth's field induces longitudinal magnetization in the ship.

magnetization. If the ship starts at the south magnetic pole and steams north, the longitudinal component of induced magnetization starts at zero at the south magnetic pole, increases to a maximum at the magnetic equator, and decreases to zero at the north magnetic pole. Hence, for a constant heading the longitudinal component of induced magnetization changes when the ship moves to a position where the horizontal component of the earth's magnetic field is different—that is, when the ship changes its magnetic latitude.

If at a given magnetic latitude the ship changes heading from north to east, the longitudinal component of induced magnetization changes from a maximum on the north heading to zero on the east heading. When the ship changes heading from east to south, the longitudinal component increases from zero on the east heading to a maximum on the south heading. On southerly headings a north pole is induced at the stern and a south pole at the bow, which is a reversal of the conditions on northerly headings when a north pole is induced at the bow and a south pole at the stern. The horizontal component of induced magnetization also changes, to some extent, as the ship pitches.

The ATHWARTSHIP INDUCED MAGNETIZATION changes when either the magnetic latitude or the heading changes, and when the ship rolls. When a ship is on an east heading, a north pole is induced on the port side and a south pole on the starboard side (fig. 11-3), which is the athwartship component of induced magnetization. The magnitude of the athwartship magnetization depends on the strength of the horizontal component of the earth's magnetic latitude. This component is maximum at the magnetic equator and zero at the magnetic poles.

The athwartship component of induced magnetization also changeswhen the ship's heading changes. The magnitude of the athwartship component of induced magnetization is maximum when the ship is headed magnetic east or west, and zero when the ship is headed magnetic north or south. The athwartship component of induced magnetization also changes, to some extent, as the ship rolls.

WEST

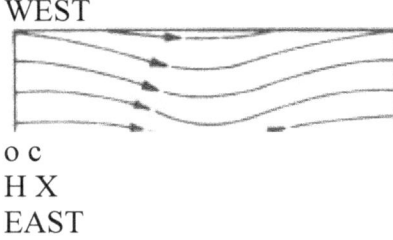

o c
H X
EAST

Figure ll-3.-Athwortship component of earth's field induces athwartship magnetization in the ship.

Degaussing Coils

As stated previously, the ship's permanent magnetization can be resolved into the (1) vertical permanent magnetization, (2) longitudinal permanent magnetization, and (3) athwartship permanent magnetization components. Similarly, the ship's induced magnetization can be resolved into the (1) vertical induced magnetization, (2) longitudinal induced magnetization, and (3) athwartship induced magnetization components. Each of the six components of the ship's magnetization produces a magnetic field in the vicinity of the ship.

Thedistortion of the earth's field caused by the superimposing of the six components of the ship's field is neutralized by means of degaussing coils. The degaussing coils are made with either single-conductor or multi-conductor cables. Coils with single-conductor cables have only one turn or a small number of turns. Single-conductor coils or a combination of some single-conductor and some multiconductor coils are usually specified for large vessels. Coils with multiconductor cables have the conductors in one or more cables connected in series to give a considerable number of turns. Multiconductor coils are usually specified for the smaller ships. The

586

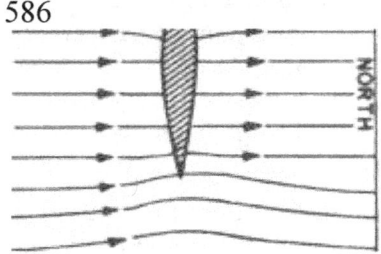

coils must be energized by direct current, which is supplied from 120-volt or 240-volt, d-c ship's service generators or from degaussing power supply equipment installed for the specific purpose of energizing the degaussing coils.

The degaussing coils consist of coils of cable wound on the ship, each having the required location and the number of turns to establish the required magnetic field strength when energized by direct current of the proper value and polarity. The coils will then produce magnetic field components equal and opposite to the components of the ship's field. Each coil consists of the main loop and may have smaller loops within the area covered by the main loop, usually at the same level. The smaller loops oppose localized peaks that occur in the ship's magnetic field within the area covered by the main loop. The resultant effect of the degaussing coils when the currents are properly set is to restore the earth's field to the undistorted condition around the ship.

The various components of the ship's magnetic field are compensated by means of standardized arrangements and combinations of degaussing coils. A degaussing installation consists of one or more of these degaussing coils, depending on the degree of safety required for the particular ship (fig. 11-4).

M COIL.—The M or main coil (fig. 11-4, A) encircles the ship in a horizontal plane, which is usually at the waterline. The function of the M coil is to produce a magnetic field, which counteracts the magnetic field produced by the vertical permanent and the vertical induced magnetization of the ship. If the M coil field were everywhere exactly equal and opposite the field produced by the vertical magnetization, the resultant of the two magnetic fields everywhere would be equal to zero. In practice, it is not possible to attain a perfect match, but the resultant is much less than the field produced by the vertical magnetization of the ship.

The permanent vertical magnetization of a ship is constant, but the vertical induced magnetization varies with the magnetic latitude and with the roll and pitch of the ship, but not with the heading. The resultant of the permanent and induced vertical magnetization will also vary with the magnetic latitude and with the roll and

"M" FIELD

1

~ ~ M COIL

A M COIL

Q-COIL OR QI AND OP COILS

F-COIL. OR FI AND FP COILS

B F AND Q COILS

••L" FIELD

L COIL LOOPS

C L COIL

•L" FIELD

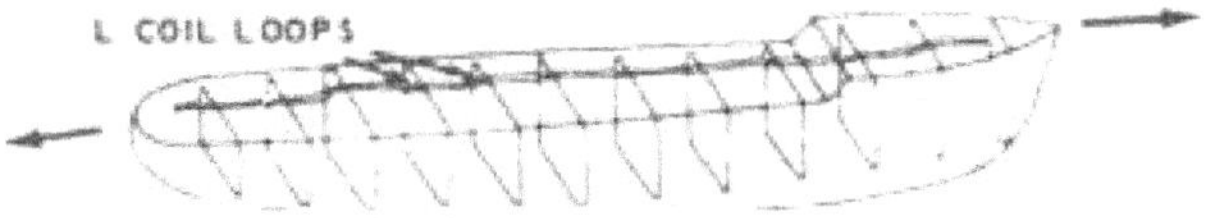

L COIL LOOPS

•A" COIL LOOPS

" COIL LOOPS   "A" FIELD

0 A COIL

Figure 11-4.-Types of degaussing coils.

pitch of the ship. Hence, the M coil strength must be changed when the ship changes magnetic latitude in order to keep the M coil field as nearly as possible equal and opposite the field produced by the ship's vertical magnetization. The change in vertical magnetization caused by roll and pitch of the ship is relatively small, and only

in special cases is it necessary to compensate for this change.

F AND Q COILS.—The F or forecastle coil encircles the forward one-fourth to one-third of the ship and is usually just below the forecastle or other uppermost deck; whereas, the Q or quarterdeck coil encircles the after one-fourth to one-third of the ship and is usually just below the quarterdeck or other uppermost deck (fig. 11-4, B). The function of the F and Q coils is to counteract the magnetic field produced by the ship's longitudinal permanent and longitudinal induced magnetization. The resultant F and Q coil strengths are then equal to the algebraic sums of their permanent and induced current requirements. The shape of the magnetic field produced by the F and Q coils is somewhat different from the magnetic fieldproduced by the ship's longitudinal magnetism, but the two fields are, in general, oppositely directed below the bow and stem of the ship. The F and Q coils are not used when it is required to compensate for the pitch of the ship.

The ship's longitudinal permanent magnetization is constant but the longitudinal induced magnetization changes with the heading and magnetic latitude. Therefore, the F and Q coil strengths must be changed when the ship changes its course or magnetic latitude in order to keep the coil strengths at the proper values to counteract the changes in the ship's longitudinal induced magnetization. Two adjustments are necessary, one to change the F coil strength and one to change the Q coil strength.

FI-QI AND FP-QP COILS.-In many installations the conductors of the F and Q coils are connected to form two separate circuits designated as the FI-Q coil and the FP-QP coil (fig. 11-4, B). The FI-QI coil consists of an FI coil connected in series with a QI coil so that the same current flows in both coils. The FP-QP coil is similar.

The FI-QI coil is used to counteract the magnetic field produced by the ship's longitudinal induced magnetization. The longitudinal induced magnetization changes when the ship changes heading or magnetic latitude, and the strength of the FI-QI coil must be changed accordingly.

The FP-QP coil is used to counteract the magnetic field produced by the ship's longitudinal permanent magnetization. The longitudinal permanent magnetization does not change when the ship changes heading or magnetic latitude, and no change is needed in the strength of the FP-QP coil. In contrast to the F and Q coils, the FI-QI and FP-QP coils require only one adjustment of the coil strength instead of two when the ship changes heading or magnetic latitude.

L COIL.—The Lor longitudinal coil (fig. 11-4, C) consists of loops in vertical planes that are parallel to the frames of the ship. The function of the L coil is to counteract the magnetic field produced by the ship's longitudinal permanent and longitudinal induced magnetization. The L coil is more difficult to install than the F and Q coils or FI-QI and FP-QP coils, but provides better neutralization because it more closely simulates the longitudinal magnetization of the ship. The L coil is often used in minesweeper vessels.

The longitudinal induced magnetization changes when the ship changes heading or magnetic latitude, and the strength of the L coil must be changed accordingly. When compensation for pitch is required, the L coil strength must also be changed as the vessel pitches.

A COIL.—The A or athwartship coil (fig. 11-4, D) consists of loops in vertical fore-and-aft planes. The function of the A coil is to produce a magnetic field that will counteract the magnetic field caused by the athwartship permanent and athwartship induced magnetization.

The athwartship induced magnetization changes when the ship changes heading or magnetic latitude, and the A coil strength must be changed accordingly. When compensation for roll is required, the A coil strength must be changed as the ship rolls.

COIL STRENGTH.—The magnetic field produced by a degaussing coil (coil strength) is proportional to the ampere turns (NI), which are the product of the number , of turns in the coil and the current in amperes flowing through the coil. A specified number of ampere turns can be obtained by using one turn and a current numerically equal to the required ampere turns, or by using more turns and a correspondingly smaller current.

The coil strengths of all degaussing coils except the FP-QP coil must be changed when the ship changes magnetic latitude zones. In addition, the coil strengths of the F, Q, FI-QI, L, and A coils must be changed when the ship changes its heading. When it is required to compensate for pitch and roll, the M coil strength must be changed when the ship pitches or rolls; the L, F, Q, or FI-QI coil strength must be changed when the ship pitches; and the A coil strength must be changed when the ship rolls. Coil strength is changed by changing the ampere turns, which can be accomplished by (1) changing the current, (2) changing the number of turns, or (3) changing the current and number of turns. In degaussing installations, the coil strength is changed by changing the value of current through the coils. This can be accomplished by means of manual or automatic degaussing control equipment.

Marking System

The degaussing installations in all types of naval vessels are marked in accordance with a standard marking system to facilitate maintenance by the ship's force. All feeders, mains, and other cables supplying power to deguassing switchboards, power supplies, and control panels are designated and marked as specified for power and lighting circuits. The system of markings and designations of conductors applies specifically for a multi-conductor system, but it is also applicable to single-conductor type installations.

Degaussing cable identification tags are made of metal. The cables are tagged as close as practicable to both sides of decks,bulkheads, or other barriers. Degaussing conductors are marked by hot stamping (branding) insulating sleeving of appropriate size. Each end of all conductors are so marked and correspond to the marking of the terminal to which they connect inside the connection box or through box. The sleeving is pushed over the conductor so that the marking is parallel to the axis of the conductor. The following letters are used for designating and marking cable tags for degaussing-coil cables and circuits.

D—Degaussing system A—Athwartship (A) coil AX—A auxiliary coil CC—Compass compensating coil F—Forecastle (F) coil to correct for permanent and induced magnetization FDR—Feeder

FI—F coil to correct for induced magnetization FP—F coil to correct for permanent magnetization I—FI-QI coil used in conjunction with feeders, compass-compensating coil, and indicator light leads IL—Indicator light L—Longitudinal (L) coil LX—L auxiliary coil M— Main (M) coil MX—M auxiliary coil P—FP-QP coil used in conjunction with feeders, compass-compensating coil, and indicator light leads

Q—Quarterdeck (Q) coil to correct for permanent

and induced magnetization QI—Q coil to correct for induced magnetization QP-Q coil to correct for permanent magnetization SPR—Spare conductor

DEGAUSSING-COIL CABLES.-The marking of degaussing-coil cables, feeder cables, and connection and through boxes is illustrated in figure 11-5. A detailed description of the marking for degaussing-coil loops, circuits, conductors, and cables and for degaussing feeder cable and feeder-cable conductors is contained in Chapter 81, Section m of the Bureau of Ships Manual.

CONNECTION AND THROUGH BOXES.-Connection and through boxes are similarly constructed watertight boxes, but they are usedfor different purposes (fig. 11-5).

A CONNECTION BOX is a watertight box with a removable cover used to connect loops together, to connect conductors in series, and to reverse turns. The power supply connection for a coil and all adjustments of ampere-turn ratios between loops are made within connection boxes. The power supply cable and interconnecting cable for the FI-QI and FP-QP coils terminate in connection boxes. The marking of the conductors in the

»»'1W'TCM
MO' «0u»«

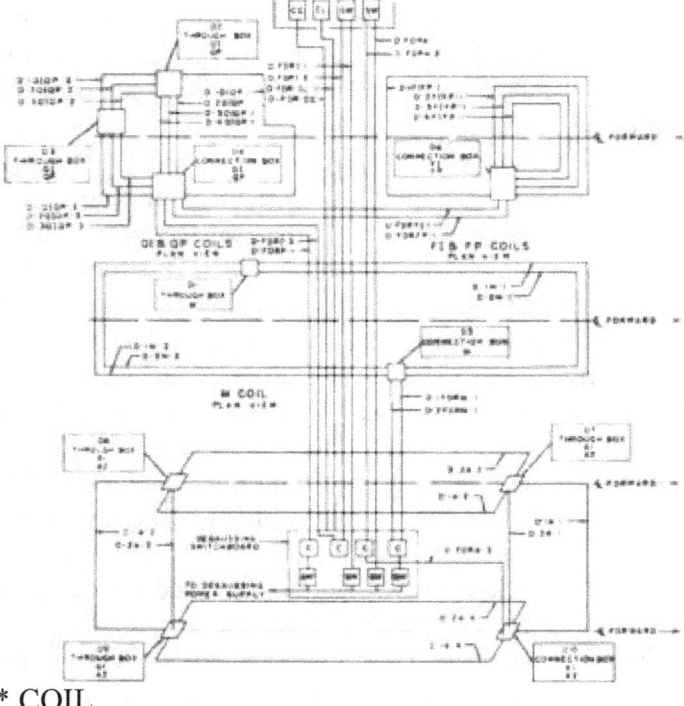

* COIL

Figure 11-5.—Marking scheme for degaussing coil cables, feeder cables, and connection and through boxes.

single-loop M coil connection box (Box D5 of fig. 11-5) is illustrated in figure 11-6.

A THROUGH BOX is a watertight box with a removable cover used to connect conductors without changing the order of conductor connections. Also, a through box is used when it is necessary to connect sections of cable. In some cases splicing in lieu of through boxes is used.

Connection and through boxes in the M,MX, F, Q, FI-QI, and FP-QP coils are considered to be in one group. They are NUMBERED D1, D2, D3, etc., in sequence, starting at the bow and going around the ship in a counterclockwise direction when viewed from above (fig. 11-5).

«QM«A.

t i<;i«u

M*t- MA 7 .

_»S1

JttL

JSC*.

M8»0-

a * ?

two c* tcm. coNcx>c T a*>s pw>»4.i.clio 'Ode'**"

mTm* *«>*»Ticm.»R StRitS 0« PARALLEL c«cm'

WCllVt 'Ml SAK CIXOUCOW OtSKlNAKO*. AS AN

EQUIVALENT SiNCt £ CONDUCTOR WOULO MECIIVC l? CONOUCTOO CABt. t use D '0« " CO*.

Figure 11-6.-Marking of conductors in the single-loop M coil connection box.

Boxes for the A coil have the next higher numbers after the numbering of the M, F, FI,

FP, QI, andQPboxes. The A coil boxes are numbered in sequence, starting with the highest forward box and continuing aft on the upper limbs of both the Al and A2 loops, then down to and forward on the lower limbs, and then up (fig. 11-5).

Boxes for the L coil have the next higher numbers after the numbering of the A coil boxes. The L coil boxes are numbered in sequence, starting with the highest forward box on the port side and continuing around the vessel in a counterclockwise direction when viewed from above. Boxes on the center line are considered to be on the port side for the assignment of numbers.

Connection and through boxes are provided with IDENTIFICATION PLATES that include:

1. Degaussing box numbers (Dl, D2, D3, etc.).
2. Connection box and/or through box as applicable.
3. Coil and loop designations (Ml,M2, FI2, QI2, etc.). For example,

Dl

Connection Box Through Box

Ml Fl M2

identifies the No. 1 degaussing box serving as a connection box for the Ml and M2 loops and as a through box for the Fl loop.

A wiring diagram of the connections in the box is pasted on the inside of the cover and coated with varnish or shellac. The wiring diagram for connection boxes should (1) designate the conductors that may be reversed without reversing the other loops, (2) indicate the arrangement of parallel circuits so that equal changes can be made in all parallel circuits when such changes are required, and (3) show the spare conductors. Spare conductors should be secured to connectors and should be connected to form a closed or continuous circuit. All conductors in a connection box should be 1-1/2 times the length required to reach any terminal within the box. Connection boxes should also be fitted with drain plugs to provide accessibility for periodic removal of accumulated moisture from the boxes.

Coil-Current Control

Degaussing installations, as previously stated, consist of different combinations of degaussing coils and manual or automatic degaussing equipment for control of the current in these coils. The selected combination depends on the size and the intended use of the particular ship.

The currents in the degaussing coils must be changed when there are changes in the ship's heading, magnetic latitude, or both. The control of the coil currents is accomplished manually in the older installations and automatically in all new installations.

MANUAL CONTROL.—Degaussing installations that have the coil currents manually controlled are energized from constant voltage d-c generators or from variable voltage motor-generators and are called (1) rheostat and (2) motor-generator installations, respectively. Reversing switches are used to change the polarity of the coils, except where provision is made to accomplish this change in the design of the rheostat for the constant voltage supply or in the generator control for the variable voltage supply.

In a RHEOSTAT INSTALLATION the power for the degaussing coils is supplied by a constant-voltage d-c generator, and the coil currents are controlled by adjusting a rheostat connected in series with the coil and power supply. An installation in which rheostats are used to control the M, FI-QI, and FP-QP coil currents is illustrated in figure 11-7. Manually operated rheostats are used in some installations and motor-operated rheostats in other installations. In the manual type, the rheostat is adjusted locally at the degaussing switchboard by turning the rheostat handle. In the motor-operated type, the rheostat shaft is turned by a motor controlled from a

120 OR 2*0
VOLT OC SUPPLY BUS
r r -oi coil
REMOTE REVERSING
SWITCH
OP-COIL

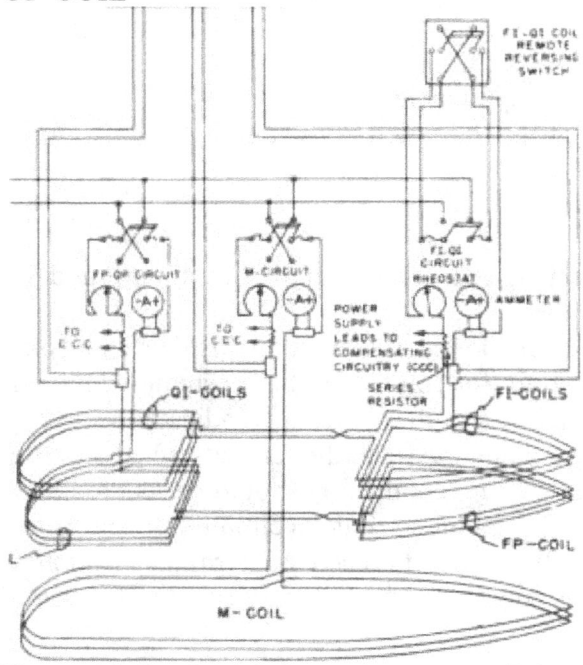

Figure 11-7.—M, FI-QI, and FP-QP coil degaussing installation with rheostat control.

remote station by means of a push-button. Motor -operated rheostats are equipped for manual operation in the event of an emergency.

In a MOTOR-GENERATOR INSTALLATION the power for the degaussing coils is supplied by a motor-generator, and the coil currents are controlled by adjusting a rheostat in the generator field circuit to vary the output of the generator. A single-line diagram of this type installation is illustrated in figure 11-8. The rheostats in the generator fields can be operated manually or by means of a motor to change the generator voltage and thereby adjust the coil current to the desired value.

An installation in which a combination RHEOSTAT and MOTOR-GENERATOR is used to supply the coil currents is illustrated in figure 11-9. The M coil current is controlled by means of a variable-voltage motor-generator,

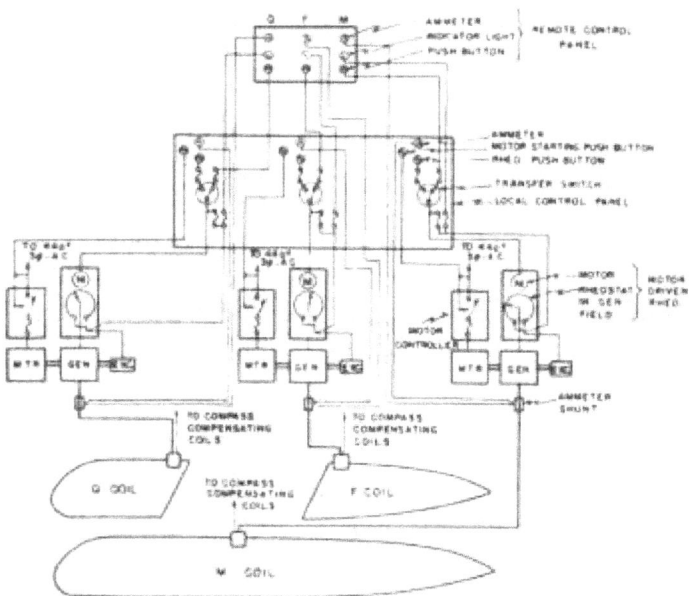

Figure 11-8.—M, F, and Q coil degaussing installation with motor generator control.

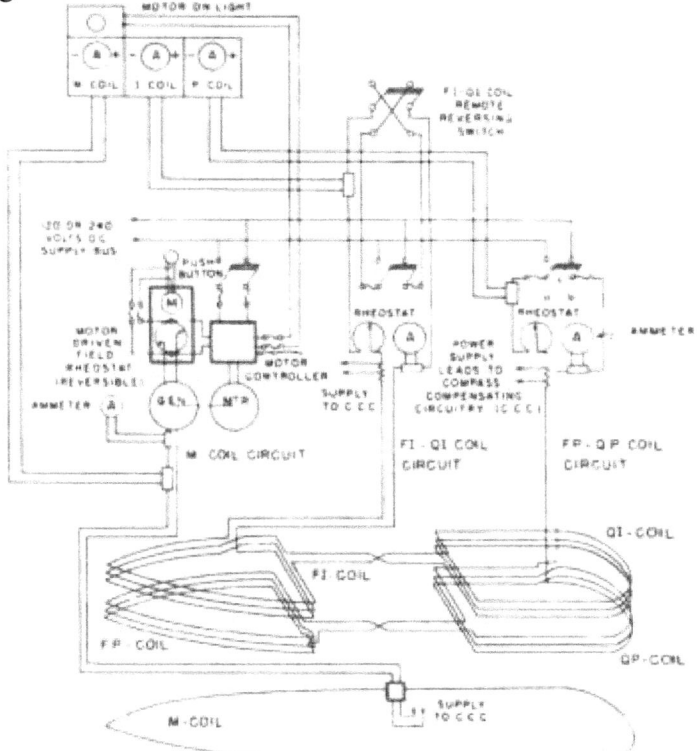

Figure 11-9.-M, FI-QI, and FP-QP coil degaussing installation with combination rheostat and motor-generator control.

and the FI-QI and FP-QP coil currents are controlled by means of rheostats.

The Degaussing Folder aboard each ship contains the Bureau of Ordnance DPG Forms, which show the number of coils and type of circuit installed in the ship. Further details of the degaussing installation are shown on the degaussing plans carried aboard each ship.

AUTOMATIC CONTROL.—Automatic degaussing control equipment is now installed in all new construction naval vessels equipped with degaussing coils that require

598

changes in the coil currents with changes in the ship's heading.

The majority of automatic degaussing control equipment now installed sets the coil currents automatically to compensate only for the change in induced magnetization caused by a change in the ship's heading. However, in some installations such as minesweepers, the concurrents are also automatically adjusted to compensate for the changes in induced magnetization caused by pitch and roll.

A block diagram of a type SM automatic degaussing control equipment for HEADING COMPENSATION only is illustrated in figure 11-10. A heading signal is obtained from the ship's gyrocompass system. For an L coil or an FI-QI coil, the type SM control equipment derives a signal that is proportional to the cosine of the ship's heading. This signal is amplified and fed to the appropriate power supply so that its output is proportional to the cosine of the ship's heading. This is the variation needed to compensate for changes in the longitudinal induced magnetization caused by changes in the ship's heading. For the A coil, the type SM control equipment maintains the coil current proportional to the sine of the ship's heading.

A block diagram of a type GM automatic degaussing control equipment for HEADING, PITCH, and ROLL COMPENSATION is illustrated in figure 11-11. The additional signals needed to compensate for pitch and roll are obtained from the gyrocompass stabilizer system and fed to the automatic degaussing control equipment. A computer element in the control equipment utilizes the heading signal and the pitch and roll signals, and causes the degaussing coil currents to vary so that they compensate for heading, pitch, and roll.

All types of gyrocompass-controlled automatic degaussing control equipment require manual settings for magnetic latitude and magnetic variation. Also, ail types of degaussing control equipment are equipped with manual controls for use if the automatic controls should become inoperative.

Other types of automatic degaussing control equipment are the FM and RM types. The FM is a magnetic amplifier type of control that controls the exciter field of the

PAUL T WGmaL CIRCUIT
CO«T ITOL - (Ml ■* TO*

UCNU.
\rcc«c it*ci
VCNAL
Man ine«
r3*t h amPl >T1 f l
cont *cto«
■ AG«f TtC V a • i AT 10**
>1 NCMtO
i
ma&mC • C Hf ADi»C inOICaTO*
«fL AT
r -

con.

re 11-10.-Block diagram of type SM automatic degaussing control equipment for heading compensation.

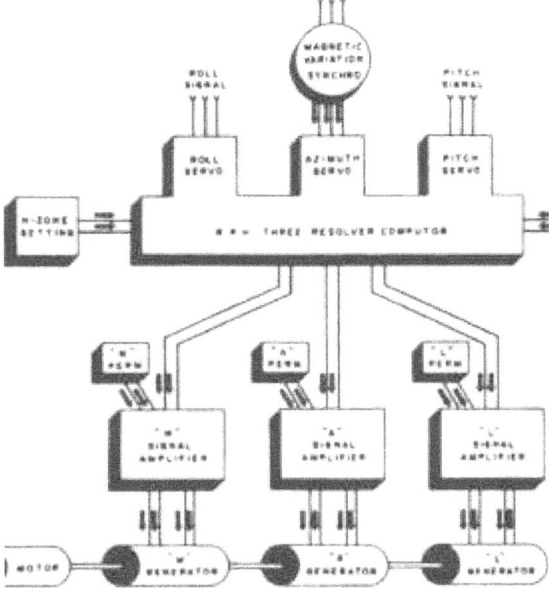

Figure 11-11.—Block diagram of type GM automatic degaussing control equipment for heading, pitch, and roll compensation.

degaussing motor-generator. The RM is also a magnetic amplifier type of control that controls the motor of the motor-operated rheostat. The rheostat is in series with the degaussing coil that is connected across the ship's constant-voltage d-c power supply. Refer to the manufacturers' instruction books furnished with the specific equipment for more detailed information concerning automatic degaussing control equipment. Also, the Electrician's Mates 1 and C training course contains additional information on automatic degaussing control equipment.

Operation

The operation of the degaussing installation should be in accordance with the instructions in the Bureau of Ordnance Degaussing Folder, which is carried aboard each ship equipped with a degaussing installation. The Degaussing Folder contains current setting charts that are prepared by the Degaussing Range when the ship is ranged and gives the degaussing coil currents to be used for any position on the earth's surface and any heading of the ship.

In most degaussing installations now in use the degaussing coil currents are set to the required values by an Electrician's Mate or other authorized personnel. The current should be checked hourly and readjusted to the correct value when necessary. Degaussing coil currents will change and require readjustment because of changes in degaussing-coil resistance caused by changes in cable temperature and changes in the voltage of the power supply for the degaussing coils.

POLARITY.—The polarity of the degaussing coil currents is of particular importance. If the polarity of any of the degaussing coils is incorrect, the ship may very likely be in much greater danger from magnetic mines than if no degaussing were installed. The polarity of a coil should be checked by observing whether the pointer of the ammeter for the coil is on the POSITIVE (right) or NEGATIVE (left) side of the zero-center ammeter. Also, the polarity of a coil can be checked by observing whether the positive or negative plate glows in the neon indicator light for the coil. For positive polarity the right-hand electrode glows, and for negative polarity the left-hand electrode glows.

The direction of current in a degaussing coil is normally indicated by zero-center ammeters located on the degaussing control panel. The pointer will deflect to the right for positive (direction) current through the coil and to the left for negative current. The direction of current can be checked by a deguassing polarity indicator or a small compass. The polarity indicator dial is marked to denote the direction of current. When taking polarity readings with the indicator or small compass, move the device toward the degaussing coil until a good deflection is obtained, and no closer. The needle in the indicator or compass will reverse its magnetic polarity if the device is held too close to the coils. The indicator or compass should be checked after each test to be certain that the magnetic polarity of the needle has not reversed.

CHANGING COIL CURRENTS.-The Degaussing Folder for each ship gives the current needed for each coil for all positions on the earth's surface and for all headings. One or more of the coil currents must be changed when one of the following conditions occurs:

1. When the ship passes from one Z zone to another. (See Degaussing Chart No. 1 in the Degaussing Folder.) The vertical intensity (Z component) of the earth's field, which is maximum at the magnetic poles and zero at the magnetic equator, is divided into a number of Z zones. The number of Z zones will vary, depending on the amount of compensation provided by the particular degaussing installation. Degaussing Chart No. 1 (fig. 11-12) illustrates the Z zones for the Atlantic and Indian Oceans. The reverse side of this chart contains the same number of Z zones for the Pacific Ocean. The coil settings are filled in for the various zones after the ship has been ranged.

2. When the ship passes from one H zone to another. (See Degaussing Chart No. 2 in the Degaussing Folder.) The horizontal intensity (H component) of the earth's magnetic field, which is maximum at the magnetic equator

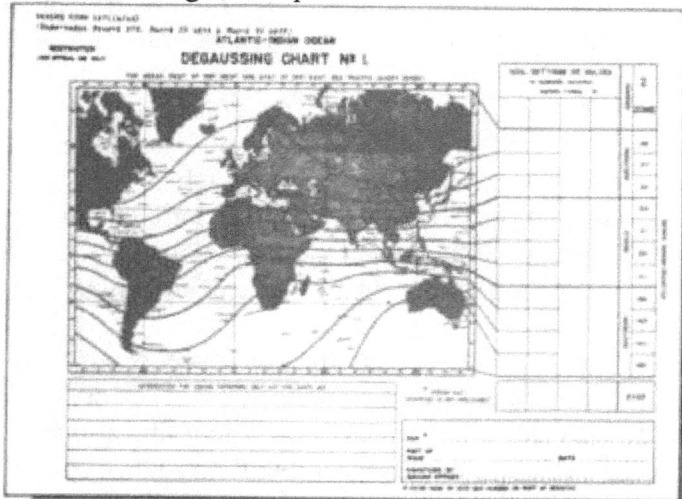

Figure 11-12.-Degaussing Chart No. 1 60S

and zero at the magnetic poles, is divided into a number of H zones. Similar to the Z zones, the number of H zones will vary, depending on the degree of compensation provided by the degaussing system. A Degaussing Chart No. 2 (fig. 11-13) illustrates the H zones for the Atlantic and Indian Oceans and the reverse side contains the same number of H zones for the Pacific Ocean.

3. When the ship's heading changes from one sector to another. The entire range of headings from 0° to 360° is divided into a number of sectors, each covering apart of the whole range of courses. (See the Degaussing Course Correction Setting Diagrams No. 1 and No. 2 and the Degaussing Course Correction Setting Tables No. 1 and No. 2 in the Degaussing Folder.)

.The Degaussing Course Corrections Setting Diagrams No. 1 for the FI-QI coil and No. 2 for the FI-QI and A coils are illustrated respectively in figures 11-14 and 11-15.

The Degaussing Course Correction Setting Tables No. 1 for the F and Q coils and No. 2 for the F, Q, and A coils are illustrated respectively in figures 11-16 and 11-17.

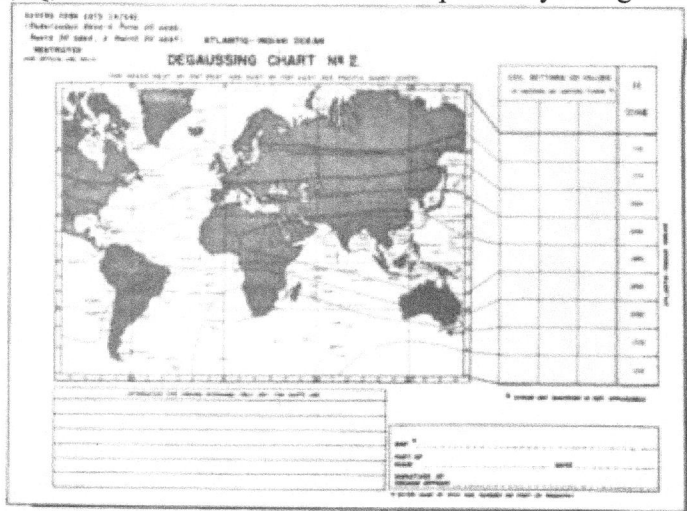

Figure 11-13.-Degaussing Chart No. 2. 604

DEGAUSSING COURSE CORRECTION SETTING DIAGRAM No. I. FI-QI COIL HEADINGS ARE MAGNETIC

N

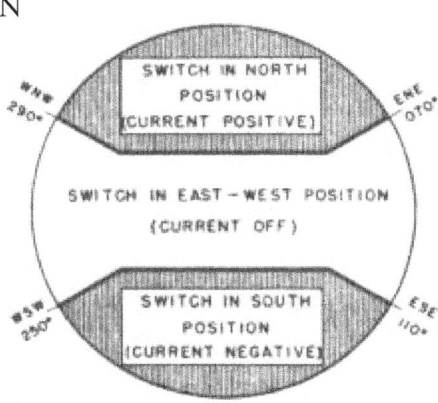

S

THE CURRENT IN THE Fi-QI COIL IS TURNED ON I POSITIVE) FOR NORTHERLY COURSES. OFF FOR EASTERLY AND WESTERLY COURSES , ANO REvERSEO (NEGATIVE) FOR SOUTHERLY COURSES BY MEANS OF A SWITCH ON THE BRlOSE OR IN THE CHART HOUSE

WHEN THE SHIP IS ON A NORTHERLY HEA0IN6 SET THE CURRENT IN THE Fl -01 COIL AT THE VALUE SHOWN IN 0E6AUSSING CHART No 2 FOR THE LOCATION OF THE SHIP AT THE TIME WHEN THE COURSE IS SOUTHERLY THE CURRENT SHOULO IE THE NEGATIVE OF THIS VALUE

CHANGE THE CURRENT VALUE IF THE SHIP MOVES INTO A DIFFERENT

ZONE

IN EMERGENCY . WHEN COURSE CHANGES ARE TOO RAPID TO FOLLOW . SET SWITCH IN EAST - WEST POSITION

Figure 11-14.-Degaussing Course Correction Setting Diagram No. 1.

None of the degaussing coil currents is changed as long as the course remains in one sector (if the Z zone and H zone remain unchanged). The FP-QP coil current is NOT changed when the ship's heading or the ship's position changes. The following changes are necessary when changing from one sector to another or from one zone to another:

•»en fom mi <i/«?) >»*.•. t ..p n»-(m i »»o.i m ,u

DEGAUSSING COURSE CORRECTION SETTING OIAGRAM No 2.

HEADINGS ARE F I - 0 I COIL MAGNETIC A COIL.

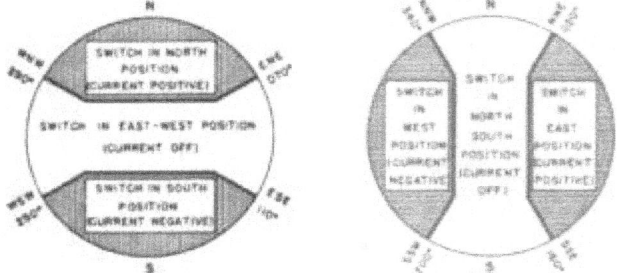

THC CURRENT IN TMC H - Oi COIL IS TURNED ON (POSITIVE) FOR NORTHERLY COURSES,OFF FOR CASTCRLY AND WEST CRLV COURSES, ANO REVERSED (NEGATIVE) FOR SOUTHERLY COURSES IV MEANS OF * SWITCH ON THE BRIDGE OR IN THE CHART HOUSE

WHEN THE SHIR IS ON A NORTHERLY

HEADING,SET THE CURRENT IN THE FI-OI COIL AT THC VALUE SHOWN IN DEGAUSSING CHART No 2 FOR THE LOCATION OF THE SHIP AT THE TIME WHEN THE COURSE IS SOUTHERLY THE CURRENT SHOULO RE THE NEGATIVE OF THIS VALUE

CHANCE THE CURRENT VALUE IF THE SHIR MOVES INTO A DIFFERENT ZONE

IN EMERGENCY , WHEN COURSE CHANGES ARE TOO RARIO TO FOLLOW, SET SWITCH IN EAST - WEST POSITION

THE CURRENT IN THE A COIL IS TURNED ONIPOSITIVEI FOR EASTERLY COURSES,OFF FOR NORTHERLY AND SOUTHERLY COURSES, ANO REVERSED NEGATIVE FOR WESTERLY COURSES BY MEANS OF A SWITCH ON THE BRIDGE OR IN THC CHART HOUSE

WHEN THC SHIP IS ON EASTERLY HEAOiNG, SET THE CURRENT M THE A COIL AT THE VALUE SHOWN IN OCCAUSSiNG CHART No 2 FOR THE LOCATION OF THE SHIP AT THE TIME WHEN THE COURSE IS WESTERLY THE CURRENT SHOULD BE THC NEGATIVE OF THIS VALUC

CHANGE THE CURRENT VALUE IF THE SHIR MOVES INTO A DIFFERENT ZONE

IN EMERGENCY , WHEN COURSE CHANGES ARC TOO RAPID TO FOLLOW. SCT SWITCH IN NORTH -SOU T H POSITION

Figure 11-15.-Degaussing Course Correction Setting Diagram No. 2.

1. The M coil current must be changed when the ship moves from one Z zone to another. The M coil current is not changed when the ship moves from one H zone to another or when the

heading changes from one sector to another.

2. The F, Q, FI-QI, L, and A coil currents are not changed when the ship moves from one Z zone to another, but must be changed if the ship moves to a different H zone, or if the heading changes to a different sector.

• itoto row I5'«i«/«s)

«t>I»lCtlO

DEGAUSSING

COURSE CORRECTION SETTING TABLE N& I.

(F AM) 0 COILS)

* I Th» T«fct« n to be u»*4 tn ALL f»Q**i~|

MSTRUCTtONS rOM USC

•y te DefewM*** O

1 M t MtMt tttt cwcm 4i«fr«M MM w*c» c*"»»p»«d I• IM H NM '+

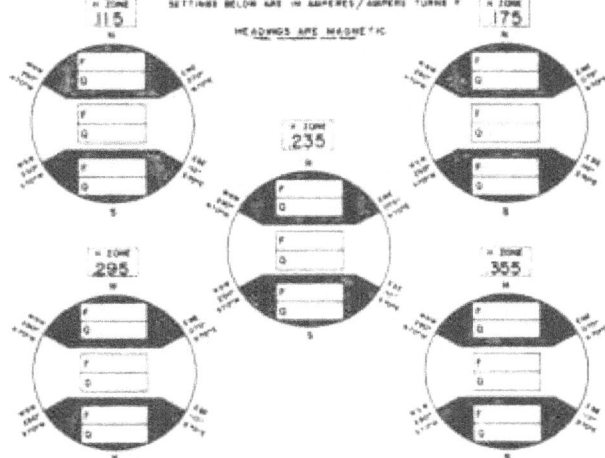

Figure 11-16.-Degaussing Course Correction Setting Table No. 1.

In a few ships, exceptional conditions may require a departure from the foregoing changes. In all cases, the Degaussing Folder will show the currents to be used.

RHEOSTAT INSTALLATION.-The degaussing coils in rheostat installations are energized by closing the main degaussing feeder switch and the individual coil

•«on roin mili'iil

DEQAU3SIN9

COURSE CORRECTION SETTING TABLE NS2.

(F.Q.ANO A COILS)

T»i« T«M » n M Ml 0"L» •••• hm •»■» •una

T)M T«M« II U M «•*« I" ALL '«)'•"•

1STRUCTIONS) FOR USC

•f nlmi It 0*MMIWMj C*«fl M| f MtM« IM C-(« *-»|- -*-<* W*tl l|l»l H *l " t«M I NM MNi*W* •* 1*« T <M VClaM HM.«|| N) h« *•«! ««« I*** « !*• IHrW>4<l Mm

MWI

SC TTINGS KlO* «»€ IN mPf »(S / *t»t ■ r TUMNS • h[»0'»SS MA6MITIC

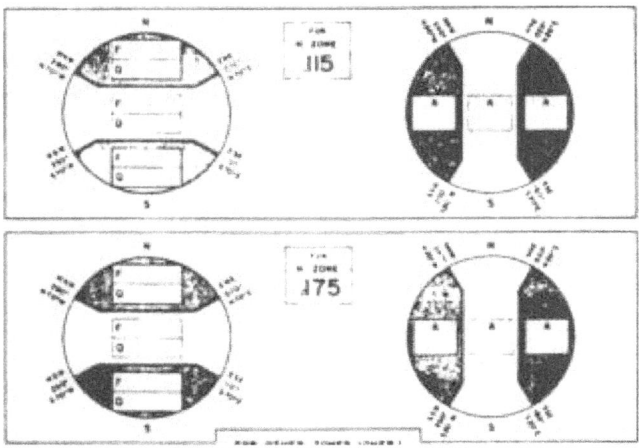

ro« OTMC". 70NCS IOVt»)

tn* it nrrtn

Figure 11-17.-Degaussing Course Correction Setting Table No. 2.

switches in the proper polarity position with all rheostats set for maximum resistance. Set the current in each coil by cutting out resistance.

Adjust the rheostats for the M and FP-QP coils (if provided) until the M and FP-QP coil currents have the values specified in the Degaussing Folder for the ship's position. Adjust the rheostats for the F, Q, FI-QI, L,

fin ft

and A coils (if provided) until the coil currents have the values specified in the Degaussing Folder for the ship's position and heading. Check the coil currents periodically, and change the coil currents as required by changes in the ship's position and heading.

In installations employing motor-driven rheostats the polarity is changed only when the current is zero either by means of a motor-driven polarity changer and pushbutton, or by operating the rheostat through and beyond the position of maximum resistance. At this point, reversal of current will occur automatically because of the internal cross -wiring of the rheostat buttons on either side of this point, or by means of a motor-driven polarity changer actuated by the rheostat arm.

Motor-driven rheostats can be operated manually by turning the emergency handwheels in the event of failure of the motor-drive system. Rheostats having polarity changes operated by means of pushbuttons are provided with small handwheels for turning the polarity changer if the system should fail. Always reduce the current to zero before operating the polarity changer; otherwise, the contacts will burn out from operation under load.

In installations employing manually operated rheostats, the degaussing switchboard is usually located in the engineroom. This switchboard includes the disconnect switches for the F, Q, FI-QI, L, and A coils; reversing switches for the M and FP-QP coils; series resistors; fuses; and ammeters. A remote degaussing panel is located in the pilot house or chart house to permit changing the F, Q, FI-QI, L, and A coil currents with heading. This panel includes ammeters or polarity indicator lights for all coils and rheostats (rheostat pushbuttons); and reversing switches for the F, Q, and L coils or reversing switches for the FI-QI and A coils. The M and FP-QP rheostats, ammeters, and switches on some installations are also included on this remote panel.

Telephone communications are provided between the remote and main panels. When the degaussing system is to be energized, the remote position (pilot house) may call the main degaussing switchboard station and have all switches closed and have those currents normally controlled from the degaussing switchboard set by means of rheostats. The remote position will then set those

currents normally controlled by the rheostat from that position and will have the responsibility of maintaining the correct current values at all times. It is the responsibility of the main panel position to correct any subsequent voltage change and current decrease caused by temperature rise of the cable.

MOTOR-GENERATOR INST ALL AT ION. -The degaussing coils in motor-generator installations are energized by closing the disconnect switch to the motors with the generator field rheostats set for the maximum resistance, and operating the pushbutton to start the motor. The degaussing switchboard is usually located in, or in the vicinity of, the engineroom near the degaussing motor -generators. This switchboard includes ammeters, pushbuttons for starting the motors, and the motor-driven field rheostat for the generator. In many installations in which the motor-generators are not located together, a separate switchboard is provided for each group of motor -generators.

After the motor has started, the currents in the coils are set by the pushbuttons that control the motor-driven field rheostats for the generators. The pushbuttons are located on the remote control panel in the pilot house or chart house, or on the degaussing switchboard. Zero-center ammeters mounted above the pushbuttons are used to determine whether the positive or negative pushbutton should be operated to obtain the specified currents. The positive pushbutton will operate the motor-driven generator field rheostat to increase the current in the positive direction (decrease the current in the negative direction). Conversely, the negative pushbutton will operate the motor-driven rheostat to increase the current in the negative direction (decrease the current in the positive direction).

Maintenance

The degaussing installation should be carefully maintained particularly to prevent deterioration resulting from moisture gradually entering the cables. Since the degaussing system consists of electrical cables, rheostats, ammeters, connection boxes, motor-generator sets, and costly automatic control equipment it must be

maintained in conformity with the instructions for the maintenance of electrical equipment given in applicable chapters of the Bureau of Ships Manual and applicable technical manuals for the equipment installed. All equipment defects, failures, and replacements must be reported by form NavShips 3621 (Material Analyses Data). This report assists the cognizant Bureau in supplying the fleet with improved equipment.

PREVENTIVE MAINTENANCE.-When the system is not in normal use, all coils should be energized at least once a week to the maximum current specified in the Degaussing Folder and operated for at least four hours with the current in one direction. Then reduce the current to zero and energize the coils momentarily with maximum current in the opposite direction.

Check each degaussing coil for grounds by insulation resistance measurements using a 500-volt megger. When the source of supply for degaussing coils is from rectifiers, the rectifiers must be disconnected prior to using a 500-volt megger on the degaussing coils to prevent damage to the rectifiers. These measurements should be made with the same degaussing equipment connected in the circuit and as closely as possible under similar conditions of temperature and humidity. The date and reading of the insulation resistance to ground should be recorded each week on the Ship's Force Degaussing Maintenance Record in the Degaussing Folder. These measurements should be taken between the coil disconnect switch and ground. In the F, Q, and A or I, P, and A coils that have a reversing switch on the coil side of the disconnect switch, the measurement should be made with the reversing switch in the CLOSED position. Also, this measurement should be made just before the weekly energizing of the coils so that the comparative readings are obtained under similar conditions. By comparing these weekly readings it is possible to detect abnormal decreases in resistance and take the necessary corrective action before failure occurs. When abnormal decreases are indicated, the various components of the circuit should be isolated and the insulation resistances checked individually to determine the cause of the low reading.

The connection boxes may accumulate considerable amounts of water due to condensation and leakage through

fin

improperly seated gaskets. The moist atmosphere In the box will gradually force moisture into the cable ends and reduce the insulation resistance. If this condition is not checked in time, the cable may be ruined and require replacement. Therefore, remove the connection and through-box drain plugs once a month to allow any accumulated water to run out. The date that each box is drained should be recorded on the Ship's Force Degaussing Maintenance Record (NavShips Form 1009) in the Degaussing Folder. Boxes that have abnormal accumulations of water should be opened, dried out, and the gaskets checked and replaced if necessary.

Remove corrosion on rheostat contact buttons with fine sandpaper and coat the buttons lightly with a graphite lubricant. Remove accumulations of dirt and dust from automatic degaussing control equipment to facilitate the natural flow of air around the components, and thus eliminate the possibility of overheating. Use a vacuum cleaner or bellows; do not use compressed air.

Observe the preventive maintenance requirements for automatic control equipment and motor-driven rheostats, as outlined in the applicable technical manual. Also, observe the maintenance of electrical equipment and electrical safety precautions, as required by Chapter 60 of the Bureau of Ships Manual.

Conduct weekly linearity checks, as required on ships equipped with automatic degaussing units in accordance with the applicable technical manual. Make weekly entries on the Ship's Force Degaussing Maintenance Record (NavShips 1009) in the Degaussing Folder.

CORRECTIVE MAINTENANCE.-Corrective maintenance consists principally of locating and eliminating grounds in the various coils, circuits, and components of a degaussing system.

Grounds in a degaussing conductor can be located by breaking the conductor into its component sections by opening connections at connection and through boxes, and testing each section of the conductor individually. Grounds in FEEDERS and control equipment CIRCUITS can be located by isolating the different components. When making these tests, reconnect all ungrounded connectors to their terminals immediately after they have tested clear, to prevent misconnection later. To locate grounds

in degaussing circuits, proceed as follows (observe precautions required when using a 500-volt megger on rectifier circuits):

1. Open the supply switch for the degaussing coil and close the reversing switch in either position.

2. Disconnect the feeder conductors from the degauss -ing coil at the feeder connection box.

3. Measure the insulation resistance to ground for each of the feeder conductors. If the insulation resistances are satisfactory in accordance with the manufacturer's technical manual

and Chapter 60 of BuShips Manual, the feeder conductors and the equipment connected to them are clear of objectionable grounds. If the insulation resistance to ground is unsatisfactory for either one or more of the feeder conductors, proceed in accordance with the method for locating grounds in feeder conductors, as described in the instruction book accompanying the installation.

4. Measure the insulation resistance from the degaussing coil to ground. If the insulation resistance of the degaussing coil is satisfactory in accordance with Chapter 60 of BuShips Manual, no further insulation tests on the coil are necessary.

Degaussing Definitions

DEGAUSSING COIL.-One or more loops of degaussing cable; each loop encircles a different total area and is connected in series with other loops to form the required coil.

PARALLEL CIRCUIT.-An arrangement of conductors connected such as to provide two or more complete and independent electrical paths through a coil or loop and having a common source of supply.

COIL SECTION.—A section of a degaussing coil cable is the length of cable between two successive connections or through boxes.

PERMANENT MAGNETISM.-That portion of the magnetic field of a ship which is proportional to the strength of the earth's field through the ship.

INDUCED MAGNETISM.-That portion of the magnetic field of a ship which is proportional to the strength of the earth's field through the ship.

EFFECTIVE TURNS.-The number of turns which are effective in producing a magnetic field under the ship. The difference between the number of series conductors in a coil or loop in which the current flows in a counterclockwise direction and the number of series conductors in which the current is flowing clockwise (called reverse turns) when viewed in accordance with polarity convention for each coil.

AMPERE TURNS.-The product of the current flowing through the coil or loop and the number of effective turns.

LOOP.—One or more turns of the conductors of a degaussing cable encircling a specific area. Different loops connected in series form a degaussing coil, for example: FI, FI2, QI, and QI2 loops are connected in series to form a FI-QI coil. In conserving cable, the conductors used for different loops encircling the same area may be contained in the same cable, for example, an FP and FI loop may both be contained in an F cable.

TURN.—One complete lap around the perimeter of a specified area by a coil or loop. One turn consists of one conductor or 2 or more conductors connected in parallel.

SERIES RESISTORS.—A resistor connected in series with the degaussing coil and used to limit the coil to its maximum design current. Where series resistors are used, the power for the compass compensating coil is usually provided by utilizing, in part, the voltage drop across this resistor and thereby varying the effect of the CC coils proportionally with the degaussing coil currents.

"THREE-COURSE EMERGENCY."-A manual three-step course correction, accomplished by a switch or switching system which provides for changes in the polarity and magnitude, of the current in the FI-QI, L and A coils to compensate for changes in induced magnetism due to changes in the heading of the ship. For this purpose, three values of ampere turns are used. With an FI-QI coil, for example, zero ampere turns are used for headings in the general direction of magnetic East, a positive value for heading in the general direction of Magnetic North and a negative value of equal magnitude for heading in the general direction of magnetic South.

REMOTE CONTROL PANEL.-This unit contains the necessary meters, electrical circuits, and controls to

enable an operator to monitor all degaussing coil currents and to control manually by Three-Course Emergency the degaussing coil currents normally controlled automatically by the AUTODEG control unit.

AUTOMATIC DEGAUSSING—CONTROL UNIT.-This unit, by utilizing a signal from the ship's gyro compass system (heading only) or (heading, roll, and pitch) varies the ampere turns with changes in heading, or heading, roll, and pitch automatically. This equipment incorporates a "Three-Course Emergency" control and generally an additional manual control.

COMPASS COMPENSATING CONTROL BOX.-A watertight enclosure having a removable cover and containing three sets of control resistors. Each set of control resistors consists of fixed resistors, a variable resistor, and jumpers. By means of adjusting the control resistors, the compass compensating coil currents supplied by the voltage drop across the DG coil, the series resistor or switchboard, is regulated to give the desired compass compensation.

DEGAUSSING FOLDER.-The degaussing folder is an official ship log, NavOrd Form 1547. It contains instructions for operation of the degaussing system, degaussing charts and values for coil settings, Installation Certificates, and a log section showing all pertinent details of action taken on the ship's degaussing system for the information of degaussing authorities. The Degaussing Folder is issued to a ship by the Degaussing Facility which renders the initial magnetic treatment.

INSTALLATION CERTIFICATES.-Installation Certificates are official NavOrd Forms showing plan or profile views of degaussing coils as installed, identification of equipment, a schematic electrical diagram, circuiting data, test results and other pertinent information as required for installation. There are also certificates of modification which describe all modifications to the degaussing installation.

ANCHOR WINDLASS

The anchor windlass installed aboard destroyers consists of a multispeed motor directly connected through

reduction gears to a vertical shaft on which are mounted a capstan and a wildcat (fig. 11-18). The CAPSTAN and WILDCAT are located on the weather deck, and the electric motor and the across-the-line starter are located in the windlass room on the next deck below. The windlass is designed to operate in both directions to raise or lower either the starboard or port anchor.

FRICTION BRAKE HANDWHEEL

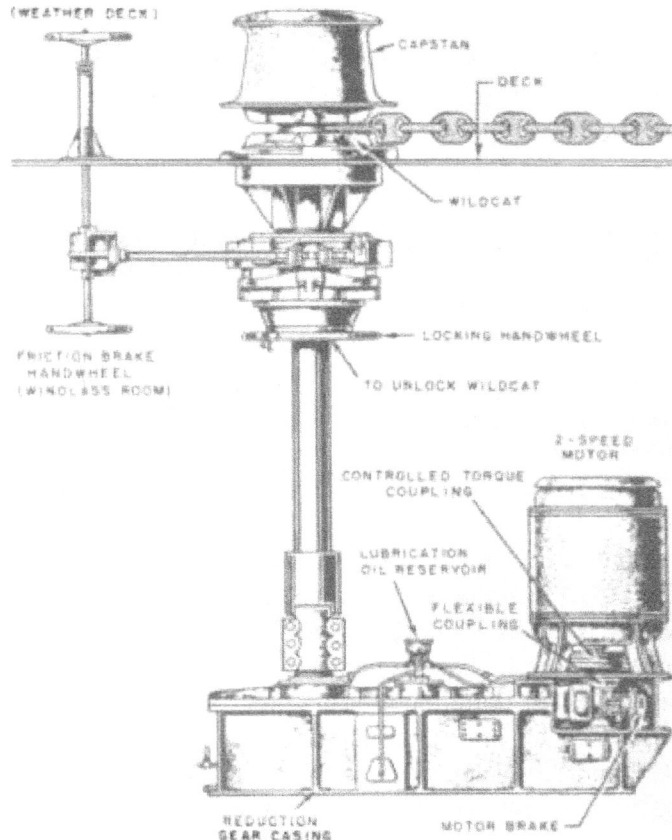

Figure 11-18.-Anchor windlass.

Construction

The windlass is driven by a 30/7-1/2 hp, two-speed (1650/400 rpm), 3-phase, 440-volt, 60-cycle motor connected to the reduction gear by a controlled torque coupling. The controlled-torque coupling prevents excessive dynamic stresses if the anchor should be hauled into the house pipe when the motor is running at high speed.

An ELECTRIC BRAKE, mounted just below the controlled-torque coupling, will release when power is applied and will set when power is disconnected or fails. The electric brake is designed to stop and hold 150 percent of the rated load when the anchor and chain are being lowered at maximum lowering speed, in the event of power^failure.

The wildcat is designed to hoist one anchor and 60 fathoms of 1-1/4-inch dielock chain in not more than 10 minutes on the high speed connection without exceeding the full-load rating of the motor. On the low-speed connection the wildcat is designed to hoist the anchor and 60 fathoms of chain without overloading the motor, and to exert a pull on the chain at least three times that required to hoist the anchor and 60 fathoms of chain.

The capstan is designed to heave a 6-inch circumference manila line at a speed of 50 feet per minute with a line pull corresponding to the full-load motor torque. The capstan is keyed to the vertical shaft. The wildcat and sleeve run free on the same shaft until connected to the shaft by a locking head located below the weather deck. The capstan can be run independently for warping by disconnecting the locking head and holding the wildcat by means of the brake band on the brake drum. The locking head is disconnected by revolving a locking handwheel located below the brake drum. The handwheel can be pinned in the LOCKED or UNLOCKED positions and should always be fully locked or fully unlocked.

A handbrake is provided on the wildcat shaft to control the anchor handling. It is designed to operate in either direction of rotation of the wildcat and to stop and hold the anchor when dropped from a depth of 45 to 60 fathoms. The brake is operated by means of a handwheel located on the weather deck and a duplicate handwheel in the windlass room.

Operation

The windlass is operated by a drum master switch on the weather deck and a duplicate switch in the windlass room (fig. 11-19). It is important to remember that if the windlass is run with the locking handwheel in the LOCKED position, the wildcat will revolve. In this case, if the chain is engaged in the whelps on the wildcat, the chain should be free to run. Exercise care to select the proper direction of rotation and be certain that the windlass is properly lubricated.

The motor can be operated from either master switch No. 1 (on the weather deck) or from master switch No. 2 (in the windlass room), but master switch No. 1 predominates (fig. 11-19). When the associated on-off switch located on master switch No. 1 is operated to the ON position, master switch No. 1 takes over control from master switch No. 2 (if both switches are operated simultaneously.

The anchor windlass is used alternately to handle either the starboard or the port anchors. The windlass is operated by a reversible motor in either of two directions. These directions may be hoist for the starboard anchor (lower for the port anchor) and hoist for the port anchor (lower for the starboard anchor). However, only one anchor can be handled at a time. The nomenclature on the controllers refers to the hoist directions for both anchors.

The motor starter is equipped with four thermal overload relays to protect the motor against overloads. Overload relays 1SOL and 2SOL are in the slow-speed motor circuit and overload relays 1FOL and 2FOL are in the fast-speed motor circuit. If an overload occurs in the slow-speed or fast-speed circuit, the SOL or the FOL relays will operate to trip the slow-speed, S, or the fast-speed, F, contactors respectively. The motor can be operated in the event of an emergency by holding either of the EMERG-RUN pushbuttons down and operating the master switch in the usual manner. To reset the overload relays, press the OVERLOAD RESET pushbuttons in the event of an overload or voltage failure. The master switch must be returned to the OFF position to restart the motor.

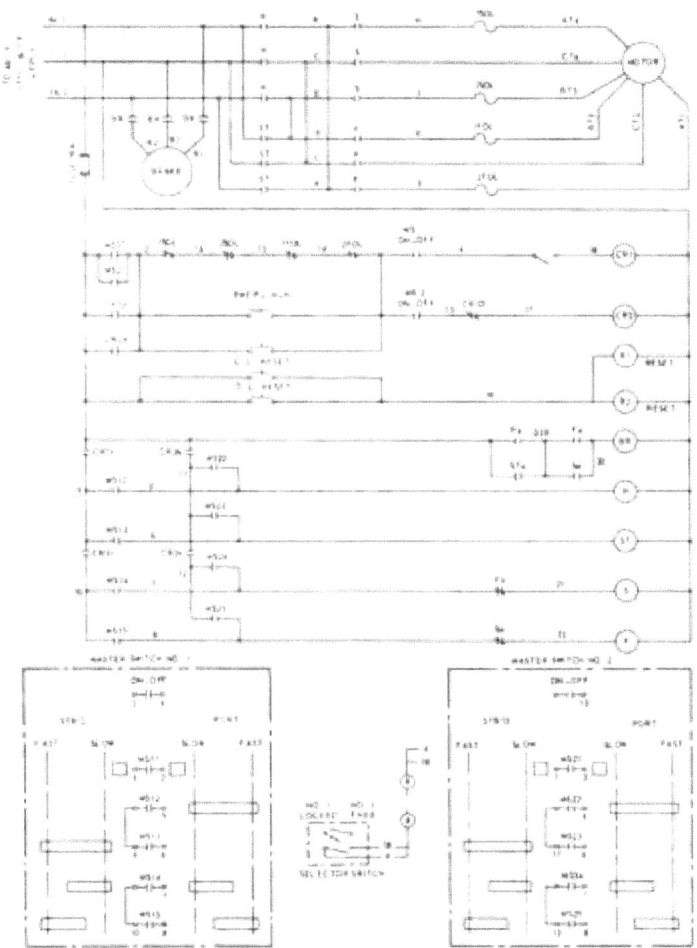

Figure 11-19.-Reversing across-the-line starter for two-speed anchor windlass.

To start the motor in the port (hoist) direction for slow speed by master switch No. 1, operate the associated on-off switch to the ON position and move the controller handle forward to the SLOW PORT (hoist) position. This action closes contacts MS 11 momentarily to energize the

operating coil of relay CR1 and to close its contacts CR1a to provide a holding circuit for relay CR1, and to open its normally closed contacts CR1d to prevent operation of relay CR2 at this time. At the same time, contacts CR1b close to prepare the circuit to controller contacts MS12 and MS13, and contacts CR1c close to prepare the circuit to controller contacts MS 14 and MSI5.

When the controller handle is moved further toward the SLOW PORT position, contacts MS11 open; controller contacts MS12 close to energize the operating coil, P, and close the port contactor in the motor starter; also, contacts, Pa, close to provide the circuit to the motor brake relay, BR. At the same time, controller contacts MS14 close to energize the operating coil, S, and close the slow-speed contactor in the motor starter. Contacts Sa close to energize the brake relay, BR, and close its contacts to release the motor brake. Also, the normally closed contacts, Sb (in the circuit to the operating coil, F, of the fast-speed contactor), open. The motor is now connected for hoisting the port anchor at slow speed.

When the controller handle is moved further to the FAST PORT position, contacts MS 15 close and contacts MS14 open. Contacts MS15 close before contacts MS14 open so that the operating coil, S, is kept energized through the normally closed contacts, Fb. When contacts MS14 open, operating coil, S, deenergizes and closes contacts Sb to energize the operating coil, F. This action opens contacts Fb to deenergize the operating coil, S, and open the slow-speed contactor. When the slow-speed contactor opens, contacts Sb close to complete the circuit to the fast-speed contactor in the motor starter. The motor is now connected for hoisting the port anchor at fast speed.

To hoist the starboard anchor, the same sequence occurs, except that controller contacts MS13 energize the operating coil, ST, to close the starboard contactor instead of controller contacts MS12 energizing the operating coil, P, to close the port contactor.

If it is desired to operate the motor by master switch No. 2, operate the associated on-off switch to the ON position and move the controller handle to the PORT or STARBOARD SLOW position. This action closes contacts MS21 momentarily to energize the operating coil of relay CR2 (if relay CR1 is not energized). The sequence of operation for master switch No. 2 is the same as that for master switch No. 1, except that contactors P, ST, S, and F are energized through the CR2 contacts instead of through the CR1 contacts. Master switch No. 1 can be locked out by turning the selector switch to the No. 1 LOCKED position. In this position the selector switch opens the circuit to relay CR1 and prevents its operation.

Maintenance of the electrical components of the anchor windlass should be in accordance with the instructions listed in chapters 5 and 8 of this training course. More detailed information concerning the maintenance of this equipment is contained in the manufacturers' instruction books furnished with the specific equipment.

ELECTRIC ELEVATORS

The elevator installation aboard aircraft carriers usually consists of hydraulic and electric types. The hydraulic type includes inboard and deck edge elevators for handling airplanes, and the electric type includes bomb, freight, mine, torpedo, and ammunition elevators. Only electric elevators are described in this training course.

Construction

The platform of an electric elevator is raised and lowered by groups of cables that pass over sheaves and drums connected to the hoisting machinery. The hoisting drums are coupled together and driven through a reduction-gear unit by an electric motor. The motor is a 2-speed (full -speed and 1/4-speed) type induction motor.

The 2-speed electric motor is controlled through a system of contactors, relays, limit switches, and selector switches (fig. 11-20). Automatic operation is obtained by selecting the levels between which the platform is to run. The start pushbutton can then be used to close contactors through safety switches to accelerate the platform from low speed to high speed. Just before reaching the desired level, the control transfers the motor to the low-speed winding through the action of cam-operated limit switches. On reaching the desired level, the control circuit is disconnected by a cam-operated stop switch, thus releasing the contactors and setting the brake to stop the platform.

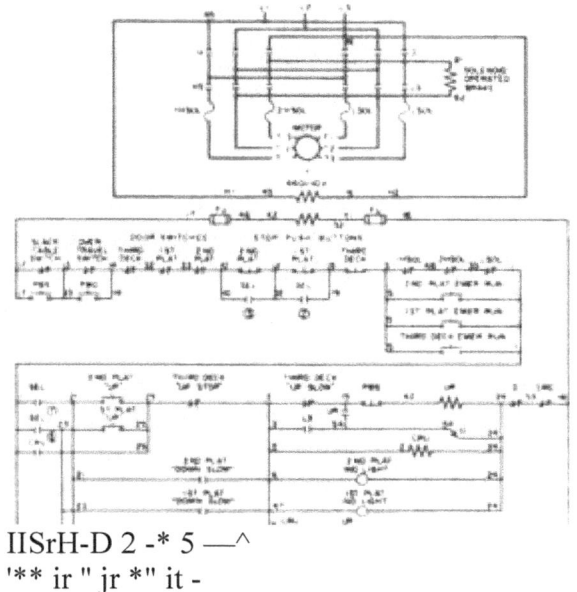

IISrH-D 2 -* 5 —^

'** ir " jr *" it -

D . »■ : TO* it LO«VflC c>»<'0«

i%x . n i "■-■•> * o*e mi *■ *«c ovia rami ■» »iju«io<

«l . Ill ICTOt twtO* W

,■:*-■*..".« \iou«MCf

r I I - *.*C« CAM f ft ' S« Jmjoi)" M<> »«00v««i0*0»Il»T» CU-w A <tC . CDNTMX Bfl Mi. U» » D0«*»

«OTO« U>*<C TON TAfcl

Figure 11-20.-Schematic diagram of electric bomb elevator.

For safety in operation, all doors at each level are interlocked to prevent operation unless they are closed. Also, all hatch covers are interlocked to prevent elevator operation unless they are fully opened. The protective features incorporated in the control are (1) slack-cable switches to prevent operation of the elevator if any cable should become slack, (2) emergency stop switches at each level served, (3) overtravel switches to stop the elevator if it should fail to stop at the uppermost level, and (4) overload protection.

Elevator controllers are designed with a double-break feature that prevents maloperation if any one contactor, relay, or switch should fail to function properly. Pushbuttons are interlocked to prevent operation of the elevator unless the platform is at the same level as the pushbutton. Some elevators are equipped with hatchway door mechanical interlocks to prevent opening the door unless the platform is at the same level.

A governor-actuated safety device is provided under the platform to grip the guide rails and stop the platform in the event of overspeed in the DOWN direction. Also, spring bumpers are provided at the bottom of the hatchway to prevent mechanical damage to the hull or platform due to overtravel in the DOWN direction.

Operation

The operation of the elevator depends on the position af the selector switch, which determines between which decks the elevator will run. This switch also renders ill master switches inoperative, except those pertaining :o the selected levels.

Suppose the selector switch is set in the second plat-:orm to the third deck position (fig. 11-20). In this posi-:ion the control is set up for the elevator to operate between the second platform and the third deck and closes :ontacts (1), (2), (4), (5), and (7). (The third deck is ibove the second platform.) Contact (2) shorts out the irst platform stop pushbutton, contact (1) places the hird deck pushbutton station in the circuit, and contacts 4) and (5) short out the first platform DOWN-STOP and he first platform DOWN-SLOW switches, respectively. Contact (7) places the second platform pushbutton station n the circuit.

If the overtravel, slack cable, door switches, stop pushbuttons, and overload relay contacts are in their normally closed positions, the control circuit is energized and set up for operation. When the second platform UP pushbutton is momentarily pressed, the up auxiliary relay, UR, is energized. This action closes contacts UR1, which energizes the up contactor, U, in the across-the-line starter. The up auxiliary relay, UR, also closes contacts UR2, which energizes the high-speed contactor, HS. Contactor HS applies voltage to the motor and energizes the brake-release solenoid. The elevator moves upward until it mechanically operates the UP-SLOW limit switch located at the third deck. Operation of the limit switch deenergizes the up auxiliary relay, UR. This action closes contacts UR3 and energizes the LS coil, which transfers the motor from the high-speed, HS, to the low-speed, LS, contactor. The elevator continues upward at low speed until it mechanically operates the UP-STOP limit switch located at the third deck. Operation of the limit switch deenergizes the up contactor, U, which de-energizes the brake-release solenoid and operates the motor brake to stop the motor. An indicating light shows when the elevator reaches the selected deck.

The elevator can be stopped at any time by pressing the stop lever at the pushbutton station located on the selected level, or, in this case, the third deck. To restart the elevator, press the UP pushbutton lever at the second platform or the DOWN lever at the third deck.

In the event of an overload, one of the overload relays will open the control circuit, set the brake, and de-energize the motor. The overload relay must be reset for normal operation by pressing the reset button that projects through the door of the enclosing case.

In the event of supply voltage failure, the line contactors will open and deenergize the motor. To resume operation of the elevator, an UP or DOWN pushbutton lever must be pressed.

As mentioned before additional protection is provided through a system of series-connected interlocks in the control circuit consisting of door, slack cable, and over-travel switches. If a cable should become slack or if the elevator should overtravel, the elevator can be operated by holding in the SLACK CABLE bypass pushbutton.

PBS, or the OVERTRAVEL bypass pushbutton, PBO, located inside the controller case. When either one of these pushbuttons is operated, the motor will run only on low speed.

If an overload should occur, the elevator can be operated (in case of an emergency) in the usual manner if the EMERG-RUN lever of either pushbutton station is held in the depressed position.

The up and down current control relays, CRU and CRD, respectively, are provided to ensure proper operation in the event of malfunctioning of other relays or contactors.

Maintenance

Adjustments to the machinery and electrical equipment of elevators should be in accordance with the procedures contained in the appropriate manufacturers' instruction books. In addition to these maintenance procedures, the following items should be checked during each shipyard overhaul.

1. Inspect wire cable for wear, corrosion, and broken strands.
2. Check equipment for worn parts and replace those parts that are excessively worn.
3. Inspect and test all safety mechanism to ensure that they are in operable condition.

4. Perform operating tests of the elevator to ensure that it is in satisfactory'ope rating condition.

5. Check the enclosures of all limit switches (except explosion proof) to be certain that the 1/4-inch drain holes are open to permit draining of condensed water.

6. Inspect the brake for evidence of excessive wear on the linings and replace as necessary. Adjust the brake in accordance with the manufacturers' instructions. When the brake is adjusted, check the leveling of the platform and correct any inaccuracy in leveling by moving the stop limit switches.

7. Inspect limit switches to determine that the arms are adjusted properly. Adjust each limit switch arm so that approximately one-half of the travel of the arm is taken up by the application of the cam to the switch arm roller. This adjustment will prevent damage to the arm or prevent failure of the switch to operate properly due to the shifting of the platform in the hatchway or movement of the guide rails with the working of the ship's structure.

ELECTROHYDRAULIC STEERING GEAR

The steering gear is one of the most vital auxiliaries aboard ship. It must be thoroughly dependable and have sufficient capacity for maximum maneuverability. The types of steering gear listed in the sequence of their development are the (1) steam, (2) electromechanical, and (3) electrohydraulic types. The electrohydraulic steering gear was developed to meet the excessive power requirements of naval vessels having larger displacements and higher speeds with the attendant increase in rudder torques.

The majority of steering gear installations in new construction naval vessels are of the electrohydraulic type.

Construction

The electrohydraulic steering gear installed in a destroyer (fig. 11-21) consists essentially of a (1) ram unit; (2) power unit; and (3) remote control system.

RAM UNIT.—The ram unit is mounted athwartship and consists of a single ram that operates in opposed cylinders. The ram is connected by links to the tillers of the twin rudders and is moved by the oil pressure built up in either of the cylinders, the oil from the opposite cylinder flowing to the suction side of the pump.

The tie rods that connect the two cylinders also serve as guides for a sliding bracket attached to the ram to prevent the ram from rotating. The bracket also provides mechanical limits to the ram travel at 42° of rudder angle. At this position the bracket contacts copper facings on the stop collars and prevents further movement.

A rack is attached to the ram and engages two gears, the rotation of which is transmitted to the respective differential control boxes through the follow-up shaft.

POWER UNIT.-The power unit consists of two pumping systems, which include two motor-driven pumps, a

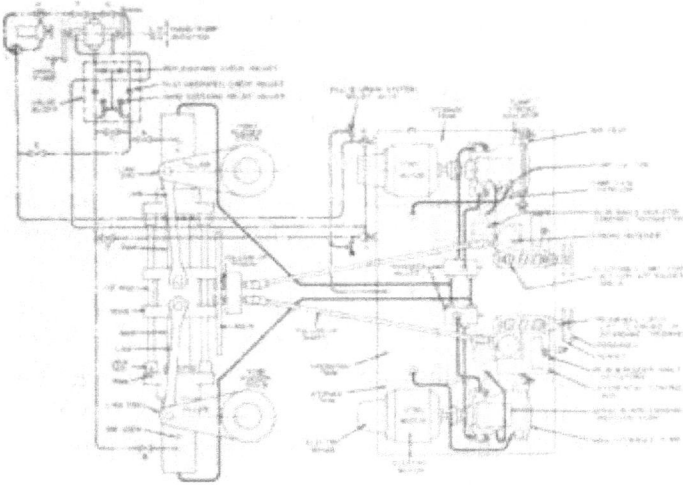

Figure 11-21.—Electrohydraulic steering gear.

hand pump, two 4-way transfer valves with operating gear, relief valve, two differential control boxes, two "trick" wheels, and a hand emergency gear all mounted on a bedplate, which is the top of an oil reservoir. Steering power is derived from either pumping system acting alone. The system not in use serves as a standby source in case of emergency.

The two pumps (port and starboard) are identical in size and design, and are of the variable delivery axial piston type. Each main pump unit includes a built-in vane-type servo pump, pressure control and replenishing valves, and two main relief valves. Each main pump is stroked through a rotary servo control.

Each pump is driven by a 20-hp, 1200-rpm, 440-volt, 3-phase, 60-cycle induction motor through reduction gears and a flexible coupling. A disk-type electric brake, which is automatically secured when deenergized, is provided on each motor.

The pumps of the power unit are connected to the ram cylinders by a high-pressure piping system. The two 4-way transfer valves are interposed in this piping, and their positions determine which pump is connected to the
cylinders in the ram unit. The hand lever, which moves both valves simultaneously, is located on the power unit between the "trick" wheels, and has three latched positions. The latched positions are marked P, N, and S, which denote port pump connected, neutral (ram locked), and starboard pump connected, respectively.

The transfer valves, which are of the ported piston type, are mounted on the power unit bedplate between the motors. When the valves are in the neutral (RAM LOCKED) position, the ports of the pipes to the ram cylinders are blocked, and the two pipes from each pump are connected through ported passages in their respective valves. If either or both pumps are started and put on stroke, oil will circulate through the valve and back to the pump. Movement of the transfer valves in either direction connects the selected pump to the ram cylinders, and the opposite pump remains bypassed.

The drain pipes from the ram cylinders lead to the reversible hand pump, which provides a means for emergency steering under limited rudder torques. The relief valve in the emergency steering system is set at 500 psi. The oil flow is through pilot-operated blocking valves that prevent the ram from overhauling the pump and kicking back the handles. When moving the ram with the hand pump, the main transfer valves must be set in the RAM LOCKED position, and the drain valves beneath the cylinders must be open.

The stroking lever and output shaft on each differential control box actuate a rotary servo valve that controls the associated main pump on the power unit. In response to movements transmitted through the control box, the pump discharge is varied between zero and maximum, and in either direction of flow.

In normal steering from the pilot house, the input shaft of the differential control box is turned by the synchro receiver (mounted on top of the box) through the remote control system. The synchro receiver is geared to the LOWER bevel gear in the differential control unit, and its rotation is transmitted through gearing to a cylindrical cam. A follower roller, which engages a groove in the cylindrical cam, is mounted on an arm keyed to the output shaft of the control box.

For example, assume that the rudder is amidship and it is desired to obtain 20° right rudder. The rotation of the lower bevel gear causes rotation of the cylindrical cam, which, in turn, imparts a motion to the servo control valve through the cam roller. This motion is transmitted to the servo control valve, which puts the pump on stroke. Oil pressure is then applied in the port cylinder, forcing the ram to starboard to give right rudder.

A rack, attached to the ram, rotates the follow-up shaft, which is geared to the UPPER bevel gear in the differential control unit. The movement of the ram and rudders (in response to the stroking of the pump) causes the upper bevel gear to rotate in a direction opposite to that of the lower bevel gear. This action rotates the cylindrical cam in the opposite direction, tending to cancel the movement of the control input and bring the servo control valve back to the neutral position to return the pump stroke to neutral and stop the pumping of oil.

Thus, the rudders are at 20° right rudder, the cam is returned to neutral, and the pump is returned to zero stroke until further movement of the steering wheel causes repetition of the cycle. The same sequence occurs if the control is from the trick wheel in the steering compartment.

An engraved dial, graduated in rudder degrees, is mounted on top of the differential control box. Two concentric pointers (one geared to the control pump input and the other geared to the rudder follow-up) indicate the positions of the helm and rudders, respectively. A helm-angle synchro transmitter, also mounted on the differential control box, actuates a synchro receiver in the steering console in the pilot house to indicate to the helmsman the helm angle.

REMOTE CONTROL SYSTEM.-The remote control system (fig. 11-22) provides control from the pilot house for normal steering. This system consists of a synchro transmitter mounted in the steering console and a synchro receiver mounted on each pump differential control box (port and starboard). A cable selector switch in the pilot house and a similar switch in the steering compartment permit a choice between the port or starboard steering cables. The selection of the control cable is made by the operation of the cable selector switches in the pilot house

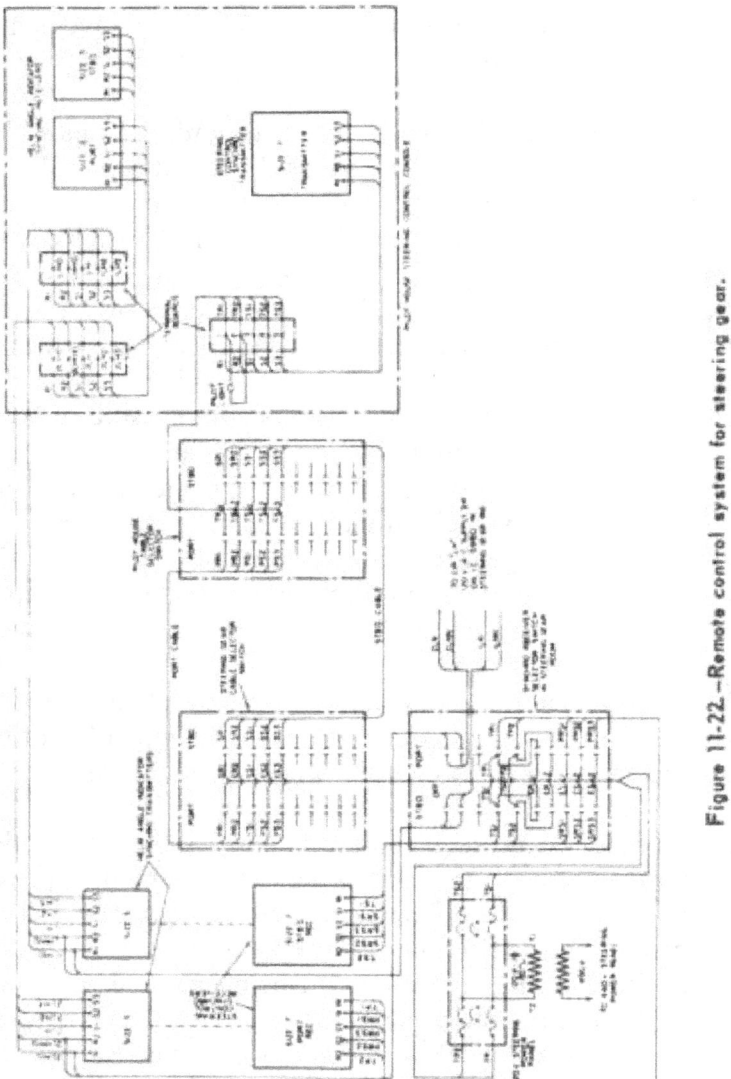

Figure 11-22.—Remote control system for steering gear.

and in the steering gear room to the desired (port or starboard) position. The synchro-receiver selector switch in the steering gear room is then set to connect the synchro receiver, on the active power unit, to the steering console synchro transmitter. The rotary motion of the synchro receiver is transmitted through gearing to the input of the differential control box (stroking mechanism) previously described.

The helm angle-indicator synchro transmitters on the differential control boxes in the steering gear room are geared to their associated steering-control synchro receivers. These transmitters under all conditions of operation actuate their associated helm-angle indicator synchro receivers in the pilot house console through their associated cables.

The 120-volt, single-phase power for the remote control system is supplied to the steering power panel through a 450/120-volt transformer from the steering power transfer switchboard. The 120-volt, single-phase power for the indicator synchro transmitters is supplied from the I. C. switchboard through circuits L (rudder order system) and N (rudder angle indicator system).

MAGNETIC CONTROLLER. -The motor of each steering gear pump is provided with a nonreversing across-the-line starter and a maintained-contact master switch (fig. 11-23). The starter is supplied with 440-volt,

39, 440 V 60X POWER
OL
STEERING POWER
TRANSFER
SWITCHBOARD

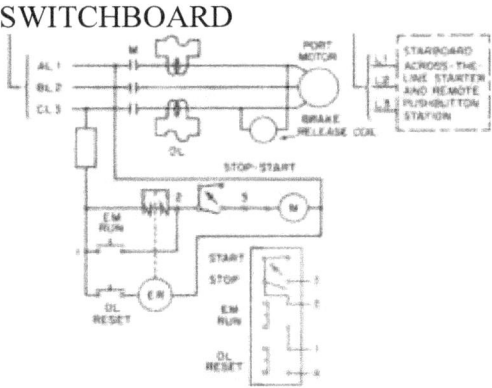

REMOTE PUSHBUTTON STATION

Figure 11-23.-Nonreversing across-the-1ine starter for steering gear pump motor.

3-phase, 60-cycle power from the steering power transfer switchboard located in the steering engine room.

The pump motor on either power unit is started or stopped by operating the maintained-contact pushbutton, on the associated pushbutton station, to the desired position. When the (maintained-contact) start pushbutton is pressed, the circuit is completed to the operating coil, M, of the line contactor. This action energizes the operating coil and closes the contactor in the motor starter to connect the motor to the line. The motor will continue to run until the contactor operating coil, M, is deener-gized because of loss of voltage, tripping of the overload relay, or by pressing the stop pushbutton.

The motor starter is provided with overload and low-voltage release protection. The overload relays are of the thermal type, similar to those installed in the previously described anchor windlass starter. The low-voltage release protection is provided by use of the maintained-contact master switch. If the operating coil, M, is de-energized due to failure of the line voltage or tripping of an overload relay, the contactor will reclose and restart the motor when voltage is restored or when the overload relays are reset by means of the reset pushbutton.

In an emergency the motor can be run (even though the overload relays have been tripped) by holding the EM-RUN pushbutton closed with the start pushbutton in the operated position. If the overload condition has not been corrected, the motor will operate only as long as the EM-RUN pushbutton is held closed.

Operation

The steering gear is normally secured with the rudders amidship and with the transfer valve shift lever (fig. 11-21) in the neutral (N) position, which hydraulically locks the ram and rudders. To operate the port power unit, using the port steering cable:

1. Set the valves and plug cocks of the hydraulic system in accordance with the operating diagram for the steering gear for normal power operation.

2. Set the transfer valve shift lever in the neutral (N) position (fig. 11-21).

3. Press the start pushbutton to start the port pump motor.

4. Engage the trick wheel and turn it until the helm indicator (pointer) corresponds with the rudder angle indicator.

5. Operate the transfer valve shift lever to the PORT (P) position to connect the port

pump unit to the ram.

6. Close the port and starboard circuit breakers on the 120-volt steering power panel to energize the synchro receiver selector switch and the cable selector switch in the steering gear room.

7. Operate the synchro receiver selector switch and the cable selector switch (in steering gear room) to the PORT positions.

8. Operate the cable selector switch (in pilot house) to the PORT position to energize the remote control synchro transmitter to steer from the bridge.

9. Disengage the trick wheel on the port unit and engage the trick wheel on the idle (starboard) unit.

To shift from the port power unit, using the port steering cable, to the starboard- power unit, using the starboard steering cable:

1. Press the start pushbutton to start the starboard (idle) pump motor.

2. Notify steering station to keep wheel motionless during change of units.

3. Allow rudder angle indicator to line up with helm indicator; then operate the transfer valve shift lever to the starboard (S) position to connect the starboard pump to the ram. Transfer the remote control from the port to the starboard unit by operating the synchro-receiver selector switch (in steering gear room).

4. Transfer the cable selector switch (in steering gear room) and cable selector switch (in pilot house) to the STARBOARD positions.

5. Disengage trick wheel- and steer from the bridge.

6. Press the stop pushbutton to stop the port motor and engage the trick wheel on this unit.

To steer from the trick wheel:

1. Select the power unit to be used and follow the procedures described previously for operating the port power unit.

2. Engage both trick wheels and steer from the trick wheel of the selected power unit.

3. Deenergizethe remote control system by operating the synchro receiver selector switch in the steering gear room to the OFF position.

To operate by hand emergency steering:

1. Set the valves and plug cocks of the hydraulic system in accordance with the operating diagram for the steering gear for hand steering.

2. Set the transfer valve shift lever in the NEUTRAL (N) position.

3. Crank the hand pump in the proper direction until the desired rudder angle is obtained.

When the hand pump is used for emergency steering, the ship's speed and rudder angles must be controlled so that the pressures in the ram cylinders will not exceed the setting of the relief valve, which is installed in the emergency steering system.

To secure the steering gear when steering under normal power operation:

1. Request steering station to put rudders on zero angle or use the trick wheel for this purpose.

2. Operate the transfer valve shift lever to the NEUTRAL (N) position.

3. Deenergizethe remote control system by operating the synchro receiver selector switch and the cable selector switch (in the steering gear room) and the cable selector switch (in the pilot house) to the OFF positions. Open the port and starboard circuit breakers on the 120-volt steering power panel.

4. Stop the pump motor by pressing the associated STOP pushbutton.

The maintenance of the steering gear must be in accordance with the instructions contained in the manufacturers' technical manuals for the equipment installed aboard your ship. Additional information concerning the maintenance of motors and control equipment is included in chapters 5, 7, and 8 of this training course.

ELECTRIC GALLEY EQUIPMENT

Electric galley equipment comprises the heavy duty cooking and baking equipment installed aboard naval

vessels and consists essentially of ranges, griddles, fry kettles, roasting ovens, and baking ovens (fig. 11-24). This equipment is supplemented by electric pantry equipment, which includes coffee urns, coffee makers, grills, hotplates, and toasters. The number and capacity of the units comprising a galley installation depend on the size and type of ship. Galley equipment is designed for operation on 115-volt or 230-volt, d-c power or 440-volt, 3-phase, 60-cycle, a-c power. A type A range is illustrated in figure 11-24, A. A roasting oven is pictured in figure 11-24, B, and a fry kettle is shown in figure 11-24, C.

Ranges

Electric galley ranges are provided in the type A (36-inch), type B (20-inch), and type C (30-inch) sizes. The ranges consist of a range-top section and an oven section assembled as a single unit, and a separate switchbox designed for overhead or bulkhead mounting (fig. 11-24, A).

The RANGE TOP SECTION (type A range) is provided with three 5-kw surface units consisting of combination griddle hotplates with cast-in, enclosed heating units or with enclosed heating units fastened to the underside. The surface units are individually controlled by a high temperature thermostat, the bulb of which is covered and clamped to the underside of the griddle hotplate. Each thermostat is connected in series with the operating coil of an associated heavy-duty contactor. When the thermostat contacts close, the contactor coil is energized and closes the contactor to energize the heating unit. The thermostat controls for the surface units are located on a panel at the front of the range-top section.

The OVEN SECTION is provided with two 3-kw enclosed heating units, one located in the top and one in the bottom of the oven compartment. The oven units are separately controlled to regulate the relative temperature at the top and bottom of the oven by individual 3-heat rotary switches located in the overhead switchbox. The oven is also provided with an adjustable temperature control (thermostat) to establish the average temperature that is to be maintained in the oven. The thermostat control for the oven units is located on a panel at the right of the oven.

A - TYPE A RANGE
B ~ TYPE 60 OVEN

C " TYPE 90 FRY KETTLE Figure 11-24.-Electric galley equipment.

A compartment at the right of the oven contains a terminal board for making the connections between the

range and the switchbox. The compartment is equipped with a removable cover for access to the interior.

The SWITCHBOX consists of a sheet-steel enclosure provided with a hinged or removable cover for access to the interior and a panel on which are mounted the two oven switches. The switchbox houses the cutout blocks, fuses, contactors, and the line terminal block. A simplified wiring diagram of a type A range is illustrated in figure 11-25.

Ovens

The type 4 and type 6 baking ovens are the older type having baking decks (the type number denotes the number of decks) that are not thermally insulated from each other. The heating units are located at the top and bottom of the oven and between each deck. Each heating unit is controlled by individual 3-heat switches located in a switch box enclosure mounted on the right-hand side of the oven.

The type 12 and type 18 baking ovens are sectional type ovens with each section constituting a separate oven that is thermally insulated and operated independently of the other sections. Each section of the type 12 or type 18 ovens has a capacity of six standard five-loaf bread pans. The type 12 oven consists of two sections and the type 18 oven consists of three sections mounted one above the other. The type numbers denote the total bread pan capacity of the oven.

Type 60 and type 125 roasting ovens are sectional type ovens with each section constituting a separate oven that is thermally insulated and operated independently of the other sections (fig. 11-24, B). The type number denotes the capacity in pounds of raw meat per section. The roasting ovens are provided in either 2 or 3 sections mounted one above the other.

The type 12 and type 18 baking ovens have the same general construction of the type 60 and type 125 roasting ovens except that the oven compartments of the roasting ovens are about four inches higher. Each oven section of both the baking and roasting ovens is provided with two heating units, one located in the top and one in the bottom (underneath the deck) of the oven compartment. The heating units are usually of the enclosed type but

RANGE
CONTROL BOX
OvEN THERMOSTAT
at
1
at
RIGHT PLATE THERMOSTAT
CENTER PLATE 'HERMOSTAT
LEFT Plate THERMOSTAT
LOWER OVEN
UPPER OVEN
Ri&ht Plate
CENTER plate
LEFT PLATE
LOWER OVEN
Switch

RELAY
uPPtR OvEN SvVCh
RELAY
RELAY
RELAY
-0>
RANGE CIRCUITS
L3
Off

MED
OVEN SWITCH CONNECTIONS
lO*
,»INEH ROD

Figure ll-25.-Schematic diagram of type A ran

some ovens have open-coil type heating units consisting of resistance wire wound on reinforced porcelain rods that are mounted in a steel frame. The oven units are controlled by individual 3-heat rotary switches (two for each section). Each oven section is also provided with an adjustable thermostat for automatic control of the average temperature. The operation of the 3-heat switches and the temperature control is the same as for the type A range previously described.

The type 60 and type 125 roasting ovens are provided with a separately mounted switch box which contains the fuses, contactors, and 3-heat switches for each section. Figure 11-26 is a wiring diagram of the type 60 oven. The types 12 and 18 baking ovens do not have a separate switch box because all control devices are mounted in a compartment at the right of each oven. The compartment is provided with a hinged, or removable, cover for access to the interior.

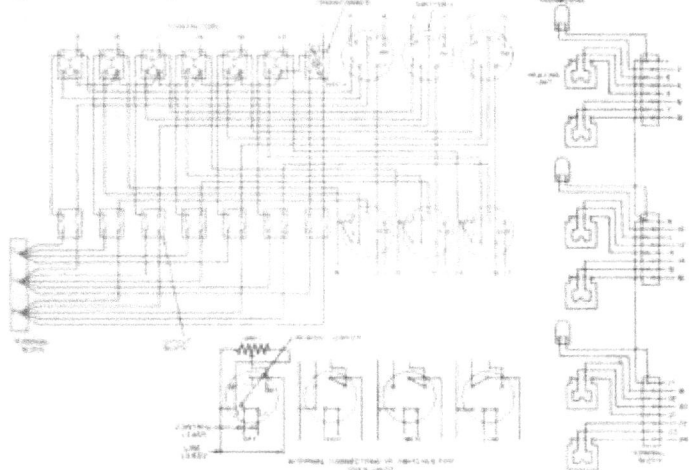

Figure 11-26,-Wiring diagram of type 60 oven.

Fry Kettles

Electric fry kettles are provided in the type 23, type 51, and type 90 sizes rated at 5 kw, 10 kw, and 18 kw,

respectively. The kettle consists of a rectangular sheet-steel enclosure that houses a polished-steel fat container (fig. 11-24, C). The fat container is also rectangular and the bottom slopes toward the center, which is provided with a sump. The sump is fitted with a suitable container to remove the sediment. The bottom of the sump is equipped with a drain pipe and gate valve for draining the fat from the kettle.

The heating units, which are of the enclosed type, are immersed directly in the fat to ensure maximum efficiency. The kettle is equipped with an adjustable automatic temperature control to maintain the fat at the desired temperature. The thermostat control, located on a panel at the front of the kettle, is provided with an OFF position and the adjustable temperature range of 250° F to 400° F is graduated on the control knob. The thermostat operates contactors, which, in turn, control the circuits to the heating units.

A rotary switch also located on the front panel is provided for disconnecting the kettle from the line. As a safety measure, the switch should be turned to the OFF position when the fry kettle is not in use.

A compartment located at the front of the kettle contains the contactors, thermostat, heating unit terminals, line terminal block, fuses, and line switch. The compartment is equipped with a removable panel for access to the interior. Figure 11-27 is a wiring diagram of the type 90 fry kettle.

Maintenance

Before starting any service work on electric galley equipment be certain that the equipment is disconnected from the power supply. Electrical circuits in the equipment should be inspected and tested at least once each month. Ground tests should be taken more frequently, preferably once each week. Be certain that wire connections and fuse clips are tight, relay contact points are bright, and the relay armature does not stick. Contact points should be kept clean and bright by polishing with fine sandpaper on a burnishing tool.

If a fault should occur in the heating units provided with electric galley equipment, the assigned Electrician's

'i-CRMOSTAT

-^/wvwwwwww-

VVWWVWWVWVW-n/WWWWWWW-

. CUTOUT

—|T1 j BL0CK

HEATING
UNITS

LINE DISCONNECT SWITCH

33

r

2

440V|

x y z

CONTACTORS

J PHASE LOADING (KV> PER PHASE) X-Y W 4S 45 90

Figure 11-27.—Wiring diagram of type 90 fry kettle.

Mate must determine the cause of malfunction. The methods and procedures for locating grounds, opens, and shorts in heating units are the same as those described in chapter 5 of this training course. When the trouble has been determined, the defective unit is removed from the equipment and replaced with a new one obtained from the spare parts box.

Refer to the manufacturers' technical manuals for instructions concerning the servicing of the electric galley equipment installed aboard your ship. These manuals also include the methods of removing and replacing the various heating units, thermostats, switches, contactors, and other components of electric cooking equipment.

nil

QUIZ

1. What is the shipboard installation called that is used to neutralize the disturbance of the earth's magnetic field caused by the ship?

2. What is the function of a Navy magnetic range station ?

3. Into what two classifications is the magnetic field of a ship divided ?

4. What is the process called that reduces a ship's permanent magnetization to a standard value ?

5. Upon what two factors does the strength of a ship's induced magnetization depend ?

6. What component of the ship's induced magnetization changes when the magnetic latitude changes and when the ship rolls or pitches, but does not change when the heading changes?

7. If at a given magnetic latitude the ship changes heading from north to east, in what manner does the longitudinal component of induced magnetization change ?

8. What is the general arrangement of the M or main degaussing coil on a ship?

9. Which degaussing coil encircles the forward one-fourth to one-third of the ship usually just below the forecastle or other uppermost deck?

10. Which degaussing coil encircles the after one-fourth to one-third of the ship usually just below the quarterdeck or other uppermost deck?

11. How is the FI coil connected with respect to the QI coil so that the same current flows in both coils?

12. Which degaussing coil consists of loops in vertical planes that are parallel to the frames of the ship?

13. Which degaussing coil consists of loops in vertical fore-and-aft planes?

14. To what is the field strength of a degaussing coil proportional ?

15. What degaussing coil is identified with the letters, CC ?

16. How is the degaussing coil connection box information presented so that the EM will know when he looks in the box (a) the conductors that may be reversed, (b) the arrangement of parallel circuits, and (c) the spare conductors?

17. How are the degaussing coil currents controlled in the degaussing installation shown in figure 11-7?

18. In the motor generator installation (fig. 11-8), how are the degaussing coil currents controlled?

19. What signal is derived from the ship's gyrocompass system for an L coil or an FI-QI coil in the SM control equipment?

20. The operation of the degaussing installation should be in accordance with what instructions?

21. How often should the EM check the degaussing coil currents ?

22. Which plate of the neon indicator light glows when the polarity of the degaussing coil current is (a) positive and (b) negative?

23. A positive (direction) current through a degaussing coil is indicated by what

deflection on the central zero ammeter located on the degaussing control panel ?

24. What degaussing coil current must be changed when the ship moves from one Z zone to another?

25. What degaussing coil currents must be changed if the ship moves to a different H zone, or if the heading changes to a different sector?

26. What provisions are made to operate motor-driven rheostats and pushbutton polarity changers in the event of a power failure?

27. Why should the current always be reduced to zero before operating the polarity changer?

28. In degaussing installations employing manually operated rheostats, where is the degaussing switchboard usually located?

29. Why is a remote degaussing panel located in the pilot house or chart house?

30. How often and for what length of time should the degaussing coils be energized when the system is not in normal use?

31. How often is each degaussing coil checked for grounds by insulation resistance measurements?

32. Where with respect to the coil disconnect switch are the insulation measurements taken?

33. In the F, Q, and A, or I, P, and A coils that have a reversing switch on the coil side of the disconnect switch, what should be the position of the reversing switch (open or closed) when the insulation measurements are taken?

34. How often should the drain plugs be removed from through boxes and connection boxes to drain any accumulated water from them?

35. Corrective maintenance on degaussing equipment consists principally of what type of work?

36. In the anchor windlass installed aboard destroyers

(a) the starboard hoist direction and (b) the port hoist direction are also used for what additional purpose ?

37. What are the actions of the electric brake on the anchor windlass equipment (a) if power is applied and

(b) if power is disconnected or fails?

38. If the anchor windlass is run with the locking hand-wheel in the LOCKED position (a) what will be the action of the wildcat and (b) what should be the condition of the chain?

39. Why is it good practice, before starting the anchor windlass motor, to press the RESET-EM-RUN pushbutton in the across-the-line starter?

40. To prevent operation of master switch No. 1 when master switch No. 2 is in operation (fig. 1 1- 19), what action must be taken by the operator?

41. For safety in operation of the electric bomb elevator (fig. 1 1 - 14), how are the doors at each level arranged?

42. What provision in the electric bomb elevator is made to guard against overspeed in the DOWN direction?

43. In order that the control circuit (fig. 11-20) be energized and set up for operation, what must be the condition of the overtravel, slack cable, door switches, stop pushbuttons, and overload relay contacts?

44. The electrohydraulic steering gear in a destroyer consists essentially of what units?

45. What is the rating of each motor that drives the port and starboard main pumps of the

power unit of the electrohydraulic steering gear?

46. What operates the follow-up shaft, which is geared to the upper level gear in the differential control unit (fig. 11-21)?

47. What function is provided by the synchro transmitter in the steering console in the pilot house (fig. 11-22) and the associated synchro receiver on each pump differential control box in the steering compartment with cable selector switches in both locations?

48. The motor for each steering gear pump (fig. 11-23) is provided with what type of starter?

49. Electric galley equipment aboard naval vessels consists essentially of what five types of units ?

50. How are the surface units in the type A electric range individually controlled?

51. How often should the electrical circuits in electric galley equipment be inspected?

52. How often should ground tests be taken on electric galley equipment?

CHAPTER

SATURABLE REACTORS

INTRODUCTION

Amplification of voltage and power can be accomplished by means of the magnetic amplifier, which employs as the controllable element an iron-core saturable reactor. The magnitude of the impedance of the saturable reactor depends upon the range of flux change that occurs in its core, and this action, in turn depends upon the magnitude of the control current. If the control current is varied a small amount, the power delivered to a load is varied through a much wider range. Herein lies the action of amplification.

The magnetic amplifier has certain advantages over other types of amplifiers. These include (1) high efficiency (90 percent); (2) reliability (long life, freedom from maintenance, reduction of spare parts inventory); (3) ruggedness (shock and vibration resistance, overload capability, freedom from effects of moisture); and (4) no warmup time. The magnetic amplifier has no moving parts and can be hermetically sealed within a case similar to the conventional dry-type transformer.

Also, the magnetic amplifier has a few disadvantages. For example, it cannot handle low-level signals; it is not useful at high frequencies; it has a time delay associated with magnetic effects; and the output waveform is not an exact reproduction of the input waveform.

The magnetic amplifier is important, however, to many phases of naval engineering because it provides a rugged,

trouble-free device that has many applications aboard ship. These applications include throttle controls on the main engines; speed, frequency, voltage, current, and temperature controls on auxiliary equipment; fire control, servomechanisms, and stabilizers for guns, radar, and sonar equipment; and pulse-forming sweep multivibrator circuits for radar and loran equipment.

Early Types of Saturable Cores

Early saturable reactors employed ordinary transformer silicon-steel cores. The amplifying qualities of these devices were not very satisfactory because of the relatively low-saturation flux density and high hysteresis loss.

To introduce the concept of controlling the magnitude of the current through a load by means of the self-induced voltage in a reactor (fig. 12-1), apply a 60-cycle, 117-volt source having a sine-waveform across a series circuit containing a variable inductance (the controlled element) and a fixed resistor (the load). The d-c control voltage is discussed later.

The circuit (fig. 12-1, A), although not an accurate analogy of magnetic amplifier action, represents the control of the magnitude of load current by utilizing the induced voltage inherent in the reactor, L.

Lenz's law states that the induced emf in any circuit is always in such a direction as to oppose the effect that produces it. Because the current in an a-c circuit is always changing, the opposition of the induced voltage is continuous. Increasing the induced voltage in the series circuit comprising r and L will reduce the magnitude of the circuit current and cause it to lag behind the source voltage by an increasing angle (fig. 12-1, B). This action reduces the circuit power factor.

The obvious advantage of controlling the circuit current with adjustable inductance is the absence of appreciable heat loss in the control element. The obvious disadvantage is low circuit power factor.

Magnitude of the inductance may be varied in a number of ways. For example, doubling the number of turns will quadruple the inductance. (Inductance varies as the square of the number of turns.) Also increasing the

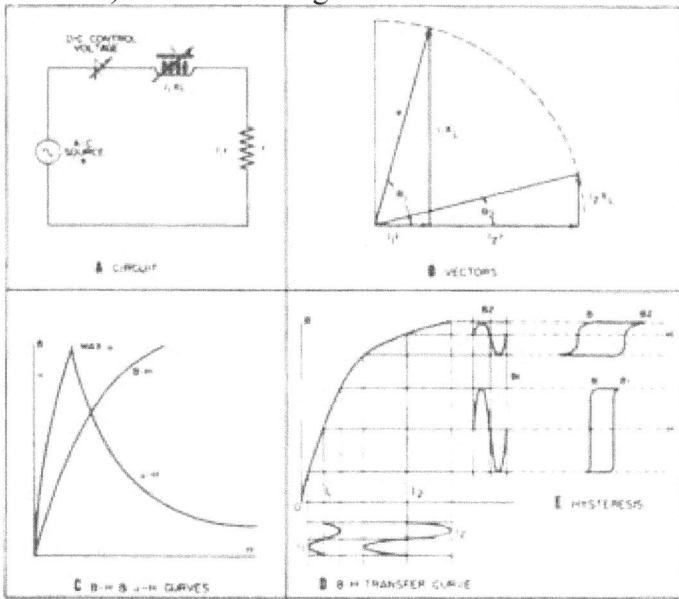

Figure 12-1.—Load current controlled by variable inductance.

permeability of the core will increase the inductance. (Inductance varies directly with permeability.)

The latter action may be accomplished in a number of ways. For example, if the reactor core is air, the permeability, n, will be unity. If a laminated silicon-steel core is gradually inserted into the coil, the permeability will increase (fig. 12-1, C) toward the point of maximum The coil impedance will increase and the circuit current and power factor will decrease.

A review of magnetism and magnetic circuits in Basic Electricity, NavPers 10086 will help in the understanding of the rest of the chapter. Another way to vary the permeability of the silicon-steel core of the reactor is to introduce a d-c control voltage (dotted battery) in series with the circuit of figure 12-1, A. Increasing the d-c voltage will increase the d-c ampere turns and H (the ampere turns per centimeter are approximately equal to H). Hence, the flux density, B (fig. 12-1, D) will increase with H and the d-c control voltage. The permeability

($M = B/H$) decreases to the right of the point of maximum, fi (fig. 12-1, C) as the d-c control voltage (and H) continue to increase.

The effect of these actions on the load current (a-c component) is represented by projecting the load current curve, ii (fig. 12-1, D) to the B-H curve and transferring the projection to the flux density curve, Bl. Thus, with low values of d-c control voltage and direct current Ii, the associated flux excursions in the reactor core are relatively large (curve Bl). The resulting induced voltage across the reactor is high and the load current, curve 11, has low amplitude.

Voltage i i r across the load for the condition of large flux change in the coil is relatively small (fig. 12-1, B), and the voltage, JiXl, across the coil is relatively large. The circuit current lags the source voltage by a relatively large angle, 9 lt and the circuit power factor is low.

Increasing the d-c control voltage and direct current from Jito 12 will partially saturate the core so that smaller flux excursions (curve B2) will occur with correspondingly reduced magnitude of induced voltage and increased load current (curve i 2). The voltage, i 2 r, across the load for the condition of small flux change in the coil is increased (fig. 12-1, B), and the voltage, I2XL, across the coil is decreased. The angle, 8 2 , by which the circuit current lags the source voltage is decreased, and the circuit power factor approaches unity.

With increased values of d-c control voltage and flux density the cycle of magnetization is reduced (fig. 12-1, E) from Bl to B2.

The load current (a-c component) in this example, is relatively insensitive to small changes in d-c control current. Also, the series reactor is never driven vary far into saturation so that its impedance never drops to a very low value. For a given value of control current, the full cycle of the corresponding hysteresis loop is completed for each cycle of source voltage. Thus, the gain of the circuit (ratio of a-c load power change to d-c control power change) is relatively low.

To obtain greater circuit gain, certain changes must be made. The reactor core is driven into saturation periodically. This action allows the load current to flow for a controlled portion of each cycle. Before saturation is

reached, the flux change prevents current flow through the load for another controlled portion of each cycle. The result of operating the reactor core in the region of saturation for a portion of each cycle will increase the circuit gain materially. This action is described later in this chapter.

The evolution of a practical magnetic amplifier has resulted from the recent development of high quality steels, gapless construction of the magnetic circuit, special low-leakage rectifiers, and self-saturating magnetic circuits. The improvement in processing magnetic materials and the successful development of dry-disk or metallic rectifiers have contributed principally to the wide use of this device as an amplifier. High-quality steels have increased the power-handling capacity. The dry-disk rectifiers convert either the entire output or part of the output from alternating current to direct current. Variations in the control current level, like the variations in the grid voltage of a thyratron, produce corresponding variations in the output.

Materials Suitable for Magnetic Amplifier Cores

Various types of nickel-iron alloys that have more suitable magnetic properties than previous ones for use as core materials for saturable reactors have been developed and are commercially available. These materials are the (1) high permeability alloys and (2) grain-oriented alloys.

High-permeability materials, such as Permalloy A, Mumetal, 1040 alloy and equivalents have low and intermediate values of saturation flux density but relatively narrow and steep hysteresis loops. These materials are used extensively as the cores in low-level input amplifier stages.

Grain-oriented materials, such as Orthonol, Deltamax, Hypernik V, Orthonik, Permeron, and equivalents, have higher values of saturation flux density and more rectangular-shaped hysteresis loops (fig. 12-2, A) than the high-permeability materials. Grain-oriented materials are referred to as square-loop materials because of the flat top and bottom of the hysteresis loop. A

conventional loop is shown in figure 12-2, B. These materials are used as the cores in high-level output amplifier stages in which maximum permeability occurs close to saturation flux density, resulting in a substantial increase in the power-handling capacity for a given weight of core material.

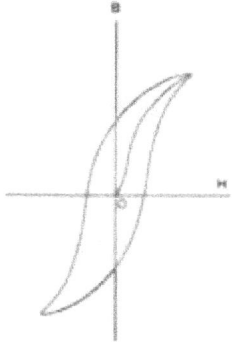

A SQUARE LOOP I CONVENTIONAL LOOP
HYSTERESIS LOOPS
Figure 12-2.-Hysteresis loops.

In the manufacture of saturable cores, the characteristics of the grain structure of the material can be altered considerably by rolling and annealing processes.

A great improvement in the magnetic properties of some materials is obtained by cold-rolling the material before it is annealed. The cold-rolling process develops an orientation of the grain in the direction of rolling. If a magnetizing force is applied to the material so that the flux is in the direction of the grain, a rectangular hysteresis loop is obtained. Thus, in some materials cold rolling produces almost infinite permeability up to the knee and almost complete saturation beyond the knee.

BASIC HALF-WAVE CIRCUIT

A description of a simple half-wave circuit (fig. 12-3, A) will be given as an example of the operating principles in general of the magnetic amplifier.

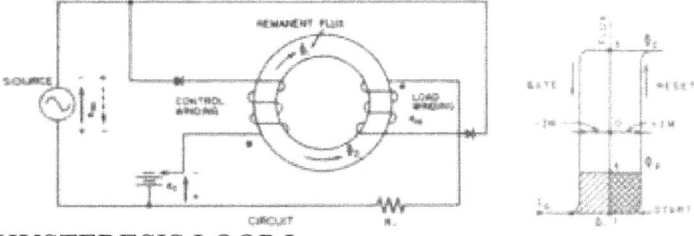

HYSTERESIS LOOP I
«S£T SATt

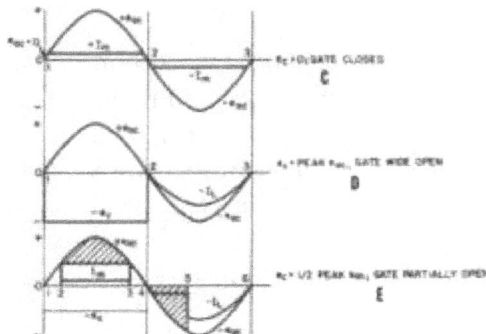

Figure 12-3.-Basic half-wave magnetic amplifier.

Windings

The magnetic amplifier contains a magnetic core made of a square-loop material upon which two windings are placed. The load, or "gating" winding, L, is connected in series with a rectifier, the load, and an a-c power supply. A second or "control" winding, C, is connected in series with a rectifier, the control signal source, and the same a-c source. The two windings have a 1:1 turns ratio. The magnetic amplifier acts in this circuit like an electrically operated contactor that gates (turns on) the load circuit periodically. A control voltage applied to the closing circuit of the contactor closes the contactor, which completes a circuit to the load. This action can be repeated periodically, for example, by introducing an a-c control voltage in series with a half-wave rectifier and the contactor closing coil.

The action of the control winding of the magnetic amplifier may be compared to that of the closing coil of

the contactor. The action of the load winding may be compared to that of the contactor's main contacts. The latter action is that of introducing a high impedance (main contacts open) for a controlled portion of each half cycle and then removing this impedance (main contacts closed) and allowing current to flow through the load during the remaining portion of the half cycle.

Polarities

In a previous study we found that transformers have polarity markings (fig. 12-4). The solid arrow at the source is marked minus at the head and plus at the tail to represent arbitrarily the positive half cycle of active source voltage. The electron flow is from the negative terminal of the source into the dotted end of the primary and returning to the positive end of the source.

<P . IS

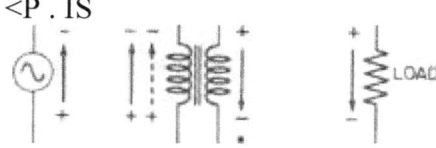

Figure 12-4.-One-half cycle of applied voltage e ac .

The solid arrow in the secondary of the transformer has the same polarity as that of the source because the secondary is the source for the load.

The dotted (dashed line) arrow in the primary winding represents the induced voltage in the primary for the first half cycle of applied voltage and has the same polarity markings as the source voltage and the secondary voltage arrows.

The direction of electron flow through the load is represented by a solid arrow, the polarity of which is opposite to that of the other three arrows. This reversal of markings is characteristic of load voltages with respect to source voltages. For voltages across loads, the arrow head is on the positive side and the tail is on the negative

side. For voltages originating in generators,transformers, batteries, etc., the arrow head is on the negative side and the tail is on the positive side.

In the example of figure 12-3, A, the solid arrow at the source indicates the direction of electron flow through the circuit during the positive half cycles of applied voltage, e^. The polarity ^markings (dots at one end of the windings) are indicative of the way the turns are wound on the core. The dotted ends of the windings of a core are assumed to always have a particular instantaneous polarity with respect to the undotted ends of the windings. Also, the dotted ends of two or more windings on a common core are considered to have the same instantaneous polarity with respect to each other. For example, in figure 12-3, A, if the voltage applied to the control winding is of a polarity at some instant to cause current to flow INTO the dot-marked end of that winding, the induced voltage of the other winding will be of a polarity (at the same instant) such as to cause current to flow OUT of the dot-marked end of that winding. The control voltage, e_c> is assumed to be a direct voltage. The rectifier arrowheads are pointed in the direction of electron flow. This direction is opposite to conventional commercial usage.

Function of Rectifiers

Rectifiers are placed in the load and control circuits to prohibit current flow in the control circuit during the gating half cycle and in the load, or gating, circuit during the reset half cycle. The magnetic amplifier is not an amplifier in the sense of a step-up transformer. Voltages generated by mutual induction (transformer action) between the control and load windings exist in these windings, but they have only a small effect on the amplifier operation under the established conditions.

During the first half cycle (solid arrow of source) of the applied voltage, the direction of the INDUCED voltages in both windings is INTO, or positive, at the polarity-marked terminals. In the load winding this action is in the forward direction of the rectifier and against e^.. Thus, the rectifier in the load circuit prevents current flow through the load and is subjected to an inverse voltage equal to the difference between e^and the mutually

induced voltage in the load winding. The time interval, corresponding to the first half cycle, is called the "reset" half cycle. The reset action is described later.

Analysis with Zero D-C Control Voltage

As mentioned previously, figure 12-3 is used in the analysis of the action of the basic half-wave magnetic amplifier. Figure 12-3, A, represents the basic circuit. Figure 12-3, B, represents the square-type hysteresis loop for the core material used in this circuit. Figure 12-3, C, D, and E, represents the waveforms of current and voltage for three conditions to be considered. The symbol representing a quantity is common to all parts of the figure. For example, the magnetizing current,! m , is represented in figure 12-3, B, C, and E. The hysteresis loop is enlarged for clarity and is not drawn to the same scale as parts C, D, and E.

RESET HALF CYCLE.-The first condition to be described is with the control voltage, e^, at zero. At the beginning of the reset half cycle, the core is assumed to possess a residual or negative saturation remanent flux level, *i (fig. 12-3, B). The direction of this flux is indicated by the arrow, +i, in figure 12-3, A. As e^ increases from 0 in a positive direction (indicated by the solid arrow at the source, fig. 12-3, A) and by the part of the sine curve, point 1 to point 2 (fig. 12-3, C), the current in the control winding establishes an mmf represented by half the width of the hysteresis loop, +1^ (fig. 12-3, B). The applied voltage establishes an mmf that acts in a direction to oppose the residual core flux, 4> if and therefore to demagnetize the core. The amount of change of flux will depend upon the MAGNITUDE of the applied voltage across the control winding and the TIME INTERVAL during which this voltage is applied.

In this example, the first half cycle of applied voltage is assumed to reverse the core magnetism and to establish its flux density at essentially the positive saturation level, 4> 2 (fig-12-3, B). This action is called reset.

As e ac increases from 0 in a positive direction in the vicinity "of point 1 (fig. 12-3, C), there is no change in core flux until the current increases to the value of +I ro , corresponding to one-half of the width of the hysteresFs

loop. Thus, with no flux change, the current rises abruptly and is limited only by the resistance of the circuit and the low value of e^. When the current reaches the value, +I ra , the core flux~starts to change from the <J>i level toward "the <P 2 level (fig. 12-3, B). The accompanying self-induced voltage opposes e ac and limits the current to a constant small value during the flux excursion from the level <t>\ to <t>2 -

This flux change continues during the time interval between point 1 and point 2 (fig. 12-3, C). As it continues, the induced voltage continues to vary in magnitude with e ac and to oppose e^ in such a manner that I,,, remains constant over the halF-cycle interval.

GATING HALF CYCLE.-The next half cycle is called the "gating" half cycle. It starts at point 2 (fig. 12-3, C), at which time the polarity of the applied voltage reverses. The direction of e^ for this half cycle is indicated by the dotted arrow (fig. 12-3, A). During this time interval, the rectifier in the control circuit blocks the flow of control circuit current. However, the rectifier in the load circuit permits current from the source to flow in that circuit. This current will magnetize the core in a negative direction—that is, in a direction to change the flux from the <t> 2 level to the <t> l level. The applied voltage is assumed to be of the correct magnitude to cause the core to be magnetized to the <h level (fig. 12-3, B). A condition of equilibrium is indicated.

The large flux change from the <t> 2 level to the <t>i level causes a self-induced voltage in the load winding, L, and a mutually induced voltage in the control winding, C. The self-induced voltage in the load winding opposes e^. The mutually induced voltage in the control winding also opposes ea C . The rectifier in the control circuit is subject to an inverse voltage equal to the difference between §a C and the mutually induced voltage in the control windingT Because of the maximum flux change in the core, maximum impedance is presented by the load winding to the circuit containing R L throughout the gating half cycle, and therefore e^wiirappear across the load winding and not across R L . For this condition, the current through the load is limited to a very small magnetizing component that is negligible compared to normal values of load current. The gate is closed.

Analysis With Maximum D-C Control Voltage

The second condition described is for the condition that ec is equal to the peak value of e^. At point 1 (fig. 12-3, D) the remanent magnetism is again at the $ 1 level (fig. 12-3, B).

FIRST HALF CYCLE.-The applied voltage, e ac , rises from 0 to maximum during the first 90° of the cycle, but has no effect on the core flux because the control voltage, §<., has a magnitude equal to the maximum value of a^, and the polarity opposite to that of £a C . Thus, the rectifier prevents the flow of battery control current during the time that e c is greater than e ac . Because there is no voltage acrosslhe control winding (the rectifier is essentially an open circuit), no flux change can occur from point 1 to point 2 (fig. 12-3, D). Figure 12-3, B, does not apply. Thus, no change in flux occurs during the reset half cycle for this assumed condition.

SECOND HALF CYCLE.-When e^ reverses its polarity (solid to dotted arrow), the rectifier in the control circuit continues to block the flow of current in that circuit. In the load circuit the polarity of e,^ during the gating interval, points 2 to 3 (fig. 12-3, D), fs such as to tend to drive the core further into negative saturation (point la, fig. 12-3, B). Because the core is already saturated, no further flux change occurs, and e ac appears across because the load winding offers no Impedance to I L .

The full value of load current flows, and its magnitude is a^/R^. The gate is wide open. The condition is analogous - to -a thyratron tube that has no grid bias. The tube fires when the plate is only slightly positive, and conduction occurs immediately and continues for essentially the full half cycle. The waveform of this current, -II, is illustrated in figure 12-3, D.

Analysis with Partial D-C Control Voltage

The third condition assumes that e^. is approximately half peak value of e^. During the reset "half cycle, voltage is applied to the control winding during the time interval from point 2 to point 3 (fig. 12-3, E). The magnitude of this voltage is e ac - e c . This voltage will be less than the

peak value of a, c , but greater than zero; and a new set of conditions will be established.

FIRST HALF CYCLE.—The reset cycle is just beginning. During the interval 1 to 2 (fig. 12-3, E), e,. is greater than e^, and the rectifier opposes any current flow in the control winding. During the interval 2 to 3, e_ac exceeds e c , and magnetizing current flows in the control windingT As mentioned previously, the extent of the change in core flux will depend on the time interval and magnitude of the voltage applied across the control winding within the half cycle. Because the time interval is very short, and the net voltage applied to the control winding is much less than e^, the core flux level is assumed to change from <Pi to the level along the line through ♦ 3 (fig. 12-3, B). During the interval 3 to 4 (fig. j 2-3, E), e^ is again less than e_c, and the rectifier prevents any further flow of control current. As in the previous examples, the rectifier in the load circuit prevents any current flow in that circuit during the reset half cycle.

SECOND HALF CYCLE.—When e^ reverses, magnetizing current flowing through the load winding changes the core flux from the level through point 3 to the level through point 1 (fig. 12-3, B). This change is assumed to take place during the interval 4 to 5 (fig. 12-3, E). The impedance of the load winding is high during this interval, and current flow through the load is restricted to the magnetizing current. However, at point 5 the core becomes saturated, and no further flux change occurs. The impedance of the core drops to zero, and current e^/R^ flows through the load during the interval 5 to 6.~The load voltage is in phase with e ac and has the same waveform as that of e ac for this partof the cycle.

Energy Considerations

The area of the portion of the hysteresis loop traversed is a measure of the energy required to complete that particular cycle of magnetization. It may be divided equally by the y axis.

For condition 1 (fig. 12-3, C) the entire right-hand area of the loop shown is figure 12-3, B, is proportional to the area under the +63,. voltage wave (which is proportional to the energy supplied), and the left-hand area is proportional to the area under the -e ac voltage wave.

For condition 2 (fig. 12-3, D), the area in both half cycles is zero (no flux change occurs), and the load, R L , absorbs the entire applied voltage, e ac . No energy is supplied to the control winding.

For condition 3 (fig. 12-3, E), the right-hand shaded area of the hysteresis loop (fig. 12-3, B) is proportional to the shaded area under (fig. 12-3, E), and the left-hand shaded area of the hysteresis loop is proportional to the shaded area of -e ac . In this case the magnitude of (e ac - e c) applied to the "control winding determines how far"the core flux is carried from negative saturation toward positive saturation, and consequently, how much of the gating half cycle will be nonconducting.

For condition 3, the energy supplied to the control winding is partially reduced. The corresponding area is reduced from the total area under +e ac for condition 1 to that indicated by the shaded portion under the +e ac curve in figure 12-3, E. The flux change is not carried to level ^{2} on the hysteresis loop, but to some point part way up the loop (level $<f> 3$), as determined by the shaded area under the -fe$^$ curve.

In other words, the time in the gating half cycle at which the core saturates (the firing angle) is determined by the shaded area under the +e ac curve or the amount of energy supplied to the control winding during the reset half cycle. Thus, as e$^$ is reduced in magnitude, the voltage, e$^$ - e c , applied to the control winding increases, and the core flux is carried further toward positive saturation during the reset half cycle.

This action increases the firing angle and delays the time in the gating half cycle when the core saturates and the load winding becomes conducting. Thus, the average load current and load voltage are reduced. They both vary with the control voltage.

By way of contrast with the relative low-circuit power factor of the example in figure 12-1, the power factor of the circuit of the example.in figure 12-3 is unity for all values of d-c control voltage between zero and maximum.

BASIC FULL-WAVE CIRCUIT

Basic circuit and waveforms for a full-wave bridge magnetic amplifier circuit are illustrated in figure 12-5.

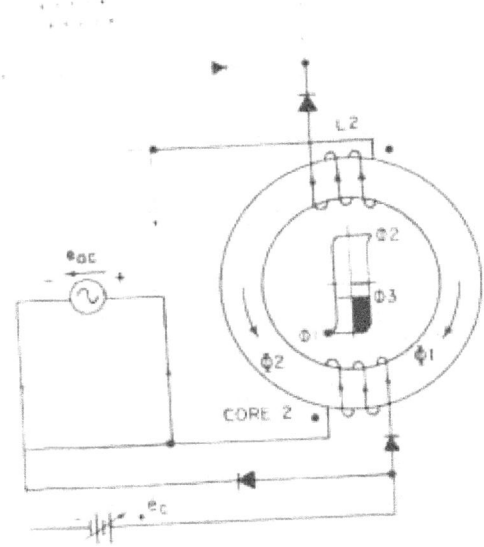

CORE 2

F CYCLE, RESET 2 GATE I

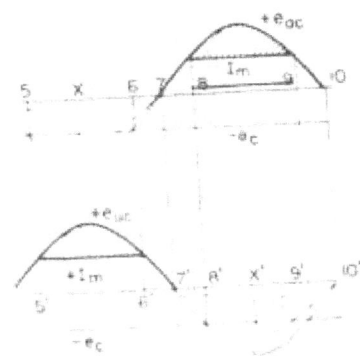

VI WHEN e c =£e ac PEAK

The core material is assumed to have the same properties as those of the rectangular hysteresis loop core employed in the half-wave circuits. Figure 12-5, A and B, represents the same magnetic amplifier. The circuits and the core hysteresis loops are repeated in order to identify the actions for each half cycle of applied voltage. Thus, the brown and red circuits (fig. 12-5, A) represent the active (conducting) circuits for the first half cycle, and the green and blue circuits (fig. 12-5, B) represent the active circuits for the second half cycle.

Brown and red portions of the sine waveform of applied voltage e ac (fig. 12-5, C) occur simultaneously with the flux changes represented by the yellow and red portions of the hysteresis loops (fig. 12-5, A) for cores (T) and ©, respectively, for the first half cycle of applied voltage, e ac .

Similarly, the green and blue portions of the sine waveform of applied voltage eac (fig-12-5, C) occur simultaneously with the flux changes represented by the green and blue portions of the hysteresis loops (fig. 12-5, B) for cores (T) and © , respectively, for the second half cycle of applied voltage e ac .

In the step-by-step analysis, two conditions are considered: (1) $e^\wedge = 0$ (fig. 12-5, C) and (2) e £ = 1/2 $e^\wedge$ peak (fig. 12-4, Dj7 Before $e^\wedge$ is applied, the remanent flux of core (f) is assumed to be $<t>i$, and the remanent flux of core ® is assumed to be $<£2$.

Analysis With e ? Equal to Zero

When $e^\wedge = 0$, the two circuits that are active (conducting) for the first half cycle of e ac are shown in brown (control, CI) and red (load, L2) of figure 12-5, A. The two circuits that are active for the second half cycle are shown in blue (control, C2) and green (load, LI) of figure 12-5, B.

POLARITIES.—The applied voltage, $e^\wedge$ (brown) is considered to be positive when the direction is OUT of the polarity-marked terminal of CI. At the same time (red) will be negative for core @ because the direction is INTO the polarity-marked terminal of L2.

Obviously, the waveforms of the source voltage cannot be both positive and negative at the same instant with respect to the direction through the source. However,

the waveforms of $e^\wedge c$ are positive with respect to CI and negative with respect to L2 at the same time because of the manner in which ev is applied to CI and L2. As stated before, $e^\wedge$ is considered to be positive with respect to CI because the direction is OUT of the polarity-marked terminal of CI at the same time it is negative with respect to L2 because the direction is INTO the polarity-marked terminal of L2.

FIRST HALF CYCLE.—As e ac increases from zero in the first half cycle (points 1 to JT, fig. 12-5, C, brown and red), the direction of $e^\wedge$ is against the rectifier in series with C2 and with the rectifier in series with CI. Therefore, no current flows in C2 and a current will flow in CI. The magnitude of the current in CI is assumed to be proportional to half the width of the hysteresis loop for core (l). The extent of the flux change depends on the magnitude of e ac ;and for the condition, e c = 0, the current flowing in CI Ts assumed to change the~flux from negative saturation 4i to positive saturation $<t>2$ - This action is said to RESET the flux in core (T), and this portion of the cycle is said to be the RESET half cycle for core (l).

During the same (first) half cycle, the direction of $e^\wedge$ is against the rectifier in series with LI and with the rectifier in series with L2 (red). Therefore, no current flows in LI, and a current will flow in L2. The path for load current includes the a-c supply and the load, R L . The magnitude of $e^\wedge$ is such that the current flowing through R L and L2 is limited to the magnetizing current, $-1m$. The magnitude of $^\wedge$ is proportional to one-half the width of the hysteresis loop, and during this half cycle the flux of core © is changed from positive saturation remanence $4>2$ to negative saturation remanence $1. This half cycle is called the GATING half cycle for core ©. As mentioned previously, it is also the reset half cycle for core (T). Because the flux change in core (2) is a maximum, the impedance developed byL2 is a maximum, and eac appears across L2. The load voltage and the load component of current are zero. This condition corresponds to the condition when $e^\wedge$ is equal to zero.

To summarize the action in the two cores during the first half cycle of applied e ac (fig. 12-5, A):

1. Magnetizing current 4T m flows in CI and resets the core (T) flux from $4>i$ to $<t>2$ (brown).

2. Magnetizing current $-1^\wedge$ flows in R L and L2 and gates (changes) the core (2) flux from $<t>2$ to$^\wedge j$.

3. The rectifiers prevent the flow of current through LI and C2.

4. The flux change through L2 is a maximum; therefore, the impedance of L2 is a maximum, and the load component of current through Bl is zero.

SECOND HALF CYCLE.—Duringthe second half cycle of e ac (points 2 to 3, fig. 12-5, C, green and blue), the polarity of e ac is opposite to the polarity of $e^\wedge$ during the first half -cycle.

As stated before, the polarities of e^c for the second half cycle are indicated in figure 12-5, B, and the paths for current flow are represented in green for the load winding, LI, of core (T) and in blue for the control winding, C2, of core ©.

As e^c increases from zero (points 2 to 3, fig. 12-5, C), the direction of $e^\wedge$ is against the rectifier in series with Cl and with the rectifier in series with C2 (blue). Therefore, no current will flow in Cl and a current will flow in C2.

The two cores are assumed to be identical; and for the condition, e c = 0, the magnitude of the current in C2 is proportional" to half the width of the hysteresis loop for core ©. The current through C2 is assumed to reset the flux in core © from negative saturation remanence

to positive saturation remanence This half cycle is the reset half cycle for core ©. (Note that during the first half cycle the core ® flux was changed from $<t>$ 2 to

(red), now it has changed back again from $<t>$i to $<t>$ 2 (blue).)

During the second half cycle of applied voltage, the direction of e ac is against the rectifier in series with L2 (fig. 12-5, B) — and with the rectifier in series with LI (green). Therefore, no current will flow in L2 and a current will flow in LI. The latter path includes the a-c supply and the load, Rl. The direction of flow through the a-c supply is opposite to that for the first half cycle, but the direction through R L is the same as it was during the first half cycle because of the arrangement of the rectifiers. As in the first half cycle, the magnitude of the current flowing through LI is limited to the magnetizing current, $-j^\wedge$, and is proportional to half the width of the hysteresis "loop. During the second half cycle, the

core (T) flux is changed from positive saturation rema-nence $ 2 t0 negative saturation remanence$^\wedge$.

(Note that during the first half cycle the core (T) flux changed from 4>i to 4> 2 (yellow), now it has changed back from $2 t0 ^1 (green).) The second half cycle is the gating half cycle for core (t). Because the flux change through LI is a maximum, the impedance of LI is a maximum, and §ac appears across LI. The load voltage and the load component of current are zero. As stated before, this condition corresponds to §c =0.

To summarize the action in the two cores during the second half cycle:

1. Magnetizing current -I ra flows in LI and gates the core (I) flux from 4> 2 to $<t>$Y (green).

2. Magnetizing current I m flows in C2 and resets the core (2) flux from $<h$ to 4>i (blue).

3. The rectifiers prevent the flow of current in CI and L2.

4. The flux change through LI is a maximum; therefore, the impedance of LI is a maximum, and the load component of current through RL is zero.

Analysis With e ? - ViQgc Peak

The second condition assumes that the control voltage, g c , is increased from zero to e c = 1/2 $e^\wedge$ (peak). The waveforms for this condition are illustrated in figure 12-5, D.

FIRST HALF CYCLE.-Waveforms for the first half cycle are indicated in yellow and red between points 1 and 4 and between points 1' and 4'. During this time the direction of e ac (fig. 12-5, A) is with the rectifier in series with CI and against the rectifier in series with C2. Therefore, CI is active; C2 is not. Similarly, the direction of e ac is with the rectifier in series with L2 and against the rectifier in series with LI. Therefore, L2 is active; LI is not. The active circuits are shown in brown and red to correspond with the yellow and red waveforms of applied voltage that appear across CI and L2, respectively.

At the beginning of the half cycle, the core (J) flux is assumed to be at negative saturation remanence ^ , and the core (2) flux at positive saturation remanence <£ 2 .

Control voltage e^ is opposed to e ac in the active circuit that includes CI. From point 1 to point 2 (fig. 12-5,

D), £ac is less than e £ , 311(1 no current will flow in CI. From point 2 to point 3, e^ is greater than e c and a magnetizing current, I^, will flow in CI. The extent of the core © flux change caused by this current flowing through CI is dependent upon the magnitude of e ac —e,. appearing across CI. The energy required to reseffhe core © flux is proportional to the shaded area (brown) under e^ between points 2 and 3 (fig. 12-5, D) and also to the shaded area (yellow) of the core © hysteresis loop between the <t> l and<f> 3 level. (It is assumed that the core © flux is reset during this interval from < t> l to <t>, .)

During the first cycle the current in L2 (red) is limited to the magnetizing component, -I m , and the core © flux is changed from positive saturation remanence <f> 2 to negative saturation remanence 4> l .

To summarize the actions occurring during the first cycle, a magnetizing current,J[m , flows in CI for a portion of the half cycle and resets the core © flux from <^ to Magnetizing current flows in L2 for the entire half cycle and changes the core (2) flux from <t> 2 to 4> l .

SECOND HALF CYCLE.—The waveforms for the second half cycle are illustrated in figure 12-5, D, between points 4 and 7 and 4' and 7'. At points 4 and 4', e^ reverses. During the second half cycle (fig. 12-5, D) the direction of e ac is with the rectifier in series with C2 and against the -rectifier in series with CI. Therefore, C2 is active and CI is not. Similarly, the direction of e^c is with the rectifier in series with LI and against the rectifier in series with L2. Therefore, LI is active and L2 is not. The active circuits are shown in green and blue to correspond with the green and blue waveforms of voltage and current acting in LI and C2, respectively.

At the beginning of the second half cycle, the core © flux is at <f>3 (green, fig. 12-5, B), and the core © flux is at <t>x (blue). From point 4 to point X (fig. 12-5, D) the current flow through LI (green) is limited to the magnetizing current -I w and the core © flux is changed from $3 to <t>\. The impedance of LI is high during this interval.

At point X the core © flux stops changing as it is driven into negative saturation along the 4>i level. Therefore, at point X the impedance of LI drops to alow value, and the current through LI and the load, R L , increases

sharply. The load current, -II, assumes the waveform of e.ac from point X to point 7 (green, fig. 12-5, D), and this Interval represents the conducting portion of LI for the second half cycle. Note that e c was increased from zero to 1/2 e^ peak at points 1 and 1' (fig. 12-5, D), and that load current, I L (green) does not begin to flow until more than 1/2 cycfe later at point X.

During the interval from point 4' to point 5' (blue, fig. 12-5, D), e^ is less than e c , and the rectifier in series with C2 prevents the flow of current in C2. From point 5' to point 6', e ac is greater than e^, and magnetizing current I m (blue) will flow in C2. During the previous half cycle Die core © flux was changed from <f> 2 to ^i (red). The core (2) flux now changes from 4>i to 4> 3 (blue). The energy required to change the core (2) flux from <t>i to <f>3 is proportional to the shaded area of the core ® hysteresis loop (blue, fig. 12-5, B) and also to the shaded area under the^ ac curve from point 5' to point 6' (blue, fig. 12-5, D). ~~

To summarize the action occurring during the second half cycle, a magnetizing current, -1,^ flows in LI for a part of the half cycle unit core (T) saturates; then the core (D flux stops changing, the impedance drops, and load current -II (green) flows through LI and R L . Magnetizing current I 5 (blue) flows for a portion of~Ehe half cycle and resets the core (2) flux from <^ to <£ 3 .

THIRD HALF CYCLE.-At points 7 and 7', e ac reverses. During the interval from point 7 to pointflu and from point 7' to point 10', the direction of e ac is the same as that for the first half cycle previously described. Therefore, the active circuits are CI and L2 (brown and red, respectively). In the first half cycle, the core (T) flux was reset from ^ to <->$; during the third half cycle the core (l) flux is again reset from 4> l to $ 3 . In the first half cycle the core (2) flux changed from <t> 2 to 6]. In the second half cycle the core (2) flux changed from $x to 0 3 . The change occurring in the core (2) flux during the second half cycle is much less than that occurring in the first half cycle. In the third half cycle the core © flux can only change from $ 3 back to <fn , and this change is completed in the early portion of the half cycle (point 7' to point X (red) fig. 12-5, D). At point X* , core © is driven into negative saturation at the ^ level

(red, fig. 12-5, A), and the flux variation ceases. The Impedance of L2 drops to a low value, and load current -1^ flows through L2 and R L . The waveform of -L follows that of -e ac from point X~to point 10'.

In subsequent half cycles of applied voltage the core © flux changes from <t> l to 4> 3 as the core ® flux changes from $ 3 to f|j then the core ® flux changes from <£ 3 to 4>i as the core (2) flux changes from 0j to <f> 3 .

Thus, in the full-wave circuit, simultaneous actions of reset in core © (brown) and gating in core © (red) occur during one half cycle. During the second half cycle, the functions reverse and gating occurs in core © (green), and reset occurs in core © (blue).

In both cores the time during which load current will flow depends upon the distance from the <f>i level along the hysteresis loops toward the 4>i level that the cores axe reset; and this action is controlled by the magnitude of the control voltage, e L . As є L is increased from zero to the peak value of e^, the reset distance along the hysteresis loops is decreased from maximum to zero, and the conducting portion of each half cycle through R L and the load windings is correspondingly lengthened". This action increases the average value of the load current from zero to the maximum value.

The amount of power supplied to the control windings is very small compared with the amount of power in the load. Hence, the power amplification is relatively large.

TRANSFER CHARACTERISTICS

In the preceding description the core material has been assumed to have a rectangular hysteresis loop with high remanence (residual magnetization), and the rectifiers have been assumed to possess no leakage current in the direction opposite to rectification. The transfer characteristics are indicated in figure 12-6 for the ideal magnetic amplifier at (1) and for the magnetic amplifier that falls short of the ideal at (2) and (3).

Effect of Low Remanence

Low remanence permits the core material to reset partially without regard to the magnitude of the control

POWER OUTPUT

© IDEAL CHARACTERISTICS

(D CHARACTERISTICS WITH LOW REMANENCE
© CHARACTERISTICS WITH RECTIFIER LEAKAGE
0
CONTROL VOLTAGE
e_c

Figure 12-6.-Transfer characteristics of magnetic amplifiers.

voltage during the reset half cycle. This action is erratic; it results in the load winding absorbing a greater part of the applied voltage during the gating half cycle, with less applied voltage available across the load when e_c is maximum. This effect is most prominent at the full-load output end of the transfer characteristics at (3) in figure 12-6.

During the reset half cycle, rectifier leakage in the load circuit will allow a current to flow through the load winding in a direction that will partially reset the core. This action will produce similar detrimental effects, at full load, to those caused by cores having low remanence. At reduced output the voltage induced in the load winding by transformer action from the control winding during the reset half cycle is in a direction to reduce the back voltage and therefore to reduce the leakage current through the rectifier. At full load the induced voltage in the load winding during the reset half cycle is zero, and the inverse voltage applied to the rectifier is the full value of . Therefore, at full load the amount of rectifier leakage and reset (flux change due to leakage) is greatest because the back voltage across the rectifier

Effect of Rectifier Leakage

fifiR

is greatest. Thus,the load voltage is reduced when e_c is maximum. However, if the back leakage current isless than the magnetizing current required in the rectangular loop material, theoretically according to this analysis, reset will not occur.

During operation there are other factors in addition to the reverse-current leakage in rectifiers that tend to reduce the output. For example, the capacitance effect of selenium rectifiers causes a frequency sensitive action to occur. Both leakage and capacity effects increase with frequency and rectifier plate area. Thus, a magnetic amplifier that operates satisfactorily at a small current rating on a 60-cycle circuit may not produce full output when operated at 400 cycles with large current rectifiers.

FULL-WAVE CIRCUIT WITHOUT CONTROL CIRCUIT RECTIFIERS AND WITHOUT CONTROL CIRCUIT $e^{\wedge}c$

In the preceding discussion of half-wave and full-wave magnetic amplifiers, the rectifiers in series with the load and control windings prevent mutually induced voltages occurring in the windings of one core from affecting the flux change of the other core. Also,e_{ac} is applied in the control circuits. Thus, in the full-wave, bridge-type magnetic amplifier circuit of figure 12-5, D, for the condition when the control voltage, e_c, is increased from zero to 1/2 e_{ac} (peak), the amplifier responds in one half cycle.

If the rectifiers in the control winding circuits and e ac are omitted, the circuit action will be complicated by the influence of the mutually induced voltage in the control winding of one core on the control current of the other core. The result of this interaction is to increase the response time.

The basic circuits for a magnetic amplifier of this type are illustrated in figure 12-7. The amplifier load circuit is equivalent to that of a full-wave rectifier employing a center-tapped transformer. Both cores employ square-loop magnetic material. The load and control windings of both cores have a 1:1 turns ratio.

Analysis With Zero D-C Control Voltage

The first condition to be described is that of = 0. The remanent flux of core (T) is at negative saturation rem-

anence $\ (fig. 12-7, A) and of core © is at positive satu-

ration remanence $ 2 . The hysteresis loop for core (T) is located in the center area embraced by core (1); the hysteresis loop for core (2) is located in the center area surrounded by core ©. Both cores are alike and both hysteresis loops are alike. The only difference is the point of operation on the loop at which the cycle begins.

FIRST HALF CYCLE.-On the first half cycle (fig. 12-7, A) the direction of e ac is with the rectifier in series with the load winding, L2, and against the rectifier in series with the load winding, LI. Therefore, current will flow in L2 and will not flow in LI. The direction and magnitude of the current in L2 are such as to cause the flux of core (2) to change from positive saturation remanence 02 to negative saturation <t>\ .

The change of flux in core (2) causes a self-induced voltage to be developed in L2 and a mutually induced voltage in C2. The self-induced voltage in L2 limits the current flowing through Rl and L2 to the magnetizing component, -I™ during the interval 1" to 2' (fig. 12-7, C).

The mutually induced voltage, e m2 , in C2 during the same interval, causes a current tollow through the control winding, CI, in such a direction as to reset the core (D flux from negative saturation remanence <t>\ to positive saturation remanence <t> 2 (fig- 12-7, A).

The change in flux of core (D causes a self-induced voltage in CI and a mutually induced voltage in LI. The self-induced voltage in CI limits the current in CI to the magnetizing current, I m , during the interval from 1 to 2 (fig. 12-7, C). This mutually induced voltage in LI is equal and opposed to e^ ; hence, no current will flow in LI during this interval.

SECOND HALF CYCLE.-On the second half cycle (fig. 12-7, B), e^c reverses polarity; the direction is with the rectifier in series with LI and against the rectifier in series with L2. Thus, current will flow in LI and no current will flow in L2. The direction of the current in LI is such as to cause the flux of core (T) to change from positive saturation <t> 2 to negative saturation .

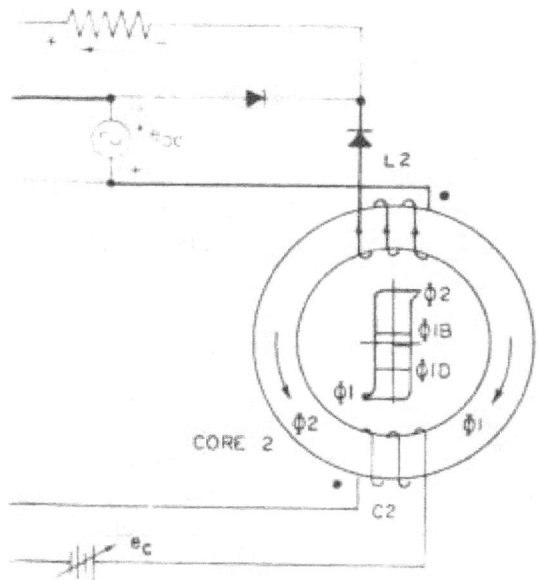

Change of flux in core (T) causes a self-induced voltage to be developed in LI and a mutually induced voltage in CI. The self-induced voltage in LI limits the current flowing through Rl and LI to the magnetizing current, -I,,, during the interval from 2 to 3 (fig. 12-7, C).

"The mutually induced voltage, e^i* in CI causes a current to flow through C2 in such a direction as to reset the core ® flux from ^ to <t> 2 (fig. 12-7, B). The change of flux in core (2) causes a self-induced voltage in C2 and a mutually induced voltage in L2. The self-induced voltage in C2 limits the current flowing in C2 to the magnetizing current, l m , during the interval from 2' to 3' (fig. 12-7, C). The mutually induced voltage in L2 is equal and opposed to e ac ; hence, no current will flow in L2 during this interval. -

A condition of equilibrium exists. During the first half cycle, e,n2 resets the core (T) flux from 4>i to<f> 2 at the same time that L2 gates the load and the core © flux changes from 4>i to <t> x .

During the second half cycle, e ml resets the core (2) flux from $i to fa at the same timeThat LI gates the load and the core (T) flux changes from <t> 2 to <f>i .

The load current is a small value, -Ig,, and e ac appears alternately across L2 and LI because of the high impedance that these windings offer to the flow of load current. The load voltage is zero, and the output power is zero.

Effect of D-C Control Voltage

The effect of a d-c control voltage on the operation of the magnetic amplifier is to unbalance the condition of equilibrium and to increase the amplifier output.

First, consider the effect of increasing the control voltage, e c , from zero to some value corresponding to half the full-load current. Because the turns ratio of both pairs of windings is 1:1, the mutually induced voltages in CI and C2 by transformer coupling with LI and L2 will not exceed e ac . The magnitude of the mutually induced voltages will "depend upon the extent of the core flux changes.

FIRST HALF CYCLE.-At the beginning of the first half cycle (fig. 12-7, A), the core Q) flux is assumed to be at negative saturation remanence 4>i and the core ©

flux at positive saturation remanence fy. The d-c control voltage acts in series with CI and C2 and is in a direction to cause both core fluxes to go toward the 0i level. This action does not occur immediately in core ©, however, because e^ is less than the magnitude required to establish a magnetizing current equal to one-half of the width of the hysteresis loop.

The direction of e^ is with the rectifier in series with L2 and against the rectifier in series with LI. Therefore, current will flow in L2 and will not flow in LI.

From point 1' to point 3' (fig. 12-7, D), -e ac is applied across L2. The magnitude of current in L2 is assumed to be proportional to half the width of the hysteresis loop for core (2). With the aid of the d-c component of ampere turns contributed by the action of e c on C2, the core ® flux will be changed from <f> 2 to ^ in the interval between points 1' and 3'. During this interval, the self-induced voltage in L2 limits the current in L2 to the magnetizing component, -Lo. At points 3' and 3 the core © flux is driven into negative saturation at the level, and the mutually induced voltage, e m2 , in C2 drops to zero. Also, at point 3' theself-inducedTvoltage in L2 ceases, and load current II flows through L2 and R L from point 3' to point 4'. The load current through R L and L2 assumes the waveform of e ac during this intervaTwhen the core (2) flux is not changing.

During the interval from 1' to 3' the core (z) flux is changing from <t> 2 to <f>i. During a portion of the same interval (points 1 to 3), the core (T) flux will be affected by the action of CI. As stated previously, the core (l) flux is at negative saturation remanence at the beginning of the first half cycle. Also, the control voltage, e c , is in a direction to hold the core (f) flux at the <t>y level?

The mutually induced voltage, e X52 > in C2 resulting from the core © flux change is applied across CI in series opposition to e c . From point 1 to point 2, -e c is greater than +e m2 , ancT the core (T) flux remains at <t>y (fig. 12-7, A). From point 2 to point 3 (fig. 12-7, D), e^ is greater than e c , and the core (T) flux changes toward the <t> 2 level. The amount of reset is proportional to the shaded area under the e^ curve. This area is proportional to the amount of energy supplied to CI during the first half cycle. Thus, the core (J) flux is assumed

to reset from to <t>i A (fig. 12-7, A) during the interval from 2 to 3 (fig. 12-7,~D). It does not reset all the way to <t>2 because e c is active in reducing the effect of e^ onCl.

In this example, the voltage applied across CI is de-rivedfromC2 (due to the core (2) flux change) and is positive during the interval from point 2 to point 3. From point 3 to point 4, e^ (dotted portion) is zero. However, the core (T) flux will not immediately change back to^i because the d-c component is not equal to one-half the width of the hysteresis loop and hence is not capable of producing a flux change.

SUBSEQUENT HALF CYCLE.-In the second half cycle (fig. 12-7, B), the core (7) flux is changed from 4> 1A back to $ i . This range is less than the full range of flux change that occurred in core © during the first half cycle \'7b<t> 2 to

). The mutually induced voltage, e ral , in CI caused by the core (T) flux change from <f> 1A to <t> 1 is applied in series opposition with the d-c control voltage, -e c . The difference voltage,^ -e c , appears across C27 The reduced average value of ^e ml (due to less than the full range of flux change in core (T)) will cause less flux change in core (2).

From point 4' to point 5' (fig. 12-7, D), -e^ is greater than e ml , and the core (2) flux remains at <f>i.~Frompoint 5' to point X', e ml is greater than -e c , and the difference voltage, e ml -e c 7 appears across C2. "The energy supplied to C2 is proportional to the shaded area under the e ral curve between points 5' andX\ This area is smaller than the shaded area under the e ra2 curve between points 2 and 3. (The larger area is portlonal to the energy supplied to CI during the first half cycle when the core (l) flux changed from <t>i to <t>^)

At point X' the core (p flux is driven into negative saturation on the $i level (fig. 12-7, B), and the mutually induced voltage, e ml , drops to zero.

With less energy supplied to C2 (point 5' to point X', fig 12-7, D), the core @ flux will reset through reduced range from <f> x to (fig. 12-7, B).

At point X' (fig. 12-7, D) the voltage, -e c , is not of sufficient magnitude to change the core (2) flux back to the <t>\ level. Thus, at the end of the second half cycle (points 7 and 7*), the core (T) flux is at the tf^ level, and the core (2) flux is at the </>ib level.

The third half cycle (fig. 12-6, A) gating occurs in core ® as the core ® flux changes from <t>\j\'7d to $1, and reset occurs in core Q) from^i to <hc. Each succeeding half cycle, the flux change becomes less and within approximately 10 cycles after e c is Increased from zero to the value corresponding to halTload, the waveforms of load current will appear, as illustrated in figure 12-7, D (point 9' to point 10' and point 11 to point 12). The corresponding control circuit currents are indicated from point 9 to point 10 and point 11' to point 12'.

If e £ is increased to the value corresponding to full-load output, both cores will, within approximately 10 cycles, approach negative saturation at 4> 1 , and no further flux change will occur in either case. Thus full-load current will flow in L2 and LI on alternate half cycles, and no voltage will be induced in either CI or C2.

If this magnetic amplifier is employed as a reactance dimmer for theater lighting, when e^ is zero the lights will be dim. As e^ is increased from zero to the value corresponding tofull - load output, the lights will gradually increase in brightness after a time lag, following the change of £c. As £c is increased, the flux change in each core occurs for a lesser part of the time in each cycle, and the range of flux change decreases during successive half cycles. Thus, the pulses of load current increase in magnitude and in length, and the lights brighten.

Disadvantages of this type of saturable reactor circuit are the time lag between the control circuit voltage change and the load response—also the instability of operation at partial loads. The tendency of the core fluxes is to go into negative saturation 4> 1 after several cycles following the increase of the control circuit voltage beyond a certain critical value.

For these reasons, this type amplifier is seldom used. The typeof magnetic amplifier shown infigure 12-4 is the preferred type.

CONCLUSION Reactors

A reactor is simply a coil connected in an a-c circuit. The flux changes direction as the current reverses, and the

opposition (impedance) to this change varies with the frequency and the amount of flux produced by a given current. Thus, if the frequency of the applied voltage is held constant, the iron-core coil having greater inductance will have greater impedance than an air-core coil. Notice, however, that if the voltage applied to the iron-core is great enough to drive the core beyond saturation, the excess voltage will "see" an impedance equivalent to that of an air-core coil. Thus, the current flowing in an iron-core coil will increase slowly with an increase in applied voltage until saturation of the core is reached. Thereafter, with an increase in applied voltage the current flow will increase rapidly.

Saturable Reactors

A saturable reactor is a reactor in which the "degree of saturation" of the magnetic core material may be independently controlled. The reactor shown in figure 12-8, A, has an output winding and a control winding. The function of the control winding is to control the degree of saturation of the core and thereby control the power delivered through the output winding to the load.

As the current supplied to the d-c control winding increases, the impedance of the a-c output winding decreases and the a-c load current increases (fig. 12-8, B). Note that the polarity of the d-c control current may be either positive or negative without altering the effect.

Because the core flux in the reactor is proportional to the ampere turns of the reactor windings, it follows that by winding many turns on the control winding and only a few turns on the output winding, a small control current can be made to control a large output current.

If the polarity of the d-c control current is not changed, the core-saturating mmf produced by the control winding will be always established in the same direction \'7b4> d , fig 12-8, A). The mmf produced by the a-c output winding, however, reverses its direction during each half cycle of the a-c supply voltage. The a-c mmf then opposes the d-c mmf one-half of each cycle and during this half cycle tends to desaturate the core. If this effect were eliminated by eliminating the "desaturating half cycle" of the output current, the change in control

R77

current required to produce a given change in output current would be considerably less; the "gain" or amplification would be increased.

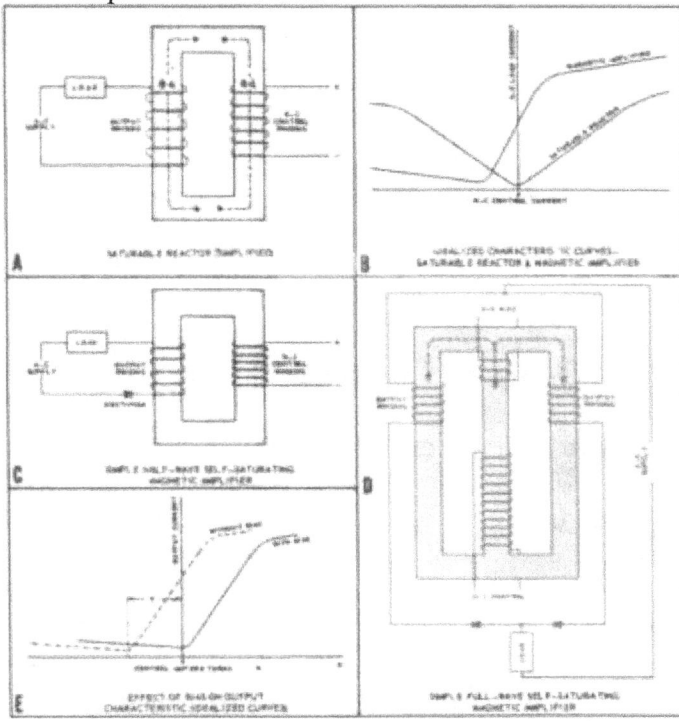

Figure 12*8.-Summary of principles of operation.

Magnetic Amplifiers

A magnetic amplifier may be regarded as a device using saturable reactors either alone or in combination with other circuit elements to secure amplification or control. A self- saturating magnetic amplifier has one or more rectifiers inserted in the output circuit to eliminate the "de-saturating" half cycle of output current. The load current is an alternating or a pulsating direct current. Figure 12-8, C, shows a simple half-wave, self-saturating magnetic amplifier.

Further refinements of the self-saturating magnetic amplifier are shown in figure 12-8, D. In this amplifier the output winding is divided into two equal sections and a bias winding is added. The output sections are arranged so as to produce alternate half-wave pulses through the output windings.

The mmf produced by the d-c bias winding, as shown by the arrows in the core, is in a direction opposite to that of the core-saturating mmf produced by the output windings. The effect of this opposition is to reduce the output current (fig. 12-8, E) corresponding to zero control current and to shift the output curve toward "cutoff." Because each output winding conducts on alternate half c y-cles, the flux in one output leg is building up while the flux in the other output leg is decaying. (The flux decays because the rectifier acts like an open switch when the source voltage is in a direction opposite to that of the rectifier.) The center leg is the common return path for the flux of each outer leg.

The result of an increasing flux in one outer leg and a decreasing flux in the other outer leg is to maintain the flux in the center leg at a constant level at the fundamental frequency of the supply voltage. Thus, no change in flux will occur in the center leg, and no voltage will be induced in the control winding at the fundamental frequency of the supply voltage.

Improved output characteristic of the amplifier is shown graphically by the upper curve of figure 12-8, B. Note that in comparison to the saturable reactor, the output current is now sensitive to the polarity of the control current.

From the characteristic curve for the magnetic amplifier (fig. 12-8, B),it may be noted that the output current is relatively high when the control current is zero. This effect is undesirable in certain applications. Compensation for this effect is made by the addition of a bias winding supplied from a separate fixed d-c voltage source with a polarity that will decrease the output current.

The effect of the bias signal is to desaturate the core and to shift the characteristic curve to the right by applying "X" ampere turns - in the negative direction (fig. 12-8, E). Note that the shape of the characteristic curve is not changed by the addition of the bias signal.

Some amplifiers are biased so that the output is a maximum when the control signal is zero. A "negative" control signal may then be used to reduce the output current. This mode of amplifier operation appears less frequently in magnetic amplifier equipment.

Feedback

Magnetic amplifiers are usually provided with several control windings, each of which has an appreciably different number of turns. The winding used as the control signal winding depends upon the signal current available and the impedance level of the signal source. The other windings may be used as additional turns for the control winding, as bias windings, or as feedback windings.

The effect of feedback in magnetic amplifiers is much the same as that in other types of amplifiers. In a self-saturating magnetic amplifier, no positive feedback (or only a very small amount) is used because the amplifier is likely to become unstable. When negative feedback is used, the power amplification and the response time are decreased, and the linearity of the control characteristic is improved.

In certain types of magnetic amplifier systems, a negative feedback signal is obtained from the output of the system and fed back to the first stage of amplification. In the first stage amplifier the feedback signal is magnetically compared with a reference signal so that the net control signal, which is called the ERROR SIGNAL, is the difference between the reference signal and the feedback signal. Thus, if the reference signal produces 10 core-saturating ampere turns, and the feedback signal produces 9 core-desaturating ampere turns, the error signal will be 1 core-saturating ampere turn. The feedback ratio is said to be "10 to 1."

If the reference signal is held constant, and, if for some reason the output of the system changes, the error signal will change in a direction that will restore the output to its original value so that it is always maintained proportional to the reference signal. When the magnitude of the reference signal is changed, the magnitude of the output signal will change proportionally, and the system will regulate about the new value of reference signal.

Conventional Schematic Representation

The schematic representations of three simple magnetic amplifier circuits are illustrated in figure 12-9. Note that the type of core on which the coils are wound is shown above each of the three schematics. Each coil is placed adjacent to the core on which it is wound. The coils and cores around which a dotted line has been drawn are usually enclosed and "potted" in a single "can"; the internal coil connections are not accessible.

ffl'

"~T—1

UTROL j =>

4-J i

r~ri

i a • i 7 i

I, •

«-c

sup»lt

A-tiNCLC com

CONTROL

IDS

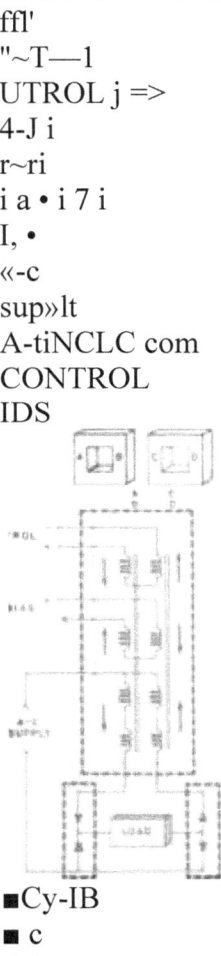

■Cy-IB

■ c

CONTROL

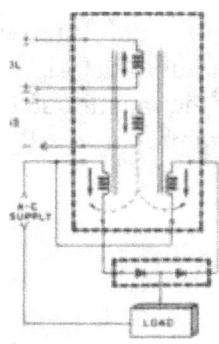

C-THREE-LEC COKE

l-TVO SINSLE CORES ONE ENCLOSURE

Figure 12-9.-Schematic representation of magnetic amplifier circuits.

The diagram shown in figure 12-9, A, is an electrical schematic representation of the circuit shown pictorially in figure 12-8, C. Figure 12-9, B, is a combination of two single cores potted in a common enclosure. The windings shown on each individual core are drawn so that, with relation to their common core, the magnetic effect of each winding is in the direction of the conventional current flow (opposite to electron flow) through the winding. This relation is shown by the arrows adjacent to the windings of figure 12-9, B. (The arrows

are not usually drawn as a part of the schematic.) The direction of the rectifiers in the figure is that of conventional current flow (opposite to the directions represented in the analysis of magnetic amplifier action, figures 12-3, 12-5, and 12-7). The conventional current flow through the output windings, as indicated by the arrows, is always in the same direction as the conventional output rectifier arrows.

Being a self-saturating magnetic amplifier, this is the direction that saturates the core and increases the output current. All windings whose current flow is in this direction are attempting to increase the output current.

Conversely, all windings whose current flow is in the opposite direction, such as the bias winding in figure 12-9, B, decrease the output current.

Figure 12-9, C, is the schematic diagram of a "three leg" reactor circuit. Two of the three-core legs are shown schematically; windings on the center leg are placed between the two legs shown as closely spaced parallel lines. The magnetic effect of the bias windings on the center leg is in a direction opposite to that of the current flow through the load windings. This effect may be visualized by following the arrows around the core, as shown by the dotted arrows in figure 12-9, C (also, see fig. 12-8, D,of which this schematic is a representation).

An analysis of the magnetic amplifier circuit shown in figure 12-9, B, reveals the following features:

1. On alternate half cycles of the a-c supply, the output windings alternately conduct.

2. The direction of current flow through the output winding is opposite to that of the bias winding. (The bias current flows in a direction that partially desaturates the core.)

3. The direction of current through the output windings is the same as that of the control windings. Both currents tend to mutually assist each other in partially saturating the core, decreasing the impedance, and overcoming the desaturating effect of the bias current. Both control and output currents can be said to be "positive"; whereas, the bias current can be termed "negative."

4. With larger positive control current, output current increases and, vice versa, decreases with decreased control current.

These features, while common to many magnetic amplifier circuits, are not necessarily used in every application.

Corrective Maintenance

When there is a casualty to a magnetic amplifier circuit, deenergize the circuit and tag it as previously described in chapter 7. Check all switches and contacts for dirt and loose connections. After inspecting the connections, check the individual components. When a winding is suspected, deenergize the circuit, then ground the winding with a tool having an insulated handle. This action will discharge any stored or static charge on the circuit. After disconnecting the unit from all other equipment in the circuit check the suspected winding for continuity and resistance, using a suitable range on the ohmmeter.

When checking the metallic rectifiers, ground and disconnect as in the case of the winding of the magnetic amplifier. A comparative resistance check (forward compared to backward) will indicate whether or not the rectifying material has shorted out. Place the leads of the ohmmeter across the terminals of the rectifier and you will read either a high (back) resistance or a low (forward) resistance. Reverse the ohmmeter leads to get the alternate reading. The high resistance should be at least eleven times the low resistance.

QUIZ

1. The magnitude of the impedance of the saturable reactor depends upon the magnitude of change of what quantity in the core?

2. State five advantages of the magnetic amplifier over other types of amplifiers.

3. State four disadvantages of the magnetic amplifier. Refer to figure 12-1 for questions 4 through 7.

4. What quantity in the winding of variable reactor, L, is utilized to control the magnitude of the load current.

5. If a laminated silicon-steel core is gradually inse rted into the coil, what will be the effect on (a) permeability, (b) impedance, (c) circuit current, and (d) power factor?

6. With low values of d-c control voltage and current, what will be the relative magnitude of the (a) flux change in the core, (b) induced voltage in the reactor winding, (c) load current, ij , and (d) circuit power factor?

7. With relatively large values of d-c control voltage and current, what will be the relative magnitude of the (a) flux change in the core, (b) induced voltage in the reactor winding, (c) load current, 12, and (d) circuit power factor?

8. Name two general classes of nickel-iron alloys that have been developed for use as core materials for saturable reactors.

9. Describe the general shape of the hysteresis loops for Orthonol, Deltamax, Hypernik V, Orthonik, Per-merson, and equivalents.

Refer to figure 12-3 for questions 10 through 17. 10. How is the load or gating winding connected with

respect to the a-c source, the load, and rectifier? 1 1. How is the control winding connected with respect

to the a-c source, the control signal source, and

rectifier ?

12. Are the rectifier arrowheads pointed in the direction of conventional current flow or electron flow?

13. To what two quantities is the magnitude of core flux change in the reset half cycle proportional?

14. For the condition e c = 0: (a) What is the range of flux change during~The reset half cycle? (b) What is the range of core flux change during the gating half cycle? (c) What is the relative magnitude of the impedance presented by the load winding to the Ri circuit?

15. For the condition e £ = e^ peak: (a) What relative magnitude of control winding current will flow during the reset half cycle? (b) What relative flux change occurs during the reset half cycle ? (c) What relative magnitude of load current flows during the gating half cycle ?

16. For the condition e c = 1/2 e^ peak: (a) Why does no current flow in the control winding until e^t exceeds e c ? (b) What is the relative change of flux in the reset half cycle? (c) Why does the load current suddenly rise at point 5 (fig. 12-3, E) ? (d) What is the relation between the right-hand shaded area of the hysteresis loop and the shaded area under the +e ac voltage curve ?

17. What is the power factor of the load circuit for all values of load current?

Refer to figure 12-5 for questions 18 through 30.

18. Before e^ is applied, identify the remanent flux as shown on the associated hysteresis loops of (a) core ® and (b) core ©.

19. Identify the active circuits (with e ac applied) (a) for the first half cycle and (b) for the second half cycle.

20. For the condition e^ = 0, during the first half cycle of e^ identify the range of (a) core (l) flux change and~(b) core (2) flux change, as indicated on the associated hysteresis loops.

21. For the condition e c = 0 during the second half cycle of e „ , identify the~range of (a) core (2) flux change and~(b) core (T) flux change, as indicated on the associated hysteresis loops.

22. For the condition e t = 0, what is the relative magnitude of the load component of current through Rl?

23. For the condition^ = 1/2 e ac peak during the first half cycle the energy required to reset the core (T) flux from the <pj level to the (63 level is proportional to which two areas ?

24. For the condition e t = 1/2 e ac peak during the second half cycle: (a) Why is the current through LI limited to the magnetizing component from point 4 to point X? (b) Why does the load current through LI and R increase at point X?

25. What are the simultaneous actions occurring (a) in CI and L2 on one half cycle and (b) in LI and C2 on the following half cycle?

26. As e c is increased from zero to the peak value of jJ ac Twhat is the effect on (a) the reset distance along the~hysteresis loops and (b) the length of the conducting portion of each half cycle through Rj and the

load windings, and (c) the average value of the load current?

27. What is the effect of low remanence on load voltage when e t is maximum?

28. What Ts the effect of reverse-current leakage in the rectifiers on the full-load voltage when e^ is maximum?

29. What is the function of the rectifiers in series with the load and control windings of one core with respect to flux changes in the other core?

30. When e t is increased from zero to 1/2 e ac peak, how long does it take for the amplifier to respond?

Refer to figure 12-7 for questions 31 through 34.

31. For the condition e £ = 0 during the first half cycle of e^ , identify the -range of (a) core (2) flux change ancT(b) core (T) flux change, as indicated on the associated hysteresis loops.

32. For the condition $e^\wedge = 0$ during the second half cycle of e ac , identify the range of (a) core (T) flux change ancTTb) core (2) flux change, as indicated on the associated hysteresis loops.

33. The magnitudes of the mutually induced voltages in the 1:1 turns ratio windings depend upon what factor?

34. What are the two principal disadvantages of this type magnetic amplifier?

35. What is the effect of an increasing flux in one outer leg (fig. 12-8, D) and a decreasing flux in the other outer leg on (a) the flux in the center leg and (b) the voltage induced in the control winding at the fundamental frequency of the supply voltage?

APPENDIX I

ANSWERS TO QUIZZES

Chapter 1

ORGANIZATION

1. Service schools and self-study courses, textbooks, and training aids.

2. Military and professional.

3. The Manual of Qualifications for Advancement in Rating (Revised), NavPers 18068.

4. One general and two emergency service ratings.

5. Record of Practical Factors, NavPers 760 (EM).

6. Training Publications for Advancement in Rating , NavPers 10052-F.

7. By application to your Information and Education Officer.

8. The success with which you stimulate others to learn.

9. Study and work should be carried out together.

10. Answering the questions at the end of each chapter.

11. (l)Man the battle stations, (2) perform basic administrative requirements, and (3) maintain continuous watches required under wartime conditions of readiness.

12. Three.

13. Maneuver and fight the ship.

14. The Watch, Quarter, and Station Bill.

15. Five.

16. Four hours.

17. At the control distribution switchboard.

18. The revised individual allowance list (RIAL).

19. The Material History Card—Electrical (NavShips 527A).

20. The Resistance Test Record (NavShips 531).

21. The Current Ship's Maintenance Project (CSMP).

22. A-C or D-C Electric Propulsion Operating Record (NavShips 3647), and Electrical Log—Ship's Service Electric Plant (NavShips 3649).

2 3. Official NavShips Forms prepared by the Bureau of Ships and ship's forms prepared by the engineering department of the individual ship.

24. Daily Ground Test Sheet.

25. The Ship's Memorandum Work Request.

26. The Equipage Custody Record (NavSandA 306A).

27. Log room, which is the office of the engineering department.

28. Custody receipts.

29. The Bureau of Ships Manual.

Chapter 2

ELECTRIC CABLE AND ACCESSORIES

1. Nonflexing and repeated flexing.

2. Double, boat, shielded, plain indicates a double conductor, small boat shielded, plain (unarmored) cable.

3. The approximate cross-sectional area expressed in thousands of circular mils.

4. The number of conductors in the cable.

5. General use and special use.

6. Shipboard General Use, Armored

7. Silicone rubber and glass.

8. As leads to portable electric equipment.

9. Heat and oil resistant, flexible.

10. (1) The total connected load current, (2) the demand factor, (3) the resultant load current, and (4) the allowable voltage drop.

11. It is the ratio of the maximum load averaged for a 15-minute period to the total connected load on the cable.

12. Unity.

13. They are directly proportional.

14. 384 feet.

15. 1.06 percent.

16. 0.535 volt.

17. Only heat- and flame-resistant armored cable.

18. (1) Splashproof, (2) spraytight (3) submersible, and (4) explosion proof.

19. Nylon, steel, brass, or aluminum alloys.

20. The single cable strap.

21. (a) Five and (b) two.

22. Fire.

23. (a) The zone in which the unit is located, (b) The number of the switchboard within that zone determined in accordance with the general rule for numbering machinery.

24. (a) The prime mover, (b) Labeling power cables between generators and switchboards.

25. (a) MG, (b) TF, and (c) RT.

26. (1) The name of the space, apparatus, or circuits served; (2) the service (power, lighting, electronics, etc.) and basic location number; (3) the supply feeder number.

27. Phases A, B, and C, respectively.

28. The positive, neutral, and negative conductors, respectively.

29. The positive and negative conductors, respectively.

30. Portable cables having a ground wire and three terminal plugs and receptacles, one terminal of which is grounded to the ship's structure.

31. Green.

32. The length of the cable in feet and the location of the cable stowage rack.

33. To preserve the insulation resistance.

34. Mica, asbestos, fiberglas, and similar inorganic materials.

35. Silicone rubber.

36. Class B.

37. Class A, B, or H materials.

38. The insulation resistance varies inversely with the length.

39. The lights should be turned off at their switches and all plugs removed from the outlets.

40. Switches or circuit breakers should be opened at the switchboard and motor controller contacts should be open.

41. 1 megohm.

42. 0.2 megohm.

43. Air tests.

44. Once each watch.

45. Quarterly.

Chapter 3

ELECTRIC LIGHTING

1. They are equal.

2. They are not necessarily equal.

3. Red, green, and blue.

4. Red.

5. The illumination on a surface that is 1 foot distant everywhere from a source of 1 candlepower.

6. (a) 25 foot-candles, (b) 4 foot-candles.

7. 8 lumens per sq ft.

8. 12.57 lumens.

9. 2.5 foot-candles.

10. (1) Incandescent, (2) fluorescent, and (3) glow lamps.

11. (1) Shape of bulb, (2) finish of bulb, and (3) type of base.

12. 300 watts.

13. 750 hours.

14. (a) 60 lumens per watt, (b) 10 lumens per watt.

15. A high resistance.

16. (1) Watertight, (2) nonwatertight, (3) pressure-proof, and (4) explosion-proof.

17. Point by point and lumen methods.

18. 9.72 foot-candles.

19. 5.7 foot-candles.

20. (1) Running lights, (2) signal lights, and (3) anchor lights.

21. (a) (1) The blue dial is illuminated, (2) the buzzer is energized, and (3) the annunciator is operated to the OUT position, (b) The reset switch must be returned to the NORMAL position, (c) Removal of the fuse in the primary filament circuit and noting the action of the components.

22. (1) Aircraft warning, (2) blinker, (3) breakdown and man overboard, (4) steering, (5) stern, (6) wake, and (7) speed.

23. (1) Light traps and (2) door switches.

24. To automatically energize the lantern when the lighting circuit fails.

25. Once each week.

Chapter 4

SEARCHLIGHTS

1. To illuminate distant objects and visual signaling.

2. (1) Stationary pedestal, (2) turn-table with arms, (3) drum with iris and signaling shutters, and (4) carbon-arc lamp.

3. To shut off the light beam without extinguishing the arc and to vary the amount of light in the searchlight beam.

4. To locate the correct position of the end of the positive carbon when it is at the focal point of the reflector.

5. To remove the gases produced by the arc and to prevent deterioration of the parts exposed to the intense heat of the arc.

6. 75 to 80 amperes and 65 to 70 volts.

7. The positive carbon.

8. The positive carbon.

9. A lens focuses light rays from the arc on the thermostat.

10. The series current regulator coil.

11. The ship's 120-volt, d-c power.

12. Both carbons should be replaced.

13. To remove the material deposited by the preceding carbons to prevent jamming.

14. To form the crater in the positive carbon and establish a normal arc.

15. After every extended run and at least once a week.

16. The arc current adjusting knob is turned clockwise.

17. It decreases the positive carbon projection.

18. The image of the tip of the positive carbon should appear on the focal line of the arc-image screen.

19. 57 ohms ± 4 ohms.

20. Usually low.

21. Lock the searchlight securely in train and elevation with the drum horizontal.

22. Signaling.

23. 1000 watts.

24. About 20 atmospheres.

25. (a) The mercury xenon lamp requires a high voltage r-f current for starting and (b) a ballast for operating it at rated output.

26. (a) 40,000 to 60,000 watts and (b) 60 to 70 volts.

27. To boost the voltage on the T2 primary 32 volts above the line supply before the arc is struck.

28. The lamp will become overheated and fail violently.

29. Use the face mask and the gloves provided with the equipment. The lamp must not be handled without enclosing it in the protective metal enclosure.

30. Minimum.

31. (a) It may burst violently, (b) In the sea.

32. Signaling.

33. (a) To ionize the xenon gas and establish an arc in the lamp, (b) To control and limit the current after starting the lamp.

34. (a) 155-160 volts and (b) 40-45 volts.

Chapter 5

MAINTENANCE OF MOTORS AND GENERATORS

1. To keep the equipment clean and free of oil, water, dirt, and other foreign particles.

2. (1) Wiping, (2) use of suction, (3) use of compressed air, and (4) use of a solvent.

3. Because it lessens the possibility of damage to insulation.

4. Do not allow drops of solder to get into the windings. Excess solder that may later

break off should be removed from the soldered joints.

5. To prevent electrolytic action between the brushes and rings or segments.

6. They should be inspected at frequent intervals to make sure they are tight.

7. Silver polish.

8. (a) They should be separated by a spreader, (b) To prevent the slings from coming into contact with the a-c rotor or d-c armature coils.

9. One-sixteenth inch.

10. One-half the diameter of the shaft.

11. Semiannually.

12. (a) Replacing one end of a screwdriver or steel rod against the bearing housing and the other end against the ear. (b) A loud, irregular grinding, clicking, or scraping noise is heard.

13. (a) A bearing housing too full of lubricant, (b) The grease becomes sticky and seals the bearing against fresh lubricant.

14. To prevent the flow of shaft currents through the bearing.

15. When they are worn down to half their original length or if the corners or edges are chipped.

16. The no load neutral.

17. Equidistant.

18. A uniform, glazed dark brown color on the places where the brushes ride.

19. Emery cloth, emery paper, or emery stone.

20. Pitting due to electrolytic action on the surface of the rings.

21. An open armature coil.

22. Practically zero.

2 3. The real grounds remain in the same bars while the
phantom grounds will shift. 24. (a) Maximum and (b) minimum.

25. (1) Disconnecting both ends of the coil and (2) installing a jumper across the risers from which the coil was disconnected.

26. (1) Cage and (2) wound.

27. An open circuit in the wound rotor.

28. An open in the shunt field winding.

29. The greatest difference in potential.

30. A small magnetic compass and battery.

31. To determine whether or not proper maintenance procedures are being carried out.

32. Daily.

33. Weekly.

34. Monthly.

35. Quarterly.

36. Semiannually.

Chapter 6

MAINTENANCE AND REPAIR OF BATTERIES

1. (1) Leclanche, (2) mercury, and (3) low temperature.

2. 1.5 volts.

3. 1.3 volts.

4. A label is attached to each battery.

5. Date of manufacture.

6. Initial capacity.

7. Delayed capacity.
8. Shelf life.
9. Six months.
10. 0.9 volt per cell.
11. Nickel-cadmium and lead-acid types.
12. Five.
13. Ten to fifteen years.
14. -65° F to 165° F.
15. A lead-acid storage battery.
16. The battery can supply 10 amperes for 10 hours.
17. 2.4 to 2.6 volts.
18. To neutralize and remove acid and thus prevent corrosion.
19. Petrolatum or cup grease.
20. The Storage Battery Tray Record, NavShips 151.
21. An explosion will occur with the sudden evolution of heat.
22. Three or four months.
2 3. Once a month or when the specific gravity drops to 1.180.
24. Daily.
25. Daily.
26. Each week.
27. Every 6 months.
28. A rheostat.
29. 5 amperes.
30. Five times the rated capacity at the 10-hour rate.
31. Because it indicates the beginning of the useful life of the battery.
32. Four.
33. 20 amperes.
34. Low (specific gravity) cells.
35. Reducing it, especially at the end of a charge.
36. An internal short circuit.
37. It may reverse.
38. 80 percent.
39. To remove the explosive mixture of hydrogen and air.
40. It should be lowered.
41. 24 hours.
42. The acid is poured slowly into the water.
Chapter 7
PROTECTIVE DEVICES
1. Because the new device will be ruined by the same defect that caused the original breakdown.
2. Delayed action.
3. They act as an arc quencher.
4. It acts as a blown-fuse indicator.
5. Fuses Fl, F2, and F3.
6. Fuses A-Al and B-Bl.
7. The circuit is continuous.

8. In fuse A-Al.

9. It is free from grounds. 10. Between 4 and 5 ohms.

1 1. Approximately half voltage.

12. Lights dim on some circuits and have full brilliance on others.

13. Simultaneous open circuits in branch land branch 2.

14. This tool is insulated so that there is less danger from electric shock while working with the fuses.

15. (1) Pushbutton, (2) knife, and (3) rotary.

16. The holding magnet will release the spring-loaded lever, which returns to the OFF position.

17. Cam-type switches.

18. By the collars, C and D, and the operating rod, E.

19. It opens the circuit and removes voltage from the operating coil, Ml.

20. (a) Bulb and (b) helix.

21. (a) Series and (b) shunt.

22. In series.

23. False operation is prevented when subjected to severe mechanical shock.

24. It is held closed by the starting current through coil Fa2, thereby shorting out the field rheostat and ensuring full-field strength.

25. A reverse-current relay.

26. 10 percent.

27. The current coil.

28. An oil dashpot mechanism gives the relay a time delay action.

29. The trip current is increased.

30. To reduce "chatter" during operation.

31. Tripping.

32. Phase-failure protective relay.

33. A resultant flux exerts a pull on the relay armature to operate the relay.

34. Seventy-eight percent of the normal root-mean-square value with no short.

35. Because with the d-c side opened, full voltage is applied across the rectox, which may cause it to break down.

36. A coiled spring in the breaker.

37. Type NQB breakers are manually operated; type AQB breakers are automatically tripped.

Chapter 8

CONTROL DEVICES

1. (a) 3 amperes, (b) 30 volts, and (c) 90 volts.

2. Up.

3. It is reversed.

4. The armature counter voltage.

5. (a) A spring and (b) an electromagnet.

6. It increases the speed of release.

7. By magnetic blowout.

8. By the sliding or wiping action after the initial contact is made.

9. Contacts 4 and 5.

10. 70 percent.

11. An electrically operated interlock.

12. In parallel.

13. A mechanical interlock.

14. The use of oil in a dashpot.

15. (1) Main line and (2) accelerating contactors.

16. A strong magnetic field in the lockout coil.

17. Shunt coil 3.

18. Pressing the START button will energize the operating coil LC of the line contactor.

19. Series relay SR drops out and closes contacts SR, which energizes coil AC.

20. F5.

21. By increasing the resistance in the shunt field rheostat.

22. Remove rust, gum, and filings from the shafts, bearings, or guides, and lubricating shafts and bearings semiannually with two drops of light oil (SAE 10).

23. Because of the large air gap, the resulting inductive reactance, X_l , and impedance are low.

24. The torque spring (19).

25. To prevent residual magnetism from holding the armature closed after the coil has been deenergized.

26. They decrease.

27. Maintained.

28. Lines 1 and 3.

29. They are mechanically interlocked.

30. To start 3-phase induction and synchronous motors, and to furnish variable voltage for test panels.

31. MSI, MS2, MS3, and MS4.

32. Because the motor is disconnected from the line during the transition from low (starting) voltage to line (running voltage).

3 3. (1) The source of voltage, (2) the load, and (3) the connecting wires between them.

34. Monthly.

35. To assume that all conductors are hot until tests prove differently.

36. All portable tools require a separate ground wire, which grounds the frame of the tool.

37. They should be removed.

38. To discharge it and thus prevent the possibility of severe electric shock by a static charge.

Chapter 9
ELECTRICAL SYSTEMS IN SMALL CRAFT

1. The lead-acid storage battery.

2. The battery-charging generator.

3. The a-c generator (alternator).

4. It is not used extensively.

5. Heavy duty.

6. It energizes the load relay, which energizes the field.

7. They are connected.

8. (a) Upper, (b) A small resistance.

9. (a) Lower, (b) Inserts resistance in series with the battery and shorts out the entire field coil.

10. In series with the charging circuit at the A terminal.

11. The lowest.

12. A loose drive belt and insufficient contact between the brushes and commutator.

13. The same (negative in this example).

14. Because of the fire hazard when flashing the field.

15. The current regulator contacts will vibrate and periodically insert resistor Rl in the generator field circuit.

16. It stops operating.

17. (a) In shunt with the field, (b) To prevent arcing across the regulator contacts.

18. The spring tension opens the contacts, which open the circuit between the generator and the battery.

19. The contacts open and the generator voltage is reduced.

20. The contacts open and the generator output is reduced.

21. An electric spark.

22. (1) The battery, (2) the ignition coil, (3) the ignition distributor, and (4) the spark plugs.

23. 25,000 volts.

24. Two.

25. The centrifugal and/or the vacuum types.

26. (a) Advanced in the direction of rotation, (b) It introduces the spark earlier in the compression stroke.

27. (a) Short and (b) long.

28. The cam (contact) angle.

29. (1) Grid resistor, (2) glow plug, (3) flame primer, and (4) ether capsule primer.

30. A low-voltage, d-c series motor.

31. 570 amperes at 2.3 volts.

32. 15 to 1.

33. (1) Bendix drive, (2) overrunning clutch, and (3) Dyer drive.

34. The clutch spring.

35. To prevent clashing gears.

36. Saturate the wicks with oil.

37. (1) Compartment lighting, (2) hand lanterns, (3) portable multipurpose signaling light, and (4) navigational lights.

Chapter 10

ELECTRIC POWER DISTRIBUTION

L. The ship's service and emergency electric plants.

2. Steam turbines and diesel engines.

3. Diesel engines and gas turbines.

4. (a) Must be capable of parallel operation with each other and (b) cannot be paralleled with any generators.

5. 450 volts, 60 cycle, 3 phase, and 3 wires.

6. To minimize the possibility that a single hit will damage more than one switchboard.

7. Load centers.

8. The generator and distribution switchboards.

9. F.

10. FB.

11. Two.

12. Bus transfer equipment.

13. (a) Three and (b) single.

14. A, B, and C.

15. A feedback tie.

16. The casualty power distribution system.

17. (1) Portable cables, (2) bulkhead terminals, (3) risers, (4) switchboard terminals, and (5) portable switches.

18. To avoid handling energized cables.

19. They can be shifted to an emergency feeder that receives power from the emergency switchboard.

20. Temporary circuits can be rigged with the casualty-power distribution system.

21. (1) Oversize exciters for the ship's service generators, (2) separate motor generator sets, and (3) metallic rectifiers.

22. The ground wire provides a conducting path between the metal housing and the ship's structure.

23. (1) Low-voltage system; (2) 120-volt, 2-wire system; and (3) 120/240-volt, 3-wire system.

24. Black, white, and red, respectively.

25. Not to ground the neutral or either leg.

26. (a) Two turbine driven ship's service generators and two diesel-driven emergency generators. (b) The ship's service generators are 400-kw, 450-volt, 3-phase, 60-cycle units. The emergency generators are 100-kw, 450-volt, 3-phase, 60-cycle units. (c) 50-kw, 120-volt, stabilized, shunt d-c generators.

27. A reverse power relay.

28. To prevent both circuit breakers from being closed at the same time so that the emergency generators cannot be paralleled.

29. An automatic bus transfer controller.

30. Both contactors cannot be closed at the same time, but both can be open.

31. (a) Drops out, (b) opens, (c) opens, (d) closes, (e) completes, (f) energizes, and (g) closes.

32. (a) Closes, (b) completes, (c) opens, (d) closes, (e) closes, and (f) opens.

33. Selective tripping.

34. The breaker closest to the fault opens first; the breaker farthest from the fault and closest to the generator will open last.

35. Greater than the short circuit current of the generator.

36. The time delay is greater for small currents than for large currents.

37. (a) The AQB load breaker, (b) the ACB feeder breaker will open almost instantaneously.

38. (a) The ACB feeder breaker, (b) time T2 (about 0.4 second), (c) time T3 (about 0.8 second), (d) time Tl (about 0.15 second), and (e) no.

39. (a) No ground exists and (b) phase A is grounded.

40. It should be checked to see that it is in good operating condition.

41. Disconnect links.

42. The ground circuit breaker will open and insert a resistor in the ground circuit.

43. (a) The alarm bell is energized and (b) the ground alarm lamp is energized.

44. They are lighted dimly.

45. A will be dark and B will be bright.

46. (a) They should be open, (b) it should cut in all the field resistance, and (c) it should contain the proper amount of oil as indicated by the oil sight gage.

47. Commutator and collector rings run true, brushes ride freely in the brush holders, and there is no chattering of brushes.

48. Close the shunt field switch and gradually cut out the field resistance until normal voltage is obtained.

49. Make certain the meter indicates the frequency of the generator being adjusted.

30. By opening the feeder circuit breakers on the power and lighting circuits.

51. It will cause that generator to take more load.

52. In proportion to their kw rating.

5 3. The governors that control the speed of the generators.

54. The voltage adjusting rheostats of the voltage regulators.

55. In the decrease direction.

56. They may start automatically because of momentary loss of ship's service voltage and be damaged while operating unattended.

57. They must not be paralleled with each other or with shore power.

58. Parallel (cross plant) or separately (split plant).

59. (a) Closed and (b) open.

60. Because a casualty to one switchboard will notaffect the loads fed from other switchboards.

61. They should be closed to energize the switchboard from one of the other switchboards.

62. Because automatic protective devices are not designed to protect the system from damage caused by careless operating practices, but to afford protection against damage caused by equipment failure.

63. At least once a year; more frequently if subjected to unusually severe service conditions.

64. Because rust on the sealing surfaces decreases the contact force and may result in overheating of the contact tips.

Chapter 11

ELECTRIC AUXILIARIES

1. A degaussing installation.

2. To measure and record ship's magnetic fields.

3. Permanent magnetization and induced magnetization.

4. Deperming.

5. The strength of the earth's field and the heading of the ship with respect to the earth's field.

6. The vertical induced magnetization.

7. Maximum on the north heading to zero on the east.

8. It encircles the ship in a horizontal plane, which is usually at the waterline.

9. The F coil.

10. The Q coil.

11. In series.

12. The L coil.

13. The A coil.

14. The ampere turns of the coil.

15. The compass compensating coil.

16. By a wiring diagram pasted on the inside of the cover.

17. By adjusting a rheostat connected in series with the coil and power supply.

18. By adjusting a rheostat in the generator field circuit to vary the output of the generator.

19. A signal that is the ship's heading.

20. The Bureau of Ordnance Degaussing Folder.

21. Hourly.

22. (a) Right-hand plate and (b) left-hand plate.

23. A deflection to the right.

24. The M coil current.

25. The F, Q, FI-QI. L, and A coil currents.

26. Emergency handwheels for manual operation are provided for both units.

27. To prevent burning out the contacts from operation under load.

28. In the engine room.

29. To permit changing the F, Q, FI-QI, L, and A coil currents with heading.

30. Once a week; 4 hours in one direction, then momentarily in the opposite direction.

31. At least once a week.

32. Between the coil disconnect switch and ground.

33. Closed.

34. Once a month.

35. Locating and eliminating grounds in the various coils, circuits, and components.

36. (a) To lower the port anchor and (b) to lower the starboard anchor.

37. (a) it will release; (b) it will set.

38. (a) It will revolve; (b) it should be free to run.

39. To reset the latch for the associated contactor in the event that the overload relay has been previously tripped.

40. Selector switch must be operated to the No. 1 lockout position.

41. They are interlocked at each level to prevent operation unless they are closed.

42. A governor-actuated safety device is provided under the elevator platform to grip the guide rails and stop the elevator.

43. They must be in their normally closed positions.

44. (1) A ram unit, (2) power unit, and (3) remote control system.

45. 20-hp, 1200-rpm, 440-volt, 3-phase, 60-cycle induction motor.

46. A rack attached to the ram.

47. Normal steering from the pilothouse.

48. A nonreversing across-the-line starter and a maintained contact master switch.

49. (1) Ranges, (2) griddles, (3) fry kettles, (4) roasting ovens, and (5) baking ovens.

50. By a high temperature thermostat, the bulb of which is covered and clamped to the underside of the griddle hotplate.

51. At least once a month.

52. Once a week.

Chapter 12

SATURABLE REACTORS

1. Flux.

2. (1) High efficiency, (2) reliability, (3) ruggedness, (4) space and weight economy, and (5) no warm-up time.

3. (1) It cannot handle low-level signals, (2) it is not useful at high frequencies, (3) it has a time delay associated with magnetic effects, and (4) the output waveform is not an exact reproduction of the input waveform.

4. The induced voltage.

5. (a) Increase, (b) increase, (c) decrease, and (d) dec rease.

6. (a) Large, (b) high, (c) low amplitude, and (d) low.

7. (a) Small, (b) small, (c) large, and (d) high.

8. High permeability alloys and grain-oriented alloys.

9. Rectangular shaped.

10. In series.

11. In series.

12. Electron flow.

13. (1) The magnitude of the applied voltage across the control winding and (2) the time interval during which this voltage is applied.

14. (a) From negative saturation <£i to positive saturation 02* (h) From the <£ 2 level to the <f>i level, (c) Maximum.

15. (a) Zero, (b) Zero, (c) Full value.

16. (a) Because the rectifier opposes it. (b) Frornnega-tive saturation <£j to_ some value <t>^ below saturation, (c) Because the core flux saturates and the impedance drops, (d) They are proportional.

\1. Unity.

18. (a) 4>i and (b) 0 2 -

19. (a) CI and L2 and (b) LI and C2.

20. (a) 0! to 0 2 - (b) ^2 to *1 •

21. (a) <f> l to <f> 2 and (b) <£ 2 to <£j.

22. Zero.

23. The shaded area of the core (I) hysteresis loop between the <f>i to 03 levels and the shaded area of e^ between points 2 and 3.

24. (a) Because the core (J) flux is changing and the impedance of LI is high, (b) Because at point X the core (T) flux is driven into negative saturation along the <f>i level and the impedance of LI drops to a low value.

25. (a) Reset in CI and gating in L2 and (b) gating in LI and reset in C2.

26. (a) Decreased, (b) increased, and (c) increased.

27. It is reduced.

28. It is reduced.

29. They prevent mutually induced voltages occurring in the windings of one core from affecting the flux change of the other core.

30. One-half cycle.

31. (a) 02 to ^1 (b) 0i to 02-

32. (a) 02 to 0i and (b) 0j to 02-

33. The extent of the core flux changes.

34. (1) The time lag between the control circuit voltage change and the level response and (2) instability at partial loads.

35. (a) No change in flux will occur and (b) no voltage will be induced.

APPENDIX II

QUALIFICATIONS FOR ADVANCEMENT IN RATING

ELECTRICIAN'S MATES (EM)

Quals Current Through Change 11 General Service Rating

Scope

Electrician's mates stand watch on motors, generators, switchboards, and control equipment; operate searchlights and other electrical equipment; maintainand repair power and lighting circuits, electrical fixtures, motors, generators, distribution switchboards, and other electrical equipment; test for short circuits, grounds, or other casualties; repair and rebuild electrical equipment in an electrical shop.

Emergency Service Ratings

Electrician's Mates P (Power and Lighting

Electricians) EMP

Operate and maintain heavy and light a. c. and d. c. power and lighting circuits and equipment aboard naval vessels; stand watches ongener-ators, switchboards, electrical propulsion motors, and control equipment.

Electrician's Mates S (Shop Electricians) EMS

Repair and rebuild electrical equipment in an electrical shop.

Navy Job Classifications and Codes

For specific Navy job classifications included within this rating and the applicable job codes, see Manual oj Enlisted Navy Job Classifications , NavPers 15105 (Re-vised), codes EM-4600 to EM-4699.

nr\A

Qualifications for Advancement in Rating

Qualifications for Advancement in Rating

Applicable Rates

EM

EMP

100 PRACTICAL FACTORS

101 Operational

1. Demonstrate knowledge of electrical safety precautions

2. Rescue a person in contact with an energized circuit; resuscitate a person unconscious from electrical shock; treat for electrical shock and burns. (Simulated conditions.)

3. Extinguish electrical fires. (Simulated conditions.)

4. Operate carbon arc searchlights; make external adjustments and replace carbons

5. Stand watch on an a. c. ship's service generator and distribution switchboard:

a. Visually scan the various switchboard meters and indicators to determine whether the proper load is being carried and the generator is operating properly 2 2

b. Take and log readings 2

c. Shift to stand-by automatic voltage regulator

d. Control voltage manually during emergency conditions 2 2

e. Maintain normal voltage 2 2

f. Maintain normal frequency 2

g. Set updist ribution board for general quarters condition 2

6. Synchronize a.c. generators for parallel operation

7. Secure a single a. c. generator which is connected alone to the bus 2

8. Secure an a. c. generator which has been operating in parallel with another generator 2

9. Set up emergency Diesel generator for automatic operation

10. Provide emergency power to main distribution board from emergency switchboard through feedback switch 2

11. Connect shore power to main distribution board 2

Qualifications for Advancement in Rating—Continued

Qualifications for Advancement in Rating

Applicable Rates

EM

101 Operational—Continued

12. Stand watch ond.c. ship's service generator and distribution switchboard:

a. Visually scan the various switchboard meters and indicators to determine whether generator is operating properly

b. Take and log readings

c. Adjust field rheostat to maintain normal voltage

d. Maintain load distribution when machines are operated in parallel . . .

1 3. Connect a d. c. generator to its bus for nonparallel operation

14. Connect d. c. generators for parallel operation

15. Secure a single d. c. generator which is connected alone to the bus

lb. Secure ad. c. generator which has been operating in parallel with another gene rator

17. Man electrical equipment at the following stations:

a. Steering engine room

b. Anchor windlass room

c. Hoist equipment and/or elevator . .

d. Degaussing switchboard

18. Read and work from electrical diagrams and sketches

19. Stand electrical watch in a turret (if so assigned)

20. Stand watch on main propulsion control switchboard if assigned to electric-drive ship

102 Maintenance and/or Repair

1. Examine motor and generator surroundings for dripping water, oil, steam, excessive dirt, and any loose gear which might interfere with ventilation or jam moving parts

2. Examine running motors and generators for cleanliness, vibration, unusual or excessive noise, heating, and condition of brushes, commutators, collector rings, bearings, and bolts

2 2

2

2

2

2

2

2 2 2 2

2

1

3 3

Qualifications for Advancement in Rating —
Continued

Qualifications for Advancement in Rating
Applicable Rates
EM
EMP
102 Maintenance and/or Repair—Continued
3. Inspect for leakage of lubricant from generator or motor bearings 3 3
4. Clean and lubricate searchlights 3 3
5. Replace storage and dry-cell batteries 3 3
6. Test running, anchor, and signal lights, and replace lamps
7. Repair portable electric tools, portable lights, fans, and appliances by:
a. Testing component parts with an ohmmeter or megger for grounds, open circuits and short circuits ... 3
b. Cleaning electric contacts and windings or elements 3
c. Replacing defective cord, plugs, switches, elements, and worn brushes
8. Locate blown fuses, using a voltage tester, and replace with fuses of proper ratings, using fuse pullers
9. Clean and lubricate electric motors and motor-generator sets 3
10. Solder electrical connections and splices
11. Replace worn gaskets and seals of watertight electrical fixtures
1Z. Stand a battery cha rging watch by giving storage batteries initial, normal, equalizing, floating, and emergency charges (excluding mixing and adding electrolyte)
13. Detect and locate grounds, open circuits, and short circuits in:
a. Lighting circuits 3 3
b. Power distribution cables 3 3
c. Motors (a. c. and d. c.) 2 2
d. Motor controllers (a. c. and d. c.) . 2 2
e. Degaussing system 1
f. Ship's service and emergency generators and associated switch gear . .
14. Service electric ranges by:
a. Taking resistance and current readings 2 2
b. Cleaning switches and relays 2 2
c. Replacing defective heating units, thermostats, and control or unit wiring
Qualifications for Advancement in Rating—Continued
Qualifications for Advancement in Rating
Applicable Rates
EM
EMP
EMS
102 Maintenance and/or Repair—Continued
15. Measure insulation resistance of ship's service and emergency generator and exciter
16. Clean generators with portable blower or vacuum cleaner and wipe with lint-less cloth
17. On motors and generators:
a. Inspect and tighten brush pigtails . .
b. Inspect and correct brush aline-ment parallel to commutator segments
c. Inspect and correct distance of brush holders from commutator. . .

d. Inspect and correct brushes to see that they move freely in holders and that holders are clean

e. Inspect and correct brush pressure

f. Inspect commutators of idle machines for commutator condition . .

g. Inspect collector rings of idle machines for evidence of corrosion . .

h. Sandpaper commutator and collector rings to smooth a slightly rough surface

18. Maintain electrical system on ship's boats:

a. Ignition (gasoline engine)

b. Starting heater circuit (Diesel engine)

c. Lighting

d. Starter

e. Generator

f. Voltage regulator

g. Batteries

19. Take such data as amount and size of wire while unwinding defective transformer coils, using wire gage to determine size of wire, and replace with same amount and size according to the following procedure:

a. Set Bakelite coil spool on coil winding machine, put wire on machine rack, and thread setting wire guide. Take several turns by hand on spool after leaving sufficient lead

2 2

2 2

2 2

2

2

2 2

2

2

2 2

2

2

2 2 2 2 2 2

Qualifications for Advancement in Rating—Continued

Qualifications for Advancement in Rating

Applicable Rates

EM

EMP

EMS

102 Maintenance and/or Repair—Continued

b. Turn on machine and allow proper number of turns to be laid on spool by observing meter and dropping lever on machine at proper time . .

c. Cover wire with cloth tape, secure cloth with Glyptal cement, and paint cloth with Glyptal paint for insulation

d. When dry, reassemble transformer and run bench, panel, and board tests

20. Repair, rebuild, or replace defective parts of storage batteries:

a. Replace or repair defective lugs, connectors, separators, cell covers, or complete cell units

b. Mold terminal lugs for use in repair work by pouring molten lead from electric lead pot into a cast aluminum mold and allowing to harden . .

21. Wind, insulate, and bake armature and field coils

22. Rewind controller solenoids

23. Turn down or take cuts on armature shafts or pins, using a bench lathe . . .

24. Replace power and lighting cable aboard ship

25. Rewind a. c. and d. c. motors

26. Test, overhaul, clean, lubricate, rewind, and replace electrical parts of electric meters

27. Test controllers and insert new contact points. Repair Bakelite panels

28. Install new power and lighting circuits when installation is duty authorized . .

29. Inspect, adjust, and repair searchlights

30. Inspect and test operation of automatic bus transfer equipment and automatic starting equipment of emergency generators

31. Mix electrolyte for storage batteries .

32. Conduct test discharge on storage batteries

33. Place new (in dry state) storage batteries in service

Qualifications for Advancement in Rating—Continued

Qualifications for Advancement in Rating—Continued

Qualifications for Advancement in Rating

Applicable Rates

EM EMP

103 Administrative and/or Clerical—Continued

4. Prepare naval shipyard and tender work requests

5. Estimate time and material needed for repair of power and lighting electrical equipment

6. Obtain replacement parts and supplies and maintain inventory

7. Conduct and supervise electrical emergency drills

8. Check electrical operating logs and maintenance records to determine if equipment is operating properly

9. Prepare reports covering power and lighting equipment failures

10. Investigate all routine repairs aboard ships alongside and determine whether repairs can be accomplished by ship's force, with the tender supplying material only, or if tender personnel must be used to make repairs

11. Prepare complete machinery histories including information on maintenance, minor repairs, and reconditioning for all motors, generators, transformers, and controllers

200 EXAMINATION SUBJECTS

201 Operational

1. Safety precautions involved in performing tasks appropriate to the applicable rates listed under 100 Practical Factors.

2. Meaning and/or significance of:

a. Conductors and insulators

b. Lines of force

c. Field intensity

d. Flux density

e. Permeability

f. Ampere-turns

g. Hysteresis and eddy currents . . . .

h. Self and mutual induction

i. Electromagnetic induction

j. Coulomb

k. Volt

C C C

C C

C C C

c c

Qualifications for Advancement in Rating—Continued

Qualifications for Advancement in Rating

Applicable Rates

EM

EMP

201 Operational—Continued 2 —Continued

1. Ampere

m. Ohm

n. Henry

o. Circular mil

p. Farad

q. Watt

r. Kilowatt

s. Horsepower

t. Power factor

u. Kilovolt-amperes

v. Reactance

w. Capacitance

x. Inductance

y. Impedance

z. Torque

aa. Frequency

bb. Cycle

cc. Phase

dd. Ambient temperature

3. Relationship of current, voltage, and resistance in d. c. circuits

4. Relationship of resistance, inductance, and capacitance in a. c. circuits

5. Types of insulating materials and varnishes

6. Relationship of reluctance, flux, and m. m. f. in a. c. and d. c. magnetic circuits

7. Relationship of the length and cross-sectional area to the resistance of a conductor

8. Relationship of resistance, temperature, and current in electrical conductor

9. Construction and types of shipboard electrical cable

10. Electric symbols used on working drawings and wiring diagrams

11. Precautions to be observed when removing paint from, or repainting electrical equipment

12. Operating principles and construction of the following:

a. Storage batteries

b. Circuit breakers

3 3 3 3 3 3 3 3 3 3 3 3 3 3 3 3 3 3 3 3

3

2

3 3 3

3 2

Qualifications for Advancement in Rating—Continued

Qualifications for Advancement in Rating

Applicable Rates

EM

EMP

201 Operational—Continued

12 —Continued

c. Generators

d. Shunt doors (d. c.)

e. Series motors (d. c.)

f. Compound motors (d. c.)

g. Stabilized shunt motors (d. c.) ....

h. Motor controllers (d. c.)

i. Transformers

j. Single-phase motors

k. Induction motors

1. Synchronous motors .

m. Motor controllers (a. c.)

n. Rotary amplifier-type motor generator

0. Automatic voltage regulators (a. c. and d. c.)

p. Magnetic amplifier

13. Purpose and intended use of the following:

a. Ohmmeters

b. Meggers

c. Ammeters (a. c. and d. c.)

d. Voltmeters (a. c. and d. c.)

e. Frequency meters

f. Wheatstone bridge

g. Wattmeter (a. c. and d. c.)

h. Power factor meter

1. Synchroscope

j. Thermocouple instruments

k. Instrument transformers

1. Copper oxide rectifiers

m. Selenium rectifiers

n. Phase-sequence indicators

14. Purpose of the following:

a. Resistors

b. Rheostats

c. Solenoids

d. Induction coils

e. Capacitors

f. Fuses

g. Switches

h. Reactors

i. Saturable reactors

15. Relationship of current, voltage, and impedance in a. c. circuits

2 2 2 2 2 2 2 2 2 2

3 3 2 2 2 2 2 2 2 2 2 2 2

3 3 3 3 3 3 3 3 2

2 2 2 2 2 2 2 2 2 2

3 3 2 2 2 2 2 2 2 2 2 2 2

3 3 3 3 3 3 3 3 2

Qualifications for Advancement in Rating—Continued

Qualifications for Advancement In Rating—Continued

Qualifications for Advancement in Rating

Applicable Rates

EM

EMP

EMS

203 Administrative and/or Clerical—Continued

3. Reports covering power and lighting equipment failures that are submitted to BuShips

4. Application of damage control principles

5. Daily, weekly, monthly, quarterly, semiannual, and annual electrical tests and inspections

6. Administrative, material, and operational readiness inspections

C

c

c c

c c

c c

c c

c c

715

Absorption, 119 Accessories, control and protective, 348-387 Advancement in rating, qualifications for, 2-6, 704-715 Air heaters, electric, 485 Allowance, combat ship, 10 Amplifier, magnetic, 678 Anchor windlass, 615-621 Armatures, 261-266 Assignments, enlisted, 11 Autotransformers, 432 Auxiliaries, electric, 579-641

Ball bearings, 241-249

corrective maintenance, 248 Batteries, 280-330

charging, 298-307

dry, 280-285

ignition system, 469-484

maintenance, 296-309

repair, 315-323

storage, 285-296

survey, 323-325

troubles, 309-315 Bearings

ball, 241

sleeve, maintenance, 236 Brakes, a-c, 422 Brushes

electric motors and generators, 249-255

maintenance, 249 Bureau of Ships

Bulletin of Information, 31

Manual, 30

technical bulletins, 31

Cable, electrical, and accessories , 35-113 casualty power, 94 connections, 81-94 designations, 35-37 ends, 81-84 fittings, 1 1 1 installation, 41 maintenance, 99-113

Cable, electrical, and accessories—continued

markings, 70

shore power, 96-99

supports, 65 Candlepower, 122 Cells

lead-acid, 287

Leclanche, 280

low-temperature, 281

mercury. 281

nickel-cadmium, 286

voltage, reversal of, 314 Circuits

breakers, 388-391

electrical; troubleshooting, 336-348

full-wave basic, 658-667 without control circuit rectifiers and without control circuit s.^, 669-676

half-wave, basic, 650-658 Coil-current control, 595-601 Coils, 427

a-c stator, 271, 272

field, 268-271

in contactors, 411

shading, 426 Collector rings, 255-257, 260, 261

Commutators, 255-260 Complement, combat ship, 10 Conductor s

identification, 79

lacing, 85-87 Contactors

a-c, 427-432

d-c. 403-427

series lockout, 413-417

shunt, 413 Control

and protective accessories, 348-387

devices, 395-443 maintenance, 438

Control—continued

voltage, d-c; effect of, 673-676 Controllers

a-c, 433-436

d-c, 417

reversing, 419 Cores

magnetic amplifier, materials suitable for, 649

saturable, early types of, 646 Cosine law, 125 Current and voltage regulator, 464-469

Damage control books, 32 Darkened-ship equipment, 159-161, 175 Deck risers, 60 Degaussing

coils, 586-591

definitions, 613

installations, 579-615 Dry batteries, 280-285 Dyer drive, 494

El ectric air heaters, 485 auxiliaries, 579-641 cable, and accessories, 35-113 circuits, troubleshooting,

336-348 elevators. See Elevators galley equipment, 634-641 lighting. See Lighting plants, operation of, 557-562 power distribution, 507-573 systems a-c, 508-518 d-c, 519, 520 maintenance, 566 operation, 557 procedures, 562-566 systems in small craft, 447-503

Electrician's Mates duties, 1

scope of training course, 8, 9 Elevators, electric, 621-626

maintenance, 625 End(s)

conductor, 84

seals, 80 Engineering department, 11-21

Feedback, 680

Flashlights, 162, 172

Fluorescent lamps, 133-138

Foot-candle, 122

Full-wave circuits. See Circuits

Fuses, 332-336 cartridge, 333 delayed-action, 332 plug, 333

Galley equipment, electric, 634-641

fry kettles, 639 Generators

air coolers, 272, 273

battery-charging, 449

cleaning, 233

maintenance, 232

operation, 547-557 preparing for, 546 Glow lamps, 138

Half-wave circuit. See Circuits

Ignition system, battery, 469-484

Incandescent lamps, 126-132

Individual allowance list, revised, 21

Inspections. See Tests

Insulation electric cables and equipment, 99-102

resistance measurements, 102-112 Interlock electric, 409 mechanical, 410

Lanterns, 163-165, 171

portable flood, 172 Leadership, 5 Lighting electric, 117-178 distribution systems, 166-172

equipment, 157-165 fixtures. 139-147. 169 maintenance, 168 glare, 173

system, 500

a-c, 166-168 d-c. 168

Lights anchor, 157 colors, 121

navigational, 147-157, 502 night-flight operation, 157 principles, 117-125 running, 147 signal, 153-157 sources, 126-138 Lumen, 124, 144

Magnetic field earth's, 580 ship's, 582 Maintenance batteries, 296-309 cable, electrical, and accessories, 99-113 cor r ective amplifier, magnetic, 683 ball bearings, 248 degaussing system, 612 motor control equipment, 421,438 degaussing installation, 610 electrical installation, 438 elevators, electric, 625 galley equipment, electric, 640

generator, 232 lighting installation, 168, 169

preventive; degaussing system, 611

project, current ship's, 22

searchlight. See Searchlights

switchboards and control equipment, 566-573 Marking

cable, 70

equipment, 78, 79 Material history, 21, 22 Military requirements, 3 Motors

air coolers, 272, 273

cleaning, 233

maintenance, 232

Navigational lights, 147-157,

502

Nonflexing service cable, 37

Organization, 1-33 administrative, electrical

division, 12 Electrician's Mates, 1

Organization—continued

watch, 16-21 Ovens, galley, 637

Personnel, assignment of, 9-11 Plans, ship's, 32 Polarities, 652 Power. See Electric

Professional qualifications, 2, 3 Protective devices, 331-391 Publications, 28-33

Qualifications for advancement in rating, 2-6. 704-715

Ranges, galley equipment, 635 Reactors, saturable, 645-683

transfer characteristics, 667 Records, 21

legal, 23

maintenance, 25 Rectifiers

functions of, 65 3

leakage; effect, 668 Reflection, 119 Refraction, 120, 121 Regulators, 460

current and voltage, 464-469 Relays, a-c, 376 Repeated flexing service cable,

38-41

Rotors, a-c, 266-268

Safety precautions batteries, 285, 325-327 control devices, 440-443 electric lighting system, 173-

176 fuses, 347 searchlight, 211 mercury-xenon lamp, 221 switchboards and other electrical equipment, 565 Searchlights, 179-229 8-in., 60-cycle sealed-beam,

ZZZ-ZZ9 maintenance corrective, 202, 221. 228 preventive, 198. 219, 227 12-in. incandescent, Z1Z-ZZZ 24-in. carbon-arc, 179-212 Ship Information Book , 32 Shore power cable, 96-99 Signal lights, 153-157, 501 Sleeve bearings, 236-241

Small craft, electrical systems

in, 447-503 Starting system, engine, 447-

500

Steering gear, electrohydraulic,

626-634 Storage batteries, 285-296 Study

plan of, 6

rules of, 6, 7 Sulphation, 310-312 Switch

float. 355

knife, 349

limit, 355

master, 352

pressure-controlled, 358 rotary, 351

temperature-controlled, 358 Switchboards, 520-546 a-c, 523 d-c, 542

Tests or inspections batteries, 297, 298, 307-309 branch circuit, 338-343

Tests or inspections —continued controllers for motors and

associated equipment, 439 generators and motors, 273-

276

shipboard electrical cable installations, 113

Transfer characteristics, 667

Trouble analysis; circuit controls, motor, 436

Troubleshooting, electrical circuits, 336-348

Tubes, stuffing, 55-60

Vacuum advance mechanism,

478 Voltage

d-c control, 654, 656, 673-676

drop, allowable; determining,

44-51 regulator, 463

Watch, organization, 16-21

Wireways, 51-55